BEYOND THE ATMOSPHERE
EARLY YEARS OF SPACE SCIENCE

HOMER E. NEWELL

Introduction to the Dover edition by
Paul Dickson

DOVER PUBLICATIONS, INC.
MINEOLA, NEW YORK

WITHDRAWN

Copyright

Copyright © 2010 by Dover Publications, Inc.

Bibliographical Note

This Dover edition, first published in 2010, is an unabridged republication of the work originally published in Washington, D.C., in 1980 in the NASA History Series as NASA SP-4211 under the title Beyond the Atmosphere: Early Years of Space Science. A new Introduction by Paul Dickson has been added to this edition.

Library of Congress Cataloging-in-Publication Data

Newell, Homer Edward, 1915–
     Beyond the atmosphere : early years of space science / Homer E. Newell ; introduction to the Dover edition by Paul Dickson.
          p. cm.
     "This Dover edition . . . is an unabridged republication of the work originally published in Washington, D.C., in 1980 in the NASA History Series as NASA SP-4211"—Pref.
     Includes bibliographical references and index.
     ISBN-13: 978-0-486-47464-9 (pbk.)
     ISBN-10: 0-486-47464-X (pbk.)
     1. Space sciences—Research—United States. 2. Outer space—Exploration. 3. Astronautics—United States. I. Title.

QB500.N48 2010
500.50973—dc22

                                                                           2009048184

Manufactured in the United States by Courier Corporation
47464X01
www.doverpublications.com

Contents

Tables

Figures

The photograph of the Veil nebula in Cygnus reproduced on the inside back cover is copyrighted by the California Institute of Technology and the Carnegie Institute of Washington (1959) and reproduced by permission from the Hale Observatories.

Photos in the text are NASA photos.

Homer E. Newell (1915–1983)

Introduction to the Dover Edition

In late 2001 I was invited to the Goddard Space Flight Center in Greenbelt, Maryland to give a lecture about a book I had written on the early days of space exploration.[1] It was only a few weeks after the September 11 attacks so security was especially tight and I had to be escorted at all times by a young engineer working at the center.

As we drove past the Homer E. Newell Memorial Library on the way to the lecture hall I exclaimed that I was pleased to see such an important man honored and added that he had been particularly helpful to me as a young journalist covering space in the late 1960s.

"Never heard of him except as the name on the library," was the response. "Who was he?"

"The father of Space Science," I said.

That was the short answer. The longer answer is that Homer E. Newell (1915–1983), a theoretical physicist and mathematician at the Naval Research Laboratory from 1944 to 1958, was the science program coordinator for Project Vanguard and acting superintendent of the atmosphere and astrophysics division. He was also a key advocate and participant in the International Geophysical Year (IGY), a sixty-seven-nation effort to unlock the secrets of the physical world, which took place between 1957 and 1958 and was considered "the greatest scientific research program ever undertaken."

Newell was an early advocate of civilian, rather than military, control of space and championed the creation of what was to become the National Aeronautics and Space Administration. As it became operational he transferred to NASA to assume responsibility for planning and development of the space science program. After President Kennedy's 1961 speech pledging to put Americans on the moon, and the first flights of Project Mercury, the agency embarked on several science programs to prepare for a moon landing. Under Newell's direction in the post of Associate Administrator of the Office of Space Science and Applications between 1962 and 1968, NASA sent Ranger, Surveyor, and Lunar Orbiter spacecraft to study the Moon. These robotic vehicles provided scientists and engineers with a greater understanding of interplanetary space and lunar geography. Newell not only estab-

---

[1] The lecture was about my book *Sputnik: The Shock of the Century* which was, in part, inspired by my last interview with Dr. Newell twenty years earlier. Newell opened my eyes to the back-story of an oft-told tale that helped me find a voice for an article on *Sputnik* and then the book, which, in turn, begat David Hoffman's film *Sputnik Mania*.

lished the lunar science program and set the direction for space science at NASA, but also spurred initiatives for communications, weather, and earth observation satellites. Over the course of his career, he became an internationally known authority in the field of atmospheric and space sciences.

Newell was not only a brilliant scientist and top administrator but also a masterful speaker and writer. In 1959 his testimony on NASA's plans to explore the planets with probes and rovers left a group of Senators "speechless but impressed" according to the *New York Times,* which also said "The Senators' silence was no reflection on Dr. Newell's ability to expound on the intricacies of the new space age, for there are few more lucid speakers on the subject than he. His is a knowledge that comes from a deep understanding and love of the subject."

He was the author of numerous scientific and popular magazine articles on space and seven books, including a primer on space for children.

Beyond the Atmosphere: Early Years of Space Science is a remarkable work in that it is at once a memoir and a lively, compelling narrative on the birth and early development of space science which, as he points out in his introduction, was often more about the politics of science than about science: "The pursuit of scientific truth gets caught up in a struggle not only with nature but also with oneself and one's fellow beings. Ambition, cooperation, strife, humility, arrogance, envy, admiration, frustration, and courage undergird and overlie the scientific process, making it more important as a story of human endeavor and achievement than as a mere accumulation of human knowledge. So it was with space science; there appeared to be a continual clash of opinions over what to do first, or next, or instead of what was being done."

So if you are looking for a dry pedestrian recounting of early scientific achievement, this book is probably not for you, but if you prefer to look over the shoulders of those that made it all come together and learn what really went on in those early days, this book has no equal.

—PAUL DICKSON

Preface

From the rocket measurements of the upper atmosphere and sun that began in 1946, space science gradually emerged as a new field of scientific activity. In the United States high-altitude rocket research had developed a high degree of sophistication by the time the Soviet Union launched the first artificial satellite of the earth in 1957. That surprise launch proved that the USSR had been pursuing a similar course.

During the period between the orbiting of *Sputnik 1* and the creation of NASA, these activities—scientific research in the high atmosphere and outer space—began to be thought of as *space science*. The first formal use of the phrase that I recall was in the pamphlet *Introduction to Outer Space* prepared by members of the President's Science Advisory Committee and issued on 26 March 1958 by President Eisenhower to acquaint "all the people of America and indeed all the people of the earth" with "the opportunities which a developing space technology can provide to extend man's knowledge of the earth, the solar system, and the universe." A few months later the phrase appeared in the title of the Space Science Board, which the National Academy of Sciences established in June 1958. Use of the term spread rapidly. From the start NASA managers referred to that part of the space program devoted to scientific research by means of rockets and spacecraft as the space science program.

The researches that came under the new rubric were themselves not new. Space science initially consisted of researches already under way that the new tools of rocketry promised to aid substantially. The large number of disciplines—such as atmospheric research and meteorology, solar physics, cosmic rays, and eventually lunar and planetary science—and the recognized importance of many of the problems that could be attacked with the new tools, attracted large numbers of scientists, giving the field of space science broad support at the outset. Even in the life sciences, where the potential contributions of space techniques were less obvious than in the physical sciences, quite a few leading researchers showed a lively, if tentative, interest.

As the program unfolded, the wide range of interest became both a source of strength and a cause for tension. For those able to penetrate beneath the impersonal exterior that science so often seems to present to the outsider, the whole gamut of human emotions is to be found. The pursuit of scientific truth gets caught up in a struggle not only with nature but also with oneself and one's fellow beings. Ambition, cooperation, strife, humility, arrogance, envy, admiration, frustration, and courage undergird and overlie the scientific process, making it more important as a story of human endeavor and achievement than as a mere accumulation of human knowledge. So it was with space science; there appeared to be a continual clash of opinions over what to do first, or next, or instead of what was being done.

To the normal attraction of probing the unknown were added the excitement bestowed by roaring rockets and speeding spacecraft and the awareness that these had opened a vast new region to the presence of man. Moreover, circumstances placed space science to a considerable extent in competition with other aspects of the space program. Congressional concern over the serious questions of national defense raised by the Soviet accomplishments in space focused attention on the nation's launch capability and technological strength rather than science. Understandably, most onlookers displayed more interest in the glamour and excitement of the Apollo program to land men on the moon than in studying cosmic rays or the earth's magnetosphere. Nevertheless, partly in its own right and also as an important supporting element to other activities in space, space science enjoyed a recognized place in the program from the outset.

In telling some of the space science story—particularly the early years when it was emerging as a vigorous new field of activity—I hope to relate this new activity to the rest of the space program on the one hand and to science in general on the other. It is a multifaceted tale, ranging from the very technical to the highly political, from the intensely personal to the institutional, from the national to the international. For long periods the participants are weighed down with the routine drudgery of calculations, painstaking testing in the laboratory and the field, and seemingly endless paper work. Then comes the reward—lifting one to the very pinnacle of excitement—when a spacecraft lands on the moon and its amazing appendages dig into the ancient soil, or from a quarter of a million miles away a breathlessly awaited voice announces, ". . . the *Eagle* has landed," or when yet a different spacecraft visits a distant world like Mars and photographs craters, volcanic peaks, and huge rifts never before seen by man. The whole world—nay, the solar system—is the stage, and the drama is played now in the comfort and safety of the computer laboratory, now amid the rigors and dangers of the launching pad, at times in the whirl of intellectual challenge in the international conference hall, at times face to face with the physical challenges of the high seas, the polar wilderness, or the ominous loneliness of the lunar wastelands.

I hope to convey some of the flavor of this complex program. To do this I shall trace several main threads as they weave their way through the story. First, of course, there is space science itself, what results have been obtained from the use of rockets and spacecraft—including manned spaceflights. What progress in various scientific disciplines can be credited to what is now called space science? It is not my intention, however, to produce another textbook on space science. Such a survey would carry well beyond the planned scope of this book. I seek rather to bring out in broad perspective the main lines of advance in major areas of research, to highlight new areas of investigation, and especially to dwell on changed and changing concepts in the different disciplines. Seeking guidance in this aspect of the task, I sent a questionnaire to a number of the foremost workers in the various disciplines of space science asking for their insights as to what the most significant contributions of space science have been. More than 60 scientists responded, and their views are incorporated into chapters 6, 11, and 20, the chapters that deal with the technical side of the story. I wish to convey here my thanks for that assistance.

The flavor of the story cannot be conveyed in isolation from the context in which the research was done. It will be necessary, therefore, to trace several threads other than space science results. On one side was the relationship of space science to science in general, while on the other were its relationships to the rest of the space program and to the social, political, and economic context.

Space science, while cohering strongly as a new activity, nevertheless is quite correctly viewed as simply a continuation and extension of numerous traditional scientific disciplines. To appreciate the significance of this observation, it is necessary to pay some attention to the meaning and nature of science.

Very little of modern science can be carried out in isolation from other activities of society—certainly not space science. Much of science today requires large pools of manpower, special facilities, and expensive equipment that private sources often cannot afford. When industry does support research, its relevance to the profit-making objectives of the company is always in mind. When government furnishes the support, relevance to national needs and objectives must be taken into account. The years of political struggles to obtain approval and funding for the National Science Foundation; the National Radio Astronomy Observatory at Greenbank, West Virginia; the Enrico Fermi high-energy particle accelerator in Illinois (which the proponents first appeared to lose, but then won); and the Mohole project to drill through the earth's crust (which the scientists first won and then lost) provide classic examples of the difficulties encountered in seeking support of large-scale, expensive scientific undertakings.

So it was with space science. In the early years, when a few sounding rockets were launched, the field received but little attention and meager support. With the appearance of Sputnik, however, space science was precipitated into prominence as an important part of a broad space program that the country suddenly found itself compelled in its own interest to undertake. Almost overnight dollars to pay for the research were no longer a problem, and numerous individuals and institutions, both at home and abroad, became interested in taking part in the program.

Two other main threads in the narrative—organization and management, and institutional relationships—concern such aspects of the space science story. Organization and management include leadership, planning techniques, budget preparation and defense, and organization and management of teams of specialists. Institutional relationships include those with the many groups that played important, often essential, parts in the program. This thread will weave into the story the roles of universities, industry, the military and other government agencies, the National Academy of Sciences, and the President's Science Advisory Committee. Working with the Congress was an essential element in securing and maintaining support for the program. As with NASA's predecessor, the National Advisory Committee for Aeronautics, advisory committees played an important role in the planning and conduct of the space science program.

Paradoxically these aspects appeared as both a hindrance and a help to space scientists. The necessity to fight continually for resources, to compete with other elements of the space program, to labor on advisory committees, and to wrestle with the mountains of paper work required by management interfered with the research. On the other hand, these distractions were necessary, for without proper organization and management, substantial resources, and far-flung teams, space science could not have expanded into the intensive probing of the solar system and the universe that it became. Moreover, one sensed the possibility of a salutary effect in having continually to be on one's toes.

From the time of the earliest sounding rockets in 1945–1946 to the end of 1973 when I left NASA, I was a participant in the space science program. It was my good fortune to take part in the pioneering work of the V-2 Upper Atmosphere Research Panel, in the International Geophysical Year rocket and satellite programs, and in organizing and conducting NASA's space science program. Inevitably what I write is colored by my own experiences. Nevertheless, this book is intended to be something more than a memoir.

I have attempted to survey the field in retrospect, consulting the literature, files, and records in an effort to discern the course of the space science program in proper perspective. It has been interesting to note that sometimes matters were not as I perceived them at the time, and occasionally were quite different. Particularly in telling of controversies in which I was among the disputants, I have tried neither to overemphasize nor to shortchange my own views. Writing the text in the third person has helped, I believe, to be objective.

There are, of course, many vantage points from which a valid picture of the space science program could be drawn. One appropriate aspect would be the story as seen by scientists outside of NASA—in the universities, the National Academy of Sciences, and elsewhere—for these comprised the greater part of the scientists in the program. A different and significant view would be obtained from scientists in other countries who took part in NASA's international cooperative program. The engineers could be expected to tell it quite differently from the scientists. A revealing, not always flattering, picture would likely come from those in industry whose talents provided much of the hardware and operations that lay at the heart of the country's space science missions. Within NASA itself a writer from the research and development centers and field stations—where lay most of the technical and operational strength of the agency—would certainly provide a decidedly different slant from that which a headquarters writer would find natural. I have, of course, tried to capture some of the flavor of all these points of view, but the principal orientation is that of headquarters management.

Another aspect of that vantage point must be considered. Most of this book deals with NASA and its program. How legitimate is this? After all, the military services, particularly the Air Force, were active in space research—through sounding rocket research, study programs, and exploratory development—before there was a NASA, and continued after NASA was established to expend billions of dollars a year on space. Also there were those other participants alluded to in the preceding paragraph. And very much in the foreground has been the Soviet Union, which precipitated worldwide interest in space by launching the Sputniks.

I believe the concentration on NASA to be legitimate, for as time went on scientists in the United States came to view NASA as the prime source of support for space science, as they had come to look to the National Science Foundation and the National Institutes of Health for general support of science. The legislators who created NASA as a civilian agency intended it to be the aegis for the U.S. national space program, which, especially on the international scene, it quickly became. Other countries interested in space research turned to NASA with their proposals to cooperate on space science projects. In the United States, in only a few years "the space science program" and "the NASA space science program" became almost synonymous.

In the USSR the situation was different. There the space program and the military space program were equivalent. Space science, like all other aspects of the program, was cloaked in military security and came into view only well after the fact, when the Soviet Union felt ready to report the results of a successful mission. Yet it would be erroneous to characterize the Soviet space science program as purely military in its import. Considerable attention to lunar and planetary research contributed to an image of Soviet scientific and technological strength, but could hardly be accorded any great military value.

While it is necessary to say enough about Soviet space science to show its more important contributions, and particularly to highlight relationships to the overall field, the main emphasis of the book is on the U.S. program. Even here there is no attempt to be exhaustive. Rather, I intend to select examples that illustrate the interplay of personalities and opinions and the steady advance of ideas that characterized the program, particularly in the 1950s and 1960s.

Overlying the space science story is a most important factor, which I like to refer to as the inexorability of the scientific process, a factor that applies not only to space science, but to all science. In the turmoil of human relations—while the battles rage over funding, priorities, control of programs, personal recognition, bureaucratic despotism, and individual shortcomings—the scientific process, as long as it is sufficiently nourished, steadily, I say inexorably, adds to the store of knowledge. To appreciate the role of the scientific process, one must begin by considering, at least briefly, the meaning and nature of science and the context in which scientists pursue their profession in the modern world. Those are the subjects of part I.

I would like to express my appreciation to the staff of the NASA History Office for the extensive encouragement and assistance I received throughout the time I worked on this book. I must single out archivist Lee Saegesser; his tireless efforts deserve special thanks. Thorough reviews of the manuscript by Dr. Norman Ness and Dr. John O'Keefe of the Goddard Space Flight Center, and many personal conversations with Dr. O'Keefe, were most helpful. Indeed, a large number of reviewers generously furnished comments on parts or all of the manuscript. Their criticisms were an invaluable aid, for which I am grateful.

Alexandria, Virginia   Homer E. Newell
January 1979

BEYOND THE ATMOSPHERE
ATMOSPHERE
EARLY YEARS OF SPACE SCIENCE

Part I

Nature of the Subject

For out of olde feldes, as men seith,
Cometh al this newe corn fro yeer to yere;
And out of olde bokes, in good feith,
Cometh al this newe science that men lere.

Chaucer

1

The Meaning of Space Science

The science managers in the new National Aeronautics and Space Administration of 1958 for the most part had limited experience in the management of science programs. By comparison with the broad program about to unfold, the previous sounding rocket work and even the International Geophysical Year programs were modest indeed. Yet the evolving perceptions of these individuals as to the nature and needs of science would play a major role in the development of the U.S. space science program. At first those perceptions were largely intuitive, growing out of personal needs and experience in scientific research, although a rather extensive literature made the thoughts and experience of others available. In addition, in launching the new program the space science managers had the benefit of the wise counsel of Deputy Administrator Hugh Dryden and Administrator T. Keith Glennan, both of whom had had considerable experience in managing science and technology programs.

Because of the central role played by the concepts of science that NASA managers brought to bear—sometimes consciously, sometimes subconsciously—on the planning and conduct of the NASA space science program, some of those concepts are set forth here at the outset. Moreover, the reader should bear in mind that these concepts are implicit in the author's treatment of space science in this book. The exposition below, while a substantial elaboration of a summary presented to Congress in the spring of 1966, is still highly condensed, and runs the risk of oversimplification.[1]

SCIENCE A PROCESS

A major theme throughout this book is that of science as a worldwide cooperative activity, a process, by which scientists, individually and collectively, seek to derive a commonly accepted explanation of the universe. The author recalls learning in the ninth grade that science was "classified (i.e., organized) knowledge," only to have to discard that definition years later as the very active nature of science became apparent. To be sure, organized knowledge is one of the valuable products of science, but science is far more than a mere accumulation of facts and figures.

Science defies attempts at simple definition. Many—both professional scientists and others—who have sought to set forth an accurate description of the nature of science have found it necessary to devote entire volumes of elaborate discussion to the subject.[2] None has found it possible to give in a few sentences a complete and simple definition, although James B. Conant perhaps came close: "Science is an interconnected series of concepts and conceptual schemes that have developed as a result of experimentation and observation and are fruitful of further experimentation and observations."[3]

On a casual reading, this definition may again appear to characterize science as a static collection of facts and figures. One must add to the definition the activity of scientists, their continuing exchange of information and ideas, and their penetrating criticism of new ideas, working hypotheses, and theories. A static mental construct alone is insufficient; one must include the process that constantly adds to, elaborates, and modifies the construct. All of this Conant—himself an eminently successful chemist—does actually include in what he is trying to convey in his brief definition, as is patent from the amplification he provides in the rest of his treatment. Indeed, the last clause of the quoted definition, requiring that the concepts and conceptual schemes of science be "fruitful of further experimentation and observations," clearly implies the ongoing nature of science.

The difficulty of conveying in brief the nature of science, particularly to the layman, has led in exasperation to such statements as, "Science is what scientists do." The circularity of this definition can be frustrating to one seriously trying to understand the subject—a legislator, for example, endeavoring to appreciate the significance of science for the country and his constituents, and to discern what science needs to keep it healthy and productive. Yet the definition suggests probably the best way of approaching the subject; that is, to tell just what it is that scientists do.

Scientists work together to develop a commonly accepted explanation of the universe. In this process, the scientist uses observation and measurement, imagination, induction, hypothesis, generalization and theory, deduction, test, communication, and mutual criticism in a constant assault on the unknown or poorly understood. Consider briefly each of these activities.

The scientist *observes and measures*. A fundamental rule of modern science is that its conclusions must be based on what actually happens in the physical world. To determine this the scientist collects experimental data. He makes measurements under the most carefully controlled conditions possible. He insists that the results of experiment and measurement be repeatable and repeated. When possible, he measures the same phenomenon in different ways, to eliminate any possible errors of method.

To experimental and observational results the scientist applies *imagination* in an effort to discern or *induce* common elements that may give further insight into what is going on. In this process he may discover relationships that lead him to formulate laws of action or behavior, such as Newton's law of gravitation or the three fundamental laws of motion, or to *make hypotheses,* like Avogadro's hypothesis that under the same pressures and temperatures, equal volumes of different gases contain equal numbers of molecules.* It is not enough that these laws be expressed in qualitative terms; they must also be expressed in quantitative form so that they may be subjected to further test and measurement.

The scientist *generalizes* from the measured data and the relationships and laws that he has discerned to develop a *theory* that can "explain" a collection of what might otherwise appear to be unconnected or unrelated facts. In seeking generalization, the scientist requires that the new theory be broader than existing theory about the subject. If the new theory explains only what is already known and nothing more, it is of very limited value and basically unacceptable.

The new theory must predict by *deduction* new phenomena and new laws as yet unobserved. These predictions can then serve as guides to new experiments and observations. By taking predictions and working them together with other known facts and accepted ideas, the scientist can often deduce a result that can be put to immediate *test* either by observation of natural phenomena or by conducting a controlled experiment. Out of all the possible tests, the scientist attempts to choose those of such a clear-cut nature that a negative result would discredit the theory being tested, while a positive result would provide the strongest possible support for the theory.

In this connection, it must be emphasized that the scientist is not seeking *"the* theory," the absolute explanation of the phenomena in question. One can never claim to have the ultimate explanation. In testing hypotheses and theories the scientist can definitely eliminate theories as unacceptable when the results of a properly designed experiment contradict in a fundamental way the proposed theory. In the other direction, however, the scientist can do no more than show a theory to be acceptable in the light of currently known facts and accepted concepts. Even a long-accepted theory may be incomplete, having been based on inadequate observations. With the continuing accumulation of new data, that theory may suddenly prove incapable of explaining some newly discovered aspect of nature. Then the old theory must be modified or expanded, or even replaced by an entirely new theory embodying new concepts. Thus, in his efforts to push back the frontiers of knowledge, the scientist is continually attempting to develop an acceptable "best-for-the-time-being" explanation of available data.

---

*For what scientists mean by the terms *hypothesis, law,* and *theory,* the reader is referred to Robert Bruce Lindsay and Henry Margenau, *Foundations of Physics* (New York: John Wiley & Sons, 1936), pp. 14-29.

In all this process the scientist continually *communicates* with his colleagues through printed journals, in oral presentations, and in informal discussions, subjecting his results and conclusions to the close scrutiny and *criticism* of his peers. Ideally, observations and measurements are examined and questioned, and repeated and checked sufficiently to ensure their validity. Theories are compared against known observation and fact, against currently accepted ideas, and against other proposed theories. Acceptable standing in the growing body of scientific knowledge is achieved only through such a searching trial by ordeal.

One should hasten to add that this is not a process of voting on the basis of mere numbers. Even though the majority of the scientific community may be prepared to accept a given theory, a telling argument by a single perceptive individual can remove the theory from competition. Thus, the voting is carried out through a continuing exchange of argument and reasoned analysis. Those who have nothing to offer either pro or con in effect do not vote.

This process or activity called science has developed its rules, its body of tradition, from hard and telling experience. Recognizing that the scientific process cannot yield the absolute in knowledge, scientists have sought to substitute for the unattainable absolute the attainable utmost in objectivity. The scientific tradition wrings out of final results as much as possible of the personal equation by demanding that the individual subject his thoughts and conclusions to the uncompromising scrutiny of his skeptical peers.

The above are things that scientists do, and through the complex interchanges among scientists these activities amalgamate into what is called science. But at this point one must ask what factor distinguishes science from a number of other endeavors. Observation and measurement, imagination, induction, hypothesis, generalization and theory, deduction, test, communication, and mutual criticism are used in various combinations by the economist, the legislator, the social planner, the historian, and others who today in partial imitation of the scientists apply to their tasks and studies their concepts of what the scientific method is. The distinguishing factor is fundamental: underlying the pursuit of science is the basic assumption that, to the questions under investigation, nature has definite answers. Regardless of the philosophical dilemma that one can never be sure of having found the right answers, the answers are assumed to exist, their uniqueness bestowing on science a natural, intrinsic unity and coherence. In contrast one would hardly argue that societal, political, and economic problems have unique answers.

These latter problems are concerned with the human predicament, and the human equation enters not only into the search for answers, but into the very solutions themselves. Human invention and devising are necessary ingredients of the solutions achieved. In science, however, although imagination and invention are important elements of the discovery process, the human factor must ultimately be excluded from its findings, and to this end the scientific process is designed to eliminate as much personal bias and individual error as possible. This aspect gives science its appearance of objectivity and impersonality, while bestowing a universality that transcends political and cultural differences that otherwise divide mankind.

The reader is again cautioned not to be misled by oversimplification. One must not conclude from the above orderly listing of activities and processes of thought, either that they constitute a prescribed series of steps in the scientific process or that one can identify a single scientific method subscribed to and followed by all scientists. On the contrary, individual scientists have their individual insights, styles, and methods of research. Conant is emphatic on this point:

> There is no such thing as *the* scientific method. If there were, surely an examination of the history of physics, chemistry, and biology would reveal it. For as I have already pointed out, few would deny that it is the progress in physics, chemistry and experimental biology which gives everyone confidence in the procedures of the scientist. Yet, a careful examination of these subjects fails to reveal any *one* method by means of which the masters in these fields broke new ground.[4]

While there is no single scientific method, there is method, and each researcher develops his own sense of order and line of attack. And major elements of the various methods are sufficiently discernible that they can be identified. Indeed, there is enough of method to the profession to lead John Simpson, professor of physics at the University of Chicago, to assert that even the plodder, while he may never make brilliant contributions, can through systematic effort aid in the progress of science.

Nevertheless, the role of insight and perceptiveness is crucial. The application, however, cannot be equated with induction in the Baconian tradition.[5] The inductive step from the singular to the general, while an important element in science, is far from routine. Often seemingly haphazard, this step calls into play inspiration, insight, intuition, imagination, and shrewd guesswork that are the hallmark of the productive researcher. Conant alluded to the elusive character of this phase of the scientific process: "Few if any pioneers have arrived at their important discoveries by a systematic process of logical thought. Rather, brilliant flashes of imaginative 'hunches' have guided their steps—often at first fumbling steps."[6]

Each individual has his own devices for trying to discern from the particular what the general might be. Certainly the reasoner does not approach his task with no preconceptions. To the new data he adds other facts and data already known, and he calls into play previously accepted ideas that appear relevant. Whatever the method, the ultimate test is whether it works.

A continuing task of the space science manager was to assess progress in the program, and various criteria for measuring the worth of scientific accomplishments have been used. In this regard the author finds attractive a number of concepts provided by Thomas S. Kuhn.[7]

A scientist approaches a new situation or problem with a definite mental picture of how things ought to be, what processes should be operative, what kinds of results are to be expected from different experiments. This mental picture—which, with some leeway for differing points of view, he shares with scientific colleagues working in the same field—has developed over the years from experimentation and observation, hypothesizing, theorizing, and testing. It has stood the ordeal of searching tests and has proved its value in predicting new results and in integrating what is known of the field into a logically consistent, useful description of nature.

To this shared mental construct, Kuhn gives the name *paradigm*, a substantial extension of the usual meaning of the term. Thus, the ionosphericists share a paradigm, in which each knows—or at least agrees to accept—that there is an ionosphere in the upper reaches of the earth's atmosphere consisting of electrons and positive and negative ions, varying in intensity, location, and character with time of day, season, and the sunspot cycle. He knows, or agrees, that most of the ionization and its variation over time are caused by solar radiation, and that the ionosphere has a complex array of solar-terrestrial interrelations. The ionosphere is affected by and affects the earth's magnetic field. It has a profound influence on the propagation of many wavelengths in the radio frequency region of the electromagnetic spectrum and acts like a mirror reflecting waves of suitable wavelength, a phenomenon that before the advent of the communications satellite afforded the only means of round-the-world short-wave radio transmissions. To develop thoroughly the paradigm shared by ionospheric physicists would be a lengthy proposition,[8] but the reader may find the above sufficiently suggestive.

As another example, solar physicists share a paradigm in which the sun is regarded as an average sort of star, about 10 billion years old and with some billions of years still to go before it evolves into a white dwarf. It originated as a condensation of dust and gases from a huge nebula and was heated by the gravitational energy released by the falling of the nebular material into the contracting solar ball until internal temperatures rose sufficiently to initiate nuclear burning of hydrogen, the major source today of the sun's radiant energy. And so on.[9] Workers in the field of solar studies understand each other, they have a common way of looking at things, they approach problems with a similar orientation.

Individual scientists usually share a number of paradigms with different colleagues. The paradigms of the upper atmosphere physicist and the ionosphericist overlap greatly. While an ionospheric investigator is applying his ionospheric paradigm to his work, he also has in the back of his mind that the laws of physics and chemistry must apply to the ionosphere, and when appropriate the ionospheric researcher brings to bear the paradigms of chemistry and modern physics. Likewise the solar physicist must constantly borrow from the paradigms of astronomy, astrophysics, physics, nuclear physics, and plasma physics.

The importance of the currently accepted paradigm or paradigms in guiding a scientist in his researches, in determining—and determining is not too strong a word—what he will perceive when he encounters a new situation, cannot be overestimated. Even the nonscientist, by osmosis from the press, television, and literature, in addition to his formal schooling, absorbs many significant concepts from the paradigms of the working scientists. Most of the fundamental concepts about the nature of the universe shared by modern man have derived from the scientific developments of the last two centuries. With these concepts infused into one's thinking, an enormous effort would be required to see the universe and the world as they were visualized by the medieval thinker. As Herbert Butterfield put it:

> The greatest obstacle to the understanding of the history of science is our inability to unload our minds of modern views about the nature of the universe. We look back a few centuries and we see men with brains much more powerful than ours—men who stand out as giants in the intellectual history of the world—and sometimes they look foolish if we only superficially observe them, for they were unaware of some of the most elementary scientific principles that we nowadays learn at school. It is easy to forget that sometimes it took centuries to discover which end of the stick to pick up when starting on a certain kind of scientific problem. It took ages of bitter controversy and required the cooperative endeavor of many pioneer minds to settle certain simple and fundamental principles which now even children understand without any difficulty at all.[10]

Thus the concept of the paradigm is more than a mere convenience. In terms of the paradigm one can discern several stages in the scientific process. First of all, the existence of shared paradigms in a scientific area indicates some measure of maturity of the field. In its beginning, a newly developing field tends to fumble along without any accepted conceptual framework, and each new datum or observation may seem to heighten the complexity and confusion. In time, however, discerning minds begin to perceive some order, and a workable paradigm begins to evolve. A good example is furnished by the birth of modern chemistry in the very confused, yet highly productive, second half of the 18th century.[11]

In its maturity a field of science exhibits alternating periods of what Kuhn refers to as normal science and scientific revolution. During a period of normal science, the accepted paradigm appears to work well, satisfactorily explaining new observations and results as they accumulate. It is a period in which measurements and observations tend to illuminate and expand upon the accepted paradigm, but not to challenge it. Most scientific work is normal science in this sense.

Occasionally new experimental results don't appear to fit the framework of the accepted paradigm. When that occurs, attention is directed toward finding an explanation. Generally the first efforts are to find a way of retaining the accepted paradigm, particularly if it has proved highly productive and illuminating in the past. Perhaps the paradigm can be extended or even bent to accommodate the new results. In fact, the scientist's inclination is to tolerate a considerable amount of misfit to save a particularly useful paradigm.

But when the challenge to the previously accepted paradigm becomes too severe, and acceptable modifications or extensions won't accommodate the new results, then a change in paradigm becomes necessary. Such periods, bringing a forced change of paradigms, Kuhn designates as scientific revolutions. Periods of scientific revolution are likely to be exciting (at least to scientists), highly active, with much debate and a lot of fumbling around trying to find a way out. Classical examples of scientific revolutions are furnished by the shift from Newtonian to Einsteinian relativity and from classical to quantum physics.[12] A more recent example is to be found in the upheavals of the 1950s and 1960s in geophysics and geology leading to the now general acceptance of the concepts of sea-floor spreading, continental drift, and plate tectonics as fundamental features of the paradigm that today guides the researcher in experimenting and theorizing about the nature of the earth's crust.[13]

For this book the concepts of paradigm, normal science, and scientific revolution furnish a way to trace and assess the development of space science through the first decade or so of NASA's existence. Nevertheless, the reader is cautioned that the concept of the paradigm in the scientific process—or the manner in which the concept is used—has been extensively criticized.[14] A major concern has been the difficulty of supplying the concept with any great degree of precision and the consequent fuzziness in the picture one can draw of the role really played by the paradigm in science. Critics have pointed out that Kuhn himself has used the concept in numerous different ways. Also, the simultaneous existence at times of conflicting paradigms, each receiving support from its separate group of adherents—as, for example, in the many years during the 18th century when both the caloric and mechanical theories of heat had their supporters—is pointed to as indicating that Kuhn's concept of scientific revolution is too simplistic to embrace the whole picture of how science moves and how revolutions occur in scientific thought.

In spite of the criticism the paradigm appeals to the author as useful and even fundamental; he suspects the criticism can be met. At any rate, for this book the straightforward interpretation of the role of paradigm in science will suffice and should be useful.

SPACE SCIENCE

This book is about space science. The subject is simple in concept, comprising *those scientific investigations made possible or significantly aided by rockets, satellites, and space probes.* But in its realization space science turns out to be very complex because of the diversity of scientific investigations made possible by space techniques.

Interest in the phenomena of space is not recent, its origins being lost in the shadows of antiquity. Impelled by curiosity and a desire to understand, man has long studied, charted, and debated the mysteries of the celestial spheres. Out of this interest came eventually the revolution in thought and outlook initiated by Copernicus, supported by the remarkably precise measurements of Tycho Brahe, illuminated by the observations of Galileo and the insights of Kepler, and given a theoretical basis by Newton in his proposed law of gravitation. The Copernican revolution continues to unfold today in human thought and lies at the heart of modern astronomy and cosmology.[15]

Yet, until recently outer space was inaccessible to man, and whatever was learned about the sun, planets, and stars was obtained by often elaborate deductions from observations of the radiations that reached the surface of the earth. Nor were all the inaccessible reaches of space far away. The ionosphere, important because of its role in radio communications, was not as far away from the man on the ground below as Baltimore is from Washington. Nevertheless, until the advent of the large rocket, the ionosphere remained inaccessible not only to man himself but even to his instruments. As a result many of the conclusions about the upper atmosphere and the space environment of the earth were quite tentative, being based on highly indirect evidence and long chains of theoretical reasoning. Time and again the theorist found himself struggling with a plethora of possibilities that could be reduced in number only if it were possible to make in situ measurements. Lacking the measurements, the researcher was forced into guesswork and speculation.

Small wonder, then, that when large rockets appeared they were soon put to work carrying scientific instruments into the upper atmosphere for making the long needed in situ measurements. From the very start it was clear that the large rocket brought with it numerous possibilities for aiding the investigation and exploration of the atmosphere and space. It could be instrumented to make measurements at high altitude and fired along a vertical or nearly vertical trajectory for the purpose, falling back to earth after reaching a peak altitude. When so used the rocket became known as a *sounding rocket* or *rocket sonde,* and the operation was referred to as sounding the upper atmosphere.

A rocket could also be used to place an instrumented capsule into orbit around the earth, where the instruments could make extended-duration measurements of the outer reaches of the earth's atmosphere or observations of the sun and other celestial objects. Or the rocket might launch an instrumented capsule on a trajectory that would take it far from the earth into what was referred to as deep space, perhaps to visit and make observations of the moon or another planet. The orbiting capsules were called *artificial satellites* of the earth; those sent farther out came to be known as *space probes* or *deep space probes*. Finally, the ultimate possibility of carrying men away from the earth to travel through deep space and someday to visit other planets emphasized dramatically the new power that men had acquired in the creation of the large rocket.

A language of rocketry emerged, which the news media popularized. Familiar words took on new meanings, and new terms were encountered: artificial satellite, spacecraft, space launch vehicle, rocket stages, countdown, liftoff, trajectory, orbit, tracking, telemetering, guidance and control, retrorockets, reentry—and space science.

Through all the centuries of scientific interest in space phenomena, the phrase *space science* had not gained common use. That the terminology did not come into use until after rockets and satellites brought it forth gives force to the definition of space science given at the start of this section. That definition sets forth the meaning in mind when in June 1957 the U.S. National Academy of Sciences combined the functions of the IGY Technical Panel on Rocketry and the IGY Technical Panel on the Earth Satellite Program into a single board, naming it the Space Science Board. That is the meaning implied by the discussions in the first book-length publication by the Space Science Board a few years later.[16] That is the meaning picked up by Samuel Glasstone in 1965 in his comprehensive survey of space science:

> The space sciences may be defined as those areas of science to which new knowledge can be contributed by means of space vehicles, i.e., sounding rockets, satellites, and lunar and planetary probes, either manned or unmanned. Thus space science does not constitute a new science but represents an important extension of the frontiers of such existing sciences as astronomy, biology, geodesy, and the physics and chemistry of Earth and its environment and of the celestial bodies.[17]

While the basic meaning of space science was clear and unvarying from the start, the exact nature of the activity, and in particular its relationship to the rest of science, was not always so clear. Glasstone's use in the above quotation of *space sciences* in one place and *space science* in the very next sentence reflects one question that arose often during the first years of the NASA program. Is space science a new scientific discipline* or, if not yet,

---

*A scientific *discipline* is an area of scientific investigation in which the investigators share a common paradigm or group of paradigms, embracing a common body of theory, and techniques and often instrumentation that stem from the underlying theoretical basis of the discipline.

will it in time develop into a new discipline? The question arose primarily because of the pure-science character of space science and the strong coherence that quickly developed in the field, but also because of the broad range of scientific topics to which research was addressed. The initial answer to the question generally agreed to by those in the program was that given by Glasstone: space science was not a new discipline and should not be expected to become one. The initial response was probably intuitive, but in retrospect it is seen to have been the correct answer.

Space science makes extensive contributions to geophysics; but this part of space science remains a part of the discipline of geophysics, using its techniques and instrumentation and employing and extending its basic theory—sharing its paradigm, that is. The researchers using space techniques for geophysical investigations, while perhaps thinking of themselves as space scientists, continue to call themselves geophysicists, to be members of geophysical societies like the American Geophysical Union, to present their papers at geophysical meetings and to publish them in geophysical journals.

Space science also makes numerous contributions to astronomy, but again the parts of space science devoted to astronomy remain a part of the discipline of astronomy, and space scientists using rockets for astronomical research continue to view themselves as astronomers. Their results are presented at meetings like those of the American Astronomical Society or the International Astronomical Union* and are published in their journals or proceedings.

Cosmic ray physicists find space methods advantageous in many of their researches, but continue to be cosmic ray physicists first and space scientists only incidentally. Examples can be multiplied at length.

Nevertheless, for several years following Sputnik the thought that space science might evolve into a separate discipline persisted. One can understand why. The demands imposed by rockets and spacecraft on the running of a science program were severe, giving a coherence to the field akin to that characteristic of a scientific discipline. But rockets and spacecraft did not rest upon or stem from the scientific disciplines they served. Rather, they were simply trucks to provide transportation to otherwise inaccessible places, while the genuine techniques and instrumentation of the investigations were those of the individual disciplines that benefited from the new means of transportation.

To emphasize the diverse scientific disciplines, writers sometimes chose to use the phrase *space sciences*. At other times authors used *science in space*

---

*Although much, probably most, of the space science research on the moon and planets falls into the disciplines of geophysics and geochemistry, the Planetary Sciences Division of AAS and Commission 16 (Planets) and Commission 17 (Moon)—in 1979 merged as new Commission 16—of the IAU have maintained a vigorous interest in these aspects of space science research.

to imply that space science was not separate from science on the ground and was neither more nor less than the familiar, everyday science carried out in a new arena. These initial uncertainties were reflected in the changing names given to the space science group in NASA Headquarters by the author and his colleagues. In 1958 and 1959 the division in the Office of Space Flight Development that had responsibility for scientific research in space was labeled Space Science. When NASA Headquarters reorganized under the new administrator, James E. Webb, the space science program was elevated to the level of a separate office, which called attention to the plural nature of its activity in its title: Office of Space Sciences. Finally, in the reorganization of 1963 that brought science and applications together under one head, NASA settled on *space science* as its choice for the rest of the 1960s, designating the new entity as the Office of Space Science and Applications.[18]

If space science had been distinctly separate from the rest of science, NASA might well have felt less impelled to draw in the wide participation that the agency encouraged in the program. As it was, recognizing that no single agency could reasonably expect to bring within its own halls the expertise needed for all the separate disciplines, NASA consciously sought broad participation from the outside scientific community, especially from the universities, where the greatest interest in pure science was to be found.

Within the universities the question arose in a somewhat different form. As the numbers of those entering space science research grew apace, a need to provide training for new scientists who might wish to pursue space research as a career became evident. Should this be done by setting up departments of space science in universities? The instinct of NASA program managers was not to do so, and when asked they advised against it, recommending instead that opportunities be provided within the traditional departments of astronomy, physics, geophysics, geology, etc., for taking on space-related problems as thesis topics. Most universities saw it this way, although a few decided to experiment with separate space science departments.*

The inseparability of space science from the rest of science and the broad range of disciplines to which space techniques promised to contribute gave impetus to the rapid development of science in the national space program. It must be emphasized that scientists came into the program with problems that had been under attack by other methods and that appeared to need some new approach if they were to be solved. The promise to provide that new approach drew researchers first to sounding rockets and later to satellites and space probes.

---

*One successful experiment was the Department of Space Sciences set up by Rice University under the direction of a young theoretical physicist, Dr. Alexander Dessler, who had done considerable research on the newly discovered terrestrial magnetosphere and who for some years served the American Geophysical Union as editor of the space sciences section of the *Journal of Geophysical Research*.

Writing about six years after the start of sounding rocket research in the United States, in what was probably the first book devoted to the subject, the author was able to find in the scientific literature significant results to report on upper atmospheric pressures, temperatures, and densities; atmospheric composition; solar radiations in the ultraviolet and x-rays; upper-air winds; the ionosphere and the earth's magnetic field; cosmic rays; and high-altitude photography. A year later the list was extended even further in a book reporting the papers presented at the first international conference on the subject of high-altitude rocket research, arranged by the Upper Atmosphere Rocket Research Panel (see chap. 4) of the United States and the Gassiot Committee of the Royal Society of London.[19] In 1956, just a decade after the start of rocket sounding of the upper atmosphere, the Upper Atmosphere Rocket Research Panel, extrapolating from its sounding rocket experience, turned its attention to the researches that would be possible with instrumented satellites of the earth. These deliberations were published in the first book on the subject to be assembled by persons professionally engaged in high-altitude research. [20] To the research topics listed above, the book added some new ones: meteors and interplanetary dust, particle radiations from the sun, the aurora, stellar astronomy, meteorology, and geodesy. The potential contributions to science of both manned and unmanned spacecraft were discussed in the Space Science Board's first book. While most attention was devoted to unmanned exploration of space, the ultimate potential of manned spaceflight was recognized in such words as: "The significant and exciting role of man lies in the exploration of the Moon and planets."[21]

Such scientific investigations, made possible by sounding rockets and spacecraft, came to define what is meant by space science. Much of the potential of space science was already discernible before ever a satellite had been launched, and by the end of 1960—by which time the first NASA administrator, Keith Glennan, had set the agency firmly on its course—the broad sweep of space science was fully apparent. By the end of a decade space science research had become worldwide, and a steady flow of results was pouring into the literature.[22]

2

The Context

To divorce modern science, including space science, from other pursuits of society is impossible. What scientists do obviously and pervasively affects the rest of society. Reciprocally, the complex activities of society, its motivations and changing objectives, what it chooses to develop and use of technology, as well as the specific support that society—for a variety of reasons—provides to science, determines in large measure what researches scientists undertake. A properly rounded history of space science should treat of more than the technical subject matter of the science itself.

As might be supposed, scientists are usually moved to take up their researches by a curiosity that impels them to find out how nature works. The scientist is likely to be driven by his personal fascination with his profession. He is willing to devote long, physically and mentally taxing hours to his work and to endure hardships and danger—like the astronomer in the small hours of the night at the mountain top observatory, or the atmospheric scientist wintering over through the long Antarctic darkness, or the undersea explorer—if only he be given the necessary resources for pursuing his researches.

But why should society support an individual in what so often appears to be a highly personal endeavor, particularly when the price tag today can run into millions or hundreds of millions of dollars? Those seeking support for science have to wrestle with this fundamental question constantly. The answer for science often can be quite simplistic. From the knowledge acquired through scientific investigations, it is argued, come eventually many of the technologies and their practical applications that people want and will pay for in the marketplace (like radio, television, home appliances, modern textiles, better automobiles, and boats) or need and must pay for (like improved agriculture, health care, modern communications, and transportation of food, materials, and supplies). That is the principal reason why society finds it profitable to support a considerable amount of science.

But the simplistic answer gives no hint of the complexity of the vexing questions that arise when government and industry are asked to foot the bill, particularly for what is sometimes called pure science. What applications will result? How long will it take? How much scientific research will be needed? What kind of research would be best for an optimum practical return on the investment? Where should the research be done—in industry, the universities, government laboratories, or research institutes? Who should decide what research to do?

There is no absolute answer to any of these questions, and circumstances can make some of them exceedingly perplexing. The literature on the subject is overwhelming, and any discussion of such matters demonstrates quickly that science has many aspects and complex relationships with other human endeavors. It becomes important, for example, to distinguish among science, technology, and application.

Technology is not science, nor is science technology, but there are important relationships between them. Technology is technical know-how, the knowledge and ability to do things of a technical or engineering nature, including the field of industrial arts. On the basis of considerable know-how, or technology, the Babylonians built and operated a remarkable irrigation system; equally remarkable was the technology of the ancients in construction. But neither technology derived from science as we know it. On the other hand a tremendous amount of technology does flow from the results of scientific research. Examples are to be found in electronics, synthetic materials, transportation, and medicine.

Technology also supports science. Electronics provides invaluable service to science in detection and measurement; the technology of materials is important in radiation-detection instruments; computer technology is a great boon to the theorist; and modern engineering is fundamental to the design and construction of modern astronomical telescopes, huge particle accelerators, and nuclear reactors. Rocket technology made space science possible; that technology in its turn rests on the results of considerable scientific research.

Application is the last step in the chain from technical know-how to actual use. Thus, the use of meteorological satellites for weather observations is an application of both scientific knowledge (of the atmosphere) and technology (of spacecraft construction, instrumentation, and operation).

The intimate relationships among science, technology, and applications give rise to many questions like those cited earlier. Some sort of rational response must be made to such questions when the public is asked to spend billions of dollars of tax money a year for scientific research and many more billions of dollars a year on civilian and military technical development. The need to respond to such queries has been a continuing requirement throughout the space science program, and most certainly will continue. These issues should be examined, therefore, in enough depth to understand how they influenced the space science program.[1]

Take, for example, the question: What applications will result? If the question is asked about applied research that is intentionally directed

17

toward a specific application already in the minds of the researcher and his supporters, then that specific application will be the end result if it turns out to be at all possible and economically sensible. For, as the researcher pursues his investigations, he will always be oriented toward the prescribed end. New scientific results that appear to lead in the direction of the desired application will be pursued, while avenues that appear to lead in some other direction will not be followed—though they may hold promise of answering very fundamental questions about nature, the answers to which might prove of more practical benefit than those the applied researcher feels constrained by his assignment to investigate.

Here is the crux of the matter. The uncommitted scientist will pursue the avenues that appear to offer the greatest promise of answering the most fundamental questions about the nature of matter, energy, physical laws, the universe; the committed orientation of the applied researcher will keep him always working toward the planned application. To the industrial manager, the legislator, the government administrator, the latter goal might seem preferable to get a specific job done—and very often it is. To invest funds in support of research that holds greatest promise of a specific desired application is the most easily justified and patently wise course of action.

Yet there is a pitfall. Time and again invaluable practical benefits have come from uncommitted research and could not have been foreseen or predicted. Pure science, almost by definition, precludes a clear prediction of results. It is the search for new knowledge. If the knowledge were known ahead of time, it would not be new.

The classical example, often cited, is the discovery of x-rays by Wilhelm Conrad Roentgen in 1895. Within a year of their discovery, x-rays were being put to practical use in medicine, and in time became of enormous value in medicine, industry, and scientific research. Roentgen's discovery resulted from experimenting with electron beams in evacuated tubes. Had he been directly seeking something of value for the medical profession, he would most likely have put away his electron beams and taken up some more "practical" line of investigation, and the discovery of x-rays would have been postponed.

Examples of practical returns from pure science can be multiplied almost ad infinitum; for example, James Clerk Maxwell's work on the theory of electricity and magnetism and the whole train of subsequent electromagnetic applications; Heinrich Rudolf Hertz's propagation experiments and radio; John Dalton's work on combining weights and modern chemistry; Christiaan Huygen's work on optics and the optical industry; and the years of purely scientific investigation into the atom and its nucleus that furnished the basis for the Manhattan Project, which in turn led to modern nuclear applications.[2]

The uncommitted researcher, while he cannot point to the future and say that his researches will produce this or that specific application as a payoff, can look back and point to use after use that was eventually made of the results of his kind of nonprogrammatic, nonapplied, uncommitted research. Many have argued the historical record to justify support of pure science, including support of enough researchers free from the constraints of programmatic or applied research to provide the uncommitted frame of mind that is most likely to follow up interesting new discoveries wherever they might lead.[3]

The importance of uncommitted research goes even deeper. Even the applied researcher relies on the scientific paradigms that he has inherited from decades and centuries of research, and these are based on data and results, a large part of which came from uncommitted research. The truth of this assertion was borne out by a series of studies supported by the National Science Foundation and published as *Technology in Retrospect and Critical Events in Science* ("Traces").[4] Several technologies or technological applications* were reviewed historically to identify scientific results that had been "key to the progress of research towards the innovation" under study. Without going too far afield, some of the *Traces* conclusions should be noted. The study found for each case that about a decade before the application—that is, about the time one was finally in a position to discern and define technically the potential application or technology—almost all of the basic research needed for the potential application had been done.† What was most significant, however, was that all applications depended vitally, critically, on a long history of basic research, a substantial part of which was nonmission, uncommitted research; in the cases studied more than 70 percent of the key scientific results stemmed from such research. Moreover, the sources of the critical information were international in scope, and universities, industry, government, and private activities made significant contributions.

This kind of story the defenders of the space science program had to convey to the administration and the Congress to obtain funding. There was a narrow path to tread. Space science was largely pure science, and researchers were by and large uncommitted to specific practical applications, although many of them showed a keen interest in applications of their results to such purposes as meteorology, geodesy, and earth-resources surveys. To retain a free hand for the investigator was important, but if the research appeared too irrelevant to the immediate needs of society or, more

---

*Magnetic ferrites, the video tape recorder, the oral contraceptive, the electron microscope, and matrix isolation.

†*Traces* dealt with a critical point raised by C. W. Sherwin and R. S. Isenson, "First Interim Report on Project Hindsight (Summary)," Dept. of Defense, Office of the Dir. of Defense Research and Engineering, 1966. The Hindsight report caused quite a stir among scientists and gave the National Science Foundation, NASA, and other government agencies supporting basic research trouble in the administration and on the Hill, because superficially the report appeared to show that only applied research was important for supporting the development and application of technology—in this case, military application.

narrowly, to the interests of the legislator's constituency, support would be hard to come by. So a substantial effort was made to point to the potential usefulness of the space science research that was in need of support,[5] at times to such an extent as to distress members of the scientific community. The pressure to produce useful results quickly was always there, and the scientists were mindful of Vannevar Bush's caution that "applied research always drives out the pure."[6]

The hazard was real, for if the importance of pure science for future practical uses was not communicated to the legislators funding could be difficult to obtain. On the other hand overselling could generate great expectations of immediate practical returns, with a day of accounting but a few years down the road in some future budget hearings. In general, the practical returns from pure science must be reckoned as being well into the future,[7] leaving the proponents of pure science with a very tricky selling job. An appropriate scale seems to be that the time from basic research result to its substantial, continuing use in practical applications is two or more decades. The author's view is that new knowledge begins to be applied extensively only when it has become second nature to the appliers and springs more or less readily to mind as needed. The time lag, then, is related to the period required for the new knowledge to diffuse through the field, become accepted, and enter textbooks, courses, and handbooks —to become a familiar element of the shared paradigm of the field.

In contrast, to develop a difficult, complex technology once the essential concept and underlying principles are known, a decade appears to be about the right time needed, while the final development of an actual application, once the basic research has been done and the pertinent technologies worked out, is a matter of some years. Examples of the development of applications in the space field are the meteorological and communications satellites which, relying on the research and technological development of previous decades, could be built and put into orbit in the first few years of NASA's history.

In the 19th century the time for new results to diffuse through the field and become accepted into the paradigm—in fields with developed paradigms— was about 50 years.[8] Today, with the more rapid flow of information, constantly changing study courses, and frequent revision of textbooks and handbooks, the interval is down to perhaps 20 years, with many examples of applications of new results sooner than that. It would seem, however, that some practical minimum time must remain for new knowledge to flow throughout a field, gain acceptance, and become second nature to sizable numbers of practitioners. If so, the most effective way of speeding up the realization of practical returns from newly acquired information is to speed up the process of making it second nature to potential appliers of the information. Is not this what, on a small scale, industrial research groups and applied research institutes try to do?

Space science was in the main pure science, and its researchers were uncommitted to the development of practical uses of the results they obtained. But administrations and congresses were committed intellectually and politically to the realization of genuine practical returns from investment of public money. Those who managed the program, therefore, had to strive to preserve and protect its pure science character, while making plain its ultimate practical worth, and to do this without undercutting the one aspect or overselling the other.

Part II

Taproots

Each venture
Is a new beginning, a raid on the inarticulate
With shabby equipment always deteriorating
In the general mess of imprecision of feeling.

T. S. Eliot, East Coker

3

Prophets and Pioneers of Spaceflight

The rocket apparently made its debut on the pages of history as a fire arrow used by the Chin Tartars in 1232 for fighting off a Mongol assault on Kai-feng-fu. The lineage to the immensely larger rockets now used as space launch vehicles is unmistakable.[1] But for centuries rockets were in the main rather small, and their use was confined principally to weaponry, the projection of lifelines in sea rescue, signaling, and fireworks displays. Not until the 20th century did a clear understanding of the principles of rockets emerge, and only then did the technology of large rockets begin to evolve. Thus, as far as spaceflight and space science are concerned, the story of rockets up to the beginning of the 20th century was largely prologue.

Nevertheless, well before the 1900s numerous authors showed a keen appreciation of what satellites might mean if only a way could be found to launch them. Their fictional accounts of space travel are often cited as early harbingers of the modern space age,[2] and in the light of recent achievements the long history of rockets makes exciting reading. Especially those engaged in space research and exploration find peculiar fascination in reading about Kepler's imaginary visit to the moon, described in the little book *Somnium, sive Astronomia Lunaris* published in 1634, several years after Kepler's death,[3] or in following the flights of fantasy recounted in Jules Verne's *De la Terre à la Lune* (1865) and *Autour de la Lune* (1870). Even the artificial satellite turned up in *The Brick Moon* by Edward Everett Hale, serialized in the *Atlantic Monthly* in 1869 and 1870. Launched by huge rotating water wheels, the Brick Moon, a manned satellite, was intended to serve as a navigational aid. In the first years of the space program, John Nicolaides, one of the engineers professionally interested in geodesy and navigation, took great delight in giving his colleagues copies of Hale's little story.

These and numerous other writings of the kind legitimately belong to the lore of the space age. While they predate the emergence of the serious work on large rockets that made the space program and space science possible, they nevertheless have a special significance. Such imaginings reflect the centuries-long interest of mankind in the heavens. Some men climbed mountains to set up astronomical observatories, others to measure how air pressure changes with height. No sooner had the Montgolfier brothers in 1783 demonstrated the feasibility of hot-air balloons than aeronauts began to fly in them. That same year J. A. C. Charles of France ascended in a hydrogen-filled balloon. This first grasp on the age-old dream of flight brought forth an amazing variety of ideas and experiments, and by the end of the 19th century powered balloon flight was a reality, the sausage-shaped dirigible being the most successful form.[4] In the 1920s and 1930s high-altitude ballooning was serious business, with men like Gray, Piccard, Anderson, and Stevens setting one altitude record after another. The record of 22 kilometers gained by the last two in the helium-inflated *Explorer II*, in 1935, persisted for two decades.

Where men could not go they sent their instruments, surrogates for the time being for those who would surely follow later. As long ago as 1749, three years before Benjamin Franklin's famous experiment with lightning, Alexander Wilson of Scotland sent thermometers aloft on kites to measure upper-air temperatures. By the late 19th century meteorologists were flying kites and balloons carrying thermometers and pressure gauges to investigate properties of the atmosphere. In 1898 the French meteorologist Léon Philippe Teisserenc de Bort started using such balloons to obtain reliable temperature measurements up to a height of 14 kilometers. By 1904 he had proved the existence of a stable, isothermal region above 11 kilometers, to which he gave the name *stratosphere*.[5]

During the 20th century free-flying balloons became a much used means of making remote scientific observations in the high atmosphere. The airplane, of course, made flying at great heights routine. While aircraft could not beat the balloons in altitude, the ability of the airplane pilot to control accurately the time and location along the flight path added a new dimension not afforded by the free balloon.

As soon as men could get to some hitherto inaccessible spot, they went there for a variety of motivations—curiosity, perhaps to make scientific observations, simply to overcome a challenge, to lay claim to a new dominion, or in pursuit of an inherent drive that it was psychologically impossible to deny. The record indicated that when men could leave the earth they would do so. The early stories about space travel served notice that man was indeed taking aim at the stars.

But, while the writings on space up through the 19th century pointed a prophetic finger toward the future, they could contribute little more. Mostly fiction, much of the writing was remarkably foresighted, but also much was incorrect. Jules Verne's enormous cannon used to launch his mooncraft from an underground pit in Florida would have subjected the passengers to bone-crushing accelerations and the spacecraft to searing temperatures from atmospheric friction. But his use of small reaction rockets to control the attitude of the spacecraft in flight was quite correct in concept. So, too, was Achille Eyraud's concept of a reaction motor to launch a spacecraft on his *Voyage à Venus* (published in 1875), but Eyraud was not aware that water, which he used as the propellant, was utterly inadequate for the job. Likewise, Hale's water wheels could never have imparted the necessary impulse to launch the Brick Moon into orbit.

The writers of the 19th century showed awareness of the science of their day, but not enough was known about the basic physics of rockets for them to understand clearly what was required to launch spacecraft. Not until a genuine understanding of the principles of reaction motors was attained could the large rocket needed for spaceflight be created. This understanding had to await the 20th century. Those who wrote before that time were accordingly cast in the role of prophets. After them came the pioneers.

Not that prophecy ceased. Far from it. Imaginative writing continued unabated, even accelerated, as later authors found themselves able to draw upon an expanding knowledge of rockets and principles of spaceflight to give their narratives plausibility and persuasiveness. Movies, comic strips, radio programs—and later television—picked up the theme. Perhaps the culmination of these prophetic writings was to be found in Arthur C. Clarke's delightful *The Exploration of Space*—honored as a Book-of-the-Month Club selection—in which the author was able, in layman's terms yet without sacrificing technical validity, to lay before the reader a veritable blueprint of the space program to come.[6]

Much of what Clarke wrote in 1951 drew upon the pioneering work of the previous 50 years when the large rocket had been brought into being.[7] During that period mathematicians and physicists developed a sound theory of rocket propulsion. In the United States, Germany, and Russia both amateurs and professionals experimented with the design, construction, and launching of both liquid- and solid-propellant rockets. The importance of their pioneering work went largely unrecognized; the general public was mostly unaware of what the rocketeers were up to except when spectacular tests of the new fangled devices, which often ended in mishaps, caught the attention of the newsreels and newspapers. But for the seriously interested there was a growing literature to seize upon and devour.

Three persons were particularly significant in the transition from the small rockets of the 19th century to the colossi of the space age: Konstantin E. Tsiolkovsky in Russia, Robert H. Goddard in the United States, and Hermann Oberth in Germany. It is generally agreed that priority goes to Tsiolkovsky (1857–1935), who apparently in his teens became interested in the possibility of spaceflight. He wrote of spaceflight in science fiction, but went further. Self-taught in mathematics, astronomy, and physics, he proceeded to develop the basic theory of rocket propulsion, and in 1898 submitted his now famous article, "The Investigation of Outer Space by Means of Reaction Apparatus," to the editors of *Science Survey*. The article, how-ever, was not published until 1903. For the next 30 years Tsiolkovsky continued to write both technical papers and science fiction, much of what he had to say being devoted to a favorite theme of flight into deep space, about which he wrote in 1911:

> To place one's feet on the soil of asteroids, to lift a stone from the moon with your hand, to construct moving stations in ether space, to organize inhabited rings around Earth, moon and sun, to observe Mars at the distance of several tens of miles, to descend to its satellites or even to its own surface—what could be more insane! However, only at such a time when reactive devices are applied, will a great new era begin in astronomy: the era of more in-tensive study of the heavens.[8]

In 1926 Tsiolkovsky suggested the use of artificial earth satellites, includ-ing manned platforms, as way stations for interplanetary flight, and in 1929 he put forth an idea for a multistage rocket which he described as a rocket train.[9]

Like the appearance of his first article on rocket principles, Tsiol-kovsky's influence in Russia was delayed. As G. A. Tokaty, aerodynamicist and chief rocket scientist of the Soviet Government in Germany (1946–1947), commented:

> Konstantin Eduardovich Tsiolkovsky (1857–1935), the man of "great efforts and little rewards," is . . . considered to be the "father" of present Soviet achievements in rocket technology. He gave Russia a spaceship project which was, for 1903, absolutely unique. But being what he was—a mere teacher in a remote provincial school, a technologist rather than a theoretician—his project did not attract the attention it deserved.[10]

Apparently it took the publication in Germany, in 1923, of *Die Rakete zu den Planetenräumen*[11] by the Hungarian-born Hermann Oberth to goad the Russians into action. Following the appearance of Oberth's work, in which the author elaborated in great detail the application of rocket propulsion to spaceflight, Tsiolkovsky's earlier works were sought out and avidly stu-died. Interest in rocket propulsion increased noticeably in the Soviet Union, which took special pains to assert Russian claims to priority by issuing in 1924 German translations of Tsiolkovsky's writings. That same year Friedrikh A. Tsander, Tsiolkovsky, and Felix E. Dzherzhinsky started the Society for Studying Interplanetary Communications, a major aspect of which concerned interplanetary travel.

After a period of grouping and regrouping, Soviet workers in the early 1930s settled down to serious experimenting with large rockets, with which the now familiar names of F. A. Tsander, Yu. V. Kondratyuk, and M. K. Tik-honravov were associated. As early as 1928 Kondratyuk had put forth the idea of using aerodynamic forces to slow down a rocket returning from a trip in space. Tsander designed and built a rocket motor using kerosene and liquid oxygen as propellants, which he successfully tested in 1932. In August of the following year the first successful flight of a Soviet liquid-propellant rocket took place. It was during this period that S. P. Korolev, who was to become the giant of Soviet modern rocketry in the 1940s and 1950s, began his work on rockets. His book *Rocket Flight in the Stratosphere* was published in 1934 by the USSR Ministry of Defense. But not long thereafter a curtain fell over Russian rocket activities, not to rise again until the launching of *Sputnik 1* revealed how much the Soviet Union had accomplished in the intervening 20 years.[12]

Of special interest to space scientists, during 1935 a Soviet liquid-propellant meteorological rocket designed by Tikhonravov was flown. Apparently, however, as in the United States, rocket research in the very high atmosphere had to await the availability of the more capable rockets that appeared in World War II. According to Tokaty, the exploration of the upper atmosphere with rockets of the V-2 class began in the autumn of 1947, and from 1949 on was continued with Pobeda rockets, described as greatly improved versions of the V-2.[13]

Second in priority among the rocket pioneers was the American physicist Robert H. Goddard (1882–1945), whose esteem in the United States today matches that of Tsiolkovsky in Russia. Goddard himself points to 19 October 1899 as the date when he, still in high school, determined to devote his career to the attainment of space exploration.[14] Like Tsiolkovsky and Oberth, Goddard clearly perceived the importance of rockets for high-altitude flight and astronautics. Almost immediately he began writing on the subject. Many of his papers were devoted to rocket theory, which he correctly expounded. Perhaps his most famous was the paper "A Method of Reaching Extreme Altitudes," the title reflecting his enduring interest in high-altitude and space research. The paper was originally written in the summer of 1914 and revised in late 1916 in the light of experimental results. With a few editorial changes and the addition of some notes, Goddard submitted the paper in 1919 to the Smithsonian Institution, and it was published in the *Smithsonian Miscellaneous Collections* of December 1919.[15]

Robert Goddard was set off from his contemporaries, Tsiolkovsky and Oberth, in that he by no means stuck to theory and writing as did the other two. From the start Goddard was busy with his hands, conducting experiments to check theory and devising hardware to put the theory into practice. The very year, 1914, in which he composed the first draft of the Smithsonian paper, he was awarded patents for a rocket using solid and liquid propellants, and for a multistage or step rocket. Goddard built and flew the first successful liquid-propellant rocket. Of primitive design and construction, the rocket flew 56 meters in 2½ seconds at Auburn, Massachusetts, on 16 March 1926.[16] In the course of his career he accumulated many ideas that came to be familiar features of successful large rockets— such as liquid-propellant motors, self-cooled motors, the use of gyroscopes for guidance and control, reflector vanes in the rocket jet for stabilizing and steering the rocket, fuel pumps, and parachutes for recovering a spent rocket. So prolific was his output that those who followed could hardly take a step without in some way infringing on one or more of his patents, a fact recognized by the United States government when in 1960 the military services and the National Aeronautics and Space Administration awarded $1 000 000 to the Goddard estate.

One might accordingly suppose that Goddard's influence on the space program and space science would be great, even to the extent of eclipsing that of other contributors. Unfortunately that does not appear to have been true. A suspicious nature—first aroused by adverse publicity connected with his Smithsonian paper and later reinforced by the conviction that his work was being plagiarized—led Goddard to work in isolation and for the most part to avoid open publication of his ideas and accomplishments. This secretiveness stood in the way of his contributing the leadership that he could so easily have given to the field and to the enthusiasts of the young American Rocket Society, which was founded in 1930 as the American Interplanetary Society. Years later G. Edward Pendray, one of the founders of the society, wrote plaintively: "When Goddard in his desert fastness in New Mexico proved uncommunicative, those of us who wanted to do our part in launching the space age turned to what appeared the next best source of light: the *Verein für Raumschiffahrt*—the German Interplanetary Society—in Berlin."[17]

The amateurs were not alone in their failure to join hands with the great pioneer. Members of the California Institute of Technology Rocket Research Project, established in 1936 by Theodore von Kármán, director of the institute's Guggenheim Aeronautical Laboratory, tried to persuade Goddard to join forces with them. When it was stipulated that a partnership would require mutual disclosure of ideas and projects, Goddard shied away. His reluctance to work openly with others deprived Goddard not only of the opportunity to provide leadership in the field, but also cut him off from the kind of professional assistance that he might have received from experienced engineers who could have helped put his many ideas into practice. In turning away from the Rocket Research Project, Goddard was also turning down the kind of funding support from the military that could have capped his long years of work with their hoped-for fruition. Working alone with extremely limited funds, Goddard could not match the progress being made in German rocketry, which was supported amply by the military during those years.

Goddard furnishes a tragic illustration of the importance of open publication and free exchange of ideas to the scientific process. Unpublished, the import of Goddard's ideas went unrecognized for the most part, and by the time they were widely known much of what he had done had been redone, as with the V-2. The opportunity to be the leader of the field during the course of his development work soon passed. Still, he was a man of genius and originality and the many honors later accorded him were well deserved. NASA's Goddard Space Flight Center in Greenbelt, Maryland, appropriately bears his name. Medals are awarded and symposia held in his honor. In 1958 the National Rocket Club began sponsoring the Robert H. Goddard Annual Memorial Dinner in Washington, faithfully attended by engineers, scientists, administrators, legislators, military men, industrialists—the Who's Who of rocket and space research—to pay tribute to Goddard's pioneering role. Space scientists also recognize in Goddard the first to work seriously on the problem of developing an effective means of sending scientific instruments beyond balloon altitudes into the upper atmosphere and outer space.

But Goddard never did personally achieve his dream of using rockets for upper-atmosphere research. While he continued to work in obscurity, spending his final years during World War II working in secrecy for the U.S. Navy, the CalTech Rocket Research Project—reorganized in 1944 as the Jet Propulsion Laboratory—went on to become the first group in the United States to build and launch a rocket specifically designed for upper-air research. Named the WAC-Corporal, the JPL rocket on 26 September 1945 rose to a height of about 70 kilometers, a U.S. record at the time. It is the JPL research, rather than Goddard's, from which a line can be traced directly to the space program. Writers associated with CalTech and the von Kármán group communicated the latest in rocketry to the public through scientific papers.[18] Although restricted in its circulation at the time, because of its bearing on military applications, a handbook of jet propulsion put out by JPL nevertheless reached large numbers of persons in rocket research and development.[19] More significantly for space science, the WAC-Corporal was the progenitor of a larger, improved sounding rocket, called Aerobee—in later versions capable of carrying a substantial instrument load above 200 kilometers—which became one of the mainstays of the American high-altitude research program.[20]

Neither Goddard's work nor the JPL rockets provided the initial impetus to the space science program in America. Circumstances made rocket sounding in the United States the beneficiary of the two decades of vigorous rocket development work by German experimenters that ensued following the publication of Oberth's *Rocket into Planetary Space*. Nourished by German military support, the German experimenters rediscovered and reinvented for themselves much of what Goddard was learning in the United States. Going well beyond what Goddard could accomplish in his self-imposed isolation, Walter Dornberger, Wernher von Braun, and their colleagues produced the V-2—Vergeltungswaffe-Zwei or "Vengeance Weapon

Two"—the first large rocket to see substantial service.[21] At the close of World War II, U.S. Army forces captured large numbers of these monsters at underground factories in the Harz Mountains in central Germany. Along with von Braun and key members of his team—who took the initiative to ensure that they became prisoners of American, not Russian, forces[22]—the Army took the captured V-2s to the United States. There the missiles were assembled, tested, and launched at the White Sands Proving Ground in New Mexico to provide experience in the handling and operation of large rockets.

Rather than fire the missiles empty, the Army offered to allow interested groups to instrument them for high-altitude scientific research. A number of military and university groups accepted, forming the V-2 Upper Atmosphere Research Panel, which became the aegis for the country's first sounding rocket program.[23]

4

The Rocket and Satellite Research Panel:
The First Space Scientists

As World War II came to a close, a group of engineers and scientists in the Communications Security Section of the Naval Research Laboratory in Washington began to cast about for new research problems to which to apply their talents. Long hours were spent on the subject, and the list of possibilities grew to sizable proportions. Milton Rosen, a competent, versatile, imaginative electronics engineer, suggested that the group might apply its wartime experience with missiles and communications, including television, to a study of the upper atmosphere. The suggestion became the eighth to go on the blackboard in the office of Ernst Krause, head of the section. Thereafter it was referred to as Project 8.

When the debate finally wound down, Project 8 was the clear winner. To the many physicists in the group the project offered an attractive and important field of research. The engineers could feel the challenge of instrumenting and launching the rockets that would be needed by the scientists. And because of the importance of knowledge of atmospheric properties to communications and the design and operation of missiles, it was possible that the Navy might support the project.

The director of the laboratory approved the upper-air research proposal in December of 1945, and the section became the Rocket Sonde Research Section, a name that appropriately enough also came from the originator of the Project 8 idea. No one in the section was experienced in upper atmospheric research, so the section immediately entered a period of intensive self-education. Members lectured each other on aerodynamics, rocket propulsion, telemetering—whatever appeared to be important for the new tasks ahead. The author gave a number of talks on satellites and satellite orbits. Indeed, the possibility of going immediately to artificial satellites of the earth as research platforms was considered by the group, which assimilated carefully whatever information it could obtain from military studies of the time. The conclusion was that one could indeed begin an artificial satellite program and expect to succeed, but that the amount of new development required would be costly and time consuming. The scientists could not hope to have their instruments aloft for some years to come and, anyway, were not likely to get their hands on the necessary funds. The Rocket Sonde Research Section accordingly shelved the satellite idea and turned to sounding rockets.

As they were considering what rockets—including the Jet Propulsion Laboratory's WAC-Corporal—might be available for the research they contemplated, word came that the U.S. Army would be willing for interested scientists to conduct experiments in some of the V-2s it was planning to fire at the White Sands range in New Mexico. Because of the narrow confines of the range, the missiles would have to be fired along nearly vertical trajectories and would accordingly make ideal probes of the upper atmosphere. To explore the possibilities Krause invited a number of interested persons to meet at the Naval Research Laboratory. At the meeting, on 16 January 1946, physicists and astronomers interested in cosmic ray, solar, and atmospheric research were present. Because of the potential importance of upper-air data to military applications, the services were well represented. It was plain from the deliberations that a number of groups both in universities and in the military would be interested in taking part in a program of high-altitude rocket research.

THE V-2 PANEL

Accordingly, at an organizing meeting at Princeton University 27 February 1946, a panel was formed of members to be actually engaged in or in some way directly concerned with high-altitude rocket research.[1] The original members (see also app. A) were:

E. H. Krause (chairman), Naval Research Laboratory
G. K. Megerian (secretary), General Electric Co.
W. G. Dow, University of Michigan
M. J. E. Golay, U.S. Army Signal Corps
C. F. Green, General Electric Co.
K. H. Kingdon, General Electric Co.
M. H. Nichols, Princeton University
J. A. Van Allen, Applied Physics Laboratory, Johns Hopkins University
F. L. Whipple, Harvard University

Because of his role in getting things started and because he would be devoting full time to upper-air research with rockets, Krause was elected chairman.

To Krause must go the principal credit for getting the program under way. He was a physicist, with a doctorate from the University of Wisconsin in spectroscopy, and a background in communications research. Both qualifications were pertinent to the development of techniques for the investigation of the sun and upper atmosphere. Krause's energy and drive were phenomenal, and his capacity for detail and thoroughness were ideally suited to welding all the elements needed to get a sounding rocket program off the ground. When Krause left in December 1947 to participate in nuclear bomb tests, James A. Van Allen was elected to the chair, a spot he occupied for the next decade.[2]

Van Allen is by far the best known of the original members of the V-2 panel. A physicist, at the time the panel was formed he was employed by the Applied Physics Laboratory of the Johns Hopkins University on the Bumblebee Project, a Navy missile research and development project. He brought to the panel an intense interest in cosmic ray physics, an interest that led in time to his discovery of the earth's radiation belts that now bear his name.

The panel had no formal charter, no specified terms of reference from an authorizing parent organization, a circumstance that left the panel free in the years ahead to pursue its destiny in keeping with its own judgment. The immediate task was to provide Col. James G. Bain of the Army Ordnance Department with advice he had requested on the allocation of V-2s to the various research groups. This the panel proceeded at once to do, and in fact until the end of the V-2 program in 1952 continued to direct its reports to Army Ordnance as principal addressee. Thereafter the reports were issued simply to the members and to observers who attended the meetings, with copies to a selected list of interested persons and agencies (see app. B).

The panel's program, if it may be called that, consisted of the collection of activities engaged in by its members. As a forum for discussion of past results and future plans, the panel was a breeding ground for ideas; but whatever control it might bring to bear on the program was exerted purely through the scientific process of open discussion and mutual criticism.

For some time after its first session, the panel met monthly (see app. C). There was a great deal to do, quickly; for Army Ordnance and its contractor, General Electric Company, intended to fire the rockets on a rather rapid schedule. Since the German warheads were not suitable for carrying scientific payloads, the Naval Research Laboratory undertook to provide the different groups with standard nose sections specifically designed for housing the research instrumentation. To send information to the ground from the flying rocket, NRL also furnished telemetering equipment to go into the rocket and erected ground stations at the White Sands range for receiving and recording the data-bearing signals. In short order the word *telemetering*, meaning the making of remote measurements by radio techniques, became a familiar part of the growing jargon of rocket sounding. To make the most of the large capacity of the V-2, NRL designed and built a large, complex telemeter. The first version supplied to the program could provide 23 channels of information; a later version provided 30. With characteristic preference for smaller, simpler instrumentation, the Applied Physics Laboratory developed and used a much smaller, 6-channel, frequency-modulated telemeter.[3]

Radar beacons were installed in the missile to track it, providing information on where measurements had been made. The range also required that each rocket be outfitted with a special radio receiver that could cut off the motor should the missile begin to misbehave after launch. Arrangements had to be made for building and supplying this equipment. Also, to supplement the tracking information provided by radar and radio, theodolites, precise cameras, and other optical instruments were installed at strategic locations along the firing range to furnish both visual and photographic trajectory data. It also would be essential to know the orientation of the rocket in order to interpret properly such measurements as aerodynamic pressures or cosmic ray fluxes. For this, still more instruments—including photocells to observe the direction of the sun, cameras, and magnetometers—were brought to bear.[4]

Although much, perhaps most, of the scientific data would be obtained by telemetering, some measurements would require the recovery of equipment and records from the rocket after the flight was over, such as earth and cloud pictures, photographs of the sun's spectrum, and biological specimens exposed to the flight environment. For this purpose several techniques were developed, including the use of explosives to destroy the streamlining of the rocket, causing it to maple leaf to the ground; the deployment of parachutes to recover part or all of the spent rocket; and even the application of the kind of sound ranging techniques used in World War I to locate large guns.[5]

At first, operations at White Sands were an amorphous collection of activities. During the first year of rocket sounding the procedures and issues that would have to be dealt with in even greater detail years later in the space program emerged: safety considerations, provision for terminating propulsion of the missile in mid-flight, tracking, telemetering, timing signals, range communications, radio-frequency interference problems, weather reports, recovery of instruments and records, and all that went into assembling, instrumenting, testing, fueling, and launching the rocket. To cope with the seemingly endless detail, the range required formal written operational plans in advance that could be disseminated to the various groups. A more or less standard routine evolved with which the participants became familiar.[6] In only a few years experimenters were harking back to the "good old days" when operations were free and easy and red tape had not yet tied everything into neat little, inviolable packages.

While the General Electric Company personnel, Army workers, and others labored to produce successful rocket firings, the scientists labored equally hard to devise and produce the instrumentation that would yield the desired scientific measurements. At first some of the instrumentation was tentative, even crude, as when Ralph Havens of NRL took an automobile headlight bulb, knocked off the tip, and used it as a Pirani pressure gauge to measure atmospheric pressure in the V-2 fired on 28 June 1946. But even before the end of 1946 spectrographs were recording the sun's spectrum in previously unobserved ultraviolet wavelengths, special radio transmitters were measuring the electrification of the ionosphere, and a variety of cosmic-ray-counter telescopes were analyzing radiation at the edge of space. A portion of each panel meeting was devoted to reporting on experimental results, which accumulated steadily from the very first flight of 16 April 1946. Papers began to appear in the literature and attracted considerable attention as experimenters reported on measurements that hitherto were impossible to make.[7] By the time the last V-2 was fired in the fall of 1952, a rich harvest of information on atmospheric temperatures, pressures, densities, composition, ionization, and winds, atmospheric and solar radiations, the earth's magnetic field at high altitudes, and cosmic rays had been reaped.[8]

THE NEED TO REPLACE THE V-2

But not all of the results had been obtained from the V-2. To be sure, the immediate availability of the V-2 as a sounding rocket was a boon to the program, for it meant that the scientists could start experimenting without delay. Its altitude performance of 160 kilometers with a metric ton of payload far exceeded that of any other rocket that the experimenters might have been able to use, making investigations well into the ionosphere possible from the outset. More significantly, the large weight-carrying capacity of the rocket meant that experimenters did not have to miniaturize and trim their equipment to shoehorn them into a very restricted payload, but could use relatively gross designs and construction. This capacity was a great help at the start, when everyone was learning, for it permitted the researcher to concentrate on the physics of his experiment without being distracted by added engineering requirements imposed by the rocket tool. Later, with some years of experience behind him, the experimenter would be able to take the outfitting of much smaller rockets in stride. And it was of advantage to go to smaller rockets as soon as possible.

Smaller rockets would be much cheaper, far simpler than the V-2 to assemble, test, and launch. Moreover, with the smaller, simpler rockets the logistics of conducting rocket soundings at places other than White Sands would be manageable. With such thoughts in mind, as panel members pressed the exploration of the upper atmosphere with the V-2 they also set out to develop a variety of single and multistage rockets specifically for atmospheric sounding.[9] James Van Allen and his colleagues at the Applied Physics Laboratory undertook, with support from the U.S. Navy's Bureau of Ordnance, to develop the Aerobee sounding rocket.[10] At the same time NRL took on the job of developing a large rocket—first called Neptune, but later Viking when it was learned a Neptune aircraft already existed—to replace the V-2s when they were gone.[11] At the 28 January 1948 meeting of the panel, Van Allen reported on a series of test firings of the Aerobee— three dummy rounds and one live round.[12] As soon as it was ready the Aerobee was put to work exploring the upper atmosphere and space, with firings not only from the original Aerobee launching tower at White Sands, but also from a second tower that the Air Force erected some 57 kilometers northeast of the Army blockhouse at the White Sands Proving Ground. The Air Force tower was located at Holloman Air Force Base near Alamogordo. Not content with the payload and altitude capabilities of the first Aerobees, both the Air Force and the Navy continued the development, producing something like a dozen different versions, one of which could carry 23 kilograms of payload to an altitude of 480 kilometers.[13] In its various versions Aerobee was used continuously in the high-altitude rocket research program through the 1950s and 1960s and was still in use in the mid-1970s.

In contrast, the Viking, although of a marvelous design—Milton Rosen, who directed the Viking development program, used to point out that in its time Viking was the most efficiently designed rocket in existence— found very little use. The dozen rockets bought for the development program were, of course, instrumented for high-altitude research. But Viking was too expensive. The groups engaged in rocket sounding each had perhaps a few hundred thousand dollars a year to expend on the research, and a single Viking would have eaten up the whole budget. When the supply of German V-2s began to run low, consideration was given to building new ones; but estimates placed the price per copy at around half a million dollars, which was prohibitive. It had been hoped that Viking would be much less expensive, but before the end of the development these rockets became almost as expensive as new V-2s. So Viking found no takers among the atmospheric sounding groups and would probably have been shelved had it not been chosen as the starting point for the Vanguard IGY satellite launching vehicle.[14]

The contrast between Viking and Aerobee typified a situation that has recurred in the space science program. One group of scientists would favor developing large new rockets, spacecraft, or other equipment that would greatly extend the research capability. Another group would prefer to keep things as small and simple as possible, devoting its funds to scientific experiments that could be done with available rockets and equipment. The former group could always point to research not possible with existing tools, thus justifying the proposed development. In rebuttal the latter could always point to an ample collection of important problems that could be attacked with existing means. There was right on both sides of the argument, and it was usually a standoff. As far as upper atmospheric research was concerned, however, Viking was too far ahead of its time.

While in the next decade researchers would be able to buy $1-million Scouts (chap. 10), in the early years of rocket sounding Viking cost too much.

Once the ball had started rolling with Aerobee and Viking, other rocket combinations began to appear. The experimenters sought less cost, greater simplicity, higher altitudes, more payload, and especially a capability to conduct firings at different geographic locations. Great ingenuity was displayed in putting together new combinations. Sounding rockets were taken to the California coast, to Florida, to the Virginia coast, out to sea, and to the shores of Hudson's Bay in Canada.[15] They were even launched in the stratosphere from balloons, a combination that the inventor, Van Allen, called a Rockoon.[16] In the panel meeting of 9 September 1954, Van Allen reported that Rockoon flights in the Arctic had established the existence of a soft radiation in the auroral zone above 50 kilometers height, which proved to be one of the milestones along the investigative track that ultimately led to the discovery of the earth's radiation belt.

Scope of Panel Activity

One of the most notable aspects of the panel record is the steadily increasing scope of activity. In the minutes of the organizing meeting, the secretary referred to the group simply as "the panel." By the third meeting Megerian was calling the group the "V-2 Upper Atmosphere Panel." This name continued for the next two meetings; but the appellation "V-2 Upper Atmosphere Research Panel" appeared at the sixth meeting, in September 1946, and stuck for the next year and a half. These first titles reflected the panel's participation in the V-2 program, but the group's primary business was high-altitude research, not V-2s. The panel, well aware that the supply of V-2s would be exhausted in the not too distant future, gave early attention to finding alternative sounding rockets. Prodded by the Office of the Chief of Ordnance, at its March 1948 meeting the panel dropped the V-2 from its title and began calling itself the "Upper Atmosphere Rocket Research Panel" (UARRP). This sufficed to describe activities until members had become so thoroughly involved in the International Geophysical Year scientific satellite program that another name change seemed appropriate. At an executive session, 29 April 1957, the panel adopted its final name: "Rocket and Satellite Research Panel."[17]

Throughout most of its active life, the panel remained quite small. By restricting its rolls to working members only, and also by limiting the number of representatives from any one agency, the panel kept its size down—which made for more manageable meetings. Yet there was no desire to limit interest or participation in the meetings. A loyal cadre of observers attended the sessions throughout the years and joined in the discussions. From the first, the National Advisory Committee for Aeronautics was represented among the observers—an interesting fact in retrospect, although at the time there was no reason to suspect that one day NACA might play a central role in a suddenly emerging space program. Increasing interest in high-altitude rocket research over the years is also shown by the steady growth in the list of addressees to whom panel reports were sent. The minutes of the organizing meeting went to only about 30 persons; 10 years later some 118 copies were being distributed among 73 addressees.[18] The composition of the distribution lists is illuminating (see app. B). The military was obviously interested. So, too, were other government agencies such as NACA and the U.S. Weather Bureau. The large number of university names on the list no doubt resulted from the pure-science nature of much of the panel's research.

For more than a decade the panel occupied a unique position in scientific research. In the United States its members represented all the institutions engaged in sounding rocket research. Attendees at meetings—members plus observers—comprised a substantial number of the individuals in the country who were involved. As one consequence of this unique position, the panel came to be regarded as the prime source of expertise in the field. In spite of the lack of any official charter, the panel soon acquired a quasi-official status. The National Advisory Committee for Aeronautics used data from the panel program in compiling and updating its tables of a standard atmosphere.[19] The Defense Department's Research and Development Board made a practice of turning to the panel for recommendations regarding sounding rockets and high-altitude rocket research. The board—called the Joint Research and Development Board before the establishment of the Department of Defense in 1947—boasted a sprawling, complex structure intended to correspond in one way or another to the military research and development programs.[20] From time to time its Committee on Guided Missiles took an interest in the rockets being used by the panel. When, in the spring of 1949, the Navy's Viking and the Air Force's MX774 rockets came into competition—it was not considered reasonable for the country to support two large, expensive sounding-rockets—UARRP was informed that a panel of the Committee on Guided Missiles endorsed Viking. The R&D board's Committee on Geophysical Sciences, and its subsidiary group for study of the upper atmosphere, took a continuing interest in what UARRP was up to. The subsidiary group endorsed the UARRP's research program and in November 1947, responding to a request for support, unanimously recognized "the importance of all phases of the well-coordinated V-2 rocket firings program and the grave consequences of any failure to give adequate financial support to all agencies involved in this program, since the lack of support of the program in any one agency would jeopardize the program as a whole."[21] At its April 1950 meeting, one finds the UARRP responding to a request of the R&D board for views on requirements for upper-air research vehicles.[22]

But, while the endorsement was of help, association with the military also brought problems. At its 7 May 1947 meeting, the UARRP learned that the R&D board's upper-atmosphere group was considering assigning primary responsibility to different agencies for different kinds of upper-atmosphere research. Although nothing ever came of this, the thought of dividing the research into assigned parcels conflicted with the basic research instincts of UARRP members.

More serious, however, was the question of security classification that arose periodically. In defense of the research program, panel members were accustomed to pointing out the many practical benefits to be gained from data and knowledge obtained. Over the years the list of potential benefits to the military grew, until a report issued at the start of the International Geophysical Year by a number of the panel members cited a dozen important applications:

- Design of missiles, high-altitude craft, and space vehicles.

- Determination of the reentry behavior of long-range ballistic missiles.

- Special techniques of high-altitude navigation.

- Evaluation of hazards to personnel and equipment in the high atmosphere and space.

- Improvement of weather forecasting.

- Study of climate.

- Prediction of the trajectories of biological, chemical, or radiological agents.

- Development of reliable point-to-point communications.

- Development of reliable and accurate methods of guidance, control, and delivery of missiles to their targets.

- Development of reliable and accurate methods of detection of enemy missiles and high-altitude craft.

- Development of countermeasures against enemy missiles.

- Remote detection of nuclear explosions.[23]

But to the extent the salesmanship succeeded, it also raised the question of why the sounding rocket results shouldn't be classified if they were so valuable to the military, which was paying for them.

From the outset the panel had assumed that its program, being basic research, would be unclassified. In a memorandum to the White Sands

Proving Ground, Col. H. N. Toftoy of the Army Ordnance Department had written that V-2 firing schedules, rocket design, and flight information would be unclassified.[24] This decision was important to the program, since the flight information was intimately related to the high-altitude data obtained from the rocket, and since design data were needed for interpreting measurements—for example, aerodynamic pressure curves were required in obtaining atmospheric densities from pressure measurements along the surface of the flying rocket. A serious threat arose when, at the October 1952 meeting of the panel, Earl Droessler of the R&D board announced that the military had again raised the question of classification of upper atmospheric data. The panel unanimously agreed to fight classification, citing the importance of the scientific process, in particular open publication and free exchange of information, to a basic research activity. While there was something to be gained by classifying certain specific uses of scientific information, there was much to be lost by classifying the purely scientific data. In these efforts the panel was successful, and the program remained unclassified.

The program called for a lot of work, but it was exciting. Panel meetings were enjoyable, with none of the tedium that so often weighs oppressively on committee meetings. For most of the members, after a period of preparation at home base—in Washington, Silver Spring, Cambridge, Ann Arbor, or elsewhere—there would be a period of some weeks or a couple of months working in the lonely beauty of the New Mexico desert. How exhilarating it was to send a rocket roaring into the clear blue sky, watch the missile trace a brilliant white vapor trail against the azure background, a trail the stratospheric winds soon blew into complicated twists and knots, and then to jump into a jeep and race northward to retrieve cameras and instruments! On one such day in March 1957, with the sky as bright a blue as it ever had been, V-2 no. 21 landed in the heart of the White Sands National Monument. What a glorious hunt riding up and down over the snow-white dunes of gypsum sand that stretched as far as the eye could see! At the end of the day, with a solar spectrograph, cameras, and other instruments safely stowed aboard the jeeps, the impact party, as it was called, slowly worked its way out of the barren wilderness. As the group approached the edge of the monument, where the gypsum deposit has acquired a pinkish tint from the surrounding red sands of the Tula Rosa Basin, the sun was setting. An occasional yucca growing amid the pink-white dunes provided a display of incomparable beauty, which the glowing sun transformed into a fairyland. When the white sands were finally left behind, one could feel the emotional release.

The routine was frequently broken by bits of humor. Early in the program, before the range was properly instrumented for tracking the V-2s, von Braun often watched the flying rocket as it rose above the desert, judging by eye whether it was on course. If the missile strayed, von Braun called for stopping the engines by radio. On one occasion, the eye failed to detect a tipping toward the south, and the missile landed in a cemetery in Juarez, Mexico, causing something of an international incident. Rumor had it that von Braun's lapse might have been related to his having some instruments riding on the rocket. At any rate preparations to track the missiles by instrument were accelerated.

The Naval Research Laboratory used radio signals from the flying rocket to measure the electrification of the ionosphere. For this purpose the laboratory installed ground stations uprange from the launching area. One day as the men were preparing one of the stations for an approaching flight, an Army jeep drove up, and a soldier got out and began driving a stake into the ground not more than a stone's throw from the station. Curious, the men asked what that meant. That, they were told, was the aiming point for some planned Honest John rocket tests. The men let it be known they didn't fully appreciate being made the target of rocket firings. "Not to worry," was the answer, "we never hit the target, anyway!"

Often there was frustration to struggle with. During the countdown for the firing of V-2 no. 16, something in the tail switch, which was supposed to turn the experimental equipment on after takeoff, was wrong. An effort was made to reconnect the switch there on the launch stand with the fully loaded rocket waiting to take off. After launch, however, instead of turning instruments on, the rewired switch proceeded to turn everything off. A postflight review showed that there were several ways in which the switch could have been connected to do the intended job, and only one way in which it would fail. The one and only wrong way had been chosen—an important object lesson regarding hasty, last-minute changes in the field. It turned out, however, that this rocket tumbled end over end in flight, which would have made the reduction of data an exceedingly complex matter. The scientist in charge later said it was probably a good thing that the equipment had been turned off, for otherwise the experimenters would surely have been unable to resist the temptation to try to interpret the measurements and probably would have wasted a lot of time on a futile exercise.

On another occasion, as a physicist watched a rocket carry aloft the cloud chamber over which he had labored long and hard, he remembered that he had forgotten to remove the lens cap from the recording camera. To add to the feeling of despair, the telemetering record indicated that the cloud chamber had worked perfectly during the flight.

Of course, it was always heartbreaking when the rocket failed to perform. It was difficult enough for some experimenters to reconcile themselves to the thought that the equipment they had struggled to perfect would often be destroyed on a single flight. There was consolation when the flight produced the data sought, but not when the rocket failed. After the program had been under way for some time, it was noted that the rockets bearing the simplest payloads seemed to have the best success. The Applied Physics Laboratory group, which never attempted to load rockets to full capacity, had acquired an image of almost perfect success. In contrast, the Air Force Cambridge Research Center, which tried to conduct dozens of complicated experiments on a single flight—and even lengthened the V-2 by a whole diameter to make additional instrument space—had developed an image of almost complete failure. The Naval Research Laboratory, which flew payloads intermediate between those of APL and AFCRL in complexity, succeeded about two-thirds of the time. There seemed to be an interaction between the experimenting and the launching operations, the more complex experiments tending to induce more problems with the rocket itself. The suspicion that this was actually happening was widely held, but never proved. On closer look, the evidence is not as clear as it seemed at the time, for the Princeton experiments were as simple as any, and yet all their rockets failed, which was no doubt the main reason for Princeton's early withdrawal from the program.

One cannot work with rockets without a certain amount of danger. Although the missiles were aimed away from them, the stations uprange were nevertheless exposed to some risk that the rocket might land on one of them. No direct hit ever did occur, but on a few occasions the wreckage from a falling rocket landed uncomfortably close. The greatest danger existed when the rocket was being loaded with propellants and people were still working around it, completing last minute preparations. When a spurt of hydrogen peroxide set a jeep afire, the industrial supplier was moved to assert publicly that the liquid was perfectly safe if only it were handled properly. Most distressing were accidents to personnel, as when a fuming sulfuric acid mixture being loaded into a V-2 prematurely ejected, spraying the face of a worker and endangering his eyesight. The acid mixture was used to generate visible clouds in the stratosphere, which were then tracked to measure stratospheric winds.

The author vividly remembers working with a companion on a platform 10 to 15 meters above ground, inserting live JATO* rockets into receptacles in the midsection of a fully loaded V-2. Tests had shown that JATO would ignite from the slightest applied voltage, and care had to be exercised not to generate any static electricity or to permit current to flow through the JATO igniter from the ohmmeter being used to check the circuits. Other workers had retired to a respectful distance. Slanting cables had been drawn between the work platform and the ground, down which—if things went wrong—one could slide and then run like hell to safety. The JATOs, which were intended to impart a spin to the rocket in the upper atmosphere, did not ignite during the loading. But, then, neither did they spin the V-2 in flight.

---

*Jet Assist Take Off rockets permitted heavily loaded aircraft to take off from short runways.

International Contacts

From the panel's labors gradually accumulated an array of answers to important questions that had previously been intractable. As noted earlier, published results began to attract attention in the United States. The sounding rocket program also aroused interest abroad. At the panel's 13 June 1950 meeting, Sydney Chapman, renowned geomagnetician from the United Kingdom, joined the discussions. From that time international contacts gradually broadened, as Chapman became a frequent participant and visitors from Belgium, Australia, Japan, and Canada came. In the fall of 1952 the Royal Society's Gassiot Committee—a committee concerned with upper atmospheric research—proposed an international meeting on that subject, to be held at Oxford the following August. At the conference the Europeans heard the U.S. program and results discussed in detail, while the Americans became aware of a growing interest among scientists from other countries. By publishing the proceedings in book form, the British stole something of a march, giving panel members occasion to reassess their own publication program.[25]

At this very period early plans for a "Third Polar Year"—a worldwide cooperative program of geophysical investigations—were taking shape (chap. 5). Van Allen and other panel members had already been considering the possibility of conducting rocket soundings in the vicinity of Fort Churchill, Canada. The author proposed at the panel's January 1953 meeting that a "full fledged operation of Northern latitude firings be organized for the Third Polar Year 1957–1958" and presented objectives and requirements for such a program at the following meeting.[26] In October 1953, Joseph Kaplan, chairman of the U.S. National Committee for the International Geophysical Year (the new name for the Third Polar Year), and Sydney Chapman, chairman of the International Committee for the IGY, both approved the idea of an IGY rocket program. Kaplan reported that the panel would be asked to serve as advisory committee to the National Academy of Sciences' National Research Council for the rocket phases of the IGY program, but very shortly thereafter the academy established its own Technical Panel on Rocketry.[27] To coordinate planning and preparations for firings at Fort Churchill—after some negotiations Canada formally extended an invitation to the United States to set up a rocket-launching range there—the panel formed a Special Committee for the IGY (SCIGY). Hearing of the Research Council's Technical Panel on Rocketry, the panel transferred SCIGY to the academy's technical group.[28] SCIGY's membership was then expanded slightly to the following:

H. E. Newell, Naval Research Laboratory, Chairman
J. W. Townsend, Jr., Naval Research Laboratory, Executive Secretary
John Hanessian, Jr., National Academy of Sciences, Recording Secretary

K. A. Anderson, State University of Iowa
Warren Berning, Ballistic Research Laboratories
L. M. Jones, University of Michigan
R. M. Slavin, Air Force Cambridge Research Center
N. W. Spencer, University of Michigan
W. G. Stroud, Signal Engineering Laboratories[29]

The military services permitted their employees to take part in the IGY program and undertook to provide logistic support for shipboard operations and for setting up an Aerobee tower and a Nike-Cajun launcher at Fort Churchill. The Army was in overall charge of the U.S. rocket contingent at Fort Churchill, while the three services shared the expenses. But additional funds were needed. Accordingly, Van Allen submitted on behalf of panel members a budget request to the IGY committee for more than one and a half million dollars, about 15 percent of America's total planned budget for the IGY.[30] The costs of the program were defrayed by both the military services and the IGY budget.

THE IGY SATELLITE PROGRAM

Although individual members had long been interested in the use of artificial satellites for scientific research, the panel up to this point had recommended only a sounding rocket program for IGY. But simultaneously with the planning for rocket firings, enthusiastic advocates were pressing for the launching of scientific satellites. Inevitably the panel was caught up in these proposals. To explore at length the usefulness of satellites for scientific research, the panel sponsored a symposium at the University of Michigan 26-27 January 1956. The proceedings were published in a book,[31] the sale of which generated a small treasury for the panel.*

Once aroused, interest in scientific satellites grew rapidly. Most members took part one way or another in the IGY satellite program. Gradually the idea emerged that the United States should go further and establish some kind of permanent space agency. In the summer of 1957, the author jotted down some brief notes outlining a "National Space Establishment" to be organized and funded to conduct unmanned space research and applications and manned exploration of outer space. Shortly thereafter the panel—which the preceding April had changed its name to Rocket and Satellite Research Panel—took steps to explore formally its potential interest in earth satellites and outer space. Report 47, 19-20 September 1957, records the creation of a Committee on the Occupation of Space, chaired

---

*Having a bank account was a source of some perplexity not resolved until years later, when the money was donated to a small, nonprofit activity called Science Services. The income from the gift was to provide for an annual award to a student competing in the International Science Fair. The panel suggested that the award be for excellence in the field of space exploration, space science, space engineering, or space application. Megerian, minutes of panel, rpt. 1968-1.

by the author. When *Sputnik 1* went into orbit, the panel intensified its efforts on behalf of a civilian National Space Establishment.[32]

On 21 November the group issued a paper entitled "A National Mission to Explore Outer Space." A different version, "National Space Establishment," appeared on 27 December 1957 (see app. D). The minutes of the 6 December panel meeting record that the earlier paper had been discussed with Detlev Bronk, president of the National Academy of Sciences. Copies had also been given to James Killian, the president's science adviser, and to Lee DuBridge, president of the California Institute of Technology. Dr. Killian referred the panel report to Herbert York, Emanuel Piore, and George Kistiakowsky, members of the President's Science Advisory Committee who were also exploring the question of the United States role in space.[33]

To this point the panel's policy of restricting membership to those working in the upper-atmosphere program had made good sense. But now the panel felt the need for additional weight behind its recommendations. During December 1957 the membership about doubled, adding key persons in the military research establishment, industry, the rocket development field, and the American Rocket Society (app. A). The society was also agitating at the time for the creation of a civilian space agency.[34] The two groups agreed to join forces in promoting the idea, and on 4 January 1958 issued a summary paper supporting their joint proposal for a "National Space Establishment" to have responsibility for investigating and exploring space.

In addition to preparing that paper, the Rocket and Satellite Research Panel mapped out a plan to bring its recommendations to the attention of persons who might be in a position to do something. Members visited congressmen and officials in the administration and sought help from the Academy of Sciences. A small group, chaired by the author and including Wernher von Braun and William Pickering, called on Vice President Nixon, who seemed most receptive. Through his good offices a number of meetings were arranged for the group on Capitol Hill and in the executive branch: with the commissioners and general manager of the Atomic Energy Commission, with George Allen and key figures in the U.S. Information Agency, and with the staffs of the House and Senate committees that were considering how to respond to the Soviet challenge in space. William Stroud, von Braun, and the author appeared before the Joint Committee on Atomic Energy and shocked members by asserting that the proposed space program could very likely require as much as a billion dollars a year and could become comparable to the atomic energy program once it got going.[35]

Panel members, of course, seized on whatever news they could acquire about what was going on. They heard that the space program could go a number of ways: a new agency might be created, which the panel had naively recommended; or responsibility might be assigned to an existing agency like the National Advisory Committee for Aeronautics; or the Department of Defense might get the job.[36]

Among panel members the NACA had an image of gross conservatism. In talking to Hugh Dryden, director of NACA, Whipple received the impression that NACA was "not prepared to undertake space research on the scale considered essential by the RSRP and by the American Rocket Society." Whipple had also talked with General Doolittle, NACA chairman, who declared his intense feeling that it would be a great error to set up any such organization outside of the Defense Department's jurisdiction.[37] His opinion was disturbing to panel members, who had felt the pinch of budgets for sounding rocket research competing with budgets for purely military purposes and who would like to remove the periodic vexation of the classification battle. Although members recognized that the new agency would have to depend on the military for a great deal of hardware and logistical support, to a man—including those employed by the services—the panel was determined that the nation's space agency ought to be civilian.

Doubts about the NACA did not lessen the feeling of satisfaction with the National Aeronautics and Space Act of 1958. Members were prepared to give whole-hearted support to the new National Aeronautics and Space Administration, which was to absorb NACA as its nucleus. Indeed, many joined the new agency. But the panel itself was now at loose ends. The purposes it had served for more than a decade would now be NASA's. For the next two years the panel devoted itself to colloquia on topics related to atmospheric and space research, but such colloquia could hardly serve the now explosively expanding field the way sessions of the scientific societies could. Members experienced a growing dissatisfaction where before a feeling of pioneering excitement had suffused the discussions. William Pickering submitted his resignation with a statement that he felt that the panel no longer served any real purpose.[38]

Having existed for so long without any formal charter, the panel now found time to compose a constitution, which was declared adopted by a three-fourths vote at the meeting of 17 February 1960.[39] After one more meeting, the panel suspended operations.

Thus, the panel's success in helping bring about the creation of a new agency devoted to the investigation and exploration of space also brought the demise of the panel. In contrast, the National Academy of Sciences, which the Rocket and Satellite Research Panel had drawn into the rocket research field, expanded its role in the program after the creation of NASA. The Space Science Board, which grew out of the academy's IGY panels on rocketry and earth satellites, was an immediate source of advice to NASA in its formative years, taking over the advisory role that the Rocket and Satellite Research Panel had once played. As a committee of the nation's prestigious Academy of Sciences, the Space Science Board enjoyed a vantage point that the panel never had commanded. How the academy went into space science and events leading to the establishment of the Space Science Board as one of the prime sources of advice to NASA are dealt with in the next chapter.

5

The Academy of Sciences Stakes a Claim

In the fall of 1957 the National Academy of Sciences in Washington was hosting an international conference on rockets and satellites. The mood was one of anticipation. The International Geophysical Year had begun officially on 1 July 1957 after several years of careful planning under the guidance of the Comité Speciale de l'Année Géophysique Internationale (CSAGI). Now, during the week of 30 September to 5 October 1957, CSAGI was giving special attention to the continued planning of the rocket and satellite part of the IGY program.

The International Geophysical Year—IGY for short—grew out of a suggestion made in 1950 by Lloyd V. Berkner to a small group gathered at the home of James A. Van Allen in Silver Spring, Maryland, that in the period 1957–1958 there should be a Third International Polar Year. Two previous International Polar Years, the first 1882–1883 and the second 1932–1933, had demonstrated in some measure the value of international cooperation in earth science investigations.[1] The group was heartily in favor of the idea.

No better promoter for such a project could have been found than Lloyd Berkner. A world-renowned geophysicist, he had long worked with problems of the ionosphere, the electrified region of the upper atmosphere that is responsible for reflecting radio signals around the world, making radio communications beyond the horizon possible. Berkner had been associated with G. Breit and M. A. Tuve who, at the Carnegie Department of Terrestrial Magnetism in Washington, had been among the first to measure the height of the ionosphere.[2] Berkner was interested in the institutional and international aspects of science, serving as adviser to the Department of Defense and in the State Department, and becoming very active in a number of the unions of the International Council of Scientific Unions. Most notable was his boldness of vision, which in scientific circles was fully a match for that of von Braun in rocketry.

Immediately swinging into action, Berkner and Sydney Chapman conveyed the proposal to the Joint Commission on the Ionosphere of the International Scientific Radio Union, the International Union of Geodesy and Geophysics, and the International Astronomical Union. In time the recommendation reached and was adopted by the International Council of Scientific Unions—parent body of the various individual unions—which in 1952 appointed a special committee to oversee the project. In the course of soliciting participation, the enterprise was enlarged to encompass the scientific study of the whole earth, a subject more broadly appealing than polar investigations.

Thus the IGY was born.[3] The special committee was formally designated the Special Committee for the IGY, referred to in both speech and writing as CSAGI (generally pronounced *kuh sah jee*), the acronym of its name in French. Chapman became the president of CSAGI, Berkner its vice president, and both labored tirelessly and effectively to make the project go. Eventually 67 countries joined.

Chapman gave some idea of the scope of the IGY as finally conceived by its planners and organizers in his general foreword to the first volume of the *Annals of the International Geophysical Year*:

> The main aim is to learn more about the fluid envelope of our planet—the atmosphere and oceans—over all the earth and at all heights and depths. The atmosphere, especially at its upper levels, is much affected by disturbances on the Sun; hence this also will be observed more closely and continuously than hitherto. Weather, the ionosphere, the earth's magnetism, the polar lights, cosmic rays, glaciers all over the world, the size and form of the earth, natural and man-made radioactivity in the air and the seas, earthquake waves in remote places, will be among the subjects studied. These researches demand widespread *simultaneous* observation.[4]

Responding to the international call to countries to join in the IGY project, the National Academy of Sciences, through its National Research Council, established a National Committee for the IGY and selected Joseph Kaplan as its chairman (app. E). Kaplan, a geophysicist who had acquired a considerable reputation working in the laboratory on band emissions from various atmospheric gases, was noted for an inexhaustible supply of pleasant anecdotes. His genial personality was ideally suited to working with the difficult, dark, moody, sometimes abrasive Hugh Odishaw, executive director of the committee. Odishaw was the guiding genius behind the organization of the U.S. IGY program, and his influence could be felt in the international arena as well. He had a remarkable ability to foresee the future consequences of present actions, and was invaluable in mapping out strategy and directing tactics for securing support from the administration and Congress and in dealing with the inevitable conflicts and politicking on the international scene.

In putting forth his original proposal, Berkner had cited the great advances in technology and scientific instrumentation since the early 1930s—much of it generated in the prosecution of World War II—as a compelling reason for not waiting out the 50 years that had intervened between the

First and Second Polar Years. Shortening the interval to 25 years would put the proposed Third Polar Year in the period 1957–1958, which would afford the added advantage of being a time of maximum sunspots, in contrast to the sunspot minimum of the Second Polar Year.

Among the new technologies that could be applied to the detailed investigation of the earth and sun were those of rocketry, whose applications to geophysics had become patently clear from the work of the Upper Atmosphere Rocket Research Panel, some of whose meetings Kaplan had attended. When the panel proposed to conduct rocket soundings as part of the IGY program, the U.S. National Committee quickly approved. Steps were taken at once to include a sounding rocket segment in the U.S. program for the IGY.

The National Research Council established a Technical Panel on Rocketry as part of the IGY machinery, and the National Academy of Sciences informed the CSAGI of the U.S. intention to use sounding rockets for geophysical investigations during the IGY.[5] By the time of the IGY planning meeting conducted by CSAGI in Rome during the week of 30 September 1954, rocket soundings had become an important element of the program.

But plans soon went beyond sounding rockets. When the U.S. sounding rocket program had begun in 1946, satellites were still deemed impracticable; now matters were different. The Navy, the Air Force, and other groups had continued to study the design and use of artificial satellites launched into orbit by large, powerful rockets, and by the early 1950s the feeling had developed that the satellite's time had come. Hidden by security wraps, some studies had moved fairly far along in the planning stage.[6] Von Braun and his people had convinced themselves that they could succeed in short order in orbiting a small satellite, and it rankled that official approval could not be obtained.

Members of the Upper Atmosphere Rocket Research Panel were aware of these studies, but those who were employees of the military did not feel free to press the issue. As has been seen, the panel recommended only a sounding rocket program to the Academy of Sciences. But geophysicist S. Fred Singer of the Applied Physics Laboratory, who had been conducting cosmic ray and magnetic field research in sounding rockets, felt under no restraints of military security. From some fairly simple calculations Singer concluded that it should be possible to place a modest (45-kilogram) satellite in orbit around the earth, and at every opportunity he urged that the country undertake to do so. Singer's conclusions were qualitatively correct, but his outspokenness generated some friction for at least two reasons. First, Singer's manner gave the impression that the idea for such a satellite was original with him, whereas behind the scenes many had already had the idea, and they felt that Singer had to be aware of this. Muzzled by classification restrictions, they could not engage Singer in debate. Second, being unable to speak out, those who had dug into the subject in much greater depth could not point out that Singer's estimates overshot the mark somewhat, and that his suggested approach was not as workable as others that couldn't be mentioned.

Nevertheless, Singer did great service to those he made so unhappy. By making known the present possibility of placing artificial satellites in orbit, Singer aroused interest in this kind of device for scientific research.[7] The IGY was to be the first beneficiary.

Singer gained international attention for his proposal when, in August 1953 at the Fourth International Congress on Astronautics in Zürich, he described his idea for a Minimum Orbital Unmanned Satellite Experiment—which he called Mouse. Mouse would weigh 45 kilograms and would be instrumented for scientific research.

The International Scientific Radio Union, at its 11th General Assembly in the Hague, gave its support to Singer's proposal. At the urging of both Singer and Lloyd Berkner, on 2 September 1954 the Radio Union adopted a resolution drawing attention to the value of instrumented earth satellites for solar and geophysical observations. Later that month, on 20 September, the International Union of Geodesy and Geophysics at its 10th General Assembly in Rome adopted an even stronger resolution, actually recommending that consideration be given to the use of small scientific satellites for geophysical research.[8] Both the resolution of the Union of Geodesy and Geophysics and the earlier one of the Radio Union were conveyed to CSAGI, which held its third general planning meeting in Rome shortly after the close of the Geodesy and Geophysics Union meeting. Indeed, it is unlikely that these two resolutions could have been missed by CSAGI, since many persons attended all three meetings—radio, geophysics, and CSAGI—and even more attended both the last two. Also, the CSAGI membership included representatives from a number of scientific unions, including radio and geodesy and geophysics. The combining of forces to promote programs of mutual interest is traditional among the scientific unions, where maneuvering has much in common with ordinary politics.

At any rate, on 4 October 1954 CSAGI agreed and issued its challenge to the IGY participants in the following words, which closely parallel the resolution adopted by the Union of Geodesy and Geophysics:

In view of the great importance of observations, during extended periods of time, of extra-terrestrial radiations and geophysical phenomena in the upper atmosphere, and in view of the advanced state of present rocket techniques, CSAGI recommends that thought be given to the launching of small satellite vehicles, to their scientific instrumentation, and to the new problems associated with satellite experiments, such as power supply, telemetering, and orientation of the vehicle.[9]

53

It remained for interested countries to respond to the recommendation. The United States had already announced its intention to conduct sounding rocket flights as part of the IGY program, but the complexity and expense of an earth-satellite program needed careful consideration by the agencies that would be expected to carry out the necessary development and IGY operations. Moreover, at the opening session of the Assembly of the Union of Geodesy and Geophysics Sydney Chapman as president of CSAGI had found it necessary to point out that the Soviets had not yet seen fit to join the IGY program. It was the United States and the Soviet Union, of course, that were expected to respond positively to the CSAGI proposal.

The U.S. National Committee for the IGY gave careful consideration to the proposal during the spring of 1955. Support was not immediately unanimous. Clearly the dimensions of this undertaking would be of a different order from the sounding rockets already a part of the IGY planning. Doubts were expressed over the wisdom of including the project in the IGY. Technical aspects were not the only considerations. There was also the concern about what would be the reaction of people to the launching of an artificial satellite that could easily be viewed as an eye in the sky, could well be accorded some sinister import, perhaps even be equated with some kind of witchcraft. Memories of Orson Welles's Mars invasion had by no means vanished. Most, however, favored endorsing the project. Joseph Kaplan, chairman of the committee, was especially enthusiastic and jokingly coined the phrase "Long Playing Rocket" for the satellite, by analogy with the long-playing records newly on the market. He suggested that, since sounding rockets had become familiar, the idea of a long-playing rocket would prove less disturbing than the completely new concept of an artificial satellite.

After much thought the National Academy of Sciences, sponsor of the U.S. IGY program, and the National Science Foundation, which provided the money, agreed jointly to seek approval of an IGY scientific earth-satellite program. The two agencies were successful, and the U.S. intent to launch an earth satellite during the IGY was announced from the White House on 29 July 1955.[10] A significant factor in administration support was the perceived need to develop and explore satellite capabilities for possible military applications. The Pentagon was assigned the job and, after a review of several possibilities, selected the Navy's Vanguard for the purpose.[11]

With scientific satellites now in the IGY program, the IGY committee established a Technical Panel on the Earth Satellite Program, to do for satellites what the rocketry panel was doing for sounding rockets. Richard Porter, the General Electric Company engineer in charge of the V-2 test program at White Sands, was asked to be chairman.[12]

When the Fourth Assembly of CSAGI met in Barcelona, 10–15 September 1956, the Soviet Union had joined the IGY and was prepared to say something about Soviet rocket and satellite plans. On 11 September, Prof. I. Bardin, speaking in Russian, announced to the CSAGI delegates that the USSR would have a rocket program in the IGY, details to be given later, and also would use satellites for pressure, temperature, cosmic ray, micrometeor, and solar radiation measurements.[13] Whereas the United States undertook to describe in considerable detail its sounding rocket and earth satellite plans, to aid those who wished to make correlated measurements by other techniques, the Soviet Union furnished little advance information.

Thus, when CSAGI convened the conference on rockets and satellites in Washington in the fall of 1957, there had been considerable time for work on the rocket and satellite projects, but it remained to be seen how much cooperative research could be done in association with those projects.

The subdued sense of anticipation that pervaded the sessions stemmed from the awareness that preparations had been under way for some time, that the IGY was already in full swing, and that the first artificial satellite must soon appear over the horizon. But those expectations did not diminish the surprise and dismay felt by U.S. scientists when the launching of *Sputnik 1* was announced on the evening of 4 October 1957. At the time many of the conference attendees were guests at the Soviet Embassy. The news, which had been broadcast by Moscow Radio, was brought to Berkner, who shared it at once with those present. His announcement practically wiped out the party as the U.S. scientists, in particular, scattered to their home bases to take stock of what had happened. The author had gone home that evening from the planning sessions at the academy and was about to call it a day when Hugh Odishaw called. As executive director of the U.S. National Committee for IGY, Odishaw wondered if a few people shouldn't meet at the IGY headquarters—1145 19th Street, N.W.— to discuss the turn of events. Once there, Odishaw, Richard Porter, the author, and others kept track of the Soviet satellite's course. From radio sightings as they were reported, the ground track of Sputnik was plotted on a map in the office, and in a few hours a pretty good idea emerged of the kind of orbit Sputnik was following.

As the group in imagination followed the course of the satellite across the heavens, the members tried to weigh the Soviet accomplishment against the fact that the launching of the U.S. satellite, Vanguard, was still some months away. They tried to estimate what the public reaction would be. Disappointment was to be expected, but they did not anticipate the degree of anguish and sometimes genuine alarm that would be expressed over the weeks and months that followed.

The next morning, Saturday, 5 October 1957, in the auditorium of the U.S. National Academy of Sciences, Anatoly Blagonravov took the floor to

speak at length about Sputnik. Understandable pride was evident in Blagonravov's bearing, but his words also bristled with barbs for his American
listeners. The speaker could not—or at any rate did not—refrain from chid-
ing the United States for talking so much about its satellite before having
one in orbit, and commended to his listeners the Soviet approach of doing
something first and then talking about it.

While there was some measure of justice in Blagonravov's ungracious
comments, his U.S. colleagues couldn't help feeling that he missed—perhaps intentionally—the point that much of the advance discussion of the
U.S. IGY satellite program was to provide necessary information for planning by those who wished to cooperate in the tracking or other operational
aspects of the mission. In view of the fruitlessness of CSAGI's efforts to
elicit any such accommodation from the Soviets, either at Barcelona in
1956 or at the meetings in Washington, the remarks of their Russian colleague were doubly frustrating.[14]

Nevertheless, admiration for the Soviet achievement was genuine and
universal, and his colleagues could heartily applaud when Blagonravov declared that he hoped that "this first step" would "serve as an inspiration to
scientists throughout the world to accelerate their efforts to explore and
solve the mysteries and phenomena of nature remaining to be explored."[15]

Reaction in the United States was strong and widespread. It was clear,
albeit intuitively to most, that a new dimension had been added to man's
sphere of thought and action. Equally clearly, something had to be done
about the fact that the United States had not been the first to put a satellite
in orbit. One read and heard talk about Soviet technological supremacy,
U.S. loss of leadership, the missile gap, and security and economic implications. In view of the impressively large weights of *Sputnik 1* (80 kg) and
2 (508 kg, 3 Nov. 1957), and the multiton launch vehicles that they implied, the 8½-kg payload of *Explorer 1* launched on 31 January 1958 did
little to allay such concerns. President Eisenhower attempted to downplay
the Soviet achievement, but couldn't carry it off.[16] Congress took the matter
seriously, largely through apprehension over military implications, and
began to crank up the machinery to respond to what was viewed as a
crisis. On his part, Eisenhower created the post of science adviser to the
president, elevated his Science Advisory Committee to White House level,
and asked the committee to develop a national policy on space. The result
was to be the National Aeronautics and Space Act of 1958.

By now atmospheric and space science had moved far beyond the narrow confines of the Rocket and Satellite Research Panel and had established a base from which the space science program could proceed following the creation of the National Aeronautics and Space Administration in
the summer and fall of 1958. From the membership of its technical panels
on rocketry and on the earth satellite program, the academy established a
Space Science Board in June 1958, to advise the government in what prom-

ised to be a fast-growing and important field. Lloyd Berkner was named chairman of SSB (app. F).

Events of the next three-quarters of a year after the first Sputnik launching make a fascinating and educational story as Congress and the administration cooperated and wrestled with each other to hammer out a legislative response to the crisis.[17] A number of circumstances combined to give scientists the civilian agency and open space program they favored. How this came about will be dealt with in chapter 7. But before proceeding to that part of the narrative, it is appropriate to pause and take stock of the rich harvest of scientific knowledge that a decade of rocket sounding had already produced before artificial earth satellites took on an importance that commanded the attention of the president and the Congress.

6

Early Harvest:
The Upper Atmosphere and Cosmic Rays

Scheduling V-2 flights, developing newer rockets, testing instruments, seeking financial support, fighting military classification, arguing and politicking in meetings national and international—such activities seemed to consume more time and energy than the actual science that was their ultimate purpose. But because of those subsidiary activities, which fill most of the pages of this book, the scientific research moved steadily forward. Month by month, year by year the results accumulated. By the time NASA began to operate, a rich harvest had already been reaped from sounding rockets, with several significant contributions from the scientific satellite program of the International Geophysical Year. These, especially upper atmospheric and cosmic ray research, gave NASA a running start in space science.

By the early 1960s the study of energetic particles and magnetic fields from the sun and their interaction with the earth's magnetic field had become a well integrated and coherent field of study. By then, also, satellite geodesy had begun to make its mark. But the space science program was open ended, and the harvest a continuing one. This steady advance of space science is the subject of three chapters (6, 11, 20), whose aim is to present in broad outline what the space science disciplines encompassed and to show how space techniques made notable contributions. The present chapter reviews achievements through 1958.

THE THRESHOLD TO SPACE

Thirty thousand light-years from the center of a disk-shaped galaxy, itself measuring 100 000 light-years from edge to edge, planet Earth revolves endlessly around an average star, the sun, which with its attendant planets speeds toward remote Vega, brightest star in the northern skies. Although containing billions of stars, nebulas, and other celestial objects, most of the galaxy consists of empty, or nearly empty, space. To inhabitants of Earth the threshold to these outer voids is the upper atmosphere.

One can easily show theoretically that the pressure and density of the atmosphere must decrease exponentially with increasing height above the ground, and experiment confirms this conclusion.[1] Roughly, at least for the first hundred kilometers, pressure and density fall to one-tenth their former value for every 10-mile (16-km) increase in altitude. Hence, above 30 km only one percent of the atmosphere remains, while beyond 100 km lies only one-millionth of the atmosphere.

Interest in the lower atmosphere where people live and experience the continuous round of changes in weather and climate is obvious, but one might well ask what could possibly hold the attention, even of scientists, in a region so nearly empty as the upper atmosphere? The initial impression, however, is misleading. After closer study the upper atmosphere is found to exhibit many fascinating, often practically important phenomena—such as the ionosphere, which profoundly influences radio communications, especially shortwave; the auroras; electric currents, which at times cause magnetic effects that blank out both radio and telephone links; and the ozonosphere, which during the debate over fluorocarbon-propelled aerosols gained temporary stature in the public mind as the protecting layer that shields the earth's surface from life-killing ultraviolet rays of the sun. So interesting were the challenging phenomena of the upper atmosphere that by the time sounding rockets put in their appearance, scientists had already evolved from afar a coherent, comprehensive picture of the upper atmosphere and solar-terrestrial relationships. In the mid-1940s this remarkably complete picture formed a paradigm that hundreds of geophysicists around the world shared and used in reporting their continuing researches at scientific meetings and in the literature.

The main features of this paradigm were set forth in an article on the upper atmosphere by B. Haurwitz, first published in 1936 and 1937 and reissued with some updating in 1941.[2] For those who began using sounding rockets in 1946 to explore the upper atmosphere, Haurwitz's concise review provided a helpful introduction, while a review paper by T. H. Johnson told much of what was known about cosmic rays from ground-based and balloon researches.[3] A classic paper by Fred Whipple on the use of meteor observations to deduce atmospheric densities at altitudes between 50 and 110 km was one of the best examples of the ingenuity necessary in studying a region not yet accessible to them or their instruments.[4] But the work that best described the state of knowledge of the earth's high atmosphere at the very time when the sounding rocket program was getting under way in the United States was a book of more than 600 pages, *The Upper Atmosphere*, by Indian scientist S. K. Mitra. Mitra furnished an exhaustive review of upper-atmospheric research, concluding with a chapter summarizing what had been learned and listing some of the most important problems needing further research. The very last paragraph noted that as the volume was going to press word had reached him:

that experiments are being conducted in the U.S.A. with the V-2 rockets to study the cosmic rays and the ionospheric conductivity up to heights of 150 km (August, 1946). It is hoped that the scope of these experiments will be extended and that the records obtained therefrom will, on the one hand, give direct information of upper atmospheric conditions and on the other reveal the true picture of the intensity distribution of solar ultraviolet radiation and thus help to solve the many mysteries of the upper atmosphere which till now have resisted all attacks.[5]

Mitra's hopes paralleled the motivations of the sounding rocket experimenters, many of whom were entering what was to them an entirely new field.

Following is an elaboration of the extensive paradigm that the space scientists inherited from the ground-based researchers (fig. 1). The description is based on the works cited above, especially on Mitra's treatise.

The atmosphere extended to great heights, auroras being observed on occasion to more than 1000 km. Pressure and density were calculated and observed to fall off in exponential fashion. If the temperature were uniform throughout the atmosphere, the decline in these quantities would be given by

$$p = p_0 \exp(-h/H) \tag{1}$$

and

$$\rho = \rho_0 \exp(-h/H) \tag{2}$$

where p and ρ denoted pressure and density respectively, the subscript zero indicated values at the ground, and h was altitude. The quantity H, known as the *scale height*, was given by

$$H = kT/Mg \tag{3}$$

where

k = Boltzmann constant = 1.372×10^{-16} erg/degree
T = temperature in kelvins
M = mean molecular mass of the air = 4.8×10^{-23} gm
g = acceleration of gravity.[6]

In (1) and (2) the value of g was assumed to be constant, whereas in actuality gravity varies inversely as the square of the distance from the

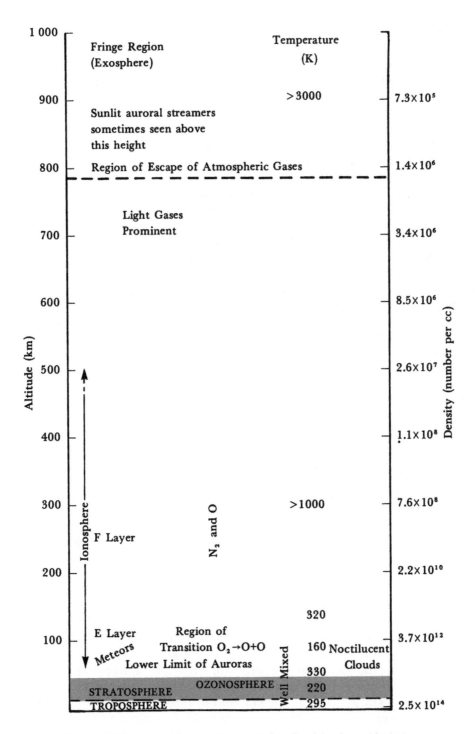

Figure 1. *The upper atmosphere as visualized in the mid-1940s.*

center of the earth. Hence the expressions for pressure and density were only approximate. More significantly, atmospheric temperature varied markedly with altitude, the scale height given by (3) varying proportionately. Thus, in regions of high temperature the pressure and density declined more slowly with height than where the temperature was low.

In regions where the atmospheric temperature was constant or nearly so, each separate atmospheric gas individually followed laws like (1) and (2) with the average molecular mass M in (3) replaced by the molecular mass of the individual gas. Thus, the corresponding scale height H varied inversely as the molecular mass of the gas, and a heavy gas like carbon dioxide fell off in density much more rapidly than nitrogen, oxygen slightly faster than nitrogen, and light gases like helium much more slowly. As a result the lighter gases appeared to diffuse upward, while the heavier gases settled out. Such considerations led one to suppose that at the highest altitudes, hundreds or thousands of kilometers above the ground, the lighter gases predominated in the atmosphere. The outermost regions were expected to consist of hydrogen or helium primarily, although no experimental evidence confirmed the supposition.

Starting at the ground, atmospheric temperature fell at a rate of about 6 K per km throughout the *troposphere* (or "weathersphere") to a value of around 220 K at the *tropopause*, or top of the troposphere, which was found at 10 to 14 km, the lower height corresponding to higher latitudes, the greater height to the tropics. Above the tropopause the temperature remained fairly constant to about 35 km. A slight increase in the proportion of helium in the air above 20 km suggested some tendency toward diffusive separation, which at one time led researchers to expect that the region above the tropopause would exhibit a layered structure—hence the name *stratosphere* for this quasi-isothermal region.

Above the stratosphere, temperature rose again, as shown by the fact that the sound from cannon fire and large explosions was reflected from these levels of the upper atmosphere. Observations on this anomalous propagation of sound waves permitted one to estimate that the air temperature was about 370 K at 55 km height. Noctilucent clouds between 70 and 90 km suggested a low temperature in the vicinity of 80 km. These extremely tenuous clouds were seen only in high latitudes and only when illuminated by the slanting rays from the sun below the horizon. With the assumption that the clouds were composed of ice crystals, the temperature around 80 km was estimated to be about 160 K. The study of meteors, investigation of the electrical properties of the high atmosphere by radio techniques, and observations of the auroras showed that temperatures rose again above 80 km to 300 K at 100 km, and to 1000 K or possibly 1500 K at 300 km, with much higher temperatures beyond. Calculations from auroral observations were, however, not always consistent with this picture, often indicating considerably lower temperatures than those deduced from radio observations.

Atmospheric composition near the ground was known to be:

(percentage by volume)

Nitrogen ....78.08 ⎤
Oxygen ....20.95 ⎬ 99.9
Argon .... 0.93 |
Carbon dioxide .... 0.03 ⎦
Neon .... 1.8×10^{-3}
Helium .... 5×10^{-4}
Krypton .... 1×10^{-4}
Xenon .... 1×10^{-5}
Ozone .... Variable, $> 1 \times 10^{-6}$
Radon (average near ground) .... 6×10^{-18}
Hydrogen .... Doubtful, $< 1 \times 10^{-3}$

Meteorological processes kept the atmosphere mixed, maintaining this composition at least up to 20 km. Between 20 and 25 km, helium increased about 3 percent above the normal value, but winds and turbulence kept the atmosphere well mixed far above stratospheric heights, up to at least 80 km.[7]

In the absence of other agents, this stirring should have kept the composition fairly uniform throughout the mixing regions. But solar ultraviolet radiation in the region from 1925 Å to 1760 Å, absorbed in atmospheric oxygen above the stratosphere, gave rise to a chain of reactions leading to the formation of ozone. Simultaneously solar ultraviolet in the neighborhood of 2550 Å decomposed atmospheric ozone. An equilibrium between the formation and destruction of ozone, combined with atmospheric motions, distributed the gas so that in temperate latitudes it showed a maximum absolute concentration at about 25 km height, and a maximum percentage concentration at about 35 km. Although never more than the equivalent of a few millimeters at normal temperature and pressure, the ozone layer shielded the ground from lethal ultraviolet rays from the sun. Ozone concentrations were observed to be higher in the polar regions than in the tropics, and tended to correlate with cyclonic weather patterns.

Above 80 km, solar ultraviolet dissociated molecular oxygen, the dissociation becoming fairly complete by about 130 km. Thus the region from 80 to 130 km appeared as one of transition from an atmosphere consisting of mostly molecular nitrogen and molecular oxygen, to one of molecular nitrogen and atomic oxygen. It was assumed that above 100 km diffusive separation of the atmospheric gases became increasingly effective, and that the dissociation of oxygen enhanced the tendency of nitrogen to settle out and the oxygen to rise. Whether nitrogen also dissociated in the higher levels was not known.

In the upper levels of the atmosphere was the *ionosphere*. The term was used in two different ways, either to mean the ionized constituents of the high atmosphere, or to mean the regions in which the ionization was found.

An ionosphere was postulated by Balfour Stewart in 1878 to explain small daily variations observed in the earth's magnetic field.[8] Later, in 1902, A. E. Kennelly in America and O. Heaviside in England suggested that a conducting layer in the upper atmosphere, which could reflect radio waves beyond the horizon, might explain how Marconi in 1901 had sent wireless signals from Cornwall to Newfoundland.[9] The first real evidence of such an ionosphere was obtained in 1925 when E. V. Appleton and M. Barnett in England detected sky waves coming down to their receiver after being reflected by a high-altitude layer.[10] Additional evidence of the Kennelly-Heaviside layer came from experiments by G. Breit and M. A. Tuve in America.[11] These experimenters sent a radio pulse upward, and observed two or more delayed pulses in a receiver a few kilometers away from the transmitter. The initial received pulse was assumed to be from the direct ray along the ground, and the other pulses to be echoes from the ionosphere. The method of Breit and Tuve became the basis for probing the ionosphere, using the reflections to determine the heights of various layers. Sophisticated formulas were worked out to explain how the various reflections observed were generated by the ionosphere. From these formulas and the experimental data, theorists could estimate layer heights, electron densities, magnetic field intensities, collision frequencies of the electrons and atmospheric particles, and reflection and absorption coefficients for the ionized media.[12]

The ionization was assumed to be caused by solar radiations, and ultraviolet was taken to be the most likely agent. Sydney Chapman showed how a monochromatic beam of ultraviolet light would generate a parabolic distribution of electron concentrations in an exponential atmosphere of molecules (like oxygen) susceptible to ionization by the radiation (fig. 2).[13] Starting with this basic theory and considering the effect of the various solar wavelength regions likely to influence the upper atmosphere, it was possible to estimate the variation with height of electron intensities and to make some guesses as to what the heavier ions might be.

From both radio observations and theory, scientists concluded that the ionosphere had two main regions of ionization, region E_1 centering on 110 km, and region F_2 centering on 275 km. The ionosphere was found to vary with time of day, season of the year, and phase of the sunspot cycle. For regions E_1 and F_2 halfway between the minimum and maximum of solar activity, the average ionization intensities corresponded to 10^5 and 10^6 electrons per cc, respectively.[14] Mainly during the daytime, regions E_2 and F_1

formed at heights of 140 km and 200 km. Region D, at some uncertain distance below the E region, was observed at times of high solar activity, and presumably because of the increased molecular collision frequency at those lower altitudes caused pronounced absorption of radio signals of medium wavelength.

At great distances from the earth, the earth's magnetic field was taken to be essentially that of a uniformly magnetized sphere; i.e., a magnetic dipole (fig. 3). Closer in, the field was observed to depart somewhat from that of a dipole, consisting of the dipole, or *regular*, part, and an *irregular* part. Some 94 percent of the earth's field, including some of the irregular field, was found to have its origin inside the earth. Of the remaining 6 percent of external origin, about half appeared to be caused by a flow of electric current between the atmosphere and the earth. The remainder, about 3 percent of the total field, appeared to be due to overhead electric currents.[15]

Such electric currents could be produced by atmospheric motions at high altitude caused by solar or lunar tides, or by nonuniform heating of the atmosphere by the sun as the earth turned. While these more or less regular daily variations could easily be accounted for by electric currents in the ionosphere, magnetic storms which occurred at times of solar activity were more likely associated with streams of charged particles from the sun. The initial increase in magnetic field observed during a storm could be explained by the arrival of charged particles from the sun, which com-pressed the earth's magnetic field slightly and thereby increased its value temporarily. The strong decline in intensity to below normal values which soon followed the initial phase might be caused by a huge ring current around the earth, fed by the particle stream from the sun, as suggested by

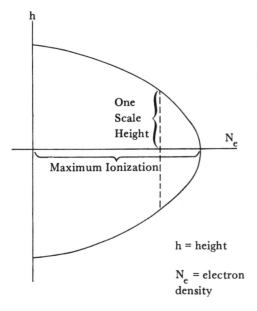

Figure 2. Chapman layer. The para-bolic distribution was estimated to be within 5 percent of the actual distribu-tion of charge densities to a distance of one scale height H (= kT/mg) above and below the level of maxi-mum ionization.

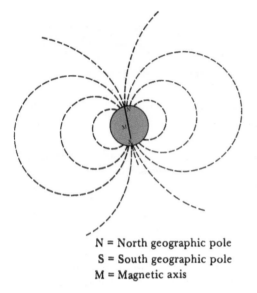

N = North geographic pole
S = South geographic pole
M = Magnetic axis

Figure 3. Earth's magnetic field. The broken lines depict the lines parallel to the direction of magnetic force. As became increasingly clear over the years, the actual magnetic field of the earth differs considerably from this idealized picture of a dipole field.

Chapman, Ferraro, and others. Then the gradual recovery from this "main phase" of the magnetic storm, as it was called, signified the gradual dissipation of the ring current and a return to normal conditions—or so it was thought.

Among the most notable of high-altitude phenomena, and among the earliest to be studied in detail, were the auroras, the northern and southern lights. These were seen most frequently at heights from 90 to 120 km, but also occurred at both lower and much greater heights. That the auroras correlated strongly with activity on the sun and appeared in an auroral belt at high latitude suggested that they must be due to charged particles from the sun. Charged particles would be steered by the earth's magnetic field, whereas neutral particles or solar photons would not be affected by the earth's field. Experimenting with cathode rays and small magnetized spheres, K. Birkeland in 1898 and others demonstrated how electrified particles approaching a magnetized sphere from a distance would be guided by the magnetic field toward the poles. Starting from Birkeland's concepts and experiments, over many years Carl Størmer developed a theory of how electrons or protons from the sun would be deflected by the earth's magnetic field into the auroral zones to produce the auroras as the particles impacted on the atmospheric molecules, causing them to glow.[16] The spectrum of the aurora was observed to exhibit primarily lines and bands of atomic oxygen and molecular nitrogen, with the forbidden green lines of atomic oxygen at 5577 Å being particularly strong.

At nighttime the high atmosphere was seen to emit a very faint light, sometimes called the permanent aurora, also consisting of the forbidden lines of atomic oxygen and of bands of the nitrogen molecule. This airglow was estimated to come from well above 100 km, perhaps from as high as 400 to 500 km, very likely from F-region ions as they were neutralized during the night. The yellow sodium D lines were also seen emanating from the lower part of the E region, and were particularly intense at twilight. From a distant cloud of material particles of some sort, the zodiacal light, with a spectrum similar to that of the sun, contributed to the light of the night sky. In the mid-1940s it was not known whether this radiation came from within the high atmosphere or from interplanetary space.

At some height, probably around 800 or 1000 km, the atmosphere was expected to cease acting like a normal gas. In this region collisions between atmospheric particles would be infrequent, and a molecule might rise along an elliptic orbit to an apogee and fall back without colliding with another molecule until returning to the denser atmosphere at lower altitudes. If the molecule had sufficient velocity it might even escape into interplanetary space. Indeed, it was supposed that hydrogen and helium had to be escaping continuously through this fringe region, even though neither had been detected in the upper atmosphere. Helium was known to be entering the atmosphere from the ground—where it was produced by the decay of radioactive elements—at a small but measurable rate; but the percentage of helium in the lower atmosphere remained constant over time. The natural conclusion was that this light gas had to be diffusing up through the atmosphere to the highest levels where the very high temperature permitted a ready escape of the gas.

Somewhere in this fringe region, or *exosphere*, the transition from the earth's atmosphere to the medium of interplanetary space was assumed to lie. One was hard put to it to define the boundary. Presumably where the atmospheric density had dropped to the few particles per cubic centimeter expected in interplanetary space the boundary must already have been crossed. But long before then the atmosphere had ceased to exist in the usual sense of the term. Across this ill-defined interface, radiations from the sun entered the earth's environs to cause the auroras, magnetic storms, ionization, and heating of the atmosphere.

Across this interface also came the cosmic rays.[17] These highly energetic particles from outer space were more the concern of the high-energy physicist than of the geophysicist. Discovered between 1911 and 1914 from balloon experiments on atmospheric ionization, cosmic rays quickly became a subject of intense interest. It was soon accepted that the rays came from outside the earth. Measurements of the ionizing power of the rays at various depths below the surfaces of mountain lakes revealed both a soft component and a hard, or extremely penetrating, component to the rays. Balloon experiments showed that the intensity of the radiation increased steadily with altitude until a maximum—called the Pfotzer maximum— was reached at about 20 km in mid-lititudes. The shape of these intensity-altitude curves is shown in figure 4a.[18] Figure 4b shows schematically that

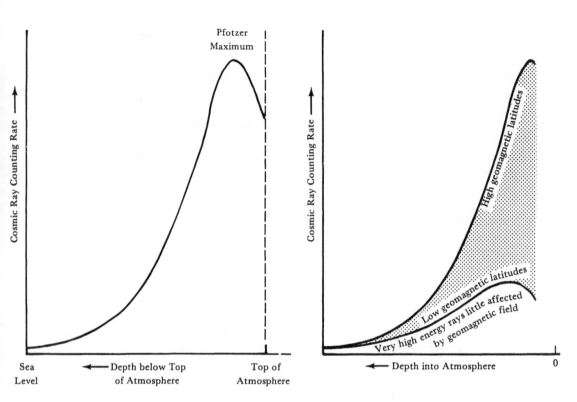

Cosmic Ray Counting Rate →

Pfotzer
Maximum

Sea Level — ←—Depth below Top of Atmosphere — Top of Atmosphere

Cosmic Ray Counting Rate →

High geomagnetic latitudes

Low geomagnetic latitudes

Very high energy rays little affected by geomagnetic field

←— Depth into Atmosphere — 0

Figure 4(a), at left. Cosmic rays at high geomagnetic latitudes. Figure 4(b), at right. Geomagnetic effect on cosmic rays. A schematic drawing showing how, according to experimental measurements, cosmic ray intensities vary with geomagnetic latitude. See Bowen, Millikan, and Neher in Physical Review 53 (1938): 855–61.

the earth's magnetic field has a distinct effect upon the radiation, leading to the conclusion that the rays are charged particles, not photons.

The shape of the intensity-altitude curve was explained as follows. The primary rays, whatever they might be, upon striking the atmosphere produced a shower of secondary rays, which, added to the primary rays, caused the initial increase in total ionization observed at high altitude. Eventually, however, an equilibrium was reached, with the atmosphere absorbing enough energy from both the primary and secondary particles to decrease the total ionizing power with further depth into the atmosphere. Such a transition curve, as it was called, would be observed not only in air, but also in lead or other substances, the principal difference being the spatial extent of the transitions, which depended on the density and nature of the material.

The early idea that the primary cosmic rays might be high-energy electrons was soon rejected. It could be shown that to penetrate the entire atmosphere and reach the ground, electron showers would have to be caused by primary electrons with such high energy that they would be completely unhindered by the earth's magnetic field. They would accordingly not exhibit the magnetic field effect already shown to exist. In 1938 T. H. Johnson concluded that the primary radiation consisted of protons, as theorists had guessed somewhat earlier. In 1941 balloon observations revealed that the cosmic rays within the atmosphere at high altitude were mostly mesotrons (mesons), presumably generated by the primary protons.[19] No significant component of electrons was observed at high altitude, supporting the conclusion that there could be no significant component of electrons in the primary radiation. But the soft component observed near the ground was believed to be electrons, decay products of the mesons generated at high altitude.

PROBLEMS TO SOLVE

Thus, the scientific paradigm for the earth's upper atmosphere in the mid-1940s was rich in ideas accumulated over more than half a century of observation and theoretical study. It had been possible to explain to a considerable degree a wide range of phenomena, many of which proved to be extraordinarily complex; but many uncertainties, unanswered questions, and problems remained.

Consider the problem of estimating atmospheric densities in the E region of the ionosphere around 100-km altitude. In the 1920s F. A. Lindemann and G. M. B. Dobson approached this problem by using observational data on the heights of appearance and disappearance of visual meteors. Intuitively it seemed reasonable that the density of the gas traversed by a speeding meteor should play a role in determining where the meteor would glow and be visible. The challenge was to develop a suitable theory to relate the observed meteor trails to the atmospheric density. Lindemann and Dobson assumed that as the meteor rushed into the atmosphere, a hot gas cap formed because of compression of the air. Heat from the gas cap was transferred to the meteor, and if the object were small enough it became incandescent. Making a number of assumptions about how heat was transferred from the gas cap to the meteor and using kinetic theory, Lindemann and Dobson derived expressions for ρ_a, the density of the atmosphere at the height of appearance, and ρ_d, the density at the height of disappearance of the meteor. The equations are reproduced here to emphasize the large number of quantities involved, uncertainties in which could cause errors in the derived atmospheric densities.

$$\rho_a = \frac{16}{3} \cdot \frac{\rho_m s \, T_2 \, r \cos \chi}{k v^2} \cdot \frac{g M_0}{R T_0} \tag{4}$$

and

$$\rho_d = \frac{24 \, r}{V_1 - V_2} \cdot \frac{\ell \Delta h}{v L} \cdot \rho_m \cdot \frac{g M_0}{R T_0} \tag{5}$$

where

ρ_m = density of the meteor
s = specific heat of the meteoric material
T_2 = temperature of the surface of the meteor
r = radius of the meteor
χ = angle of the meteor path to the vertical
g = acceleration of gravity
M_0 = molecular weight of the air
k = $(V_1-V_2)/3v$ = calculated efficiency factor of heating
V_1 = velocity of the compressed gas molecules in front of the meteor
V_2 = velocity of the gas molecules at the temperature of the meteoric surface
ℓ = latent heat of vaporization of meteoric material
v = velocity of the meteor, assumed constant
R = universal gas constant
T_0 = temperature of the atmosphere, assumed isothermal throughout the range of consideration
L = total length of the meteor trail
Δh = projection of L on the vertical.[20]

From the apparent brightness of the meteor the rate at which energy was being emitted could be calculated, which multiplied by the time of visibility gave the total amount of energy radiated. Setting this equal to the kinetic energy $\tfrac{1}{2}mv^2$ yielded the mass m of the meteor. If one then assumed that the meteor was iron and essentially spherical, one got from the expression

$$\text{mass} = \text{density times volume}$$

$$m = \rho_m \cdot (4\pi r^3/3) \tag{6}$$

which gave the radius r. The other quantities in the expressions for the atmospheric density could be either measured directly or estimated from plausible assumptions, thereby giving densities at two altitudes, that of appearance and that of disappearance.

The chain of reasoning was lengthy, with many assumptions. The results obtained by the investigators immediately put some of the assumptions into question. For example, the air densities obtained proved three times too high to correspond to an isothermal atmosphere at the stratospheric temperature of 220 K, requiring instead temperatures around 300 K. Between the stratosphere and the E region of the ionosphere, then, there had to be a significant variation in temperature. Moreover, other observations indicated that it was not even likely that the temperature would be constant in the E region. Experiments with the anomalous propagation of sound mentioned earlier showed that atmospheric temperatures rose markedly between 30 and 55 km to between 336 K and 350 K at the latter altitude. Noctilucent clouds, on the other hand, strongly suggested very low temperatures at 80 km.[21] The conclusion was forced, then, that the atmosphere was not isothermal, having temperatures which rose sharply above the stratosphere to somewhere at or above 55 km, fell again to very low values around 80 km, and then rose once more between 80 and 100 km.

Disagreements also arose over how the meteors became incandescent. One investigator objected to the idea of a gas cap, preferring to assume that the meteor was heated by direct impact with the air molecules.[22] In the early 1940s Fred Whipple obtained very accurate photographic records of meteor trails from which he could deduce decelerations. He developed an elaborate theory of how the properties of the upper atmospheric gases, the deceleration of the observed meteor and its heating to vaporization and incandescence, and its physical properties were all interrelated. Then, making some suppositions about properties of the incoming meteors and measuring deceleration and luminosity from the photographs, Whipple finally deduced the densities of the atmosphere along the trail.[23] Again there were assumptions and corresponding uncertainties in the results.

For the ionosphericist the theoretical maze was even more complicated. The prober's principal tool was the radio wave. A signal sent into the ionosphere would be bent by the ionized medium, and if the charge density were great enough would be reflected downward again. For a simple layer in which the strata of equal ionization were horizontal, the condition for total reflection of a signal propagated vertically was:

$$(4\pi \; N_e^o \; e^2)/mp^2 = 1 \tag{7}$$

where

N_e^o = value of the electron density at the point of reflection
e = the electronic charge
m = mass of the electron
p = angular frequency of the radio signal.[24]

Thus, a radio signal of low enough frequency sent into the ionosphere would continue upward until it reached a level at which the electron density was great enough to satisfy equation (7). At that point the wave would be reflected, returning to the ground after a delay corresponding to its flight along the upward and downward paths. As the wave frequency was increased, the wave would penetrate farther into the layer before being reflected, and the delay in the ionosphere would be increased. If the layer had a maximum electron density, when the signal frequency exceeded the value (called the *critical frequency*) for which that maximum charge density would produce total reflection, then the wave passed through the layer and no return was observed at the ground.

By sweeping the signal frequency from low values to higher ones, one could generate a record of signal returns which could be displayed as shown in figure 5, curve E. The critical frequency could be read from the figure, from which the charge density at the point of reflection could then be calculated, using equation (7). With a little additional calculation, the height of the point of reflection could also be estimated, showing where the reflecting layer existed.

If, in the charge density, other maxima lay above and exceeded the initial maximum, then as the wave frequencies were increased new reflections would be observed, corresponding to the higher-altitude, more intensely ionized layers, as shown in curve F of figure 5. From the critical frequencies for these higher layers, estimates could be derived for the charge densities and heights of the upper layers.

By using an appropriate theory like that of Chapman concerning the formation of ionized layers by solar radiations (fig. 3), one could then estimate charge densities above and below the maxima obtained from the radio propagation measurements, and thus construct a continuous curve of charge densities versus altitude.

The concept was simple, but enormous complications entered when all the pertinent factors were considered. First, the ionosphere was by no means as simple as assumed in the foregoing example, and at times the propagation measurements indicated gross inhomogeneities. Moreover, one had to take into account the earth's magnetic field, collision frequencies among the particles in the ionosphere, and the fact that the ionization consisted not only of electrons but also of both positive and negative ions. The earth's magnetic field produced double refraction of the radio signals used to probe the ionosphere, splitting the signal into what were called *ordinary* and *extraordinary* rays, which followed different paths, had different points of reflection and different delay times, and were differently polarized—that is, the electric vectors of the two rays vibrated in different planes. When there were several ionospheric layers to deal with, and particularly under disturbed conditions, the problem of identifying properly the various return signals could become next to impossible. In addition, when the signal had to traverse a region in which the collision frequencies were high, as in a strong D region during times of high solar activity, the signal could be greatly attenuated or even blanked out. Not knowing the ions in the ionosphere simply added to the complication.

The mathematical expression of how all these factors affected the propagation of signals through the ionosphere was far more complicated than the simple expression of equation (7), and applying it to the determination

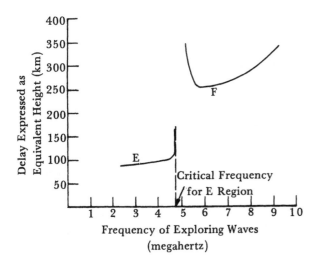

Figure 5. Radio wave reflections from the ionosphere. The time required for a signal to go to the ionosphere and return to ground gives a measure of the reflecting layer's height.

of charge densities in the ionosphere put great demands on ingenuity and insight.[25]

These two examples of how investigators restricted to working with observations obtained at or near the ground had to wrest the information they sought from long chains of supposition and theoretical reasoning illustrate the sort of opportunity that befell the rocket researchers, who expected to make direct measurements in situ. Since much, even most, of what went on in the upper atmosphere was caused directly or indirectly by energy from the sun, a most important contribution the rocket sounder could make was to measure the solar spectrum both outside the appreciable atmosphere and as affected by altitude within the atmosphere. Knowing the former would let the theorist know what wavelengths and intensities were generating ionization, various photochemical reactions, and ultimately heating in the atmosphere. Knowing the latter would immediately tell where the different wavelengths were having their effect. The importance Mitra put on this vital information is seen in his assertion that "the greatest obstacle in the study of the upper atmosphere, is undoubtedly the lack of our direct and precise knowledge of the energy distribution in the near and extreme ultraviolet radiation of the sun. For, conditions in the high atmosphere are almost entirely controlled by the sun."[26]

Many data the sounding rocket could obtain apparently could be obtained in no other way. In addition, many quantities that could be estimated from ground-based studies contained serious uncertainties which could be removed or lessened by rocket measurements. These circumstances made it possible for a number of young rocket experimenters in short order to compete respectably in upper-atmosphere research against much more knowledgeable scientists of many years' experience. The ways in which newcomers could contribute may be illustrated by listing some of the problems that in the mid-1940s still awaited solution.[27]

Diurnal, seasonal, and other temporal variations in atmospheric pressure, temperature, and density were needed.

A correct description of atmospheric composition at all altitudes would be invaluable. One could determine the distribution of ozone in the upper stratosphere and middle atmosphere and find the level at which most of the ozone was formed. Knowing the composition would also allow one to know definitely to what altitudes the atmosphere was completely or nearly completely mixed, and at what altitudes diffusive separation played an important role. In particular one would want to know where oxygen began to dissociate into atomic form and at what altitude the dissociation had become complete, and whether at some altitudes nitrogen also dissociated. At what level would lighter gases like helium become an appreciable or even dominant component of the air?

With respect to the ionosphere, radio sounding could not determine the ionization intensities in a region lying above one of higher charge density. One had to rely on theory to try to fill in the missing information. But in situ measurements might remove this lack. Moreover, if the precise nature and concentrations of both the positive and negative ions could be determined, a better understanding could be developed of how the balance between those agents creating the ionosphere and those tending to destroy it was established. One would then be in a better position to determine the specific causes of the temporal and geographic variations in the various ionospheric layers.

There was little doubt that excitation, dissociation, and ionization of atmospheric constituents, as well as various energy transfer and recombination processes, were responsible for the night sky radiations; but there were various possibilities among which to choose. Moreover, there were gross uncertainties in the altitudes from which many of the radiations were thought to arise. Again in situ measurements should help to resolve the difficulties, not only by pinning down altitudes, but also by providing additional insight into the recombination coefficients and other fundamental parameters involved.

As for magnetic field effects, a prime target would be to locate the electric currents that were responsible. One would hope, too, to be able to detect and identify the particles that caused the auroras.

With regard to cosmic rays, the precise composition of the primary radiation needed to be determined; for this purpose, measurements in outer space well above the atmosphere of the earth should be useful. Additional information on the effect of the earth's magnetic field upon the cosmic rays would be interesting, but more fundamental would be data on whether the radiation was isotropic or anisotropic in free space. An intriguing question was how many of the cosmic rays coming to the earth were from the sun and how many were from outside the solar system.

THE HARVEST

Such were the problems to which the rocket experimenters addressed themselves. Once started, the results of their research flowed in a steady stream into the literature, contributing to a growing understanding of upper atmospheric phenomena. A concise summary of some of the more important results from the first dozen years of high-altitude rocket sounding appears in the author's book *Sounding Rockets*.[28] A deeper, more detailed insight into what had been achieved may be had from volume 12 of the *Annals of the International Geophysical Year*.[29] The following brief review is derived from these and other sources.

It is not surprising that the first questions taken up by the rocket experimenters were those considered the most significant by the ground-based researchers. Naval Research Laboratory investigators built spectrographs and sent them aloft to photograph the solar spectrum at high altitude. On 10 October 1946 Richard Tousey and his colleagues obtained the first photographs of solar spectra from above the ozonosphere.[30] This event marked the beginning of many years of intensive research on the structure and energy content of the solar spectrum in both the near and far ultraviolet and eventually in the x-ray region, using a variety of techniques including spectrographs, photon counters, and photosensitive phosphors.[31] Experimenters at the Applied Physics Laboratory of the Johns Hopkins University quickly followed up the NRL achievement with spectrographic experiments of their own, obtaining highly detailed spectrograms.[32] In March 1947 the Naval Research Laboratory workers obtained additional spectra at various altitudes reaching to 75 km, and in June 1949 more spectrograms were recorded.[33] In the years that followed, both University of Colorado and Navy workers developed pointing devices to keep rocket-borne spectrographs aimed at the sun, and with these obtained more detail and continually extended the spectra to shorter and shorter wavelengths. Using the pointing control, the group at the University of Colorado in 1952 flew a spectrograph to about 85 km. In addition to the by now familiar ultraviolet spectrum from 2800 Å to about 2000 Å, there was a strong emission line at 1216 Å. This was quickly identified with the Lyman alpha line of the neutral hydrogen atom.[34] Between 1952 and 1955 both the Naval Research Laboratory and Air Force groups confirmed the presence of other emission lines between 1000 Å and 2000 Å. In 1958 the University of Colorado team used a specially designed spectrograph to photograph the solar spectrum from 3000 Å all the way to 84 Å in the extreme ultraviolet.[35] About 130 emission lines were measured and their intensities roughly estimated. The resonance line of ionized helium at 304 Å was found to be very strong. In the years following, the Colorado workers, those at the Naval Research Laboratory, and a group at the Air Force Cambridge Research

Center in Massachusetts contributed much detail on the solar spectrum in the far ultraviolet.

As had been anticipated, the ultraviolet spectrum of the sun, which proved to be very complex, did not correspond to a simple black body radiating at a 6000 K temperature as in the visible part of the spectrum. This finding was dramatically shown in a comparison of actual intensities obtained by NRL on 7 March 1947 with the 6000 K blackbody curve, shown in figure 6.

On 5 August 1948 in an Aerobee rocket flight to 96 km, T. R. Burnight detected what appeared to be x-rays in the upper atmosphere. Burnight did not follow up on his discovery, however, and it was left to others to pursue the subject.[36]

These data on the solar spectrum below the atmospheric cutoff at about 2800Å supplied theorists with much of the missing information to explain how and where the sun's radiation produced different atmospheric layers. The workers at the Naval Research Laboratory and the Applied Physics Laboratory used observations on the change in solar ultraviolet intensities with altitude to determine the distribution of ozone in the upper atmosphere.[37] It was established that the level of maximum ozone production lay in the vicinity of 50 km, hence that the higher concentrations of ozone at lower altitudes had to be due to atmospheric circulations.

Solar ultraviolet could be tied with confidence to the E region of the ionosphere. The intense Lyman alpha radiation of the neutral hydrogen atom penetrated to 70 km and influenced the lower E region and upper D region of the ionosphere. But x-rays in the vicinity of 2 Å penetrated deep into the D region and were far more efficient in producing ionization in the D layer than was hydrogen Lyman alpha.

Atmospheric structure—that is, the variation of pressure, temperature, and density with altitude—also received the early attention of the rocket experimenters.[38] Almost every flight carried gauges to measure these fundamental parameters. Signal Corps and University of Michigan groups adapted anomalous sound propagation techniques to the rocket by sending explosive grenades aloft to be set off at high altitude; the sound waves could be used to determine both air temperatures and winds up to 60 km or higher.[39] Those measuring x-ray intensities used the observed absorption of x-rays in the ionosphere to estimate air densities there.[40] As a result of many rocket observations, in the early 1950s the Rocket and Satellite Research Panel was able to issue an improved estimate of upper-atmospheric structure for use by geophysicists.[41] By the time Sputnik went into orbit, the groundwork had been laid to describe the structure through the F region of the ionosphere and to give a considerable amount of information about both geographical and temporal variations of these quantities.[42]

The ionosphere was also receiving immediate attention in the sounding rocket program. Among the early experimenters, J. Carl Seddon under-

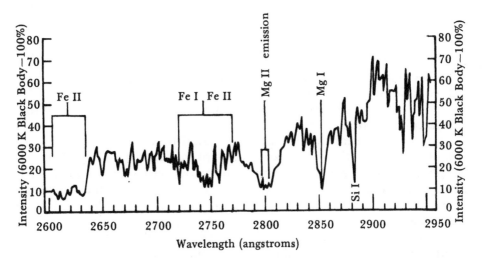

Figure 6. Solar spectrum. Solar intensities above the ozonosphere at White Sands, New Mexico. Spectrum of 7 March 1947, replotted on a linear intensity scale relative to the intensity of a black body at 6000 K. Durand et al, Astrophysical Journal *109 (1949): 1–16. Illustration courtesy of the* Astrophysical Journal, *published by the University of Chicago Press, copyright 1949, The American Astronomical Society. All rights reserved.*

took to adapt the propagation techniques of the ground-based probers to the rocket. He used the influence of the ionosphere on radio signals from the flying rocket to deduce charge densities existing in the atmosphere.

The phase speed c, wavelength λ, and frequency f of a steady-state radio signal satisfy the equation

$$c = \lambda f, \tag{8}$$

while the relation between c and c_0, the phase velocity in free space, is

$$c = c_0/n, \tag{9}$$

where n is called the index of refraction of the medium in which c is the phase velocity. If the signal source is in motion relative to the observer, a shift in frequency, the well known doppler shift, results:

$$\Delta f = - f(v/c) \tag{10}$$

$$= - fnv/c_0 \tag{11}$$

The original transmitted frequency could be carefully fixed in an experiment, Δf and v could be measured, and c_0 would be a known constant. Hence n could be calculated. Since n depended on the electron and ion concentrations, their collision frequencies, and the strength and direction of the magnetic field, one could thus get an equation relating the very quantities to be determined.[43] Seddon arranged his experiment so as to get several such equations, which could be solved simultaneously to give electron densities as a function of height, and sometimes some of the other quantities such as collision frequencies.

Although transmitting the probing signal from the flying rocket was supposed to reduce the complexity, many of the difficulties experienced by the ground-based probers remained. Inhomogeneities in the ionosphere, multiple reflections of the propagated wave, splitting of the signal into ordinary and extraordinary rays, and not knowing the identities of the ambient ions made the reduction and interpretation of the data a challenge. Nevertheless, Seddon was able to improve upon electron density curves obtained from the ground and to furnish some information about the low-density regions that had been hidden from the probing of the ground-based investigators. Figure 7 shows a curve of electron density changing with altitude, drawn by John E. Jackson from a composite of NRL data and measurements by other groups.

Other experimenters preferred to avoid the problems inherent in propagation experiments by using various kinds of ionization gauges. Even though the rocket introduced complications of its own, such as exuding gases carried from the ground and distorting the ambient electric field,

Figure 7. Ionospheric charge densities. Summary as of August 1958 of ionospheric data corresponding to summer noon, middle latitudes, and sunspot maximum. Courtesy of J. E. Jackson, CSAGI meeting, Moscow, 1958.

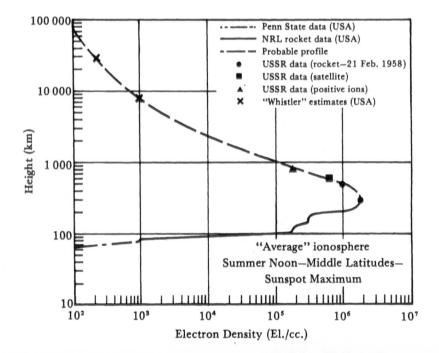

such gauge measurements were felt to be more "direct" than those obtained from the propagation experiments. Both techniques made their contributions, with the result that ground-based experimenters were provided with a standard, one might say, against which they could calibrate the methods of deducing results from their cheaper, more widespread observations.

Since the vexing question of composition continually entered into discussions of the upper atmosphere, particularly special regions like the ionosphere and the exosphere or fringe region at the top of the atmosphere, investigators soon tackled the problem of identifying atmospheric constituents as a function of altitude. At altitudes up to the bottom of the E region, workers from the University of Michigan tried sampling the air by opening evacuated glass vials or steel bottles in the upper atmosphere and immediately resealing them to lock in the sample before the rocket descended. It was tricky, because one had to ensure that the bottles weren't sampling gases carried by the rocket itself and also that the sampling procedure was not somehow altering the composition of the sample. While these experiments provided some hints of diffusive separation of helium over limited ranges above the stratosphere, by and large they confirmed that the atmosphere was thoroughly mixed, up to the E region.[44]

The most powerful technique to be brought to bear upon the problem of atmospheric composition was that of the mass spectrometer.[45] This device separates out the atmospheric particles in accordance with their molecular masses—or, more properly, in accordance with the ratios of these to their charges in the ionized state in which they are fed to the spectrometer's analyzer. While there can be some ambiguity, one can feel considerable confidence in the identifications achieved. With such an instrument John W. Townsend, Jr., and his colleagues at the Naval Research Laboratory produced a considerable amount of data on upper atmospheric composition above White Sands, New Mexico, and over Churchill, Canada.[46] They confirmed that there was little diffusive separation below 100 km; but above 120 km separation processes, at least as indicated by the separation of argon relative to nitrogen, became quite effective. The changeover from molecular oxygen to atomic oxygen appeared to be slower than had been supposed. Neutral nitric oxide, NO, was shown to be a negligible constituent of the E region and above, since its presence would have been apparent in a pronounced absorption in the ultraviolet. No such absorption was observed in rocket solar spectrograms. On the other hand, NO^+ turned out to be a major positive ion in the E region of the ionosphere. In northern latitudes, during the daytime above Fort Churchill, as altitude increased from 100 to 150 to 200 km the relative abundances of positive ions changed from (O_2^+, NO^+) to (NO^+, O_2^+, O^+) to (O^+, NO^+, O_2^+). In the United States above White Sands the results were similar except that the nitric oxide ion NO^+ was the predominant ion in the E region. In all cases O^+ was the predominant positive ion above 250 km, while according to Soviet data N^+

was never more than about 7% of the O^+ for altitudes up to more than 800 km. On several flights the negative nitrogen dioxide NO_2^- was detected in the E region.

Some of the uncertainties concerning the heights of emission of the night airglow were removed by rocket experiments.[47] The atomic oxygen green line at 5577 Å was found to have its maximum at about 95 km, to show a sharp lower cutoff at 90 km, and to trail off at 120 km on the upper side. The sodium D lines (5890 Å – 5896 Å) came primarily from the region between 75 and 100 km, peaking at about 90 km. The red oxygen lines (6300 Å–6364 Å) came from above 163 km, while the 6257 Å Meinel hydroxyl, OH, band was emitted in the region from 50 to 100 km.

Although some measurements were made of the earth's magnetic field at high altitude and of associated current flows,[48] this aspect of the high atmosphere received less attention during the first decade of rocket sounding than it would later when satellites became available. Cosmic rays were, however, a matter of intense interest to a few researchers. Of the many aspects of this fascinating subject to pursue, two topics in particular stood out: (1) What was the cosmic ray intensity above the atmosphere? (2) What was the composition of the cosmic rays? James A. Van Allen tackled these questions in a rather straightforward way. By sending a single geiger counter into the upper atmosphere, he was able to trace out a counting rate curve that rose to a Pfotzer maximum at a height of about 19 km, above which the remaining atmosphere corresponded to about 56 g/cm² of material (fig. 8).[49] With increasing altitude beyond that level the counting rate declined until it leveled off at a constant rate at and above 55 km. After several flights Van Allen was able to estimate the vertical intensity of cosmic rays at high altitude above White Sands to be 0.077 ± 0.005 particles per sec-cm²-ster,* close to the value that workers at the Naval Research Laboratory obtained. With rather poor statistics the rocket experimenters estimated that the primary cosmic rays consisted of protons and alpha particles in the ratio of about 5 to 1, with less than one percent heavier nuclei.[50] These figures differed somewhat from better measures being obtained from balloon observations.

The energy spectrum of the cosmic rays had suggested a distinct lower bound for the cosmic ray particles. Van Allen began to investigate the lower energy end of the cosmic ray spectrum. He sent counters aloft at latitudes ranging from the geomagnetic equator to the polar regions. During these investigations, Van Allen noted a pronounced increase in the numbers of soft radiation particles encountered above the stratosphere in the auroral zone, particles that were not found at either lower or higher latitudes.[51]

---

*Ster, short for steradian, a common measure of solid angle.

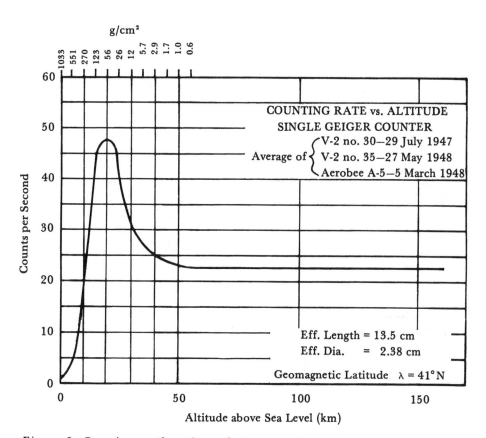

Figure 8. Cosmic ray flux. Smoothed composite curve of Applied Physics Laboratory single-counter counting rates above White Sands, New Mexico, geomagnetic latitude 41°N. Gangnes, Jenkins, and Van Allen in Physical Review 75 (1949): 57–69; *courtesy of J. A. Van Allen and* Physical Review.

SIGNIFICANCE

The foregoing nontechnical description is merely illustrative. A review of a length appropriate to this book cannot cover in detail 12 years of work by hundreds of scientists. Nor can the description convey to the reader the many subtleties and innumerable interrelationships with which both experimenters and theorists concerned themselves. Nevertheless, brief though it is, the summary shows how the rocket sounding work contributed to atmospheric and cosmic ray research.

The new tool, the high-altitude research rocket, had indeed made it possible to obtain data hitherto unobtainable and to solve problems hitherto intractable—as anticipated. The rocket results enabled ground-based observers to improve their techniques and to obtain better results from their measurements—that is, to calibrate their experiments. Whereas at the start some had expressed grave doubts as to the wisdom of using rockets for high-altitude research, a decade later the importance of the sounding rocket to the field was universally recognized. It is natural, then, to ask whether sounding rockets had revolutionized the field of upper atmospheric research.

In the excitement of new discoveries amid a continuing flow of important data from a long list of topics—solar physics; atmospheric pressure, temperature, density, composition, and winds; the ionosphere; magnetic fields; the airglow; the auroras; and cosmic rays—the rocket experimenters liked to think and speak of their work as revolutionizing the field. But it is clear in retrospect that the first decade of high-altitude rocket research was normal science, not revolution. Put otherwise, the results from those years of research elaborated and expanded upon the already accepted paradigm, but did not force any fundamental changes in it.

Were one to compose a schematic diagram of the upper atmosphere based on what was known immediately following the launching of Sputnik, the picture would probably look much like the drawing of figure 9. A comparison with figure 1 drawn from information set forth in Mitra's book of 10 years earlier shows a striking similarity in overall concepts. In both, the atmosphere is visualized as consisting of a number of characteristic layers—troposphere, stratosphere, ozonosphere, ionosphere, and exosphere—at essentially the same altitude levels. In both, temperatures vary markedly with altitude, and these variations are associated with the different atmospheric layers. Solar radiations are considered to be the cause of photochemical processes going on in the atmosphere, affecting composition, giving rise to the night airglow, and forming the ozonosphere and ionosphere. Heating of different levels in the atmosphere is ascribed to incoming solar energy, which in a series of stages ultimately degenerates into heat. There is little doubt that the auroras are caused by charged particles from the sun, and that in some way such particles are also responsible for changes in the earth's magnetic field during magnetic storms.

Clearly the two paradigms, before and after, are essentially the same. The expert will, of course, see a new richness of detail in the later picture, but nothing that the earlier paradigm could not accommodate once the facts were known. Thus, space science's first decade, the sounding rocket period, must be characterized as extremely fruitful normal science. Nevertheless, in that early harvest were the elements of some remarkable discoveries.

The soft radiation that Van Allen had detected in the auroral regions presaged the discovery of a largely unsuspected aspect of the earth's environment. Following up his interest in these soft radiations, Van Allen instrumented the first American satellite, *Explorer 1*, with counters to probe further the incoming cosmic rays. Unexpectedly high radiation intensities were found above the atmosphere and, after additional measurements in *Explorer 3*, Van Allen on 1 May 1958 announced the discovery of a belt of

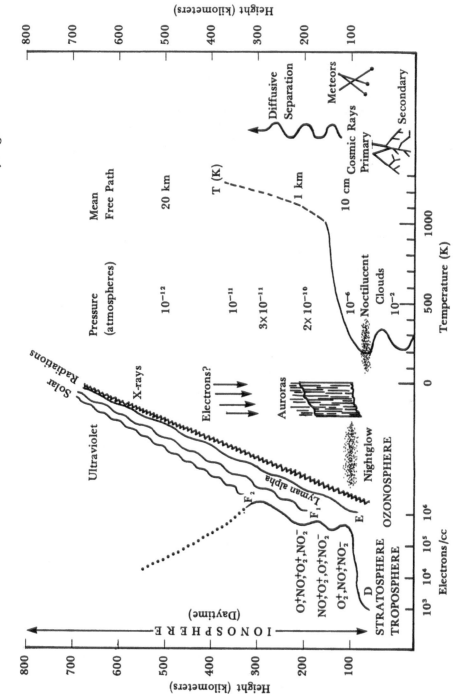

Figure 9. Upper atmosphere as visualized in 1958. The general features are similar to those of figure 1, corresponding to the mid-1940s, but there is much more detail.

radiation surrounding the earth, which at once became known as the Van Allen Radiation Belt.[52] The discovery set in motion a long chain of investigations that in the course of the next several years forced a revision of the picture scientists had developed of how the sun's particle radiations affect the earth's atmosphere. The new features of the geophysical paradigm will be presented in chapter 11.

The second discovery came from the rocket investigations of the sun's short-wavelength spectrum. The discovery that x-rays were an important variable in the solar spectrum suggested that x-rays might also be important in other stars and celestial objects, which later proved to be correct.[53] When the experimenters in the Naval Research Laboratory group turned their ultraviolet and x-ray detectors toward the stars, they initiated a new field of rocket astronomy, which will be described more fully in chapter 20.

In the meantime, the early harvest from rocket sounding of the upper atmosphere was convincing evidence of the rich returns that could be expected from a program of scientific research in space. In this aspect of space, at least, the United States could consider itself fully competitive with any rival.

Part III

Setting the Course

The heavens themselves, the planets, and this center
Observe degree, priority, and place,
Insisture, course, proportion, season, form,
Office, and custom, in all line of order.

Shakespeare, Troilus and Cressida

7

Response to Sputnik: The Creation of NASA

How brightly the Red Star shone before all the world in October of 1957! Streaking across the skies, steadily beeping its mysterious radio message to those on the ground, Sputnik was a source of amazement and wonder to people around the globe, most of whom had had no inkling of what was about to happen. To one nation in particular the Russian star loomed as a threat and a challenge.

In the United States many were taken aback by the intensity of the reaction. Hysteria was the term used by some writers, although that was doubtless too strong a word. Concern and apprehension were better descriptions. Especially in the matter of possible military applications there was concern, and many judged it unthinkable that the United States should allow any other power to get into a position to deny America the benefits and protection that a space capability might afford. A strong and quick response was deemed essential.

Actually, as has been seen in chapters 3 to 5, the United States was not far behind. A full decade of pioneering work had brought into being a respectable stable of rockets and missiles and still more powerful ones were under development, some of them nearing completion. Tracking and telemetering stations were operating, and a number of missile test ranges were functioning. Sizable teams of persons with capabilities pertinent to space research and engineering were available in both government and industry. And more than 10 years of sounding rocket research combined with open publication of results had given the United States a definite edge over the USSR in space science, in spite of its priority in the satellite program. Without question the United States was competitive in space even as the country deplored its loss of leadership.

Leadership was the key word. To be competitive was not enough. In an age when technology was vital to national defense, essential for solving problems of food, transportation, and health, and important to the national economy, technological leadership was an invaluable national resource not to be relinquished without a struggle. It was technological leadership that would generate a favorable balance of trade for the United

States and afford strength in international negotiations. Moreover, the *appearance* of leadership—while in no way equivalent to genuine technological strength in importance—nevertheless had a strong bearing on the international benefits to be won. President Eisenhower was right when he asserted that the country's position in rockets and missiles was a strong one, which the launching of a small scientific satellite by Russia could not substantively weaken, but in the mood of the times people were not disposed to listen.

One heard of a race with Russia, a topic that would be debated often in the years to come. While many would deny the necessity to run a race—and some would even contend that no race existed—for most, competition with the Soviets was serious business. Even those in a position to appreciate the strength of the U.S. position did little to bring it out, most likely because they, too, were persuaded of the importance of recapturing leadership in space, especially in view of the military implications. National leaders were worried about the obvious great size and lifting capacity of the Soviet missiles. Also, insertion of a satellite into orbit proved that the USSR had mastered the final ingredient of a successful intercontinental ballistic missile, guidance. Moreover, all this had occurred while the U.S. Atlas missile was still under development, far from deployment.

During those formative months of late 1957 and the first half of 1958 the broad spectrum of forces impacting on space matters—at once synergistic and conflicting—began to become apparent. In the face of the Soviet challenge academic, industrial, and political forces merged in a common conviction that the country must put its space house in order. The mutually reinforcing effect of these disparate interests all pushing for a properly organized, unified national space program led eventually to the creation of the National Aeronautics and Space Administration; their continued cooperation during the ensuing years produced the broad spectrum of achievements in manned exploration, science, and applications in outer space with which the world has become familiar. But the individual motivations—political objectives, commercial goals, professional aspirations—and the differing philosophical backgrounds of the industrialist, academician, legislator, administrator, soldier, scientist, and engineer set up cross currents and conflicts of varying intensity that run through the early years of NASA's history.

During the months preceding the passage of the National Aeronautics and Space Act of 1958, space science and potential military applications were already established areas of space activity that contended to attract the various interest groups that had or might have a stake in the nation's future in space.[1] Industry gravitated toward the military, with which it already had a close and profitable association. Professional societies, the Rocket and Satellite Research Panel, the Academy of Sciences, and the President's Science Advisory Committee naturally pursued the scientific side of the matter. It was the military implications of space, the bearing that future space developments might have upon national defense and security, that imparted a sense of urgency to the deliberations in the executive and legislative branches. Communications satellites, weather satellites, and earth observations from space for intelligence were all important to the military, the last named being of special significance in the Cold War. These considerations caught and held the attention of the legislators. But, though the military implications were deemed the primary concern of the country, circumstances elevated the scientific aspects to a position of considerable influence.

A majority of those who would finally make the decision soon became convinced that the most effective way of proving U.S. leadership in space would be to demonstrate it openly.[2] Moreover, a space program conducted under wraps of military secrecy would very likely be viewed by other nations as a sinister thing, a potential threat to the peace of the world. A cardinal point in the U.S. military posture had always been that the development and maintenance of U.S. military strength was peaceful, not intended for aggression, but for self-defense and to enable the country to help maintain stability in a world in which weakness too often provided the occasion for trouble. It was an important thesis for the U.S. public to continue to believe and to sell to the rest of the world and, in a matter as portentous as space seemed to be, special efforts were needed to present the proper image. It seemed important, therefore, that the U.S. space program be open, unclassified, visibly peaceful, and conducted so as to benefit, not harm, the peoples of the world.

A logical conclusion of this reasoning was that the program should be set up under civilian auspices. Thus, although the military had by far the greatest amount of experience pertinent to conducting a space program, it was by no means a foregone conclusion that the Pentagon would be assigned the principal responsibility. To be sure, the Army, Navy, and Air Force had been among the earliest to study the usefulness of space to support their missions.[3] Military hardware afforded the only existing U.S. capabilities for space operations.[4] Moreover, the services had provided the funding and much of the manpower for the rocket-sounding program of the Rocket and Satellite Research Panel, many of whose members were civilian employees in military research laboratories, as shown in appendix A. Yet so powerful was the conviction that the program must project an image of benevolence and beneficence that the otherwise overriding military factors were themselves outweighed.

Reinforcing these views were President Eisenhower's own convictions. Already distressed over the enormous power and unmanageability of what he later called the military-industrial complex, Eisenhower was not disposed to foster further growth by adding still another very large, very costly enterprise to the Pentagon's responsibilities. Moreover, at the time the Pentagon did not enjoy the best of relations with Capitol Hill. One heard talk of a "missile mess" and interservice rivalry in the Pentagon. Such concerns led, during the very period when the administration and Congress were deciding America's role in space, to the appointment of a new secretary of defense, the creation of the Advanced Research Projects Agency, and passage of the Defense Reorganization Act, which among other things set up the Office of the Director of Defense Research and Engineering. Such considerations, plus Eisenhower's not seeing in Sputnik the crisis for national defense that others considered it to be, predisposed him to favor a space program with a strong scientific component under civilian management.

The scientists were united in their desire to have a strong scientific component in the space program. The greatly expanded federal funding of science in the years following World War II had declined. Members of the President's Science Advisory Committee and James Killian, special assistant to the president for science and technology, saw in the space program an opportunity to renew national support of science. Under the circumstances Killian and PSAC had a considerable influence in the creation of NASA, pressing for a space program under civilian management with a strong scientific flavor.[5] In this they were supported by the Rocket and Satellite Research Panel; the National Academy of Sciences, where President Detlov Bronk took a personal interest and where the Space Science Board was set up; by the American Rocket Society; and by other groups of scientists who felt impelled to speak out on the issue.[6]

Against this background the debate on how precisely to respond to the Soviet challenge proceeded. A deluge of proposals descended upon various congressional committees. In the Department of Defense the administration had set up the Advanced Research Projects Agency, approved by Congress in February as a temporary holding operation.[7] But, as pointed out, there were cogent reasons for a space program under civilian auspices—in which case provision would also have to be made to meet the vitally important military needs. Among the civilian possibilities was the creation of a new agency—which some of the scientists had recommended—but to those who knew what was involved, that was a horrendous undertaking. Alternatively one could assign the responsibility to an existing agency, or build a new agency around an existing organization as nucleus. With the application of nuclear power to rocket propulsion in mind, the Atomic Energy Commission was interested in taking on the job, as both the commissioners and the Joint Committee on Atomic Energy on the Hill made plain to members of the Rocket and Satellite Research Panel when they called to enlist support for the creation of a National Space Establishment.[8] But, for a number of reasons, the choice finally fell on the National Advisory Committee for Aeronautics (NACA).

The NACA would not have been the choice of most scientists. As a highly ingrown activity, the agency did not enjoy a particularly great esteem in scientific circles, being thought of more as an applied research activity serving primarily industry and the military. Members of the Rocket and Satellite Research Panel, in particular, were skeptical of the ability of an agency almost entirely oriented toward in-house research and with no experience in the management of large programs to take on all the research, development, and operational tasks of a space program that some members thought would soon entail $1 billion a year. For most of its life NACA had managed at most a few tens of millions of dollars a year. In fact, the annual budget had not exceeded $1 million before 1930 and had not passed the $10-million mark until World War II. It took the construction and operation of the large wind-tunnels of the 1950s to push the budget toward $100 million. Skepticism within the panel was not lessened by the cautious attitude NACA management had displayed through the years toward letting NACA people take part, even in a small way, in the sounding rocket program. Doubts about the choice of NACA were increased in the months following the launching of Sputnik by conversations between panel members and NACA's Director of Research Hugh Dryden and Chairman James Doolittle.

The views of the scientists probably carried little weight. More telling was the disenchantment with NACA on the part of its own clients, the Air Force and industry. The agency had started in 1915 as an advisory group, as its name implied, but became gun shy when its advice began to generate at least as many enemies as friends.[9] As a consequence the NACA soon turned away from advising and toward research. Even here it was necessary to keep from treading on the toes of either industry or the military, and as a consequence the agency gravitated toward aerodynamic and wind-tunnel research, in which both clients were happy to have help. Over the years the agency had acquired a reputation of caution and conservatism. This conservatism may have caused NACA to miss out on a number of important aeronautical advances, the most significant of which was jet propulsion, where Britain and Germany took the lead. At any rate, because of such missed opportunities, NACA in the 1950s no longer had the unqualified endorsement of the military and industry that it once had, and in the view of at least one historian might well have died had not the space program come along to revive it.[10] Under the circumstances the agency was available, and it was a case of assigning responsibility for the space program to an organization whose future was otherwise in doubt.

NACA pursuit of this opportunity was something less than sparkling. At the urging of younger members of the agency, Dryden and his staff developed a number of papers on the subject of space research. On 14 January 1958 the so-called "Dryden Plan" was made public.[11] The title, "A National Research Program for Space Technology"—rather than a name referring to the exploration and investigation of space—reflected the agency's characteristic caution and narrowness of outlook. The plan was a hodgepodge born of a desire to keep much of NACA's old way of life while embracing the interests of both military and civilian groups. Under the plan the national space program would be a cooperative effort among the Department of Defense, the NACA, the National Academy of Sciences, the National Science Foundation, and various private institutions and companies. The Department of Defense would be responsible for military development and operations, the National Academy of Sciences and the National Science Foundation would have responsibility for the scientific experiments to be conducted, mostly by the outside scientific community, while NACA would be responsible for research and scientific operations in space. This cautious approach—which courted everyone and satisfied no one—was endorsed in a resolution passed by the Main Committee of NACA on 16 January. On 10 February 1958 the agency issued an internal document giving details of the expansion of NACA that would be required to support the Dryden plan.[12]

In spite of the negative feelings about NACA, the availability of the agency, coupled with doubts about its future in the field of aeronautics and the desire to put the space program in civilian hands, eventually made NACA the prime candidate for the job. On 5 March Chairman Nelson Rockefeller of the President's Advisory Committee on Government Organization, Director Percival Brundage of the Bureau of the Budget, and Special Assistant for Science and Technology James Killian jointly delivered to President Eisenhower a memorandum recommending that "leadership of the civil space effort be lodged in a strengthened and redesignated National Advisory Committee for Aeronautics." The memo listed a number of liabilities, but stated that these could be overcome by enacting appropriate legislation. The NACA would be renamed the National Aeronautical and Space Agency, and the 17-member governing committee—which NACA insisted was the kind of buffer a research agency needed at the top to shield it from external forces—would remain, but the membership would be changed and its power reduced.

That same day President Eisenhower decided to build "a civilian space agency upon the NACA structure."[13] From that point matters moved rapidly within the executive branch. The Bureau of the Budget prepared draft legislation with assistance from Killian's office and NACA. The pace with which this was accomplished left little time for coordination with other agencies such as the Department of Defense, a matter that aroused considerable criticism during the congressional hearings on the bill. On 2 April 1958, Eisenhower submitted his proposal to Congress. The Bureau of the Budget had insisted on a single responsible head for the new agency, one who would be advised by a board of experts but would not be responsible to and shielded by such a board. NACA leaders disagreed, and according to Arthur Levine some members of the agency sought help from friendly congressmen to preserve the traditional NACA organizational pattern.[14] But although the administration's bill was considerably tighter than the diffuse approach of the Dryden plan, and although the presentation of the bill served to channel the congressional deliberations into the course that led to the passage of the National Aeronautics and Space Act of 1958, both committee members and witnesses found much in it to criticize.

> The original bill lacked provisions dealing with Congressional oversight and control, international cooperation and control, patents, indemnification, limitation of liability, conflict of interest, definition of terms, ceilings on salaries, relations with the Atomic Energy Commission, formal liaison committees, and over-all policy determination and coordination. There was no provision for nuclear propulsion, or even any recognition of its importance in this new field. Vagueness regarding the delineation of military and civilian activities in outer space was charged by many. There was no formal provision for determining agency jurisdictions in space research or settling of jurisdictional disputes. There was much criticism of the lack of clarity in the size and makeup of the board proposed in the Administration bill. Concern was voiced over the lack of substantive provisions backing up various aims put forth in the declaration of policy.[15]

In the end Congress adopted a bill which, while it accepted much of what the administration had proposed, nevertheless introduced substantial changes to meet the various criticisms.

The remarkable congressional response to the Sputnik crisis has been analyzed by a number of authors.[16] Even before President Eisenhower showed any willingness to take the matter seriously, Congress had begun to probe the subject of the nation's missile and satellite programs.[17] The Preparedness Investigating Subcommittee of the Senate Committee on Armed Services opened hearings on 25 November 1957, continuing through 28 January 1958, accumulating more than 7000 pages of printed testimony largely devoted to how the United States and the Soviet Union compared in science and technology in general and rockets and missiles in particular. In his opening remarks the chairman, Lyndon B. Johnson, set a tone of bipartisan, nonpolitical searching for the best possible national response to the Russian challenge, a tone that was to characterize the entire process of the next half year leading to the passage of the NASA Act. The unanimous report from these first hearings called for quick and vigorous action. Indeed, the clear determination of the Congress to do something about the crisis had much to do with goading Eisenhower into action to develop an administration proposal.

At first congressional investigation and study, while extensive and much to the point, showed little agreement on how to proceed. Numerous resolutions and bills were offered, some of them proposing the establishment of a permanent space organization.[18] For a while there seemed to be too many cooks, but in February 1958 matters began to gel. Senate Resolution 256 on 6 February created a Special Committee on Space and Aeronautics to frame legislation for a national program of space exploration and development. On 10 February, 13 senators, comprising a powerful representation of the Senate leadership, were named to the committee.[19] The membership included the chairmen and ranking minority members of all major Senate committees concerned. On 20 February, in a rare break with tradition, the majority leader, Senator Johnson, was elected chairman.

The House soon followed suit and on 5 March established its own blue ribbon group, the Select Committee on Astronautics and Space Exploration, to which 13 members were appointed.[20] As in the Senate, the House regarded the matter as sufficiently important to set aside tradition, and Majority Leader John W. McCormack was named chairman. Minority Leader Joseph W. Martin, Jr., was picked as vice chairman.

Meanwhile the administration had been preparing the draft legislation. The appearance of the administration bill drew congressional activity into focus. On 14 April Senators Johnson and Bridges introduced the bill as S. 3609. The same day McCormack introduced it in the House as H.R. 118811, with identical bills being put forth by eight other representatives.[21] The House committee began hearings the next day, 15 April, and continued them through 12 May. Not having conducted a previous inquiry, as had the Senate, the House hearings were thorough and extensive. In contrast, the Senate committee directed its inquiry more narrowly at the proposed draft legislation. The Senate hearings covered six days, opening 6 May and closing 15 May.

Many complex issues were debated: the organization and salary structure of the new agency and its location in the executive branch; the matter of policy guidance at the top, and how to provide coordination and liaison between the civilian space agency and numerous other activities—like the Department of Defense and the military services, the Weather Bureau of the Department of Commerce, the National Science Foundation, the Atomic Energy Commission, and the Department of Health, Education, and Welfare, for example—which had legitimate and important interests in space research and applications; and the matter of ensuring the military the necessary freedom of action to pursue applications of space that were deemed of military significance.[22] The last-named issue was of great concern and brought in by implication the question of how to divide responsibility in the space program between a civilian agency and the military establishment. Numerous other issues also had to be ironed out, such as organization within Congress and how to provide for congressional oversight, policy on information and publicity, and how to handle international matters such as cooperation in space.[23] Both committees felt that the administration proposal failed to cover adequately many of the important issues. As a consequence, the bill finally passed differed considerably from that initially proposed.[24]

Most significant for space science, Congress did not prescribe the specific content of the space program with which the NASA Act was concerned.

> In the end the legislative formulation of a detailed space program was bypassed. The legislation set up an agency, created its machinery, and provided for coordination and cooperation between it and other branches of the Executive.[25]

The Congress had found it impossible to divide the program between the military and the new civilian agency:

> It rapidly became evident that it was the use made of it and not the satellite itself which might well determine whether it would be of a military or a peaceful nature. For example, a reconnaissance satellite could be used to map and photograph the surface of the earth for purposes of defense or attack. It could also provide vastly improved means for the study and exploration of the universe.[26]

Intelligence gathering was generally conceded to be entirely military, space science essentially civilian—although the military would necessarily be interested in certain aspects of space science. All else was contested: manned spaceflight, launch vehicle development, and applications like communications and meteorological uses of satellites. In the face of this dilemma the legislators chose to provide a framework that would give both the military and the civilian space agencies the necessary freedom of action, while requiring coordination and mutual assistance. Having established the framework, Congress would leave it to the two agencies to work out between them the appropriate division of labor and responsibility—precluding, of course, unwarranted duplication of effort.

The lack of a specifically prescribed program gave the first administrator of NASA a wide degree of latitude in selecting projects and missions to undertake, a freedom of choice that was but little curtailed by guidance that James Killian supplied in the summer of 1958, assigning manned spaceflight, meteorology, passive communications, and science to NASA, and active communications and reconnaissance to the Department of Defense. The latitude NASA enjoyed permitted the development of a broad-ranging program of science and exploration, and the accompanying development of technology and the application of space techniques to practical uses. During the first several years this situation was entirely in keeping with the spirit of the times, and on the Hill there was more questioning of whether NASA was being bold enough than there was concern about overstepping any bounds. In fact, it was conservatism within the administration that led to considerable moderation in building up the program.

The climate was ideal for the growth of a space science program. Not being prescribed in detail—as far as science was concerned the NASA Act simply called for "the expansion of human knowledge of phenomena in the atmosphere and space"[27]—the science program could be permitted to unfold in keeping with the scientific process. Relying on the nation's scientists, including the National Academy of Sciences, NASA proceeded to attack the scientific problems of the atmosphere and space that the scientists themselves deemed most important and most likely to produce significant new information. The organization of the space science program, the establishment of advisory committees, the agency's funding requests, and the means by which individual scientists, universities, and other research organizations were invited to participate—all were designed to make the space science program a creature of working scientists, in the conviction that such an approach would produce the best possible program for the country.

In many ways, although it didn't always seem so to the scientists, space science occupied a favored position. As a means of diverting attention from the military overtones of the Sputnik crisis, President Eisenhower had favored a national space program with a scientific complexion. During the months of discussion on the Hill, there never arose the slightest question but that space science would be an essential element of the national space program. Long lists of scientists were called as witnesses, or their opinions sought by letter as to what to do. The importance of science to the program and the importance of a civilian arena for science, plus the international character of science, contributed to the argument for placing the space program in the hands of a civilian agency. Reinforcing such considerations in the minds of congressmen and senators was the image of success science had acquired in the International Geophysical Year that had brought forth the Sputnik challenge.

Of course, the freedom that the first administrator of NASA enjoyed in developing the civilian space program had also been accorded the military services in pursuing military interests in space. As already mentioned, it was the military potential of space that aroused the concern and held the attention of many legislators, and that virtually guaranteed a formally designated national space program. But the broad overlap of common interests that had stymied the legislators in their efforts to effect a satisfactory division between the civilian and the military in the first place was a potential source of conflict between the new agency and the military services. Such conflict the National Aeronautics and Space Council and especially the Civilian-Military Liaison Committee, called for in the NASA Act, were intended to handle.

Another feature of the NASA Act that was of importance to space science was the provision of a single responsible head for the agency. Under the pressure of a national clamor to close the gap with the Russians in space—a pressure continually reinforced by the urging of Congress to get on with the task—NASA had its best chance to break away from the conservatism that had characterized its predecessor. To continue the old NACA structure, as NACA officials had urged, with an advisory board determining policy and shielding the director from many of these outside pressures, might well have had a greater impact on science than on other aspects of the space program. Boards and committees tend to be conservative. Paradoxically, scientists as a class are quite conservative. As a group they would doubtless have been content to move more slowly, more cautiously, less expensively, making the most of the tools already developed in preference to the creation of larger, more versatile—and more expensive—tools. Exposed directly to the outside pressures to match or surpass the Soviet achievements in space, NASA moved more rapidly with the development of observatory-class satellites and the larger deep-space probes than the scientists would have required (chap. 12). Some of the most intense conflicts between NASA and the scientific community arose later over the issue of the small and less costly projects versus the large and expensive ones—a conflict that NASA's vigorous development of manned spaceflight exacerbated.

Of course, the scientific community is not monolithic, and there were so many widely differing opinions on these matters as to make speaking of a single position of the scientific community nonsense. Nevertheless it seems clear that the new organizational structure prescribed for NASA not only helped NACA people drop much of their conservatism, but also had an impact on the space science program in effecting a faster development of more advanced space tools than many leading scientists would have called for.

The National Aeronautics and Space Act of 1958 was a remarkable piece of legislation, and the process which produced it even more noteworthy. The thoroughness with which the subjects of space and its potentials and implications were investigated and studied, the thoughtfulness given to the issues raised, and the care taken in responding to the crisis precipitated by Sputnik provide a model that could well be commended as a pattern for the handling of legislative matters. As a practical matter, however, it is not likely that the Congress could find the time and resources to devote such attention to more than a select few of the issues that come before it. Also, few other issues are so free of partisan concerns and vested interests.

At any rate, the act provided an effective framework for both the civilian and military components of the nation's space research and exploration. In the course of time, some changes were found desirable.[28] Perhaps the most telling were those in coordination, the area in which Congress had displayed so much concern and on which so much time had been spent. President Eisenhower made little use of the Aeronautics and Space Council and did not provide a permanent staff for it, so it was left to NASA and the Bureau of the Budget to do the staff work. In April 1961 the NASA Act was amended to place the National Aeronautics and Space

Council in the Executive Office of the President, to replace the president with the vice president as chairman, to decrease the size of the council, and to broaden its functions to include cooperation "among all departments and agencies of the United States engaged in aeronautical and space activities."[29] Also the Civilian-Military Liaison Committee proved ineffective from the start. In September 1960 NASA and the Department of Defense jointly established an Aeronautics and Astronautics Coordinating Board, cochaired by the deputy administrator of NASA and the Defense Department's director of defense research and engineering. Because it worked, the AACB rapidly took over the functions of the Civilian-Military Liaison Committee. The new board succeeded because its cochairmen and members were in positions of authority in their respective agencies, where they could personally put into effect agreements arrived at in the board. No longer of any use, the liaison committee was abolished by reorganization in July 1965.[30] There were some other changes, and additional authorities were acquired from related legislation—such as the authority to award grants in support of basic science.[31] But, all in all, the strength and effectiveness of the NASA legislation lay in the original act of 1958.

Under its provisions NACA prepared to move out on its new career—as NASA. Dryden was not chosen as the first administrator. In retrospect it is easy to see why. The cautious and diffuse approach of the NACA with which Dryden was identified, and Dryden's conservative views on the budget needed by the new agency, did not jibe with the legislators' sense of urgency in space matters.[32] Instead of Dryden, T. Keith Glennan—president of Case Institute of Technology, former head of the Navy's New London Underwater Sound Laboratories, and for two years a member of the Atomic Energy Commission—was chosen. In spite of the difficulties with Congress, Dryden had an undiminished reputation for technical and administrative competence which led Glennan to ask specifically for him as his deputy.

After a brief preparatory period, Glennan officially opened NASA's doors on 1 October 1958. Space science was one of the first of NASA's programs to flourish. Nevertheless it was not the Sputnik crisis that brought space science into being. What Sputnik did achieve was to break out much of the U.S. space program, including space science, from under the military wing where it had resided during the pioneering years. Had it not been for the shock generated by Sputnik, the American space program would probably have evolved into one largely devoted to military objectives—with space science as an adjunct. Under such circumstances, in spite of the commendably enlightened policies of the U.S. military establishment regarding support of basic research, the free play of the scientific process would have been difficult to maintain. Pressures would have been in the direction of supporting research with military applications and imposing security classification on some of the results. With the program in NASA, the scientific community was in a stronger position to impress its brand on American space science and to work openly with foreign colleagues when that seemed appropriate.

Yet it is of interest that the members of the Academy of Sciences and of the President's Science Advisory Committee who had worked so hard to push the space program in the direction of science and toward the civilian arena were not those who proceeded to carry out the space science program. As leaders of the scientific establishment, they continued to be beset by the problems of maintaining adequate appreciation and support for science in general; and as soon as the space program was launched they returned to these broader matters. Rather, it was those who had already been engaged in rocket and satellite work, especially those working on projects connected with the International Geophysical Year, who began to develop the nation's space science program. These individuals, with years of experience behind them in industry, on the Rocket and Satellite Research Panel, and in the IGY program, naturally had proprietary feelings about space research; and it was easy for them to regard the space science program as very much their own creation. But the academy, from its association with IGY, and PSAC from its role in laying the legislative foundation for NASA, also had certain proprietary feelings about the program. There arose accordingly a tension—constructive for the most part—between NASA managers and advisers in the academy and on PSAC. The issues of what the space science program should be, how it should be carried out, and who should make the decisions arose early and recurred continually throughout the 1960s and into the 1970s.

8

NASA Gets under Way

None of the traditional conservatism of the National Advisory Committee for Aeronautics was evident in the autumn of 1958 when the National Aeronautics and Space Administration got under way. Rather, the industry, care, and thoroughness that had earned for NACA the respect of Congress over the years could be sensed as the new agency geared up for the challenges ahead. A seemingly endless list of matters had to be taken care of in the first few months after NASA was formally opened by Administrator Glennan on 1 October 1958, and everyone had his hands full.

The agency showed no inclination to take its role in the nation's space program for granted. The debates during the previous year about the importance of the space program and the country's poor position relative to the Soviet Union demonstrated that Congress would take a deep interest in what NASA did. Also, the significance of the choice of a new man, T. Keith Glennan, as the first administrator, rather than Hugh Dryden, the director of NACA, was not lost upon former NACA employees. Even though the National Aeronautics and Space Act of 1958 had given NASA extensive authority, the agency still felt the need to sell itself. As the staff prepared for NASA's first budget hearings, Abe Silverstein, director of spaceflight programs, admonished his people with words like the following: "Remember, it is not the program we have to sell. That has already been bought. What we have to prove is that we are the right ones to do it!"[1]

That was the mood of NASA as it bent to the tasks ahead. If anything stood out at the time, it was that everything seemed to be happening at once. In the white hot light of public interest, NASA had to establish its organization, expand its staff, acquire new facilities, find contractors for the work to be done, carry out Vanguard and the projects transferred from the Advanced Research Projects Agency, work out its relations with the military and other agencies, develop a budget, prepare for the first congressional hearings, and plan for the future—all while attempting to get a program immediately under way. Again it was Silverstein who put it into words: "Two years. It will take two years to get things really under control. After that you can begin to take it easy." As a prophet, Silverstein was half right. It did take about two years to set NASA on the course it would follow for the next decade.

The jumbled character of NASA's first years is readily apparent in Robert L. Rosholt's review of the period;[2] but in the midst of all the scramble, things were getting done. From hour to hour, and from day to day, NASA managers would move from topic to topic, keeping things moving on all fronts. Gradually the program began to take shape. Space science, even though it had the advantage of a head start from the previous sounding rocket work and the scientific satellite program of the International Geophysical Year, shared in the growing pains of the new agency. In addition, problems peculiar to a scientific endeavor had to be solved.

The following pages take up a number of subjects that the National Aeronautics and Space Administration had to address itself to for all its programs, but here they are considered in the light of their bearing on space science. Although discussed under several topical headings, these matters were inextricably interwoven and were being worked on simultaneously.

ORGANIZATION

President Eisenhower's decision of 5 March 1958 to build a civilian space agency around NACA set in motion the train of events that led to the establishment of NASA. On 2 April, when the administration's draft legislation was sent to the Hill, the president instructed NACA and the Department of Defense to work out the necessary plans. For its part NACA set up an Ad Hoc Committee on NASA Organization, under Ira Abbott, NACA assistant director for aerodynamic research, which made a preliminary report in May.[3]

The committee's suggested organization showed four major divisions: Aeronautical and Space Research, Space Flight Programs, Space Science, and Management.[4] The last named stemmed from a recognition that the prospective program would require substantial management attention, requiring, among other things, contracting for development and operations as well as for research. Aeronautical and Space Research would cover the advanced research of the NACA plus that pertinent to the investigation and exploration of space. The large development projects and operations required for the space program would be handled by Space Flight Programs.

Space Science remained a separate box on the organization chart through the tentative plan of 11 August 1958. In keeping with the plan that Dryden had proposed in January,[5] it was specially noted on the charts that the space sciences program would use the services of the scientific community, including the National Science Foundation and the National Academy of Sciences. On 19 August, Administrator Glennan met with key

NACA officials to go over the planning, and a provisional organization chart was issued on 21 August 1958, from which the space science box had disappeared. About this time the author began negotiations with Abe Silverstein for a number of the space scientists at the Naval Research Laboratory to join NASA. As an outcome of these negotiations, John Townsend, John Clark, and the author transferred to NASA Headquarters on 20 October 1958. A few days later, on 24 October, a tentative organization chart again showed a box for space sciences, but this time in the Office of Space Flight Development, under Silverstein. Glennan's first official organization plan in January 1959 retained space science in the Office of Space Flight Development.[6]

According to Glennan, one should not read too much into the shifting position and status of space science, which simply reflected the fact that "'space sciences' was only one of many organizational elements to be fitted together." Moreover, the administrator looked to Deputy Administrator Hugh Dryden to ensure that science was "accorded its appropriate role and status in the NASA family."[7]

Had the 21 August chart persisted, it is safe to say that the scientific community would have been most distressed. As it was, making space science a subsidiary of spaceflight development did not sit too well with key scientists, who did not hesitate to characterize science as one of the major purposes of the space program. At the 18 December 1959 meeting of the Space Science Panel of the President's Science Advisory Committee, for example, Chairman Purcell closed by declaring that "space science was the backbone of the American space program, the foundation of what we can do in applications."[8] Space science may have been put where it was in the fall of 1958 because the scientists from the Naval Research Laboratory and elsewhere who came into NASA were unknown quantities to Dryden and Silverstein.

Space science again disappeared from the organizational nomenclature when in February 1960 the author was listed as deputy to Silverstein. In its place were two titles: Satellites and Sounding Rocket Programs, and Lunar and Planetary Programs. Almost two years later, in November of 1961, the second administrator, James E. Webb, announced his first major reorganization of NASA; at that time the author became director of a newly created Office of Space Sciences, giving science the kind of visibility in the NASA organization that the scientific community felt it should have.[9]

An often repeated statement of NACA people was that the strength of NACA lay in its centers.* That was where the trained people, who represented the research and technical competence of the agency, lived. The same would be true of NASA. But from the outset Hugh Dryden was especially

---

*Langley Aeronautical Laboratory in Virginia, Ames Aeronautical Laboratory and the High-Speed Flight Station in California, and Lewis Flight Propulsion Laboratory in Ohio.

concerned that the research character of the NACA centers—on which NACA's reputation in aeronautical and aerodynamic research had rested—be preserved and protected against encroachment by the development and operational demands of the space program. Thus, the Office for Space Flight Programs had the dual purpose of providing new capability for space research and development, while leaving the old centers free to pursue the advanced research and technology that were their forte. This policy, which appeared in the earliest planning, persisted throughout the evolution of NASA, but weakened with the passage of time. Thus, to avoid overloading the Jet Propulsion Laboratory (which was already struggling with the Ranger, Surveyor, and Mariner projects), in the summer of 1963 management of the Lunar Orbiter was assigned to the Langley Research Center.[10] This sizable project was followed in the latter half of the decade by the even more demanding Viking.[11] When the Centaur rocket stage needed special attention to pull it through its development difficulties, the project was assigned to the Lewis Research Center.[12] At the Ames Research Center, studies of an astronomical satellite undertaken in 1958 and 1959 became the basis for much of the planning for NASA's Orbiting Astronomical Observatory.[13] Later Ames became the management center for the Pioneer projects.[14] The urge to take part in the space portion of NASA's program, the need for additional support to important projects, plus the argument that a modest development work would provide insights into technological needs that would benefit advanced research, militated against keeping the research centers "pure." It eventually became a matter of keeping the development work at a modest level.

Given the policy of protecting the research character of Langley, Lewis, and Ames, an entirely new capability for the unmanned and manned space programs had to be built. On the 29 January 1959 organization chart, Glennan listed under the Office of Space Flight Development, as space project centers: Jet Propulsion Laboratory, Beltsville Space Center, Wallops Station, and Cape Canaveral.[15] Wallops Station had been an arm of the Langley Research Center and would now be devoted to a variety of test projects, including the launching of sounding rockets and the Scout satellite-launching vehicle.[16] Cape Canaveral, of course, was the site of the Air Force's East Coast missile launching facilities, which would be expected to support NASA, as well as military, programs. The Jet Propulsion Laboratory was transferred from the Army to NASA by executive order on 3 December 1958, giving NASA a substantial capability for spaceflight development.[17] By mutual agreement JPL was steered in the direction of lunar and planetary exploration. The Beltsville Center—which took its temporary name from its location on surplus government land near the Beltsville Agricultural Center in Maryland—grew out of planning that had started before NASA was activated. This center was to provide a satellite research and development arm for the agency.[18] On the first of May 1959, just a week after construction had begun, Glennan announced that the new center would be called the Goddard Space Flight Center in honor of Robert H. Goddard. By September the first building was fully occupied. The center was dedicated on 16 March 1961. Goddard, the Jet Propulsion Laboratory, and Wallops Island were to become the principal NASA centers in the space science program, although as mentioned earlier, other centers contributed substantially.

STAFFING

NASA's task of assembling a team for the space program was helped immeasurably by being able to build on the NACA team. Abe Silverstein, brought to Washington by Hugh Dryden, played a key role in the pre-NASA planning and in getting the space program under way. His imprint was to be found on most aspects of the program, including space science.

Abe Silverstein was a hard-nosed, highly practical, boldly innovative engineer, with a solid conviction—consistent with NACA tradition—that all research had to have a firm justification in practical applications to which it would ultimately contribute. Abe had come from the position of associate director of the Lewis Propulsion Laboratory in Cleveland to NACA Headquarters to help plan the new NASA. He remained a few years to head NASA's Office of Space Flight Development and later the Office of Space Flight Programs, in which NASA's space science division was then located. With him from Lewis he brought a number of persons who were to play key roles in NASA's management structure: Edgar M. Cortright, for many years deputy in the Office of Space Science and Applications and later director of the Langley Research Center; DeMarquis D. Wyatt, a leading figure in programming and budgeting for the agency; and George M. Low, who took over the Apollo Project Office at the Manned Spacecraft Center after the tragic Apollo fire at Cape Kennedy. Still later, Low became deputy administrator of NASA. Abe also drew upon other centers in NACA, selecting Robert Gilruth of Langley to manage a manned flight Space Task Group, which evolved into the Manned Spacecraft Center, subsequently renamed the Johnson Space Center. From the Ames Research Center he chose Harry Goett to take over the directorship of the Goddard Space Flight Center after John Townsend had that enterprise well under way.

The space science team grew largely from researchers who flocked to NASA from other agencies. The author's upper-air-research colleagues at the Naval Research Laboratory comprised an appreciable number of these. Soon after transferring to NASA, John Townsend, who had been head of rocket sonde research at NRL, was given the task of bringing the Beltsville Space Center into being. Negotiating with the director of the Naval Research Laboratory, Townsend worked out the details of the transfer of additional scientists and engineers from his former branch, 46 of whom were placed on the NASA rolls on 28 December 1959.[19] From NRL Townsend also secured temporary housing for the new NASA group. With the members of Project Vanguard who were transferred en masse by President Eisenhower on 1 October 1958, these employees accounted for most of the original staffing of the center. The manned flight Space Task Group at Langley was administratively assigned to the Goddard Space Flight Center for a while, but before any physical transfer took place the group was sent to Houston in 1961 as the nucleus of the Manned Spacecraft Center.

When Robert Jastrow, a physicist interested in properties of the upper atmosphere, transferred from the Naval Research Laboratory on 10 November 1958, he immediately set to work helping to plan the future space science program. An attractive, able scientist, Jastrow quickly earned the support of the administrator's office. He took the lead in developing for NASA a theoretical space sciences group, from which eventually came both the Theoretical Division and the Institute for Space Studies of the Goddard Center. Through both of these activities Jastrow was instrumental in drawing a high level of scientific talent into the agency, either onto NASA rolls or as visiting scientists.

Remaining at headquarters, John Clark and the author worked with Morton Stoller, Edgar Cortright, and other NACA people to build up a space sciences staff. Nancy Roman was enticed to leave the Naval Research Laboratory radio astronomy group to put her hand to developing an astronomy program for NASA. To help plan lunar and planetary programs, Gerhardt Schilling shifted over from the Academy of Sciences, where he had been associated with the International Geophysical Year and Space Science Board staffs. Robert Fellows, a chemist, came from Sprague Electric Company to join in planning and directing the upper-atmosphere research program.

Such was the pattern, but by no means the full accounting, of the early space science staffing of NASA. Those who had been pioneering in space research and development swelled the rolls of workers in the space program, both within and outside of NASA. And a great many of these were scientists interested in taking part in the space science program.

Program

The purpose of an organization and staff is to do something. Ideally one should know what that something is before trying to shape an organization or to hire people, for the program should determine how and with whom to go about it. In practice the ideal can hardly ever be attained. In NASA much of the planning had to be done as the agency organized itself and hired staff. But not entirely, because NACA people—and others—had thought a great deal about what should be included in the space program.

In anticipation of the use of near-earth satellites for geodesy, during the 1950s a number of groups had been busily engaged in preparatory

Table 1
Pre-Sputnik Ideas for Space Projects

Project	Advocates
Sounding rocket research	Robert Goddard. Wernher von Braun. Upper Atmosphere Rocket Research Panel. Russians.
Explorer-class satellites	Upper Atmosphere Rocket Research Panel. International Geophysical Year rocket and satellite groups.
Geodetic satellites	Dirk Brouwer. Luigi Jacchia. R. K. C. Johns. John O'Keefe. American Geophysical Union Committee on the Geodetic Applications of Artificial Satellites. Upper Atmosphere Rocket Research Panel.
Biosatellites	Heinz Haber.
Astronomical satellites (solar and stellar)	Lyman Spitzer. Fred Whipple. Upper Atmosphere Rocket Research Panel.
Weather satellites	Harry Wexler.
Communications satellites	Arthur Clarke.
Manned space stations and orbital bases	Konstantin Tsiolkovsky. Hermann Oberth. Wernher von Braun. Willy Ley. H. E. Ross.
Interplanetary manned space flight, including manned lunar missions	K. E. Tsiolkovsky. Wernher von Braun. A. V. Cleaver.

studies and analyses. Among them were John O'Keefe of the Army Map Service, Luigi Jacchia of Harvard, Dirk Brouwer of Yale, and their colleagues. For the International Geophysical Year the Smithsonian Astrophysical Observatory had taken the lead in developing and putting into effect plans to use data from the IGY satellite camera network. As an outgrowth of an informal committee organized at the suggestion of R. K. C. Johns, the American Geophysical Union Committee on the Geodetic Applications of Artificial Satellites kept geodesists informed of the possibilities of the new tools. A committee report issued in September 1958 shows that considerable thought had been given to the subject.[20]

As early as 21 November 1957 the National Advisory Committee for Aeronautics had voted to establish a Special Committee on Space Technology.[21] Chaired by H. Guyford Stever, dean of the Massachusetts Institute of Technology, the committee included several members from the Rocket and Satellite Research Panel, such as James A. Van Allen, its chairman; Wernher von Braun, leader of the German missile experts working for the Army Ballistic Missile Agency in Huntsville, Alabama; and William Pickering, director of the Jet Propulsion Laboratory. The committee, assisted by a number of specialized subcommittees, formulated a space research program. Although the report, completed on 28 October 1958 after NASA was already operating, never was published, NACA people had had the benefit of the thinking of the committee and its various subgroups on a wide variety of subjects, including technology, space applications, the physical and life sciences, and manned spaceflight.[22]

With respect to manned spaceflight, both NACA and military agencies had been very active. NACA's prospective thinking in February of 1958 had envisioned the travel of man to the moon and nearby planets. During the summer of 1958 the Advanced Research Projects Agency was besieged with requests for support of Air Force and Army manned spaceflight proposals. But shortly after the passage of the NASA Act of 1958, President Eisenhower assigned the new agency the responsibility for manned spaceflight. In mid-September Administrator Glennan and Roy Johnson, head of the Advanced Research Projects Agency, agreed that the two agencies should work together on a man-in-space program, and to coordinate the activity they established a joint Manned Satellite Panel.[23] Robert Gilruth, who was to be a key figure in the Mercury, Gemini, and Apollo programs, was chairman. By virtue of this spade work, NASA was ready only one week after its opening to proceed formally with Project Mercury, the nation's first manned spaceflight mission.[24]

The roots of the space science program went at least as deeply into the past as did those of the manned spaceflight program. Taking over the Vanguard program, assuming responsibility for much of the nation's sounding rocket research, and acquiring the Pioneer deep-space probes from the Air Force, NASA had an ongoing space science program from its first day. Building on these activities and drawing upon their own experiences of the past decade, the rocket research scientists who had come into NASA were able to put together a plan that described in a general way what NASA would be doing in space science for the next two decades. The shape of the emerging program was evident in NASA's first hearings before Congress.[25]

To trace the evolution of the space science program in the thinking of the NASA planners is interesting. A sheaf of working papers in the NASA files for January 1959 gives an overall summary of the space science program as envisioned at that time.[26] The program was described in terms of the different scientific disciplines to which space research could contribute—for example: particles and fields, astronomy, atmospheres, and ionospheres. A more formal document, 10 February 1959, elaborated further, listing atmospheres, ionospheres, energetic particles, electric and magnetic fields, and astronomy as areas of research within the NASA space science program.[27] Within the larger categories was a detailed breakdown.

Atmospheric research had been a major part of the sounding rocket work of the past decade (chap. 6). It investigated the principal properties of the earth's atmosphere, such as the pressure, temperature, density, and composition of the atmosphere as a function of height and geographic location. Particularly important were the variations of these quantities with time, especially the variations caused by solar influences. With the prospect of sending spacecraft to other planets the comparative study of different planetary *atmospheres* would be immeasurably aided. Among the important solar influences to study were those causing atmospheric molecules to become ionized forming electrified regions known collectively as the ionosphere. Again, with the possibility of including other planets, the plural *ionospheres* was used. *Energetic particles* referred to planetary radiation belts, radiations in the interplanetary medium, and cosmic rays. *Electric and magnetic fields* in the upper atmosphere and space would be important aspects of relationships between the sun and the earth, and presumably other planets as well. There was considerable interest in studies of *gravitational fields* in connection with geodesy and the celestial mechanics of the solar system and particularly for investigation of various theories of relativity, for which it was agreed that satellites should be very effective. The areas of *astronomy* were the familiar ones of sun, moon, planets, and stars. With satellites measurements could be made in ultraviolet, x-ray, and other wavelengths that could not be observed at the ground.

Amplifying the general description of the program were several pages devoted to specific problems ripe for attack. For example, under atmospheres one asked what were the primary sources of energy affecting the high atmosphere, and what was the relationship of the Great Radiation Belt to the heating of the upper atmosphere. A question that was still not answered a decade later concerned the precise relations between the earth's surface meteorology and the upper atmosphere. Of particular importance was the problem of learning in detail exactly how the sun exerted its influence on the atmospheres of the earth and planets. For ionospheric studies it was regarded as important to obtain details on the structure of the highest regions. Of especial interest, one looked forward to the study of other planetary ionospheres. Typical among the problems under energetic particles and magnetic fields were those in pinning down exactly how solar particles behaved in the vicinity of planets and, in particular, how the aurora was generated. Under gravitational fields the first problem was to get a detailed analysis of the earth's field into its various components, since from this would flow the solution of many other problems. Likewise one looked forward to determining the gravity fields of other planets. In astronomy, the first problem was to learn the spectral emissions of the sun, stars, and interstellar medium. This selection of existing problems is only illustrative.

Even more detail could be found in a second document of February 1959 entitled "The United States National Space Sciences Program," which added biosciences to the list of space science areas.[28] Of particular interest was the indication that working groups would be established immediately in half a dozen program areas, others later as needed. The evolution of the idea of working groups as a means of drawing outside scientists into the program will be discussed in some detail in chapter 9.

A five-page paper of 30 March 1959 elaborated on plans for a biosciences program, including the establishment of working groups in the area. By mid-April a paper titled "National Space Sciences Program," plainly a derivative from the 4 February document, showed clearly the directions the program was taking.[29] Although this was only half a year after NASA began, the agency would follow those directions for the next decade.

This early thinking of NASA was reflected in the report the National Academy of Sciences made to the international Committee on Space Research at its second meeting held at The Hague, 12–14 March 1959. In this report the breakdown of NASA's February working papers was followed.[30]

Quite properly the planning began with a consideration of the scientific objectives to pursue, a listing of the important areas of research, and an assessment of the significant problems to attack. One can see this approach in the NASA documents of the first half of 1959. But a program—by which is meant a long term, continuing endeavor of rather broad general objectives—must be carried out in discrete steps. In NASA parlance such discrete steps were called projects. Thus, the astronomy *program*, which was to investigate the universe from above the atmosphere, might be expected to continue as long as space techniques could produce significant new information, and there was no foreseeable end to that. But a satellite *project* to measure gamma rays from the depths of the galaxy—which most certainly would further the astronomy program by shedding light on energetic processes in the galaxy—would be of limited duration, long enough to prepare, fly, and operate the satellite, and interpret the data obtained.

Even as NASA's program plans were being developed, the agency started numerous projects to conduct investigations ranging over the different areas of space science. In reporting on the program, then, it soon became possible to list ongoing projects in which the nation's scientists were participating. In April 1960 NASA's report on the space science program could go far beyond the generalities of the papers developed in early 1959.[31] The various parts of the program were integrated under a few broad objectives; namely, to learn more about sun-earth relationships, the origin of the universe and the solar system, and the origin of life. The specific disciplines supporting the broad objectives were given as aeronomy, ionospheres, energetic particles, magnetic fields, astronomy, gravitational fields, lunar sciences, planetology and interplanetary sciences, and micrometeorites and cosmic dust. The similarity to the earlier breakdown is evident, although a few new terms appear. Replacing atmospheres was *aeronomy*, a term coined by Sydney Chapman during the 1950s from the Greek words meaning "laws of the air." The term *planetology* had been introduced to mean the study of the body of a planet as distinguished from the investigation of its atmosphere and ionosphere. It was felt that dust in space might be an important factor in studying the origin and evolution of planetary systems and in designing spacecraft for interplanetary flight, hence the prominence given to it. Micrometeorites were simply cosmic dust particles that struck another body such as the earth or a spacecraft.

In the April 1960 paper a great deal of detail could be given on projects to support the various programs. Under each of the above disciplinary categories were listed physical parameters to be measured, instruments to be used, experimenters responsible, and sounding rockets, satellites, or space probes to be employed. For example, under aeronomy the 30-meter sphere of thin metallized plastic constructed by William O'Sullivan of Langley Research Center was listed as a sounding rocket experiment to measure upper-atmosphere densities. The sphere would be carried aloft, ejected at altitude, and inflated. Accurate measures of the air drag on the falling sphere would give the desired densities. O'Sullivan's sphere was the forerunner of the Echo passive-communications satellite launched on 1 August 1960, which observers around the world were able to follow with the naked eye.

Under energetic particles the document had a long list of experiments on radiation belts and the magnetosphere, some of them having already been carried out successfully. For the related area of magnetic fields, *Pioneer 5* was shown as having been launched into deep space in March 1960. Radio, gamma-ray, solar, and stellar astronomy projects were in the works, and a solar observatory satellite was actually under development. A considerable amount of work was indicated as under way in gravitational fields, relativity, and geodesy. A series of lunar probes was also listed.

From this point on, the space science program evolved pretty much along the lines already established. A continuing effort to keep a spark of life in the planning recast the program objectives in different words from time to time. For example, in March 1961 the principal areas in the space science program were grouped under the headings: the earth as a planet, the earth's atmosphere, solar activity and its influence on the earth, origin and history of the moon and planets, and the nature of the stars and galaxies.[32] By the mid-1960s, when NASA made a special effort in its authorization hearings to present the broad perspective of the space science program, the principal categories had been reduced to two: exploration of the solar system and investigation of the universe. Elaborating on these objectives the author stated to the Senate Committee on Aeronautical and Space Sciences:

> The first category includes the investigation of our Earth and its atmosphere, the Moon and planets, and the interplanetary medium. The nature and behavior of the Sun and its influence on the solar system, especially on the Earth, are of prime importance. With the availability of space techniques, we are no longer limited in direct observations to a single body of the solar system, but may now send our instruments and even men to explore and investigate other objects in the solar system. The possibility of comparing the properties of the planets in detail adds greatly to the power of investigation of our own planet. Potentially far-reaching in its philosophical implications, is the search for life on other planets.

> The fundamental laws of the universe in which we live are the most important objects of scientific search. Space techniques furnish a most powerful means of probing the nature of the universe, by furnishing the opportunity to observe and measure from above the Earth's atmosphere in wavelengths that cannot penetrate to the ground. There is also the opportunity to perform experiments on the scale of the solar system using satellites and space probes to study relativity, to delve into the nature of gravitation, including the search for the existence of gravitational waves.[33]

However expressed, the basic substance of the program was remarkably stable.

In contrast to the overall program, one would expect the projects to change considerably as the years went by. But even here many projects had their origins in the thinking of the first few years. Table 2 lists the major space science, or science-related, projects up to mid-1968. For each project the dollar symbol indicates the first fiscal year in which money was specifically charged against the project, although money from supporting research or other general sources most likely had been spent earlier in exploratory work on the project. In some cases an asterisk is inserted to show how much earlier serious consideration of such a project had been under way. Of the 25 projects named, 22 (or 88 percent) were under way by mid-1962 in the sense that costs were being formally charged to them. More than three-quarters of the projects were begun or were being seriously considered during Glennan's tenure.

Considering the rapid development of the NASA program—on all fronts as well as in space science—and the wide range of projects, including launch vehicle development to be discussed later, that were set in motion during Glennan's term of office, it seems clear that the first administrator must be given the credit for setting NASA on the course that it followed for the next decade. Nevertheless, as he himself stated, Glennan was

Table 2
First Recorded Direct Obligations
to Space Science or Related Projects

(by fiscal year)

Project	1959	1960s 0	1	2	3	4	5	6
Sounding rockets	$							
Vanguard	$							
Explorers	$							
Physics and astronomy advanced research	$							
Lunar and planetary advanced research	$							
Bioscience advanced research			$					
Manned space science advanced research						$		
Orbiting solar observatories	*	$						
Advanced orbiting solar observatories				$				
Orbiting astronomical observatories		$						
Orbiting geophysical observatories		$						
International satellites		$						
Pioneer	$							
Ranger	$							
Surveyor			$					
Surveyor Orbiter						$		
Lunar Orbiter				$				
Mariner			$					
Voyager†			*				$	
Scout development	$							
Centaur development	$							
Delta development	$							
Mercury	$							
Gemini				$				
Apollo	$							

SOURCE: Jane Van Nimmen and Leonard C. Bruno with Robert L. Rosholt, *NASA Historical Data Book, 1958-1968*, vol. 1. *NASA Resources*, NASA SP-4012 (Washington, 1976), pp. 136–48. By the time a specific project appears as such in the financial records, generally a considerable amount of time (sometimes years) has been spent on advanced planning and research to lay the groundwork for the project.

* = already under consideration.
$ = first record of direct obligations.
† = never finished.

no "space cadet." His was just the right balance of conservatism and interest in space to make him congenial to President Eisenhower and acceptable
to the Congress. "Thus, with strong support, when it was needed, from
Eisenhower the Administrator's Office . . . with pushes—strong pushes—
from Abe Silverstein and believable and solid advice from Newell and
others—and strong pushes from the Congress—set the pace."[34] It may be
said that Glennan set a strong but measured pace.

The effect can be discerned in the methodical way in which the space
science program was made to unfold. NASA's first and natural step was to
extend the sounding rocket work, and the Pioneer deep-space investigations already under way. The modest step from those to solar and astronomical observatories came next, although the Orbiting Astronomical Observatory proved to be a much bigger bite than the space science managers
had imagined. The investigation of the moon would be much more demanding and costly than near-earth missions, and a serious commitment
to a lunar science program came more slowly, even though Harold Urey,
Nobel Laureate and renowned student of the moon and planets, had begun
to press for such a program in the first few months of NASA's existence.[35]
Also, to maintain what he considered the right pace, Glennan for a while
showed a reluctance to discuss planetary missions except as plans for later,
for the more distant future.

But "later" was not long in coming, as table 2 makes clear. Before
Glennan left office NASA was engaged in space science projects that took
in not only the earth and its environs, but also the moon and planets, the
sun, and even the distant stars. One may surmise that Glennan, exposed to
the pressures from both within and without the agency, and perhaps himself caught up in the enthusiasm of those around him, moved more rapidly
than he had originally intended. At any rate, he turned over to his successor, James E. Webb, a well rounded program, well under way.

By the time Webb took office, the course of the space program for the
next decade has been set. Even Apollo, under study since the start of NASA,
had been commended to President Eisenhower in the last months of the
Republican administration. Though Eisenhower did not approve, the ideas
were there ready to be seized when President Kennedy came to feel that the
successful accomplishment of an extremely challenging space mission
would be important to U.S. prestige. The renewed sense of urgency that
the Apollo decision bestowed on the space program made Webb's task one
of loosening the shackles imposed by the previous administration and stepping up the pace. But the program content was already there. Thus, Apollo
and Gemini may be looked upon as super projects designed to pursue an
already existing program with greater vigor.

In this climate the space science managers put together plans to expand their program. On 22 May 1961 the Space Sciences Steering Committee, which consisted of NASA's principal space science program managers,

met with selected consultants to review the proposed expansion.[36] The consultants represented a cross section of the disciplines of space science: Dirk Brouwer of Yale (astronomy), Joseph W. Chamberlain of Yerkes Observatory (atmospheric sciences and planetary astronomy), Robert A. Helliwell of Stanford (radio physics), Harry H. Hess of Princeton (geophysics and geology), Bruno B. Rossi of the Massachusetts Institute of Technology (high-energy physics and x-ray astronomy), and Harold C. Urey of the University of California at San Diego (lunar and planetary science). The group endorsed the expansion of the program proposed by NASA and emphasized a number of exciting researches to pursue, including the moon's gravity, the almost nonexistent lunar atmosphere, solar radiations, the vicinity of the sun as close as 16 000 000 km from the solar surface (one-tenth the distance from the sun to the earth), and micrometeoric particles in space. In characteristic fashion the scientists favored large numbers of small spacecraft for investigating the vicinity of the earth and heartily endorsed small grants from NASA to large numbers of universities for basic research. They recommended that serious consideration be given to a proposal from General Motors for obtaining by unmanned methods a sample of material from the moon.

By the following autumn NASA had moved forward substantially in the expansion of its program and was beginning to feel the need for a full-scale exchange with the scientific community on the content and course of the program. The Academy of Sciences was requested to organize a study of the space science program, which the Space Science Board agreed to do.[37] The study would be conducted during the summer of 1962; the program to be reviewed was described by the author at the NASA management conference held at the Lewis Research Center on 11 January 1962.[38] In the program were sounding rockets, satellites, and space probes. The Scout, Delta, Agena, and Centaur rockets, to be discussed in chapter 10, were included. Spacecraft, also to be taken up in chapter 10, included a variety of Explorers; solar, geophysical, and astronomical observatories; the lunar spacecraft Ranger and Surveyor; and Mariner planetary spacecraft. Some advance thinking about a spacecraft for a bioscence program was mentioned. The scientific fields were those already mentioned. Geodesy was described as important but stymied by difficulties over classification. An international cooperative program including many of the disciplines was well under way. What would be of special interest to the summer study participants was the university program, which in January of 1962 was rapidly increasing. The author's report listed university program funding as $3 million for fiscal 1959, $6 million for fiscal 1960, and $14 million for fiscal 1961 (the fiscal year beginning July 1 preceding the corresponding calendar year). In fiscal 1962 plans were to use $28 million on research projects in universities, largely flight experiments, plus $12 million for support of graduate training in space related fields, research facilities on university campuses, and grants for research of a more general nature than the specific flight projects. Space science managers projected a university program growing in the future to $100 million a year in projects and about $70 million a year in the broader grant program—a growth that was only about half realized in the 1960s.

In short order NASA's new team of leaders, which included many of the NACA's top people, remade the organization and activities acquired from the National Advisory Committee for Aeronautics into a National Aeronautics and Space Administration as called for in the NASA Act of 1958. As the new agency organized, developed its staff, and built its new facilities, NASA started the space science program along the lines that would be followed for the next decade. The rapidity with which this was done was both a tribute to the NASA team and convincing evidence that a strong base had existed on which to build in a number of areas, including space science. The thoroughness of the early work would be attested to by the fact that during the next decade—though the pace would be increased—little that was new would be added to the program.

9

External Relations

To provide an organization for handling relations within NASA was not enough. In a program that clearly would involve more people outside NASA than in, and in which numerous other agencies would have strong interests, NASA had to devise mechanisms for a variety of external relations. During Glennan's 29 months as administrator a considerable amount of management time was devoted to such matters.[1] Some of the arrangements worked out concerned space science.

DEPARTMENT OF DEFENSE

Congress had shown great concern over how to ensure proper coordination between the civilian and military space programs. Space science was one of the areas of mutual interest between NASA and the Defense Department. Sounding rocket research had been supported by the military services during the 1940s and 1950s, and the services had participated in the scientific satellite program of the International Geophysical Year. The potential military applications (p. 41) were adequate motivation for such activity, and there was no reason to suppose that the creation of a civilian space program would end military interest.

Most of the space scientists who came to NASA in the fall of 1958 had been associated with the Army, Navy, or Air Force rocket or satellite research programs. Their long-standing personal associations with people in the Department of Defense made coordinating the two programs relatively easy. The Civilian-Military Liaison Committee was too far removed from the day-to-day action to be as effective as informal personal contacts were. These personal contacts gradually led to a more formal arrangement. On 4 May 1959 a meeting on the subject was held in the office of Herbert York, director of defense research and engineering. Attending, in addition to York, were Samuel Clements and John Macauley of Defense, and John Clark, N. Manos, and the author from NASA. The participants agreed to set up a 10-man group with 5 members from each organization and to meet monthly to exchange information. Named the NASA-DoD Space Science Committee, the coordinating group under the chairmanship of the author held its first meeting at NASA Headquarters on 11 August 1959.[2] A month later, when Defense and NASA established the Aeronautics and Astronautics Coordinating Board, the Space Science Committee was renamed the Unmanned Spacecraft Panel of AACB.[3]

The deliberations of the Unmanned Spacecraft Panel were anything but monotonous. The panel quickly developed a mechanism for routinely tabulating, updating, and exchanging a great deal of information on space projects in the civilian and military programs.[4] Since it was unnecessary to devote the time of the panel meetings to routine coordination, attention could more easily be given to special problems. The problems varied widely in substance and seriousness. One of the first was the question of how much space science the military would do. Many NASA members felt that the military services should look to NASA for their space science needs and devote themselves to researches specifically related to military applications. With this position the services violently disagreed, insisting that they had to be working in science to make the most effective applications of the science results. The author agreed with this position and had to take a bit of flak from his own colleagues, because they feared that arguments over duplication of effort might compromise the NASA program. Dryden agreed that it was not reasonable to try to stand in the way of a Defense Department space science program, particularly because of the benefit to military applications. The dispute was eventually turned in the direction of ensuring, by careful coordination, that the military and civilian space science work did not bring—in the jargon of the day—"wasteful duplication."

More serious were the disputes over questions of military classification. Such problems arose in connection with accurate observations of the earth's surface and in geodesy. Earth observations were directly related to military interest in reconnaissance and surveillance, and intelligence agencies were sensitive about revealing either their interest or national capabilities in the field.[5] Applications people ran into this problem first in connection with weather pictures of rather gross resolution that were obtained from the Tiros weather satellites. There was concern over possible international sensitivity to U.S. satellite photography of foreign territories—even at resolutions of no better than 400 to 800 meters.[6] Some feared that international reaction might precipitate a confrontation that could compromise U.S. ability to pursue legitimate defense interests in earth observations. This controversy was heightened in the late 1960s when NASA and other agencies began to push earth-resource surveys of much finer resolution.[7]

The space scientists also had their bouts with classification problems. The most knotty had to do with geodesy, the science of measuring the earth. The accuracy with which the gravitational field could be measured and analyzed into its various components—or harmonics, as they are called—was important in determining the size and shape of the earth, the distribution of mass in the earth's crust, and stresses within the mantle below the crust. But such data were also essential for accurate guidance of long-range missiles. To the scientists the precise location of different points on the earth's surface relative to each other was vital for checking newly emerging theories about the movement of the earth's crust. But to the military those data would determine the position of potential military targets relative to missile launching areas. The conflict was fundamental. The scientists needed such information for their research and during the International Geophysical Year had entered into worldwide, multinational, cooperative programs for making geodetic measurements from observations of IGY satellites. The IGY program had naturally extended into the NASA program and along with it went the tacit assumption that the scientific data obtained would be available to all participants. Indeed, as with all the IGY programs, the results were to be published in the open literature.

This policy was painful to the military people, who felt that data of such vital military significance should be kept under wraps and potential enemies forced to expend similar efforts to obtain the information. A muddle of exchanges began between NASA and DoD on the subject. Geodetic scientists complained about footdragging. At the March 1960 meeting of the Space Science Board, George Woollard urged NASA to start at once on the preparation of a satellite specifically for geodesy. A little over half a year later, the Aeronautics and Astronautics Coordinating Board was still discussing how NASA might obtain geodetic data for the scientific community.[8] On 14 November the Department of Defense announced that the Army, Navy, NASA, and Air Force were jointly building a geodetic satellite.[9] In that same period Deputy Administrator Hugh Dryden was seeking clarification from the Academy of Sciences as to exactly what international commitments regarding open publication of geodetic data the United States had entered into. In reply he received a pile of paper three centimeters thick showing that, internationally, there was a general understanding that the United States would publish data from its IGY satellites that could be used for geodetic studies, with the necessary information on the precise location of tracking stations.[10]

The joint satellite, which acquired the name Anna from the initials of the cooperating agencies, did not end the controversy over classification. The rumblings reached the ears of Congressman Joseph Karth, chairman of the Space Science and Applications Subcommittee of NASA's authorizing committee in the House of Representatives. He plunged into a series of hearings on the subject. The Karth hearings, and pressure the president's science adviser received from the scientific community, forced a decision very much like apartheid. It was finally agreed that the scientific geodetic program would continue, with open publication of results on the NASA side. Likewise, the DoD program would continue, and when appropriate the two agencies would cooperate, as with the Anna satellite. But DoD would decide unilaterally on the disposition of the data and results from its part of the program. Because of the knotty problems in this area, NASA, DoD, and the Department of Commerce—where the Coast and Geodetic Survey was located—set up a special Geodetic Satellite Policy Board for the difficult problem of coordination.[11]

It would be unfair to leave the impression that all the struggles with questions of classification were caused by the military. Within NASA a pressure arose to classify launch schedules. Some of the pressure came from the use of military hardware and launching ranges, but much of the desire to classify stemmed from the poor showing that NASA had made in its early attempts and from an embarrassing tendency of schedules to slip because of technical problems. One could not properly use classification to avoid embarrassment to the agency, but the argument was put forth that it was important to protect the already damaged national reputation in space exploration from any further damage.

After a year's experience Administrator Glennan felt it unwise to publish schedules with specific launch dates too far in advance. Past and imminent launches could be given by date, but Glennan suggested that launches over the next two years be announced only by quarter, and only by year thereafter.[12] In March 1960 the author wrote to Ira Abbott, chairman of a committee dealing with matters of security classification, citing numerous problems that would arise in the space science area if blanket classification were applied to NASA launch dates.[13] It did not seem appropriate to classify sounding rocket firings in which many universities participated—and for which schedules had been unclassified for more than a decade of the Rocket and Satellite Research Panel program. For planetary shots the timing was specified by the celestial mechanics of the solar system and, if the existence of the mission was known, its date was more or less obvious. Even where nature did not reveal the date of a prospective mission, NASA had other problems to work out. A large part of the space science program was carried out by researchers in the universities, who did not ordinarily have security clearances. Also, the civilian, peaceful character of the national space program would appear to be compromised if an effort were made to operate under security restrictions. Baker-Nunn optical tracking stations, operated by the Smithsonian Astrophysical Observatory, would not be welcome in countries like India and Japan, which opposed classified activities on their soil. It would be difficult or impossible to work with volunteer groups providing supporting observations of satellites if schedules could not be issued in advance. The same problem would arise with groups assisting in telemetering satellites and space probes—various universities and the Jodrell Bank Radio Astronomy Observatory in England, for example.

Concern about this aspect of classification continued through NASA's first two years, but policy developed to meet the need. Those participating in a mission were furnished the necessary information for planning and meeting schedules; and, in space science missions, while experimenters generally did not have to wrestle with problems of security classification, they were expected to handle schedule information discreetly.

ACADEMY OF SCIENCES

Most important for space science were relations with the National Academy of Sciences and the Space Science Board. It was assumed without question that NASA would look to the Space Science Board for advice on scientific questions. Accordingly NASA joined the National Science Foundation and the Department of Defense in providing funding for the board. In the fall of 1959, when time approached for the National Science Foundation to renew the annual contract with the board, Dryden sent to Alan Waterman, director of NSF, a work request that NASA would like to see incorporated in the new contract.[14] The contents of the request, a copy of which was sent to the National Academy the same day it went to NSF, had been discussed in advance between the author and Hugh Odishaw, executive director of the Space Science Board.[15] NASA sought assistance from the board on (1) long-range planning, (2) specific planning for the separate scientific disciplines, (3) international programs, and (4) the handling of space science data and results. The first two were straightforward, but care was taken to emphasize planning, and NASA took this opportunity to turn back an incipient interest on the part of the board in getting into operational matters like the review and selection of experiments for space science missions.

A major point under (3) concerned U.S. representation on the international Committee on Space Research (COSPAR). At the invitation of Lloyd Berkner, who was then president of the International Council of Scientific Unions, the author had convened the organizing meeting of COSPAR in London 14 November 1958. Subsequently the question arose as to whether America's permanent representative should come from NASA. The Academy thought not, and Dryden and the author agreed. It was traditional and appropriate that the country's representation on international scientific, as opposed to political, bodies should fall under the aegis of the Academy of Sciences. NASA supported this view and further agreed to pay America's annual subvention to the Committee on Space Research.

The final item in the work request on data and results was fuzzy, not at all clear at the time. Since the Academy had been involved during the

International Geophysical Year with the operation of world data centers, which archived and distributed data and information derived from the IGY science program, it was thought that the Academy might continue this function for the national space program. After all, there had been a Data Center on Rockets and Satellites, so what could be more direct than to have that center expand its responsibilities? There were subtleties to the problem, one of which surfaced in a meeting 9 December 1959, held at Boulder, Colorado. Hugh Odishaw asked if NASA would support a center devoted to data from all upper-atmosphere and solar research, not just those obtained from rockets and satellites.[16] NASA representatives equivocated and, after prolonged discussion with the Academy, established the Space Science Data Center at the Goddard Space Flight Center.[17] Although the new organization did undertake to archive a great deal of data that were not obtained from space experiments, in general such data were selected because they would increase the value of the space data.

During the period that NASA was developing its working relations with the Space Science Board, the agency was also feeling its way toward some mechanism to provide broader and closer contacts with the scientific community than could be expected from the Space Science Board alone. For the most part unaware of the extensive and skillful use NACA had made of committees to keep in touch with thinking outside the agency, NASA space scientists began to move in a similar direction. Internally a Space Sciences Steering Committee was established in April 1960, with responsibility for recommending space science programs and projects to the director of spaceflight programs, Abe Silverstein. The steering committee also recommended the selection of experiments and experimenters for space science missions.[18] Subcommittees were formed for the scientific disciplines.[19] Unlike the steering committee, however, which consisted solely of NASA employees, about half the members of the subcommittees were outside scientists. From this group of disciplinary subcommittees the NASA advisory structure in space science evolved over the years. Advisory committees became a major element in NASA's relations with the scientific community and in planning and conducting the space science program. This subject will be discussed in detail in chapter 12.

PRESIDENT'S SCIENCE ADVISORY COMMITTEE

In space science at least, NASA's relations with the President's Science Advisory Committee grew out of the central role it had played in the formation of the agency. From the start the Space Science Panel of PSAC took a close interest, frequently reviewing what was being done and offering advice. When astronomers could not agree on specifications for the orbiting astronomical observatory, and NASA found itself in the middle, the Space Science Panel and its chairman, Edward Purcell, pushed NASA to resolve the difficulties. Of NASA's desire to be cooperative, Glennan years later would write: ". . . no major operating agency ever gave more consideration to the very much less than objective cries of the 'scientists[.]' Within the Administration—that is, NASA [—] we had solid and often brilliant scientists who were able to plan a truly 'NATIONAL' science program in spite of the often controversial advice and complaints so freely given by the Scientific Community!"[20]

By the spring of 1962 the space science group in NASA Headquarters had settled on policies to use in developing the program and in working with the scientific community. These policies were described to the Space Science Panel in April 1962 and appeared to have the panel's blessing.[21] The policies, together with the NASA management instruction on responsibilities of principal investigators in the flight program, provided the framework for the conduct of the space science program during the 1960s.[22]

It is worth dwelling a bit on these policies, since they colored all of NASA's relations with the scientists. The agency undertook, with the best advice it could get, to determine the most important areas of research— clearly a subjective matter, which the agency sought to handle as objectively as possible. Then NASA tried to support competent scientists working on what were thought to be the most important problems in each area. No attempt was made to saturate any area with researchers, in the belief that high quality could best be achieved by supporting only those investigations that seemed most fundamental and most likely to yield significant new information. When funds were ample, this policy could be followed without difficulty; but when money became tight, difficult choices would have to be made, and perhaps an entire area of research might have to be curtailed to afford adequate funding for the remaining areas. Such situations did arise later on. For example, in the budget squeezes of the late 1960s NASA chose to decrease ionospheric and magnetospheric research in order to maintain adequate support for solar system research and space astronomy. Although the Space Science Board endorsed this choice, the board had to face dissension in its ranks from the particles-and-fields workers who were hard hit by the cutbacks.

NASA tried to provide continuity of support to researchers. It was recognized that a single experiment usually was but a step in an investigation and that it was important to enable a scientist to complete the entire investigation. For example, a single sounding rocket flight could yield interesting data on ion densities in the ionosphere, say at White Sands at noon on a summer's day. But to understand the processes in ionospheric behavior, geographic and temporal variation, and the relationship of solar activity to the ionosphere—an immeasurably broader and more significant objective than to know the state of the ionosphere at only one time and location— would take years of research and many experiments.

Continuity of support was a genuine worry to non-NASA scientists. In this regard they felt at a disadvantage with respect to scientists in the

NASA centers who could count on being supported continuously by their agency. Moreover, the NASA scientists clearly had an inside track in placing their experiments on NASA spacecraft; many outsiders worried that NASA would take care of its own scientists first and assign the leftovers to outside experimenters. To allay such fears, the author informed the Space Science Board that NASA would pick experiments on the basis of merit and would assign most of the available payload space on NASA science missions to outside scientists.* When, in November 1959, Lloyd Berkner, as chairman of the Space Science Board, sent a lengthy criticism of the space science programs to George Kistiakowsky, the president's science adviser, Berkner found few things to praise. One was the stated policy of reserving no more than 20 to 25 percent of the payload on science missions for NASA personnel.[23]

This policy did not have universal support within NASA, where there was much sympathy for the idea of taking care of one's own. After all, it was argued, NASA people had undertaken to create and operate the necessary space tools for scientific research, to defend the program before the administration and the Congress, and to do a lot of the drudgery needed to keep a program going. For this they should be guaranteed first rights over those who chose to remain in the academic world with all its niceties and privileges. In sympathy with the NASA laboratories, Silverstein himself voiced such views, and the author at times found himself in the middle. Nevertheless, a genuine effort was made to adhere to the stated policy, and for a while the proportion of outside scientists finding berths on NASA spacecraft increased. But the limited amount of payload space available, along with the increasing numbers of applicants who wanted to take part, militated against reaching the ideal. Responding to renewed criticism, in March 1960 John Clark, in a NASA memorandum discussing relations between the agency and the Space Science Board, reiterated: "It is still the NASA objective that the larger part of the scientific work will be done outside of the NASA organization. . . . about 60 percent of the present space science work is being done outside NASA, compared to 40 percent in-house."[24]

During the 1960s, except on some individual flights, NASA never quite achieved the stated, admittedly arbitrary goal. While occasionally a cause for grumbling, the matter did not become serious again until the early 1970s when tight budgets once more seemed to put research groups in universities at a decided disadvantage relative to those in NASA centers.

NASA also recognized that it was not enough to pay only for flight experiments. A certain amount of related research had to be supported, particularly that required to lay the groundwork for experimenting in

---

*NASA scientists pointed to this statement of policy as logically inconsistent. How could the policy be adhered to if, on the basis of merit, the center proposals surpassed all those from outside?

space. During its first year, however, NASA appeared to be neglecting this important aspect of space science in its concentration on getting space-flight projects going. In his critical letter of November 1959 to Kistia-kowsky, Berkner unleashed a lengthy critique of the program as he saw it at the end of its first year. Berkner dwelt on a number of concerns scientists repeatedly returned to throughout the years. Along with worries about relative amounts of money going into manned spaceflight—Mercury at the time—Berkner expressed the interest of scientists in having large numbers of small vehicles in the program in preference to a few larger ones. He also registered complaints about the domineering attitude of NASA project engineers toward experimenters and about the difficulty outside scientists had in competing with NASA scientists unless the necessary engineering facilities were provided to enable the outside scientists to compete. Berkner considered the question of support for long-term, space-related research a major issue, averring that NASA had to provide support, since the National Science Foundation was not likely to do so.[25] Responding to Berkner's criticism, Administrator Glennan wrote to Kistiakowsky on 3 December 1959 agreeing among other things that NASA should support the long-range basic research important to space science.[26] In this vein, NASA's university program office later devised a method of step-funding research projects so as to assure a university scientist of at least three years continuous support (chap. 13).[27]

While recognizing its own responsibilities toward experimenters, NASA also asked principal investigators to assume considerable responsibility on their part—specifically for the preparation, calibration, installation, and operation of their instruments. This policy, which was somewhat fuzzy at the start, grew in clarity as time passed, until it was articulated in April 1964 in a formal NASA issuance.[28] Basically, a principal investigator was given a place on a satellite or space probe for his instruments, was assigned the necessary electrical power, telemetering, and other support from the spacecraft, and was promised a certain period of time after the flight during which the data obtained would be reserved to him for analysis, interpretation, and publication. In return, the investigator was expected to work as a member of the project team, meet all relevant schedules, and ensure that his equipment was properly constructed, passed prescribed tests, and was available in operating condition for installation in the spacecraft at the appropriate time. In addition to using the data for his own research, the experimenter was expected to put them into a suitable form for archiving in the data center, so that later researchers could use them for further studies.

The policy was simple in concept, but problems arose from time to time. Not infrequently the scientists would feel that too much prominence was being given to engineering, as opposed to scientific, requirements, and that the project manager did not appreciate that the scientific experiments were the purpose of the project. On his part, the project manager often would feel that the scientists did not understand the difficulties in getting an operating spacecraft aloft, and the importance of meeting schedules and test requirements. Such conflicts often seemed in the nature of things, for the engineer was trained in disciplined teamwork, while the scientist's stock in trade was highly individualistic questioning of authority. The engineer would find the scientist's propensity for last minute changes to make an improvement in the experiment baffling, while the scientist would find the engineer's insistence on prescribed routine frustrating. Yet the scientists and engineers could and did work out their differences, though sometimes at the expense of management time.

More subtle was the question of what was meant by the experimenter's taking responsibility for his experiment. If the investigator interpreted that as giving him control over the project manager, the scientist-engineer clash was enhanced, and management had to make clear that the project manager was in charge. That the investigator should take responsibility for ensuring that all phases of his experiment were being properly taken care of did not mean that the scientist had to do them all himself, though sometimes there was confusion about this. The investigator was expected to work out with the project manager how the investigator would meet his responsibilities. Often a contractor would be engaged to construct the scientific equipment. Perhaps NASA would agree to do part of the work. Whatever arrangement was made, it was still the investigator's responsibility to be aware of how things were going and when necessary, perhaps with the project manager's help, to see that steps were taken to correct deficiencies.

Granting an investigator the exclusive use of data for a specified period was important to both NASA and the investigator.[29] To the scientist the opportunity to publish his results and earn the acclaim of colleagues for what he had accomplished was a substantial part of his reward for conceiving and carrying out an experiment. Were NASA unable to grant a scientist the necessary time to claim his reward, the best researchers would surely have sought other scientific fields to plough. Yet from time to time this policy came under attack by Congress. The argument was that the taxpayer was putting out enormous sums for space research and, therefore, had a right to the data as soon as acquired. Most often this argument flared up when the data were spectacular pictures of the moon or Mars. Then a clamor from the press to issue the pictures at once would be echoed by members of Congress, no doubt inspired to speak out by a few well placed phone calls from enterprising reporters. Little concern was expressed over release of ionospheric measurements or data on magnetic fields in space.

NASA held its ground on the basic policy. Setting aside the question of the scientist's pay for contributing his original ideas and carrying out the experiment, NASA pointed out that only the scientist who had conceived the experiment and had personally struggled with the intricacies of calibrating the measuring instruments, could reduce the data properly to remove ambiguities and errors that would otherwise make the data useless to other researchers. In return for the exclusive use of data for a mutually agreed time, NASA required an experimenter to put his data in suitable form for archiving and use by other researchers. This was the taxpayer's quid pro quo; without such an arrangement, the taxpayer would not be getting his money's worth.

The time required to put data in order varied from case to case and was negotiated between the agency and the experimenter. For a simple experiment, perhaps a repetition of a previous one, a few months might suffice. A more complicated, more subtle experiment might take the investigator a year or more to work up the data and publish his first paper. As an illustration, NASA could point to the ionospheric experiment devised by J. Carl Seddon and colleagues at the Naval Research Laboratory for sounding rocket experiments at White Sands.[30] Simple in concept, the experiment ran into tremendous difficulties in practice. The idea was to measure the effect of the ionosphere on radio signals from a flying rocket and to use that effect to deduce the electron densities in the ionosphere. But the influence of the earth's magnetic field, the splitting of the radio signal into separate components, and reflections of the signal from inhomogeneities in the ionosphere required many years to decipher. Until that was done, the data would have been useless to other researchers. But once the various physical processes were understood and could be unraveled, the analysis of data from a new set of measurements could be accomplished in a few months.

Pictures were a special case. That was where the greatest public interest lay, and NASA adopted a policy of releasing pictures as soon as they could be put in suitable form. Often this was virtually immediately, as with much lunar photography. But pictures of Mars received with low signal-power usually took a great deal of electronic processing to bring out all available detail and it could be many weeks or months before they were ready for release.

PUBLICATION OF RESULTS

Important to the scientific community was the question of where scientific results from the space program would be published. Publication in the open literature is, of course, a fundamental aspect of the scientific process. Both the outside scientists and those who had joined the agency were dedicated by training and habit to open publication. In this they ran head on into NACA tradition and practice of issuing research results in series such as NACA Reports, Technical Notes, and Technical Memoranda.[31]

NACA papers were highly respected in the field of aeronautics and aerodynamics. They were carefully critiqued and severely edited within the agency before being widely distributed to aeronautical centers, appropriate military offices in the United States and elsewhere, and industrial and academic libraries around the world. It was NACA's position that the procedure ensured both high quality in its publications and provided for getting them to those who needed them in their work. Moreover, the existence of such series of NACA publications was the best possible advertising for the agency.

NACA was not alone in this practice. Both the Bureau of Standards and the Bell Laboratories put out journals of their own; and, during the Rocket and Satellite Panel days, the Naval Research Laboratory had issued much of its rocket-research results in NRL reports.[32] In the space science field, the Jet Propulsion Laboratory began putting out a Technical Report Series under the imprimatur of JPL and the California Institute of Technology.[33] In academic circles Gerard P. Kuiper, noted astronomer of unbounded energy and wide-ranging interests and head of the Lunar and Planetary Laboratory at the University of Arizona, put out a series entitled *Communications of the Lunar and Planetary Laboratory*, listing the University of Arizona as publisher.[34] In the *Communications* Kuiper and his colleagues published a great deal of excellent material, much of it from research supported by NASA. But Kuiper was severely criticized by his scientific colleagues for using this means of bringing his work to the community. Their reasons for criticising were fundamental, deeply rooted in the scientific process. First, it was pointed out, the usual scientific journal accepted an article for publication only after it had been given a careful review by one or more impartial experts in the field addressed in the article, whereas a scientist publishing in what amounted to his own journal could hardly subject his own work to the same kind of review. Secondly, the limited distribution of a publication to a selected list of recipients was bound to miss persons who had not only a legitimate, but often a significant, interest in the material, for how could one individual or a small group hope to be aware of all such interests? This point was particularly pertinent in a rapidly growing field with imprecise and fluctuating boundaries. In contrast, regularly published journals, open by subscription to all who were interested, were widely known in the scientific community; a scientist from another discipline could quickly find his way to material of importance to his work. Although the NACA had had a very large organization to draw upon for reviewing papers before publication, the same sort of criticism had been leveled at the NACA publication policy.

For NASA's first year, the question of publication remained in the background, with the NASA scientists assuming that the policy was to publish the results in the open literature, and former NACA people tending to expect a collection of NASA publications to evolve. Harry Goett, director of the Goddard Space Flight Center, precipitated a confrontation when in May of 1960 he proposed to issue NASA papers that had been given at a meeting of the international Committee on Space Research in a NASA series.[35] When the proposal reached Thomas Neill, an employee in the Office of Advanced Research and Technology who had carried over from the NACA the responsibility for overseeing the publication of in-house reports, Neill refused to permit the COSPAR papers to go out as NASA technical reports. Neill's position was that the papers had already been published in the COSPAR sphere and to put them out now in a NASA series would be wasteful duplication. It was an understandable position, but it stood squarely in the way of those who wanted to build up NASA's own fine "fourteen foot shelf" of space science literature, as Abe Silverstein described it.

There was a great deal of discussion of this issue during the spring and summer of 1960. The scientists, recognizing the intense desire of the NACA people to build up a library of NASA publications along the NACA lines, favored dual publication. A check with a number of scientific societies revealed they would be willing to accept papers for publication that had previously been put out under a NASA cover, since they did not regard the latter as genuine publication. This was the view of Lloyd Berkner, president of the American Geophysical Union, when the author called him on 19 May 1960. For AGU's own publication, the *Journal of Geophysical Research*, Berkner was sure there would be no problem, and he thought there should not be any difficulty for the *Physical Review*—which was later confirmed by the editors.[36] Several other journals took the same position; of those queried only the American Chemical Society expressed disapproval. Taking smug satisfaction in the considerable evidence they had gathered that NACA or NASA reporting was not generally viewed as genuine publication, the NASA scientists persevered in urging a policy that space science results would be published in the open literature, but that where desired duplicate NASA publication would be permitted. Dryden approved the idea and asked that an appropriate paper be drawn up articulating the policy, which led to more discussions but no clear statement of policy that could be given formal approval.

Instead the policy was established by practice. Space science results were published in the open literature, and management issuances pertaining to the program presumed such a policy. In international, cooperative space science projects, implementing agreements called for publication of results in the open literature.[37] Simultaneously in-house publications took a variety of forms. The Jet Propulsion Laboratory report series has been mentioned. From time to time the Goddard Space Flight Center issued bound collections of reprints of published papers by Goddard authors.[38] In September 1959 Abe Silverstein was considering establishing a NASA journal, much like that of the Bureau of Standards, which was cited as an example.[39] But such a NASA journal did not materialize. Instead there evolved the NASA Special Publications, an aperiodic series, generally book length, devoted to the whole spectrum of NASA's activities. The Special Publications were an excellent means of publishing under the NASA imprimatur integrated reviews of a topic or field, but were not usually suitable as an outlet for original scientific research. They were in fact accorded the same sort of mild disdain the academic community reserves, not always with justification, for most government publications.

UNIVERSITIES

NACA had had a rather small involvement with the universities.[40] What university research NACA did pay for usually was tied into research projects going on at the NACA laboratories. For NASA, however, relations with universities would be more extensive and different. This was especially true in space science, where the number of disciplines encompassed in the program dictated that a great deal of the work would have to be done outside and largely in the universities. Much of this would be an extension of a university's own research, with the addition of new tools— rockets and spacecraft. NASA would accordingly be funding university research as a major part of a broad space science program rather than as specific support to in-house projects. By undertaking to carry out a substantial part of the national space science program, the universities became allies of NASA.

But when NASA also decided to create space science groups at the Goddard Space Flight Center, the Jet Propulsion Laboratory, and other centers, the universities found themselves in the role of rivals to NASA. For, the in-house groups would inevitably be in competition with those outside for funding of their research and for accommodation on scientific flights, as mentioned earlier. A number of the mechanisms that NASA devised for working with the scientific community were influenced by the need to moderate the tensions that soon appeared. For this reason the responsibility for selecting space science experiments and experimenters was kept in headquarters even during periods when there was a general attempt to decentralize authority by transferring to the field many functions previously handled by headquarters.

Work with the universities was sufficiently important to the space program—particularly to the space science program—that NASA established an organizational unit specifically for handling university relations.[41] The university office guided NASA's work with the academic community, not hesitating to experiment with new ideas on government-university relations. More attention is given to the NASA university program, particularly as it bore on space science, in chapter 13.

SCIENTIFIC SOCIETIES

In seeking to bring the scientific community into the space science program and in insisting on publication of results in the open literature, NASA could hardly escape a close association with the scientific societies. The societies afforded the most common meeting ground of the scientists, and their journals formed much of the open literature.

A number of scientific societies soon became involved. The American Astronomical Society's interest was at first tentative, although a number of its leading members were fully committed to space astronomy—like Richard Tousey of the Naval Research Laboratory, Leo Goldberg of the University of Michigan, Gerard Kuiper of Yerkes Observatory, and Lyman Spitzer of Princeton. Spitzer had been among the first, in the mid-1940s, to write about and advocate the use of satellites for astronomical research. In the sounding rocket program of the 1940s and 1950s, Tousey had been one of the pioneers in rocket astronomy. And no sooner had NASA opened its doors than Leo Goldberg was urging support of a solar astronomical satellite project which the McDonnell Aircraft Company had designed with advice from University of Michigan astronomers. Under the pressure of such widespread interest, the American Astronomical Society's participation grew steadily throughout the 1960s. Papers appeared in its journal and at its meetings, and the society began to promote important aspects of space astronomy. The spectacular results of planetary missions, particularly in 1969 and early 1970s, helped dispel the disdain and lack of interest with which astronomers had regarded the planetary field for decades.

Among the first learned societies to show strong interest were the American Physical Society and the American Geophysical Union. In April 1959—six months into NASA's first full year—the Physical Society sponsored, along with NASA and the National Academy of Sciences, a symposium on space physics, which was well attended.[42] Anticipating the importance of space science for extending geophysics to other planets, the Geophysical Union went even further. In November 1959, AGU officers considered the question of providing a home for space science. Encouraged by the show of interest, NASA's Robert Jastrow and Gordon J. F. Mac-Donald, a brilliant young geophysicist, on 10 December 1959 wrote to President Lloyd Berkner recommending that the union create a section on planetary physics.[43] After consulting with AGU officers, Berkner responded by inviting the author to become chairman of a Planning Committee on Planetary Science, with members Jastrow (secretary), Leroy Alldredge, Joseph W. Chamberlain, Thomas Gold, MacDonald, Hugh Odishaw, Alan Shapley, Harry Vestine, Harry Wexler, Charles Whitten, and Philip Abelson (and later Walter Orr Roberts), all of whom had had important roles in the International Geophysical Year program. For the next two years the committee organized sessions on space science for the union meetings, and promoted the interests of space science within the union.

For the summer of 1960 committee members prepared a series of papers reporting on progress in the planetary and interplanetary sciences for publication in the *Transactions of the American Geophysical Union*. The President's Page in the *Transactions* for September 1960 carried a note from the author pointing out the importance of space science to geophysics and calling attention to the existence of the Planning Committee on Planetary Sciences.[44]

Within the union there was a steady movement toward the creation of a new section on the planetary sciences. But space science was itself but an extension of the traditional disciplines, and there was opposition to the proposed action. The argument was that the existing sections of AGU could provide the desired home for the new activities in space. The section on meteorology, for example, could accommodate satellite meteorology. Any section dealing with an aspect of the earth sciences could house that same aspect of the planetary sciences. In fact, some feared that a separate section on the planetary sciences would become another little union within the overall union. Even members thoroughly involved in the space sciences—like John Simpson, experimenter on Pioneer and Explorer satellites, theoretical physicist Alexander Dessler, and Harry Wexler, director of research for the U.S. Weather Bureau—were opposed. Nevertheless, the strong coherence in the space sciences, generated by the peculiarities and demands of the space tools, sparked the push for a new section. The spring of 1961 saw a great deal of discussion of the matter, and at its 22 April 1961 meeting the council of the union approved in principle the formation of a new section—by a margin of one vote! The council asked that the entire organizational structure, activity, and nomenclature of the union be reviewed as a precaution against intolerable dislocations within the society from addition of the new section. The review concluded that no other changes were required, and on 25 April 1962 the council gave final approval for the formation of a section on planetary sciences (which later in the decade divided into several groups). The author became the first president of the section, and Jastrow its first secretary, thus symbolizing the close relations that NASA had developed with the American Geophysical Union.

The examples given here are only illustrative. The breadth of the space sciences generated an important association with many scientific and technical societies and institutes. The interest of the American Rocket Society and the Institute of Aeronautical Sciences—which soon merged into the new American Institute of Aeronautics and Astronautics—was an obvious one, as was that of the American Astronautical Society and the International Astronautical Federation, although their concern tended more toward the engineering and technology side of the picture. More directly concerned with space science were the Optical Society of America, the International Astronomical Union, the American Meteorological Society, the Geological Society of America, the American Institute of Biological Sciences, and a long list of others. For some of these, interest in space science flared up at the very start, while for others the interest gradually emerged as the program unfolded.

Inheriting so much from the International Geophysical Year, NASA had an international program from the outset.[45] There were two main arenas, that of the international scientific circles such as the International Council of Scientific Unions and its newly formed Committee on Space Research, and that of a political nature, falling generally in the sphere of the United Nations. There were numerous political considerations relative to space, and NASA was immediately drawn into United Nations deliberations on space matters.

But the natural arena for space science was the international scientific community, and from the start NASA gave strong support to the Committee on Space Research. Among the unions of the council represented on COSPAR were the International Union of Scientific Radio and the International Union of Geodesy and Geophysics, which had first recommended the use of scientific satellites during the International Geophysical Year. Following the organizing meeting convened by the author in London in November of 1958, COSPAR held its first full-scale business session in The Hague, 12–14 March 1959.[46] At that meeting Richard Porter of the Space Science Board, U.S. representative to COSPAR, asked the author whether the United States might offer to launch space science experiments for COSPAR members. In a phone call to Washington, the author obtained Hugh Dryden's approval to inform the meeting that NASA would be willing to do so. Porter then wrote to President H. C. van de Hulst, saying that the United States would accept single experiments as part of larger payloads, or would launch complete payloads prepared by other countries.[47] The response to the U.S. invitation was immediate, and before the year was out a number of cooperative projects had begun. With the Soviet Union, genuine cooperation proved to be difficult during the 1960s, less difficult in the 1970s climate of detente. These subjects are discussed at length in chapter 18.

As the leaders of NASA worked to reshape the NACA into an aeronautics and space organization, they also laid the foundation for the many relationships with other government agencies, industry, and the scientific community that played an essential role in planning and conducting the program. But none of this would have been of any avail without the principal tools, the rockets and spacecraft essential to the investigation and exploration of space. A first order of business was to provide for these tools. That NASA set about to do, striving to overcome as soon as possible the visible gap that lay between the United States and the Soviet Union in propulsion capabilities and launchable spacecraft weights. Because of the central importance of launch vehicles and their payloads, the next chapter is devoted entirely to them.

10

Rockets and Spacecraft: Sine Qua Non
of Space Science

Even as NASA was being formed, the stable of American sounding rockets was impressive. There were small (Deacon, Cajun, Arcon, Arcas), medium (Aerobee, Aerobee-Hi), and large (Viking) rockets. Viking had been designed to replace the V-2, which was no longer used after the test program ended in 1952. There were rockets using solid propellants (Deacon, Cajun, Arcon, Arcas) and rockets using liquid propellants (second stage of Aerobee, Viking). Multistage combinations (Nike-Deacon, Nike-Cajun, Aerobee) achieved higher altitudes than could economically be attained with single-stage rockets. Rockets had been launched from balloons, from aircraft, and from launchers aboard ships at sea. These sounding rockets and the high-altitude research program that went with them provided NASA with an immediately on-going component of its space science program.[1]

A similar situation existed with respect to the larger vehicles needed for launching spacecraft into orbit. The reentry test vehicle Jupiter C—which launched America's first satellite, *Explorer 1*, and which used the Redstone missile as its first stage—gave rise to a first group of what were called Juno space launch vehicles. Later versions of Juno used the more powerful Jupiter intermediate-range ballistic missile as the first stage.[2] The Redstone, which was created for the Army by the von Braun team and in which one could detect a distinct V-2 ancestry, was on hand and was used for America's first suborbital manned flights.[3] The Vanguard IGY launch vehicle, which used derivatives of the Viking and Aerobee sounding rockets as its first and second stages, was also available.[4] NASA took over Vanguard from the Naval Research Laboratory and completed the program, after which the Vanguard first stage was retired; but the upper stages were combined with the Air Force's Thor to create the Thor-Delta, or simply Delta, launch vehicle, which from the very start was one of the most useful of the medium-sized combinations.[5] In 1958 the Air Force's Atlas was the most powerful U.S. rocket that could be quickly pressed into service as a space launcher. To it was assigned the launching of the first American astronauts to go into orbit.[6] Atlas eventually became the main stage of Atlas-Agena and Atlas-Centaur, multistage launch vehicles used to put multiton payloads into space.

The imposing presence of the Soviet Union in space following the launching of the first Sputniks and the substantial lead it apparently had over the U.S. in payload capability generated a sense of urgency to develop very large payload capabilities. But with the variety of vehicles already in its stable or imminent, the United States clearly was not going to have to start from scratch. Indeed, had the country been willing to use the von Braun rockets for the IGY satellite program, the first satellite in orbit could well have borne a "made in America" stamp. At any rate, even this partial survey of the situation at the time NASA got going shows how deep in the rocket and missile work of the 1950s lay some of the roots of the subsequent space program.

Of course, along with the missiles and rockets available to NASA and the military were associated facilities and equipment already in operation. Launch ranges existed in Florida, California, New Mexico, and Canada. Tracking and telemetering stations, strategically located in the U.S. and elsewhere, were working. Vanguard Minitrack network of radio-tracking and telemetering stations for operating with the IGY satellites spanned the globe and provided a nucleus on which to build for an enlarged program of the future. IGY optical tracking stations also girdled the globe and were immediately available for photographic and visual tracking of spacecraft that were large enough to be detected by such means. To produce all these a substantial component of American aerospace, electronics, and other industry had been employed, generating hardware and acquiring an experience ready to be used for tackling the challenges that lay ahead.[7]

One of the first tasks facing NASA in the fall of 1958 was to determine what additional launch vehicles would be required to accomplish the space missions planned for the program. While most of the launchers would derive from military hardware, some, especially for the manned spaceflight program, would have to be built from scratch. So, too, would the spacecraft for the science, applications, and manned spaceflight missions.

LAUNCH VEHICLES

It is not necessary for understanding the relationship of launch vehicles to the space science program to delve deeply into how they were developed, but a few principles should be understood. First, a number of different vehicles were required. One might have supposed that a single launch vehicle, which could do everything the program required, would be ideal. With only one manufacturing line, one kind of assembly, test, and launch facilities, one kind of operational equipment and procedures, and basically one launch team, a substantial background of experience would quickly build up for that vehicle. Engineers and technicians would become thoroughly familiar with its characteristics and idiosyncrasies, so that a high degree of reliability could be ensured.

But in an era when a launch vehicle was expended for each firing, the economics would not be favorable. For, to be acceptable, a single vehicle would have to be able to accomplish both the simplest and the most difficult of the missions required—from small, near-earth satellite missions to manned flights to the moon. On the most difficult missions, the launch vehicle would presumably operate most efficiently, and the costs would be commensurate with the accomplishments. But to use such a vehicle for less demanding missions would be most inefficient; indeed, the cost of the launch vehicle could overwhelm the cost of the spacecraft. To mitigate this problem of cost there would, of course, be pressure to fly many small missions on a single launch vehicle, or to let small missions ride piggyback on larger ones, thus, reducing the cost per spacecraft; but then different kinds of complications would enter in. Some of these would be fundamental, as when one set of experiments required a circular orbit, another set an eccentric orbit, still another a polar orbit, and a fourth an equatorial orbit.

For expendable launch vehicles such considerations led to the conclusion that the most efficient approach would be a graduated series running from a small, inexpensive vehicle to the very large, very expensive ones. The gradation between vehicles would be large enough to yield a substantial increase in payload and mission capability, but small enough to avoid having to use vehicles too costly for their assigned missions. Obviously the way in which these requirements were met was a matter of judgment, and in some respects arbitrary. The subject was constantly under study by both the military and NASA.[8]

Second, the basic physics of rockets dictated that major launch vehicles should be multistage, or step, rockets—that is, combinations of two or more rockets, called stages, which burn one after the other. As soon as the first stage has used its propellants, it is discarded, after which at an appropriate time the second stage is ignited. When the second stage has burned out and been discarded, the third stage is ready to fire. And so on. Multistaging is important for rockets that must work against the force of gravity, for otherwise the propellants must supply the energy needed to propel the entire rocket structure against the pull of gravity for the whole launching phase. But with staging, in which portions of the structure are discarded as soon as they are no longer needed, only a small fraction of the entire vehicle need be propelled into the final orbit or space trajectory. The early rocket pioneers recognized the importance of staging, a point that Robert Goddard elaborated in his famous Smithsonian paper.[9] In the space program two-stage and three-stage vehicles became common, four- and five-stage combinations not uncommon.

In the scramble to put together a national launch vehicle capability after the formation of NASA, it was natural that whatever vehicles were available or could be assembled from the existing military programs would be used. In 1959 six out of the seven vehicles that were available to NASA came largely from the missile program.[10] The seventh was Vanguard, which had been built by the Navy for the IGY.

To expand the national capability, additional vehicles would be developed (fig. 10). Scout, a four-stage, solid-propellant vehicle, would provide an inexpensive means for launching 70 kilograms to 185 (or even 550) kilometers. Vega and Centaur, the latter using the high-energy propellants liquid hydrogen and liquid oxygen, would substantially extend the launch capability of Atlas-based vehicles. Juno V vehicles and Nova were intended to support a variety of manned missions. Nova was expected to generate a thrust of 6 000 000 pounds—almost 27 000 000 newtons—and would be required for launching men in a direct ascent from the ground to the moon's surface.

Fifteen years later the situation was entirely different. By then a bewildering variety of rocket stages and launch vehicle combinations had been developed, along with an extensive literature.[11] What the vehicles could do for the various space missions can be deduced from the performance figures for the launch vehicles in figures 10 through 14.

As seen in figure 11, by 1962 the Hustler and Vega had been eliminated. Likewise, Juno II, which had not proved particularly useful, had been dropped. Nova plans called for doubled thrust, and Saturn was to rely on liquid-hydrogen, liquid-oxygen engines for its upper stages. By 1966 Nova had disappeared because of NASA's decision in July 1962 to use lunar-orbit rendezvous instead of direct ascent for the Apollo missions.[12] Several versions of Saturn would support NASA's manned spaceflight programs; the largest, Saturn V, would be used for the manned lunar missions.[13] The Department of Defense preferred not to be tied into the expensive and highly experimental Saturn, so in following years the Titan III line of launchers was introduced into the stable to support large-scale military missions. Titan III additions can be seen in the display for 1972 (fig. 13), at which time the only Thor-based vehicle remaining was Delta.

These launch vehicles, with the sounding rockets discussed at the start of this chapter, made up the backbone of U.S. capability to explore and investigate space. They resulted from joint planning by NASA and the military to serve their respective needs.[14] Other rockets and rocket stages were put together for special purposes, usually by the military, but their existence did not change the overall picture.

In the 1970s a basic change was initiated with the commitment to the Space Shuttle, which in the 1980s would supplant most of the expendable boosters for launching spacecraft into near-earth orbit. The launch vehicle line-up in the fall of 1976 (fig. 14) shows how the elimination of the

Saturns, following the completion of the Apollo and Skylab programs by the mid-1970s, and the prospect of the Space Shuttle by the 1980s had thinned out the stable. Only five vehicles remained: Scout for small payloads, Delta and Atlas-Centaur for medium and large payloads, and two Titan III combinations for the very large payloads.

Throughout the entire evolution of the launch vehicle stable, both Scout and Delta remained. Relatively inexpensive, able to support a substantial number of the researches that scientists wanted to do, these launch vehicles had great appeal. Even with the Shuttle in operation, Scout at least was likely to remain, for even the Shuttle might not prove economical for small missions with a wide variety of special trajectory requirements.

Scout and Delta also illustrate another feature of the national launch vehicle program. As the group of vehicles improved as a family over the years, performance of the individual vehicles also improved. In 1962 Scout could put 100 kg into a near-earth orbit; by the 1970s this performance had doubled. In the same period Delta's performance had shown an even greater growth. In 1962 Delta could send several hundred kilograms into a near-earth orbit or 25 kg to Mars or Venus; by 1976 Delta could loft 2000 kg into a 185-km orbit or 340 kg to the near planets. The increased performance, which most vehicles experienced over the years, was brought by continuing programs of improving and uprating the vehicles. While improvement programs were the pride of the vehicle engineers, they were sometimes the bane of top management, which would often have preferred to settle upon some acceptable level of performance and then stop spending any more money on vehicle development.

Like the United States, the Soviet Union developed a launch vehicle stable.[15] During the 1960s, however, the Soviet Union appeared to rely on fewer kinds of vehicles than did the United States, perferring to use a single model for a wider variety of missions. The Russian preference may be prima facie evidence that the economic factors of using large vehicles for small-payload missions were not as prohibitive as American planners felt they were; but comparing Russian economics with American is a risky business. The USSR may simply have decided to pay the extra cost for the convenience afforded.

Only the United States and the Soviet Union developed extensive launch capabilities. Other nations interested in space research and applications turned largely to the United States for assistance in launching spacecraft or individual experiments, as will be explored in chapter 18. Some nations, however, desiring to lessen this dependence, proceeded to develop vehicles of their own. Among those were Britain, France, Japan, People's Republic of China, and the European Launcher Development Organization, a coalition of countries that pooled resources to develop a launcher approaching the Atlas-Agena capability.[16] Italy set up an equatorial launching facility in the Indian Ocean off the coast of Kenya, but used the American Scout as launch vehicle.[17]

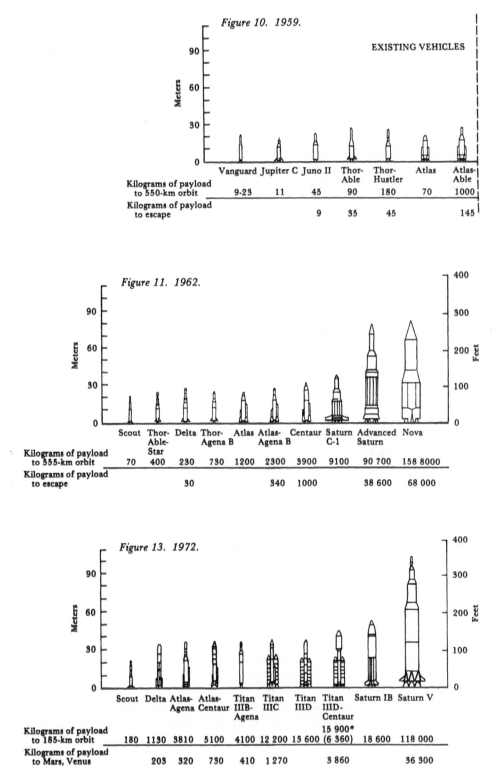

Figure 10. 1959.

EXISTING VEHICLES

	Vanguard	Jupiter C	Juno II	Thor-Able	Thor-Hustler	Atlas	Atlas-Able
Kilograms of payload to 550-km orbit	9-23	11	45	90	180	70	1000
Kilograms of payload to escape			9	35	45		145

Figure 11. 1962.

	Scout	Thor-Able-Star	Delta	Thor-Agena B	Atlas Agena B	Atlas-Agena B	Centaur	Saturn C-1	Advanced Saturn	Nova
Kilograms of payload to 555-km orbit	70	400	230	730	1200	2300	3900	9100	90 700	158 8000
Kilograms of payload to escape			30			340	1000		38 600	68 000

Figure 13. 1972.

	Scout	Delta	Atlas-Agena	Atlas-Centaur	Titan IIIB-Agena	Titan IIIC	Titan IIID	Titan IIID-Centaur	Saturn IB	Saturn V
Kilograms of payload to 185-km orbit	180	1130	3810	5100	4100	12 200	13 600	15 900* (6 360)	18 600	118 000
Kilograms of payload to Mars, Venus		203	320	730	410	1 270		3 860		36 300

*With modification

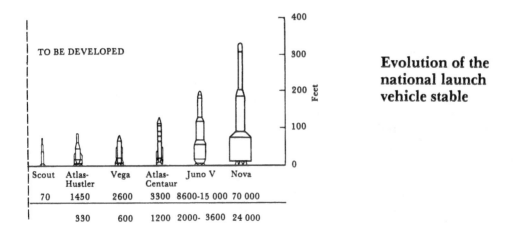

TO BE DEVELOPED

	Scout	Atlas-Hustler	Vega	Atlas-Centaur	Juno V	Nova
	70	1450	2600	3300	8600-15 000	70 000
		330	600	1200	2000- 3600	24 000

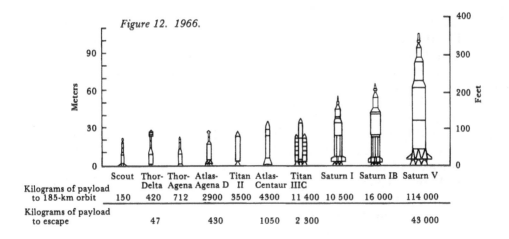

Figure 12. 1966.

	Scout	Thor-Delta	Thor-Agena	Atlas Agena D	Titan II	Atlas-Centaur	Titan IIIC	Saturn I	Saturn IB	Saturn V
Kilograms of payload to 185-km orbit	150	420	712	2900	3500	4300	11 400	10 500	16 000	114 000
Kilograms of payload to escape		47		430		1050	2 300			43 000

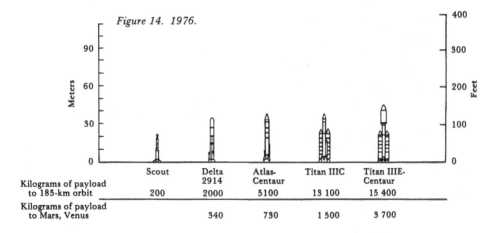

Figure 14. 1976.

	Scout	Delta 2914	Atlas-Centaur	Titan IIIC	Titan IIIE-Centaur
Kilograms of payload to 185-km orbit	200	2000	5100	13 100	15 400
Kilograms of payload to Mars, Venus		340	730	1 500	3 700

By 1966—when Centaur became fully operational—the United States could at last launch spacecraft for just about any space mission the country might want to undertake, except the very demanding ones requiring the Saturn or Titan still under development. Although the debate over whether the United States could or could not match Russian launch capability still arose occasionally, the subject no longer had the importance it once did. As long as the United States could carry out the scientific investigations and make the space applications it desired—sometimes using miniaturization techniques to overcome limitations on total payload weights—the country could compete with the Soviet Union on essentially equal terms, and the comparative sizes of rockets were then an artificial criterion on which to judge Soviet and U.S. space prowess.

The ability to launch objects and men into space rested on a substantial investment in manpower and facilities—design, engineering, construction, assembly, and test facilities in both industry and government. Most visible were the launching ranges, which along with the launch vehicles themselves symbolized the nation's space capability. In the United States the principal facilities were the Eastern Test Range, with launch areas at Cape Canaveral and Merritt Island in Florida; and the Western Test Range, for which the launching areas were at Vandenberg Air Force Base in California.[18] A smaller launch station was used by NASA for firing Scouts from Wallops Island on the Virginia coast.[19] Supplementing these were sounding rocket ranges at the White Sands Missile Test Facility in New Mexico, Wallops Island, Point Mugu in California, and Fort Churchill in northern Canada. Occasionally shipborne launchers were used for special missions.

The Soviet Union operated a number of major launch ranges out of Tyuratam, Kapustin Yar, and Plesetzk.[20] A few other countries established satellite launch ranges.[21] By invitation American sounding rockets were fired at numerous ranges around the world—for example, at Thumba in India, at Woomera in Australia, and at Andoeya, Norway.[22]

SPACECRAFT

What the United States could do with its launch capabilities was revealed by the spacecraft the country built and launched. As with the many rockets and rocket stages, one could easily become totally immersed in the subject of spacecraft, which displayed a bewildering variety of shapes, sizes, and purposes. A partial listing of U.S. and foreign spacecraft is given in tables 3 and 4.[23] As with launch vehicles, it is not necessary to know all the spacecraft in intimate detail to understand their role in the space science program. A few general concepts suffice. By dividing spacecraft into a few representative classes, one can understand their various functions better.

First, there was the sounding rocket, a rocket instrumented for high-altitude research and fired along a vertical or nearly vertical trajectory. To

Table 3
United States Spacecraft

(Partial list, excluding commercial and Defense Department spacecraft)

Earth Satellites	
Vanguard	International Geophysical Year, Explorer class.
Explorer	Small satellite for near-earth missions.
Interplanetary Monitoring Platform (IMP)	Explorer-class satellite to explore cislunar and lunar space.
Orbiting Geophysical Observatory (OGO)	Observatory-class satellite for geophysical research.
Orbiting Solar Observatory (OSO)	Observatory-class satellite for solar studies.
Orbiting Astronomical Observatory (OAO)	Observatory-class satellite for stellar astronomy.
High Energy Astronomical Observatory (HEAO)	Very heavy, observatory-class satellite for studying shortwave and high-energy phenomena in cosmos.
Biosatellite	Observatory-class, recoverable satellite for life sciences.
Pegasus	Observatory-class satellite for micrometeoroid studies.
Echo	Applications satellite. Large metallized sphere for passive communications studies.
Relay	Active communications satellite.
Syncom	Active communications satellite in synchronous orbit.
Applications Technology Satellite (ATS)	Platform for a variety of applications technology researches, particularly in synchronous orbit.
Tiros	Large Explorer-class weather satellite.
ESSA	Operational version of Tiros.
Nimbus	Observatory-class weather satellite.
ERTS	Earth Resources Technology Satellite. Applications satellite devoted to earth resources research.
Synchronous-orbit Meteorological Satellite (SMS)	Applications satellite in synchronous orbit for meteorological research.
PAGEOS	Passive geodetic satellite.
GEOS	Active geodetic satellite with flashing lights and radio instrumentation.
LAGEOS	Geodetic satellite with corner reflectors for use with laser beams from the ground.

Table 3—Continued

Mercury	One-man satellite.
Gemini	Two-man satellite.
Apollo (earth-orbiting)	Three-man satellite.

Space Probes

Ranger	Lunar hard lander.
Surveyor	Lunar soft lander.
Lunar Orbiter	Lunar satellite for photography of the moon and lunar environment studies.
Apollo lunar module with command module	Manned lunar lander with manned lunar orbiter.
Pioneer	Explorer-class interplanetary probe.
Mariner	Observatory-class planetary and interplanetary probe.
Viking	Observatory-class planetary orbiter plus lander.
Voyager	Observatory-class planetary and interplanetary probe for outer planet studies. (Spacecraft of the 1970s, not the Voyager of the 1960s that was displaced from the program by Viking.)

Table 4
Foreign Spacecraft

Soviet Spacecraft

Sputnik	Geophysical research satellite.
Luna	Unmanned lunar orbiter, lander, and return missions.
Vostok	First Soviet manned spacecraft.
Voskhod	Adaptation of Vostok to accommodate two and three cosmonauts.
Soyuz	Two- or three-man spacecraft, with working compartment.
Salyut	Earth-orbiting space station for prolonged occupancy and revisitation by cosmonauts.
Cosmos	Catchall name for variety of research and test spacecraft.
Venus (Venera)	Unmanned Venus probe.
Polyot	Earth satellite with onboard propulsion for changing orbits.
Elektron	Radiation belt satellite, launched in pairs.

Table 4—Continued

Zond	Lunar and deep-space probe.
Molniya	Communications satellite in 12-hour orbit with low perigee and with apogee near synchronous-orbit altitude.
Meteor	Weather satellite.
Intercosmos	Soviet international satellite.
Oreol	Scientific satellite for upper atmosphere and auroral studies.
Mars	Unmanned Martian probe.
Prognoz	Satellite to study solar plasma fluxes.
Raduga	Geosynchronous communications satellite.
Ekran	Television broadcasting satellite.

Other Foreign Spacecraft, Launched with U.S. Cooperation

Ariel	United Kingdom satellite for geophysical and astronomical research.
Alouette	Canadian satellite for ionospheric research.
ISIS	Canadian satellite for ionospheric research.
San Marco	Italian satellite for geophysical research.
French (FR-1)	French satellite for ionospheric research.
ESRO	European Space Research Organization satellite for particles and fields investigations.
AZUR	German satellite for particles and fields research.
Skynet	United Kingdom satellite for communications.
NATO	NATO satellite for communications.
CAS (Eole)	French satellite for data collection and meteorology.
Barium Ion Cloud	German satellite for geophysical research.
ANIK	Canadian geosynchronous satellite for communications.
Aeros	German geophysical satellite.
ANS	Netherlands satellite for ultraviolet and x-ray astronomy.
INTASAT	Spanish satellite for ionospheric research.
Helios	German deep-space probe for interplanetary and solar studies inside the orbit of Mercury.
Symphonie	French-German satellite for communications.
COS	European Space Agency satellite for study of cosmic gamma rays.

distinguish between sounding rockets and space probes, a limit of one earth's radius was arbitrarily set on the altitude of a sounding rocket.[24] The sounding rocket was really both launch vehicle and payload combined. Only rarely was the payload separated from the flying rocket, and when this did happen, the payload still traversed an up-and-down trajectory alongside that of its launching rocket.

Sounding rockets, which were the only high-altitude research vehicles capable of exceeding balloon ceilings before the launching of Sputnik, continued through the 1970s to be important in the U.S. space program, being launched at the rate of about 100 a year. They provided the best means of obtaining vertical cross sections of atmospheric properties up to satellite altitudes and also were inexpensive devices for trying out new instrumentation or making exploratory measurements of phenomena to be studied in detail later with more expensive spacecraft. Their relatively low cost and the speed with which a sounding rocket experiment could be prepared and carried out also made sounding rockets useful for graduate research where the student needed to complete a project in a reasonable amount of time to support his dissertation. But not only students found the sounding rocket attractive. Many professional space scientists continued to favor sounding rockets for much of their research, as opposed to the more complicated, more expensive, and more demanding satellites.[25] Through the years there was a steady demand on NASA for sounding rockets, and the agency was frequently urged to increase its budget in this area, even though by 1965 the budget had risen to $19 million a year, an order of magnitude more than had been spent per year on such research during the days of the Upper Atmosphere Rocket Research Panel.

Yet satellites were required for many space experiments, particularly for long-duration observations above the earth's atmosphere. For these, many different satellites were devised, as may be seen from table 3. Generally these spacecraft could be divided into three classes: Explorers, observatories, and manned spacecraft.

The simplest were the Explorers, whose weights usually ranged from less than 50 kilograms to several hundred (fig. 15). Each was devoted either to a single experiment or to a small collection of related experiments. Explorers were either unstabilized or used fairly simple techniques for obtaining a rough degree of stability. If the spacecraft were spun around a suitable axis, that axis would maintain its general direction in space. Long booms could be used to generate a gravitational torque on the satellite, keeping a chosen side toward the ground. Explorers usually used battery power-supplies, sometimes replenished by energy from solar cells.

Explorers were much simpler than observatory satellites. The observatory class—which included the Orbiting Solar Observatory (fig. 16), the Orbiting Geophysical Observatory (fig. 17), and the Orbiting Astronomical

Observatory (fig. 18)—consisted of very heavy, complex, accurately stabilized spacecraft. Weights ranged from several hundreds of kilograms to tons. The much greater size and weight of the observatories permitted more scientific payload and more sophisticated instrumentation. They could devote a considerable weight to what was called "housekeeping" equipment—power supplies, temperature control, and tracking and telemetering equipment. As with the Orbiting Geophysical Observatory, there might be provision for special operations such as erecting booms to hold instruments like magnetometers at a distance from the main body of the spacecraft, which otherwise might influence the measurements to be made. The Orbiting Astronomical Observatory and the Nimbus meteorological satellite were equipped with large paddles, covered with solar cells, which could be unfolded and kept facing the sun to furnish power for the spacecraft and its instruments. As a rule observatories carried elaborate systems for maintaining a desired orientation in space, so that scientific instruments could be pointed in a chosen direction. This was especially true of the astronomical satellites, whose instruments measured radiations from selected celestial bodies. The Orbiting Astronomical Observatory, for example, used specially designed star trackers which, by fixing on several chosen stars in widely different directions, could establish a reference frame for the spacecraft. With this refrence frame as a guide, the observatory could be slewed around to any chosen direction. Once properly oriented, the spacecraft could be held fixed to within a minute of arc, and telescopes within the satellite could be trained for long periods of time on their observational targets with an accuracy of fractions of a second of arc.

A variety of schemes provided the ability to alter and then maintain spacecraft orientation. Most frequently used were small intermittent jets, developed either by small chemical rockets or by releasing high-pressure gas—nitrogen, for example—through small nozzles. Other methods were also used, often supplementing the jets. Electric current passing through loops of wire mounted in the spacecraft would develop magnetic fields which, reacting against the magnetic field of the earth, would exert a torque on the satellite that could be used to alter the orientation. Rapidly spinning wheels—gyroscopes—that stubbornly resist efforts to change their orientation in space, could also help point and stabilize spacecraft. Most of the structure of the Orbiting Solar Observatory rotated to provide gyroscopic stabilization. Finally, if the mass of a spacecraft was distributed much as the material in a dumbbell, the earth's gravity could keep the satellite pointed toward the earth's surface, as in Applications Technology Satellites. Since the earth's gravitational field varies inversely with the square of the distance from the center of the earth, the end of such a spacecraft nearer the earth would experience a greater pull of gravity than would the end farther away. Thus, whenever the end facing the earth tended to

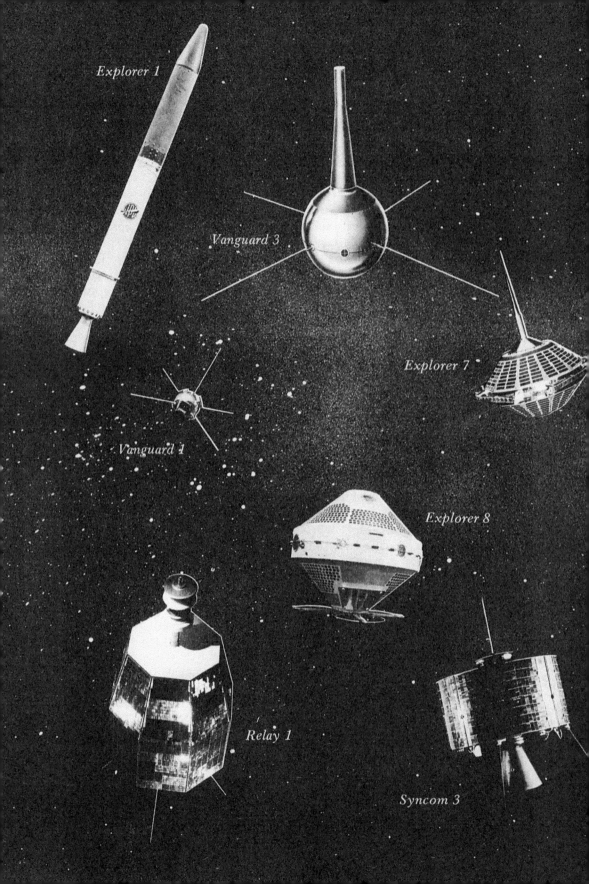

Explorer 1

Vanguard 3

Explorer 7

Vanguard 1

Explorer 8

Relay 1

Syncom 3

Echo 1

Explorer 11

Explorer 17

Explorer 20

Alouette 1

Ariel 2

Figure 15. Explorer satellites. The great variety is impressive, but underlying that variety was a common technology.

Figure 16. Orbiting Solar Observatory.
OSO was the first observatory-class satellite built by NASA. In the photo above,
OSO 3 is prepared for tests before launch
into orbit in 1967.

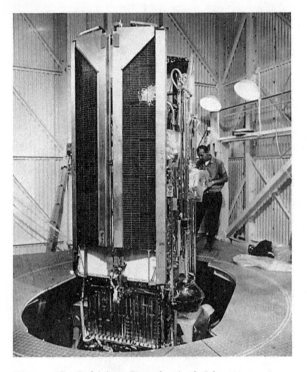

Figure 17. Orbiting Geophysical Observatory, a very capable spacecraft. Although once dubbed the "streetcar satellite," OGO never acquired the ease of
preparation hoped for it. OGO 4's booms,
antennas, and solar panels are folded in
the photo above, for vacuum chamber
tests and launch in 1967.

Figure 18. Orbiting Astronomical Observatory. OAO was the most complex and
difficult of the observatory-class satellites
initially undertaken by NASA. Above,
solar panels are mounted on OAO 3 in
preparation for its 1972 launch.

drift away, the greater force of gravity on it would pull that end back toward the earth again.

Observatory-size spacecraft were also large enough to carry additional rockets which, when fired against the direction of the satellite's motion, could return the spacecraft to earth, where its equipment and records could be recovered. Once on its way down, the spacecraft would have to be protected against heat generated by friction of the atmosphere, for which purpose retrorockets—that is, rockets fired in the direction of motion to slow the spacecraft—or parachutes, or a combination of the two, could be used.

Finally, there were the manned spacecraft (figs. 19–22). These were even larger than the unmanned observatories, since, in addition to housekeeping equipment and instruments, the spacecraft had to be completely maneuverable to return the astronaut to earth after the mission and also had to afford a suitable environment for the crew.

All three classes of satellites had their uses. The manned spacecraft introduced the element of exploration—of personal, on-the-spot investigation—into the program. With men aboard it became possible to extend laboratory research into the environment of space, as was done dramatically in Skylab and the Apollo-Soyuz Test Project.[26] When Space Shuttle development was begun in the early 1970s, one motivation was to make it relatively easy for researchers to perform experiments in space laboratories.

At the other end of the spectrum, Explorer-class satellites permitted scientists to perform a wide range of space science experiments, as shown in table 5. To the scientists, the lower costs meant that more of the funds available could be put into the scientific research itself. Also, the effort required to put an experiment into an Explorer was considerably less than that required for an observatory—one or two years as opposed to many years. When a discovery was made, an experimenter could more quickly follow up with new experiments on sounding rockets and Explorers than he could using observatories. Moreover, not having to fit his instruments alongside those of many other experimenters—with all the problems of electrical, radio, and other kinds of interference—the experimenter could exercise greater control over his experiment. These advantages account for the unvarying insistence of the scientific community that NASA continue to provide sounding rockets and small satellites. Whenever larger projects appeared to threaten the funding of the smaller ones, the scientific community rose in defense of the smaller. Over this issue the scientists came as near to unanimity as they ever did.

On the other hand, some investigations required a greater capability than the Explorers afforded. Astronomy experiments that required large telescopes and precision pointing are one example. Bioscience experiments that required the recovery of specimens after exposure to the environment of space—as in the Biosatellite—are another.[27] The Orbiting Geophysical Observatory afforded the means of conducting 20 or more experiments on

Figures 19-22. Manned spacecraft. Mercury (left) and Gemini (below) laid the groundwork for Apollo (lower left), while Skylab (lower right) grew out of the Apollo technology.

Table 5
Examples of Investigations in Explorer-Class Missions

Spacecraft	Subject of Investigation
Vanguard 1	Satellite geodesy.
Vanguard 2	Cloud-cover studies.
Vanguard 3	Survey of earth's magnetic field and lower edge of radiation belts.
Explorer 1	Charged particle radiations in space.
Explorer 6	Radiation belt; meteorology.
Explorer 7	Energetic particles; micrometeoroids.
Explorer 8	Ionosphere; atmospheric composition.
Explorer 9	Atmospheric pressures and densities.
Explorer 10	Interplanetary magnetic field near earth; particle radiations.
Explorer 11	Gamma rays from space.
Explorer 12	Magnetospheric studies.
Explorer 13	Micrometeoroids.
Explorer 14	Charged particles and magnetic fields in magnetosphere.
Explorer 16	Micrometeoroids.
Explorer 17	Atmospheric composition.
Explorer 18	Charged particles and magnetic fields in cislunar space.
Explorer 20	Probe of topside of ionosphere.
Explorer 27	Geodesy by radio tracking methods.
Explorer 30	Solar x-ray studies.
Explorer 38	Radio astronomy.
Explorer 42	Catalog and study of celestial x-ray sources.
Pioneer 5	Interplanetary charged particles and magnetic fields.
Ariel 1	Ionospheric and solar research.
Ariel 2	Atmospheric research and radio astronomy.
Alouette 1	Charge densities in upper ionosphere and radiation belt studies.
San Marco 1	Atmospheric physics.
ESRO 2B	Cosmic rays and radiations from sun.

one spacecraft, so that a variety of related phenomena could be observed simultaneously and correlated.

As with the launch vehicles, it proved impossible to find a single spacecraft carrier that would serve all needs, although some attempts were made in this direction. One spoke of building a standardized satellite, to effect economies and improve reliability. When conceived, the Orbiting Geophysical Observatory was described as a "streetcar satellite," whose continuing use would so reduce the preparation time for an experiment that researchers could get their equipment aboard at the last minute—like catching a streetcar or bus—to follow up on some recent space science discovery. Initial reaction to the streetcar concept was positive, and the idea had the blessing of the Space Science Panel of the President's Science Advisory Committee.[28] But the problems of serving so many experimenters on one spacecraft defeated the objective. For each observatory a great deal of tailoring was required, compromises had to be worked out on orbits, orientation, placement of instruments, magnetic cleanliness of the spacecraft, allocation of telemetering capacity, and operating time. Use of a common electrical power supply invited electrical interference among different experiments, and often an offending experimenter was required to provide his own power. Ionospheric and radiation belt phenomena were fundamentally related to the earth's magnetic field, making it important to measure ions, radiation particles, and magnetic fields simultaneously. But the various measuring instruments could easily interfere with each other unless care was taken. Those who wished to determine atmospheric composition at spacecraft altitudes required that the satellite and other instruments not contaminate the natural atmosphere with gases brought up from the ground—another difficult problem when large numbers of experiments were being conducted simultaneously.

Such problems defeated the efforts to produce standardized satellites in the same sense as standardized autos and auto parts or standardized home appliances. Nevertheless a considerable amount of uniformity was achieved. The basic structure, housekeeping, and orientation systems of the solar observatories were essentially the same from spacecraft to spacecraft. Even the geophysical observatories, with all the tailoring that they required, had much in common with each other. More important, the technology on which spacecraft were based acquired over the years a certain amount of standardization.

In this sense even the Explorers were standardized. Certainly no more varied looking group of satellites could be assembled than those of figure 15, which shows a large number of the Explorer satellites. Yet they were all cousins, stemming from a common, rather straightforward technology. When an engineer started out to design an Explorer-class satellite, he had pretty much in mind the kinds of structure, temperature control, tracking and telemetering devices, and antenna systems he might use. He was familiar with the kinds of vacuum, thermal, and vibration tests the spacecraft would have to pass to be approved for flight. To be sure, the technology advanced over the years as better components and materials became available, and improved housekeeping equipment was devised. But the family relationship remained. The steady, though gradual, change in Explorer technology did make the later Explorers considerably more capable than earlier ones. For example, *Explorer 35* launched on 19 July 1967, weighing 104 kg and operating from an orbit of the moon, could far outperform the first several Explorers which weighed only tens of kilograms. Yet in the evolution of the series, any given Explorer was quite similar in its technology to the immediately preceding one. The same point is illustrated by the solar observatories. These also changed gradually over the years. In August

1969 the sixth solar observatory, weighing 290 kg, went into orbit. Though looking a great deal like the first observatory, launched on 7 March 1962 and weighing 200 kg, *OSO 6* was more versatile than earlier ones, having the capability to point two telescopes at the sun to study in detail ultraviolet and x-ray spectra at any point on the solar disk.

Like the earth satellites, space probes—which were spacecraft sent away from earth into deep space—fall into several classes. The analogy is very close. Akin to the Explorer satellites were the Pioneer space probes (fig. 23). These were modest-sized vehicles, ranging from around 40 kg in the first models to the 260-kg weights of *Pioneer 10* and *11* sent to Jupiter in 1972 and 1973. They were spin-stabilized and instrumented to investigate the interplanetary medium and the environs of a planet as the spacecraft flew by. During the 1960s a number of Pioneers revolving like little planets about the sun provided a great deal of information on the solar wind and magnetic fields in space. When on the far side of the sun from the earth, Pioneer radio signals were carefully observed to find effects of the sun's gravity predicted by the general theory of relativity. On 3 December 1973 *Pioneer 10* passed Jupiter at 130000 km from the planet's surface—taking pictures, measuring its radiations, and investigating some of its satellites— and then receded along a trajectory that would eventually carry it out of

Figure 23. Pioneer. The earliest of the U.S. space probes provided the kind of flexibility for the deep-space investigator that Explorers provided for earth-satellite missions. Pioneer 8 in the photo is prepared for December 1967 launch into orbit of the sun.

the solar system. A year later *Pioneer 11* also flew by Jupiter, sending back more data on the giant planet and its satellites, leaving the planet on a course that would carry the spacecraft in September 1979 to the vicinity of the second largest planet, Saturn, famed for its rings. These missions into deep space give some idea of Pioneer's usefulness to the space scientist. As with Explorer, Pioneer technology also advanced steadily through the years, so that in the early 1970s it was possible to plan to use Pioneer spacecraft to carry orbiters and atmospheric probes to Venus in the late 1970s.[29]

Analogous to the observatory satellites were the larger lunar and planetary probes: Ranger, Surveyor, Lunar Orbiter, Mariner, and Viking (figs. 24–28). Like the observatories, these spacecraft were maneuverable and, using a celestial reference system, could be accurately oriented in space. In addition to small jets for orienting the spacecraft, they also carried a larger rocket that could be fired to alter the trajectory so as to keep the craft on course to its intended target.

Lunar Orbiter carried enough auxiliary propulsion to place it in a lunar orbit, from which the spacecraft obtained a series of photographs of the moon. These were later used to produce maps of the moon's surface and to aid in planning Apollo missions. Several Mariners carried enough propulsion to place them in orbit about Mars. Surveyor used retrorockets to slow the spacecraft for a soft landing on the moon, after which remotely controlled instruments televised and investigated the surrounding landscape. Viking combined both the orbiting and landing capability, the main vehicle first going into orbit of Mars, after which a portion of the spacecraft separated and was forced by rockets to descend to the Martian surface.

Deep-space probes had to overcome problems additional to those encountered by earth satellites. For example, a Martian probe took about two-thirds of a year to get to its destination. *Pioneer 10* and *11* required almost two years to fly to Jupiter, and had to survive those two years in the environment of space to accomplish their assigned missions. If an earth satellite operated properly for a few months, the experimenters would have a few months worth of data for their trouble; but if a planetary probe operated for only a few months, they would get no planetary data at all. Of course, one also made interplanetary measurements on planetary flights— observing the solar wind, magnetic fields in space, dust, meteor streams, and cosmic rays—but on planetary missions these were secondary objectives.

Another requirement for the great distances traveled by interplanetary and planetary spacecraft was more powerful radio communications systems. The antenna had to be pointed toward the earth. If omnidirectional or wide-angle antennas were used, the power requirements went up, sometimes prohibitively. If, to conserve power, narrow-beam antennas were used, they exacerbated the antenna-pointing requirement. Finally, spacecraft that flew toward the sun had to be protected against overheating by solar radiations, while those that flew away toward the outer solar system had to be protected against freezing.

With these larger space probes one could plan in the course of time to investigate all the planets and major satellites of the solar system, and the asteroids and comets. Spacecraft could be placed in orbit around other bodies, as was done with Lunar Orbiter around the moon and Mariner around Mars. Landers could place instrumented laboratories on the surfaces of other bodies, as did Surveyor and the Soviet Luna on the moon, Viking on Mars, and Soviet Venus probes on Venus. It was even possible to deposit roving laboratories on those bodies, as the Soviet Union did with Lunokhod on the moon, or retrieve samples of material from them as did *Luna 16* and *17*.[30]

Finally, there were the manned space probes (fig. 29). During the 1960s these consisted solely of the Apollo–Lunar Module combinations that the American astronauts flew to the moon. As with the manned satellites, these provided the added dimension of manned exploration. The successful Apollo missions yielded such a wealth of scientific data as to soften at last the years-long lament of the scientific community over the tremendous expense of the Apollo program.

Of course, operation of these spacecraft required auxiliary equipment and systems. Out of the Minitrack tracking and telemetering network of the International Geophysical Year grew a versatile satellite network for issuing instructions to satellites, determining their orbits, and receiving telemetered information.[31] To work with the deep-space probes, the Jet Propulsion Laboratory established a deep-space network using 26-m and 64-m parabolic antennas at three stations spaced roughly equally around the world in longitude, so that a distant space probe could at all times be viewed from at least one of the stations.[32] For manned spaceflight a special network was linked to the Johnson Space Center in Houston.[33] As needed these were supported by tracking-telemetering ships and aircraft furnished by the Navy and the Air Force.

The rockets and spacecraft were a sine qua non of the space program and of space science. It is not surprising, therefore, that most of NASA's activity and resources went into the creation and operation of these vehicles. The scientific researches themselves, while not inexpensive, required only a fraction of what the tools—the spacecraft and launch vehicles—cost. Since the tools were where most of the money was going, Congress spent a great deal of time probing the budgets for them, and NASA managers became accustomed to thinking of their programs in terms of launch vehicles and spacecraft. One would speak of Ranger and Mariner programs, and of the Polar Orbiting Geophysical Observatory program and the Orbiting Solar Observatory program—or rather, to the distress of those to whom acronyms are anathema, of the POGO and OSO programs. From time to time scientists would chide NASA on this habit, pointing out that as far as space

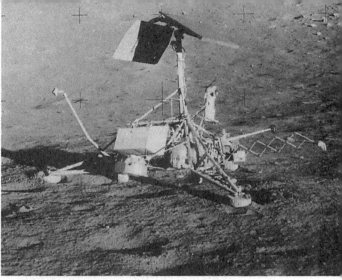

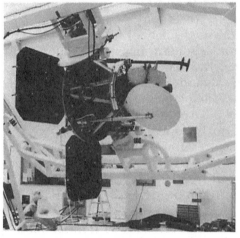

Figures 24–28. Lunar and planetary probes. These were the deep-space analogs of the observatory-class satellites: Ranger, above left; Surveyor, above right, on the lunar surface; Lunar Orbiter, lower right; Mariner, opposite at top; Viking, lower opposite.

science was concerned, the program was to investigate the magnetosphere, to probe the origin and evolution of the moon and planets, to understand the solar processes, etc. NASA managers agreed, of course, and indeed from the very outset NASA people set forth the scientific objectives, not the space hardware, as the purpose of the space science program.[34] But the shorthand was too convenient, and the practice persisted even among the scientists.

HARD-LEARNED LESSONS

The rockets and spacecraft that won NASA an image of success did not come easily at first. During NASA's first two years the launch vehicles especially produced their share of grief. There were important lessons to learn, and experience proved to be a stern teacher. Atlas-Able and Pioneer provide a good illustration.

The Atlas-Able—a launch vehicle using the Atlas missile as the first stage combined with a second stage from the International Geophysical Year's Vanguard—had been brought into the spaceflight program by Abe

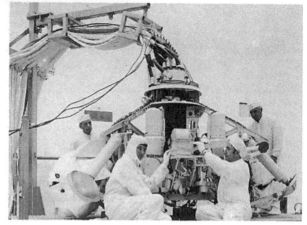

Silverstein, head of NASA's Office of Space Flight Development. NASA hoped to fill in and perhaps even to steal a march on the Soviet Union in lunar and planetary exploration while the agency pressed the development of its own space launch capability.[35] Contracting with Space Technology Laboratories for three deep-space probes—Pioneer spacecraft—to be launched by a space vehicle yet to be proved, NASA hoped to move faster in its science program than would otherwise be possible with the smaller rockets available to the agency. At first intended to launch a Venus probe, Atlas-Able was switched to lunar missions when the planetary flight appeared to be too much to attempt at the start. Hopes ran high for Atlas-Able and a number of scientists vied to help outfit the probe with scientific instruments. They had reason to be unhappy when their hard work went for naught.

Pioneer spacecraft on the Atlas-Able vehicles carried instruments to investigate interplanetary space and the moon. The Pioneer program, begun by the Air Force, was taken over by NASA when the new agency assumed responsibility for the nation's space science program.[36] Between 11 October 1958 and 15 December 1960, eight attempts were made to launch Pioneers into space.[17] Six, including all three Atlas-Able firings, were failures. *Pioneer 4*, riding a Juno II launch vehicle, achieved a limited success.

Figure 29. Apollo lunar module. Six LMs descended from lunar orbit to land men on the moon, three of the craft carrying a lunar roving vehicle. Above, *Apollo 16 LM and Rover on the moon in 1972 (with lunar dust on the camera lens showing in streaks).*

Only *Pioneer 5*, launched by a Thor-Able on 11 March 1960, could be called an unqualified success. Although instruments on *Pioneer 5* provided a great deal of information on cosmic rays and the space environment, the success could hardly erase the gloomy picture of six outright failures.

The significance of the Pioneer and Atlas-Able failures was all too clear at the time. In retrospect that significance stands out in high relief. For scientists it emphasized the sad plight that could await the experimenter who put his instruments on an as yet unproved vehicle. Long months, even years, of hard work could prove fruitless if the vehicle did not do its job properly. John Simpson of the Fermi Institute at the University of Chicago had reason to lament this hard fact of life: His group had developed new instruments to investigate the composition and energy of the solar wind and had accepted an invitation to put their instruments aboard the Pioneers, of which only *Pioneer 5* had been satisfactory.

The fiasco precipitated a long series of exchanges between Simpson and NASA in which the frustrated investigator explored ways of recouping his losses by gaining access to other NASA flights. On 16 December 1960, the day after the last Atlas-Able attempt, Simpson was on the phone urgently reviewing his situation with the author.[38] A half year later Simpson wrote that his group had participated in at least eight launchings with only two successes, a circumstance he attributed to having the Chicago experiments flown on unproved rocket systems or being assigned the role of secondary objectives.[39]

NASA had to face the issue of backups for important experiments, an issue that would always be at the back of an experimenter's mind when he signed up for a long difficult period of preparing for a space mission.[40] It soon became apparent that the time and effort required to conduct an experiment on a satellite or a space probe was far greater than that required to perform experiments in the laboratory. For the laboratory, one normally thought in terms of months, whereas space experiments could require years of hard work. Scientists began to point out that taking part in one or a few space science experiments could consume an appreciable fraction of a person's productive career, and in a "publish or perish" world the failure to get results because a rocket or a spacecraft didn't function properly could seriously affect that career. NASA was early moved by such considerations to adopt a policy of rescheduling experiments which, through no fault of the experimenter, did not succeed.[41]

Atlas-Able made plain that success on space missions would be neither automatic nor cheap. There was a price to pay, and part of that price was failure of some missions. This price NASA management would never find comfortable. The trouble with a philosophy of accepting a certain number of failures as normal and inevitable, was that even "learning failures" in an open program like NASA's, conducted under the watchful eye of a whole world, looked to the public and Congress like absolute failures. The press treated them as failures. It didn't matter that previous development projects like the V-2, Atlas, Thor, Polaris, and almost any other major rocket development one could name, had had their share of unsuccessful early firings, and that this had been accepted as a necessary growing pain. Those difficulties had been hardly visible under the cloak of military secrecy. But space program difficulties were highly visible and distressing.

The first notable application of a double standard came even before NASA was formed, with the spectacular Vanguard explosion in December 1957.[42] Before the Vanguard development ever got under way, it had been agreed at the National Academy of Sciences that if only one International Geophysical Year satellite out of six made it to orbit, that would be taken as a successful outcome of the project, so difficult was it considered. In fact, when the number of launches for the program was cut back from 12 to 6, the scientists argued strongly that they couldn't reasonably expect more than two successful flights.[43] Actually three were successfully launched before the end of the program. Moreover, out of the Vanguard development came the liquid- and solid-fueled upper stages that made the Delta launch vehicle the tremendous success that it became. In addition, the Vanguard program contributed the Minitrack and Baker-Nunn tracking nets to help get the new space program off to a good start.

By any reasonably objective measures—certainly by previously accepted standards—Vanguard was a successful development. Yet the early launch failures made the entire program a symbol for failure in the public mind.

The lesson of Vanguard was plain. NASA could not afford to regard failure as acceptable under any guise. Success had to be sought on the first try, and every reasonable effort bent toward achieving that outcome. This philosophy was epitomized in the "all-up" approach adopted by George Mueller, who had taken over direction of NASA's manned spaceflight programs in September 1963.[44] In Apollo the all-up philosophy—which called for assembling a complete launcher and attempting to carry out a complete mission even on the early test flights—was intended to produce economies as well as to preserve an image of success.

But the all-up approach had already been applied to other NASA projects, particularly in the space science and applications areas. Explorer satellites were simply expected to succeed, and did. The Orbiting Solar Observatory was fully instrumented for space science on its first launch in March 1962. The flight to Venus in August 1962 was Mariner's maiden voyage. Later, Associate Administrator Seamans did not permit the Office of Space Science and Applications to try for only a part of the project's objectives on early launches of Surveyor to the moon. He insisted on a full-scale mission on the first lunar flight attempt. It became customary to plan for only one or two spacecraft for a project, expecting those to meet the desired objectives. While this policy kept the question of backups always in view, and was a source of some uneasiness, the detailed attention required to make the policy of success-on-the-first-try work was important in building confidence in NASA's ability to do its job.

And that was the third issue highlighted by the ill-fated Atlas-Able project. While getting on with the business of systematically building up the national launch capability, NASA had more or less turned over the handling of Atlas-Able flights to the industrial contractor. It was not a procedure that the agency would follow very often. Rather, with its open program—operating in a goldfish bowl, as it were—the agency would prefer to monitor its contractors very closely, often more closely than the contractors thought necessary. Indeed, a great deal of NASA's management time would be taken up in overseeing the work of contractors on rockets, spacecraft, and other equipment needed for the space program.

Finally, the Atlas-Able missions served to emphasize the key role played by the rocket in space operations. Successful rockets opened up space to human exploration. Without rockets, space must remain ever remote and inaccessible. Moreover—and this was the point behind the Atlas-Able attempts—some rockets could do more than others. Neither the Jupiter C, which had launched *Explorer 1*, nor the Vanguard launch vehicle could send probes to the moon or planets. With such vehicles one would have been restricted essentially to small artificial satellites of the earth. The limitations of these first American launch vehicles were further emphasized by comparison of the 8-kg *Explorer 1* and the 28-kg *Vanguard 3* with the 84- and 508-kg weights of *Sputnik 1* and *2*.[45] Indeed, even more disturbing than the Soviet Union's launching of the first satellite was its obvious superiority in launch vehicles and implied superiority in long-range missile capabilities. The disparity was brought out even further by the launch of *Sputnik 3*, weighing 1327 kg, on 15 May 1958.[46]

COSTS

Building the launch vehicles and spacecraft described in previous sections was expensive, but necessary if the United States was to achieve the stated objectives in science, applications, and exploration. Moreover, costs did not end with development. Since most rockets and spacecraft of the 1960s were expended in the accomplishment of their missions, it was necessary to buy a new launch vehicle and a new spacecraft for each new mission.

Individual costs for launch vehicles are given in table 6. These are rough, order-of-magnitude figures. Actual costs varied, depending on how many special requirements were placed on the launch by the science objectives. To start, launch vehicle costs exceeded those of the payloads they carried, but in just a couple of years Hugh Dryden was informing the Congress that spacecraft costs had come to exceed those of the launch vehicles.[47] As time went on, engineers and scientists became expert in miniaturizing equipment, cramming their satellites and space probes with instruments. This practice increased the amount of research that could be accomplished in a given payload weight and space, but it also made for

Table 6
Launch Vehicle Costs

Vehicle	Cost of Hardware Plus Launching* (millions of 1970 dollars)
Scout	3.6
Delta (= Thor-Delta)	9.0
Atlas-Agena	17.8
Atlas-Centaur	19.4
Titan IIIC	28.6
Saturn IB	56
Saturn V	225

SOURCE: NASA Comptroller's Office.

*These are only order-of-magnitude costs. Moreover, they are the prices after many years experience. NASA had hoped Scout would cost about $1 million per rocket, but both inflation and programs to improve performance raised the price substantially above the target. Inflation and improvement programs also increased the cost of Delta over the years.

expensive spacecraft. From experience with Explorers, a rule of thumb developed that scientific spacecraft would cost about $20000 to $40000 per kilogram, but in time more complicated and sophisticated vehicles, such as the larger deep-space probes, were far more expensive. Typical costs for NASA scientific spacecraft are shown in table 7.

The tabulation illustrates why many scientists were wary about getting into projects using the larger spacecraft. The cost of about three-quarters of a million dollars for four Vikings could pay for at least twice as many Pioneers, or for dozens of Explorers. When the costs of the Viking program continued year after year to delay undertaking Pioneer missions to Venus, there were strong protests. For both satellites and space probes, the larger spacecraft were recognized as essential to the accomplishment of many important investigations, but generally were not acceptable at the sacrifice of the smaller missions, which gave the scientist more flexibility and personal freedom of action.

Manned spacecraft were an order of magnitude more expensive than unmanned satellites and probes. This was so not only because of the larger

Table 7
Scientific Spacecraft Costs

Spacecraft	Cost per Kilogram[*] (thousands of 1970 dollars)
Explorer	20–40
Interplanetary Monitoring Platform	50
Orbiting Solar Observatory (depending on complexity)	30–70
Orbiting Geophysical Observatory	70
Orbiting Astronomical Observatory	55
Small Astronomy Satellite	174
High Energy Astronomical Observatory	60
Surveyor	85–90
Lunar Orbiter	95
Pioneer (from no. 6 on)	175–225
Mariner	85–105
Viking Orbiter	85

SOURCE: NASA Comptroller's Office.
*Order-of-magnitude figures only.

size and greater complexity of vehicles that were to carry men, but also because every effort had to be bent to guarantee the safety of the crew. Trying to guarantee perfect performance was very costly, often requiring much redundancy. For unmanned spacecraft and launch vehicles, it did not make economic sense to try to achieve the same degree of refinement. Scout, for example, was designed to be an inexpensive launch vehicle for science and applications missions. Reasonable care and good engineering and operational practices could achieve success rates of 90 percent or better, and still keep the vehicle in the inexpensive category. To try to guarantee 100-percent success, on the other hand, would have increased costs enormously. Thus, if one required 10 successful firings for a certain program, and was willing to shoot for getting those 10 successes out of a total of 11 or 12 firings, the total program costs would be much lower than if one insisted on achieving the 10 successes with only 10 firings.

The author vividly remembers a discussion of this point before the Space Science and Applications Subcommittee of the House Committee on Science and Astronautics during hearings on NASA's fiscal 1966 budget request. Concerned that NASA might be escalating the costs of Scout by insisting on too high a degree of reliability, Congressman Weston Vivian, himself a former engineer, queried Edgar Cortright, deputy in the Office of Space Science and Applications, on the matter.[48] While acknowledging the desirability of striking a proper balance between costs and reliability, NASA people took special delight in this new twist. The normal experience was to be challenged to explain why the unmanned program wasn't striving all-out, as in the manned program, to achieve perfection.

Neither NASA nor the Department of Defense had carte blanche to spend unlimited sums on rockets and spacecraft. While desiring that the nation's space program should be first rate and that the country should regain the image of leadership in the field, the administration and Congress still were concerned that costs be kept down. While it was apparent at the outset that in time many agencies would come to be interested in applying space techniques to their work, it soon became equally apparent that these agencies could not expect to operate their own launch and spacecraft facilities. If the Weather Bureau, the Geological Survey, the Federal Aviation Administration, the Maritime Commission, the Department of Agriculture, and the Forest Service had attempted to run their own space programs, the aggregated costs would have been prohibitive. As a consequence it was expected that NASA, and occasionally the Department of Defense, would service other agencies wishing to use space methods in their own programs. As the number of space applications grew over the years, more and more of NASA's work was expected to go into providing support to others.

But manpower and money were not the only price to pay for exploring and utilizing space. As the example of the ill-fated Atlas-Able missions showed, there were failures and frustrations to endure. Also, rockets and spacecraft could at times be hazardous. Perhaps the best known illustrations were the Apollo fire, in which three astronauts were burned to death in a tragic holocaust of flammable materials in the oxygen atmosphere used in the Apollo capsule, and the April 1967 flight of *Soyuz 1*, in which Cosmonaut Vladimir Komorav was killed.[49] Because of the universal interest in the manned flight program, these tragedies received worldwide attention. Apollo, for example, became the subject of an intense, deeply probing congressional investigation.[50] But others also gave their lives in the course of the program, though with less notice from the public. Astronauts who died in accidents in their training airplanes received only momentary notice. On 5 October 1967, at Northern American Rockwell's plant in Downey, California, a hazardous mixture containing barium used in NASA sounding rocket experiments exploded, killing 2 workmen and injuring 11. The accident was thoroughly investigated by a NASA board and the procedures for handling such chemicals were revised. To the public, however, the matter appeared to pass as just another industrial accident.[51]

Just as tragic as the Apollo fire was the accident to an Orbiting Solar Observatory on 14 April 1964. In an assembly room at Cape Canaveral, a Delta rocket's third stage motor had just been mated to the spacecraft in preparation for some prelaunch tests. Suddenly the rocket ignited, filling the workroom with searing hot gases, burning 11 engineers and technicians, 3 of them fatally. An investigation following the accident showed that a spark of static electricity had probably set off the fuze that ignited the solid propellant.[52] But, whereas the Apollo fire had evoked a national outcry, the OSO accident drew little attention except from those closely associated with the project.

One measure of the difficulty encountered in a development program was the increase in cost and schedules over the original estimates. When estimates proved on the mark, engineering difficulties had been correctly estimated and the project could be carried out in the specified time and for the stated price. But when unexpected technical problems required extra time to solve, costs increased and exceeded the original estimates. These overruns, as they were called, were usual in the complex, novel developments of the space program, and special management attention was needed to keep them under control.

The Department of Defense had experienced such problems in the development and acquisition of large weapon systems. Studies showed that in the course of 12 major development projects, costs increased by an average 3.2 times and schedules lengthened by 36 percent.[53] NASA fared little better.

Although the space agency, after its initial troubles, began to develop an enviable record of successes in its numerous programs, acquiring during the 1960s a reputation of being able to do what it set out to do, nevertheless the record was not as neat as the agency would have desired. An analysis in 1969 by D. D. Wyatt, who from the start had played an important role in NASA's programming and budgeting, showed that cost increases over the life of a project were likely to rise substantially when the estimates were made before "establishing a well-defined spacecraft design and a clear definition of required experiment development."[54] Space science programs had their share of horrible examples. The Orbiting Astronomical Observatory, the Orbiting Geophysical Observatory, and Surveyor all increased in cost by about four times, as did the meteorological satellite, Nimbus. In contrast, the communications satellite projects Relay and Syncom, and the Applications Technology Satellite, for each of which a good definition of requirements was reached before estimating, showed only moderate cost increases, by between 1.1 and 1.3 times. While the manned spaceflight projects showed somewhat lower cost rises, Wyatt noted that the cost projections were made a considerable time after the projects had started and that there was evidence that estimates at the true start would have been much lower and cost increases accordingly much higher.

Wyatt's basic thesis was correct. His analysis, however, did not go unchallenged. Hans Mark, director of the Ames Research Center, wrote that the analysis failed to take into account that programs sometimes were expanded in scope in midstream, as had happened in the Pioneer program. This, of course, also added to the total cost, but such increases were not properly classed as overruns. To get a true picture, one needed to take into account intentional changes in program.[55]

The Orbiting Astronomical Observatory was a good example of the kinds of trouble one could get into by trying to force too big a technological step. Some of the required subsystems for the satellite were not far enough along to ensure a smooth development of the spacecraft. The star trackers, for example, essential for establishing the stellar reference frame against which the spacecraft would be stabilized, ran into difficulties that took a long time to resolve. The cost of solving the problem was only part of the total increase, for, while engineers wrestled with the star trackers, a far greater number of workers on the rest of the observatory project also had to be paid as they waited for the star trackers.[56]

The observatory finally proved to be a powerful astronomical facility. But in retrospect it can be seen that NASA might have done better to follow the recommendations of its advisers, who would have preferred to start with a less ambitious astronomy satellite that would have permitted astronomical observations sooner. Having the less capable astronomy satellite sooner, the astronomers would have been content to wait for the larger one, as Edward Purcell and other members of the White House's Space Science Panel had indicated.[57]

Both Ranger and Surveyor suffered from launch vehicle troubles. Five launch vehicle failures in a row impeded the development of Ranger, drawing the attention of NASA management and the Congress. When on the sixth flight the launch vehicle finally did work, sending the Ranger space-craft precisely to the intended spot on the moon, only to have Ranger's television system fail in the last seconds, the reaction was swift. Searching investigations into the engineering and management of the project were conducted both within NASA and by the Congress.[58] The performance of the Jet Propulsion Laboratory and the Office of Space Science and Appli-cations was under scrutiny, as well as that of the television contractor, Radio Corporation of America. The steps proposed to solve the problems were carefully examined by NASA management, and only the resounding success of *Ranger 7* averted what might have been drastic management changes.

Surveyor's launch vehicle troubles were of a more subtle kind. At the start the project suffered from lack of adequate management attention at both the Jet Propulsion Laboratory and the contractor, Hughes Aircraft Company. But Surveyor's difficulties were exacerbated by simultaneous development of launch vehicle and spacecraft. The Centaur was the first major rocket to use the high-energy propellants liquid hydrogen and liq-uid oxygen. The Advanced Research Projects Agency had started Centaur in 1958 before NASA was created. When NASA took over responsibility, the project had been handled more as a research project than as a serious development effort, but NASA considered the Centaur stage an important component of the launch vehicle stable.

Nevertheless, since only estimates could be given of the vehicle's per-formance, it was difficult for Surveyor engineers to pin down weight and payload requirements for their spacecraft. These difficulties were further enhanced by the large number of other spacecraft being assigned to Cen-taur, influencing specifications for the rocket. In April 1962 Surveyor, Mariner, a variety of near-earth and synchronous-orbit missions, and the Department of Defense's communications satellite, Advent, were assigned to the Centaur.[59] The synchronous-orbit missions—that is, launchings to an altitude of 36000 km, where the satellite's rate of revolution would equal the earth's rate of rotation—would be about as demanding as missions to the moon.

When the program difficulties—including failure of the first flight test of Centaur—became too intense, NASA cut the Gordian knot by asserting that the initial Centaur would be developed for Surveyor only. Once Cen-taur had proved successful, a suitable program to uprate the vehicle could be instituted to meet additional requirements. Along with this decision, NASA also moved management of Centaur from the Marshall Space Flight Center—where the demands of Saturn preempted the center's management attention—to the Lewis Research Center.[60]

But these changes did not come before Congress again had seen fit to delve into how NASA was performing. Congressman Joseph Karth's Space Science and Applications Subcommittee of the House Committee on Science and Astronautics explored all aspects of Centaur: its management, its importance to the space program, its funding, the contractor, and even the engineering principles underlying the design.[61] Such investigations into NASA failures made it perfectly plain that the agency had to bend every effort to avoid mishaps and provided much of the motivation for Associate Administrator Seaman's policy, mentioned earlier, of seeking success on the first try. A variety of management tools was used to make the policy succeed. During the fall of 1963 and throughout 1964, much attention was given to devising and applying more effective management.[62]

For the development of the Polaris missile, the Navy had devised what was called the Program Evaluation Review Technique, or PERT, system.[63] This laid out in graphical form the schedules of the different parts of a project to show how individual schedules were interrelated, and in particular to highlight the components or subsystems that were most likely to delay the whole project. Once highlighted, the critical elements of the project could be given the necessary funding and management and engineering attention needed to keep them moving in pace with the rest of the project. NASA adapted this system to its own projects.

Like the Department of Defense, NASA also experimented with special contracting devices to reduce cost overruns and delays in schedule.[64] In the development of a new launch vehicle, spacecraft, or other equipment, it was not possible for either NASA or the contractor to specify in detail the product desired. Instead, performance specifications were given, and the technical approach then agreed on between NASA and the contractor. In this situation, the contractor was in no position to sign a fixed price contract. Instead, contracts were customarily "cost plus fixed fee," that is, the government agreed to pay the actual allowable costs incurred in the development, plus a fee that was based on the contractor's original estimate of what the costs would be. There was a certain amount of incentive in such an arrangement for a contractor to finish the work within the specified time and cost, for the longer a project ran and the higher the costs went, the smaller would be the percentage represented by the fee. Also, the performance of a company was bound to influence the government's decision on the choice of contractors for other projects.

But experience showed that these incentives were not particularly effective in keeping costs down and schedules short. Therefore, additional incentives were introduced in the form of bonuses for meeting or beating schedules within estimated costs, and penalties for exceeding estimated costs and schedules.[65] The value of such incentives was difficult to estimate precisely. Perhaps the greatest benefit lay in the increased management attention they evoked.

One of the most effective ways to keep costs down was to see that schedules were met. The longer a project ran, the longer salaries on the project were being paid, hence the greater were the total costs. Time was money, and if a project manager could keep to his schedule, he probably could keep close to his original cost estimates. A striking example was furnished by the 1967 Mariner mission to Venus, preparation for which began only a year or so before the launch date. Since the date for launching a spacecraft to a planet is fixed by the celestial mechanics of the solar system, there was no leeway in the schedule and the date had to be met. One result was that the project was completed for slightly less than the original cost estimate.[66]

Most projects had several distinct phases. First was the period in which the project was being conceived. The specific objectives would be worked out—for example, to investigate the solar wind and magnetic fields in interplanetary space. The feasibility of the project would be determined. Could appropriate experiments be designed and instrumented? Did a suitable spacecraft for the experiments exist or could one be made? Was there a suitable launch vehicle that could be used? Much of the effort during this phase would be paper work, supported by limited amounts of laboratory research.

A second phase would be that in which engineering studies analyzed various approaches, seeking, if possible, a best one. During this phase performance specifications were worked out, to provide the basis for detailed engineering design carried out in the third phase. Once the engineering design had been completed, it was possible to move into the fourth, final phase, that of actual development, in which hardware was made and tested. Major costs in the project would come during the fourth phase, and if delays occurred because engineering difficulties had not been properly anticipated during the earlier phases, large overruns could accumulate.

To try to avoid such overruns, NASA instituted what it called a "phased project planning system," recognizing four phases A, B, C, and D, corresponding to those described above.[67] The plan required completing the early phases before proceeding to later ones. The costs of phases A and B would be a minor part of the project expense, and one could afford to spend time at these stages getting matters right. Similarly, in the detail design phase, C, although costs would be appreciably higher than in phases A and B, they would still be far less than those of the construction phase, D. Again, it would pay to spend whatever time was needed in phase C to avoid problems later in phase D.

The phased project planning document was used in NASA less as a bible than as a set of useful guidelines. Wyatt's study seemed to show that when the principles embodied in the phasing scheme were followed, costs and schedules were indeed kept in line.

By such devices NASA sought to sustain a high level of performance. For its own use, the Office of Space Science and Applications held monthly status reviews, just before those of the associate administrator. At these reviews progress and problems were discussed for all parts of the program.

Beginning in November 1966 these reviews, which had already been going on for several years, were documented in OSSA Management Reports.[68]

To keep a tight rein on the agency's activities, the administrator required that the program offices obtain formal approval of all programs or projects. This process evolved over the years until in the latter 1960s a signed Project Approval Document—or PAD, as it came to be called—was required for each major element of the program.[69] The PAD set forth the purpose of the project or program, outlined approaches to be taken, and gave estimates of schedules, manpower, and costs. The document described how the phases A, B, C, and D would be accomplished. Where outside contractors were to be used, a procurement plan acceptable to NASA's legal and procurement offices had to be provided. Before a program office could obtain approval to move to the later phases of an approved project, the office in charge would have to furnish a detailed project development plan showing that the groundwork had been properly laid and that the path to completion of the project had been thought through. Administrator Webb liked to refer to the Project Approval Documents as contracts between him and his program managers. They furnished written evidence, for those who might wish to probe the agency's performance, of the care that NASA took to ensure effective performance on its projects.

The wide scope of the space science program required a large number of space science PADs. Although these were primarily the concern of the program and project managers, they did affect the scientists themselves, in that experimenters were required to meet schedules, adhere to cost estimates, and furnish an appreciable portion of the documentation needed by managers to keep track of progress on their projects. Much of this was onerous to the scientists, who preferred to spend their time on their experiments. For manned spaceflight projects especially, where managers felt keenly the burden of ensuring absolute success, not only to justify the many dollars spent on the program but more importantly to protect the safety of astronauts, the schedules were quite rigid, and documentation considerably more detailed than for unmanned projects. Many scientists shied away from working in the manned program, preferring to fly their experiments in unmanned spacecraft for which the management requirements were less burdensome.

It is not likely that advanced research and development programs like those of NASA and the Department of Defense will ever be entirely free of mishaps and failures, cost overruns and schedule slips. Working at the frontiers of science and technology, the likelihood of encountering unforeseen technical difficulties must ever be present. The prescription for performing satisfactorily under such conditions is constant management attention and the most effective techniques for reducing the unforeseen to only the unforeseeable. In the space program, such management attention could produce acceptable performance. For launch vehicles, NASA's per-

169

formance improved over the years from a very poor showing in the first two years, to better than 90 percent successes in the 1970s (fig. 30).[70] With spacecraft the success-on-the-first-try policy appeared to bear fruit. During the 1960s every Explorer satellite that was properly placed in space by its launch vehicle achieved its mission. The Orbiting Solar Observatory, Surveyor, Lunar Orbiter, Mariner, and Viking worked the first time out, as did the Canadian Alouette, the British Ariel, and the Italian San Marco. The first Orbiting Astronomical Observatory failed, but the second was highly productive. Like the astronomy observatory, Biosatellite was successful on its second flight. In spite of its hectic development history, the second time Ranger reached the moon it performed perfectly.

To complete the picture, applications satellites that succeeded right away included the Tiros and Nimbus weather satellites, the Echo, Syncom, and Intelsat communications satellites, NASA's Applications Technology Satellite, and the geodetic satellite Geos. All the manned spaceflight spacecraft—Mercury, Gemini, Apollo, and the Lunar Module—worked on their initial manned flights.[71] After more than a decade of experience, both NASA and the Department of Defense came to expect spacecraft to function correctly once they were launched properly, and with the increasing reliability of launch vehicles, the probability of success on a mission was very high.

With reliable launchers and spacecraft available to carry out their experiments, scientists could conduct a variety of experiments from near the earth to the remote regions of the solar system, and the value of what they did depended very much on their own scientific competence. Recognizing this, NASA had consciously set out to interest the best of the nation's scientists in the space science program. Policies for supporting research, procedures for selecting experiments and experimenters, and arrangements for securing to NASA the best possible advice were designed to attract the best researchers into the program. Most important was the policy of letting the space science program become very much the creation of the scientists, with projects to attack the problems the scientists themselves perceived as the most fundamental and most likely to provide important new knowledge. The way in which this policy was put into practice is the subject of chapter 12. Before proceeding with the narrative, however, it is appropriate to review some of the significant space science results obtained in NASA's first three or four years.

NASA Record of Spaceflight Performance
(as of 31 December 1975)

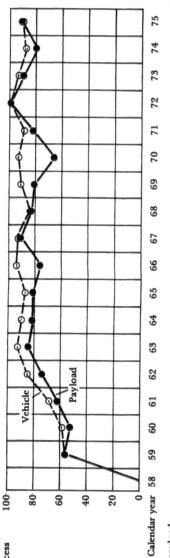

NASA payloads	58	59	60	61	62	63	64	65	66	67	68	69	70	71	72	73	74	75	Total
Attempts	4	14	17	24	27	13	30[a]	28[a]	29[b]	22	18[a]	16	9	11	10	10	5	13	300
Successes	0	8	9	15	20	11	25	23	22	20	15	13	6	9	10	9	4	12	231
Percentage successful	0	57	53	63	74	85	83	82	76	91	83	81	66	82	100	90	80	92	77
NASA vehicles (some carrying non-NASA payloads)																			
Attempts[c]	4	14	17	24	27	15	30	30	36	27	23	22	14	18	18	14	17	21	316
Successes	0	8	10	16	23	14	27	26	34	25	19	20	13	16	18	13	15	19	371
Percentage successful	0	57	59	67	85	93	90	87	94	93	83	91	93	89	100	93	88	91	85

[a] Two missions launched on a single vehicle.
[b] Five missions on two vehicles are included.
[c] These figures include NASA vehicles used for non-NASA or cooperative missions.

Figure 30. NASA success rate. After a poor beginning, NASA's success rate rose steadily, eventually bettering 90 percent. NASA, "Historical Pocket Statistics" (Jan. 1972 and Jan. 1976); Astronautics and Aeronautics, 1972, NASA SP-4017 (1974), app. B.

11

Deepening Perspective:
A New Look at the Old World

Among the most important contributions rockets, satellites, and space probes made to science was the new perspective they afforded in many areas, particularly in the earth and planetary sciences. Earth scientists, of course, had always enjoyed an advantage in being close to the object of study, living on the earth and immersed in its atmosphere, where the investigator could collect great quantities of data in situ. This was the very advantage that scientists seized upon when sounding rockets made it possible at long last to get on-the-spot measurements in the upper atmosphere. But a certain myopia was also associated with being too close to the object of study.

One of the tasks facing the researcher on the ground was developing an integrated picture of what was often a very large-scale, as well as complex, system. The meteorologist, for example, in spite of the enormous quantities of data he gathered on the weather, still found them too sparse. Even on land they came from rather widely separated stations, and there were none at all from vast stretches of the oceans. As a consequence the investigator was hard pressed to describe with any confidence the huge cyclonic systems and their interrelationships that characterized the general circulations of the earth's lower atmosphere, let alone tell what the weather was like in remote unobserved regions. But when the first weather satellite pictures became available, showing cloud patterns over both continents and oceans, the meteorologist had at hand one of the integrating factors that he needed. For, clouds, being intimately associated with pressure patterns and air circulations, showed by their distributions the major weather systems. Most of what was seen in the early cloud pictures was expected, but there were also surprises. The author can recall hearing Dr. Harry Wexler, director of research for the U.S. Weather Bureau and strong proponent of weather satellites years before any satellite had flown, exclaim that he had never expected the large-scale patterns of atmospheric vortices that stood out in many satellite photographs. When in the course of time satel-

172

lite cloud imaging was improved in resolution and supplemented by techniques for measuring cloud heights, the vertical distribution of atmospheric temperatures, and local winds, meteorology became not merely local, not merely regional, but the global science it had always aspired—and needed—to be.[1]

Meteorologists were among the most ready to take advantage of the new approach and in short order used the satellite pictures in making weather forecasts. But such pictures also showed complete ice fields, total watersheds, entire geological provinces such as volcanic fields or geosyncline basins, varying patterns of land use, and vast expanses of ocean. To many it was clear from the start that the perspective afforded by satellite observations would in time prove of immense value in these and other areas. Such has proved true.[2]

After more than a decade of rocket sounding of the upper atmosphere, space science was quite ready to benefit from the new perspective. In the first half dozen years following the formation of NASA, especially rapid progress was made in the continued study of the upper atmosphere and ionosphere, solar physics, rocket astronomy, geodesy, and the magnetosphere. Accomplishments in the last two areas provide good illustrations of the power of space techniques for scientific research and are the subject of this chapter. The contributions to geodesy were anticipated, causing a number of researchers to give serious attention to the possibilities during the 1950s, years before Sputnik went aloft.[3] In contrast, the magnetosphere emerged as something of a surprise from the early rocket and satellite work on particles and fields.

THE MAGNETOSPHERE

For want of a more appealing name the phrase *particles and fields* came into early use in the space program to denote the study of magnetic and electric fields in space and a variety of particle radiations. Among the last named were the extremely energetic cosmic rays, plasma radiations from the sun, and the electrons, protons, or whatever they were that were thought to cause the auroras. (Gravitational fields were not included, falling rather under geodesy, relativity, and cosmology, with which gravity studies were naturally associated.) The term *magnetosphere* denotes the region of space surrounding the earth where the earth's magnetic field plays a prominent, often controlling, role relative to various particle radiations found there. As will be seen, magnetospheric physics constituted an important aspect of the discipline of particles and fields.

The discovery of the magnetosphere began with Van Allen's discovery of the earth's radiation belt. At White Sands, New Mexico, Van Allen had traced the curve of cosmic ray intensity through the Pfotzer maximum to a more or less steady value at heights greater than 55 km that looked very much as though it might be the free space value of the cosmic ray intensity.[4] Cosmic rays, being charged particles, were affected by the earth's magnetic field, and fewer of them were able to get in over the geomagnetic equator than in the polar regions. The less energetic rays were the most affected by the magnetic field, making it difficult to determine what the lower end of the cosmic ray spectrum might be in interplanetary or interstellar space. Since the total energy spectrum of the cosmic radiation in space would be an important factor in trying to figure out how and where cosmic rays were generated, Van Allen took a special interest in investigating the variation of the high-altitude cosmic ray intensity with geomagnetic latitude. For this purpose he took Aerobee rockets to sea aboard the U.S. Navy's seaplane tender *Norton Sound*, which had to be specially outfitted with an Aerobee launching tower. Van Allen's sounding ranged from the geomagnetic equator off the coast of Peru to Alaskan waters.[5] The measured variations were sufficiently intriguing that Van Allen pursued the subject further with Rockoons—the small sounding rockets that he launched from Skyhook balloons in the stratosphere. These Rockoon experiments turned up a most interesting and puzzling phenomenon. In the auroral regions above 60 km was a rather soft—i.e., moderately penetrating—radiation that could be a mixture of charged particles and x-ray photons.[6] This radiation was assumed to be in some way connected with the auroras, and efforts were begun to explore the connection.

At about this time the appearance of the International Geophysical Year satellite program gave Van Allen the chance to extend these investigations to even higher altitudes. When the first Explorer was launched (31 January 1958), Van Allen's counters appeared to show a zero counting rate at certain locations, which didn't seem to make sense. Further study showed, however, that actually the counters were saturating because of ambient radiations far exceeding intensities with which the counters had been expected to cope. *Explorer 3* (26 March 1958) pursued the question.

Soon Van Allen decided that he was observing a region of intense radiation surrounding the earth at high altitude, and on 1 May 1958 he announced his discovery.[7] The region at once became known as the Van Allen Radiation Belt. Soviet measurements in *Sputnik 3* (15 May 1958) confirmed the discovery.

An explanation was quickly forthcoming. The radiations were attributed to charged particles caught in the earth's magnetic field, unable to escape because their energies were too low to allow them to cross the surrounding field lines. One thus visualized *trapping regions* within the earth's field and spoke of *trapped radiations*. Suddenly it was crystal clear that the earth's magnetic field, which could prevent some charged particles in interplanetary space from ever reaching the earth, could also prevent other particles already near the earth from leaving.

In retrospect it seemed remarkable that the existence of the radiation belt had not been anticipated long before its discovery. Workers concerned with the problem of how gases escaped from the atmosphere understood that the magnetic field would hinder the escape of ions.[8] More significantly, the experiments of K. Birkeland and E. Brüche with cathode rays aimed at small magnetized spheres and the half century of theoretical work by Carl Størmer and others on the influence of the earth's magnetic field on auroral particles and cosmic rays provided a substantial basis for predicting the existence of trapped radiations near the earth.[9] Seeking an explanation for the auroras, Størmer had developed a theory of the motion of an electron approaching the earth's dipole magnetic field from the sun. He showed that such an electron would be deflected by the earth's field away from the equator to the polar regions, an action that appeared to him to explain the existence of auroral regions or zones at high latitude.

Størmer's calculations showed that there were regions inside the earth's magnetic field which such solar electrons could not reach, to which he gave the name "forbidden regions." Birkeland, with whom the theory had originated, had already demonstrated in the laboratory that electrons would be deflected to the polar regions, a fact Størmer's calculations nicely brought out.

Later, in the 1930s and after, theorists interested in explaining the geomagnetic-latitude effect observed in cosmic ray intensities, extended Størmer's work to much higher energy relativistic particles—i.e., particles approaching the speed of light—such as were to be found in the cosmic rays.[10] Their calculations also revealed forbidden regions toward the geomagnetic equator and served to explain why cosmic rays increased in intensity with increasing geomagnetic latitude.

These investigations furnish an excellent example of how initial orientation can markedly bias an investigator's conclusions. To those seeking explanations of the auroras or the cosmic-ray–latitude effect, the orientation was from outside in. Their particles were approaching the earth from great distances. It was natural, then, that the regions which the earth's magnetic field prevented those particles from entering should be named forbidden regions. While the point was not missed, still the investigators did not focus on the fact that for a particle already within one of those regions, it could be the outside that was forbidden—in other words, a particle of too low an energy already within one of those regions couldn't get out. What were forbidden regions for particles approaching from the outside were trapping regions for some particles already there.

It was only a tiny step from this realization to the idea that these trapping regions might well be filled with trapped radiations forming a radiation belt around the earth. But no one paid any attention to this possibility until, on the eve of Van Allen's discovery, S. Fred Singer in discussing magnetic storms touched upon the possibility that regions of trapped radiations might be found at high altitudes around the earth.[11] Following Van Allen's announcement, this field of investigation blossomed forth as researchers vied with each other to learn about the fascinating trapped radiations.[12]

In the next half-dozen years a new paradigm emerged to characterize the magnetosphere and magnetospheric physics. Whereas before the spring of 1958 the space environment immediately surrounding the earth was thought to be relatively uncomplicated, it soon became clear that the recently discovered magnetosphere was extremely complex. Before the recognition of the radiation belts, there was no generally accepted picture of the space environment near the earth. Students of the earth's upper atmosphere and ionosphere tended to think of these as attenuating more or less exponentially with altitude, eventually merging at some considerable, but unknown, height with the medium of interplanetary space. Around the planet the earth's magnetic field was visualized as essentially that of a dipole, much as depicted in figure 3 in chapter 6. It was known that particles from the sun swept across the earth's atmosphere, some of them causing the auroras. Sidney Chapman, V. C. A. Ferraro, and others supposed that some of the solar particles impinging upon the earth's magnetic field would compress it, thereby causing the sudden increase in the surface field that had long been observed to follow flares on the sun. Such a theory implied, of course, that the earth's magnetic field would be distorted somewhat by the solar particles. Moreover, to explain the main phase of magnetic storms in which the field dropped well below normal for a day or more, Chapman and Ferraro thought of the cloud of solar particles as somehow setting up a ring current around the earth; the current generated a magnetic field that caused the considerable drop in field intensity an hour or so after the sudden increase of the initial phase of the storm. The cloud of solar particles was presumably a plasma; that is, a gas composed of equal numbers of positively and negatively charged particles. Thus, the plasma, though neutral in the large, would be highly conducting. Also, since the positive particles would be deflected in one direction by the earth's magnetic field, the negative particles in the opposite, one could sense intuitively how a current might be set up around the earth— although there were formidable difficulties to overcome in developing such a theory. The period of one to several days required for the field to return to normal would then be the time required for the ring current to dissipate.

Chapman and Ferraro visualized the ring current as flowing on the surface of a huge cavity which the earth's magnetic field carved out of the plasma cloud as it swept by the earth. There were, of course, two sides to this coin. From one point of view the earth's magnetic field generated a cavity in the flowing plasma. From the other point of view, however, one could think of the plasma cloud as confining the earth's field to the cavity region. The discovery of the radiation belt focused attention on the second point of view, and the region within the Chapman-Ferraro cavity became known as the *magnetosphere* (fig. 31).

Because of the intense interest in the new topic, many of NASA's early spacecraft—and those of the USSR, also—were instrumented to make measurements of the particles and fields in the vicinity of the earth and in interplanetary space. By the end of 1964 a highly detailed picture of the magnetosphere had been worked out, a picture that was still evolving.[13]

Explorer 1 measurements put the radiation belt at about 1000 km above the equator, and *Explorer 3* and *Sputnik 3* confirmed this observation. From *Explorer 4* and the space probe *Pioneer 3*, Van Allen could show that, at least for particles that could penetrate one gram per square centimeter of material, there were two radiation belts, an inner zone and an outer zone as shown in figure 32. *Pioneer 4*, which eventually went into orbit around the sun, gave additional information about the extent of the radiation belts. It appeared that the belts extended to about 10 earth radii from the center of the earth, but the exact location of the outer edge appeared to be variable.

The variability was quickly tied to conditions in interplanetary space, which in turn were controlled by solar activity. A major factor influencing the earth's space environment was shown to be the *solar wind*. In 1958 Eugene Parker had shown theoretically that the sun's corona had to be expanding continuously, and that a continuous wind from the sun should be blowing through interplanetary space.[14] Highly conducting and virtually free of collisions among the constituent particles, this solar wind should entrap and draw out magnetic field lines of the sun. Such interplanetary plasma fluxes of about 10^8 particles per square centimeter per second were measured by Gringauz on *Lunik 2* and *3*.[15] With a probe on *Explorer 10*, H. Bridge and coworkers at the Massachusetts Institute of Technology confirmed the fluxes detected by the Luniks and found that the wind came from the general direction of the sun at about 300 km per second.[16] More definitive measurements from *Mariner 2* and *Explorer 18* showed a very gusty wind, nearly radial from the sun, to be blowing at all times with velocities of roughly 300 to 500 km per second. Protons and helium nuclei appeared to be present in the wind.[17]

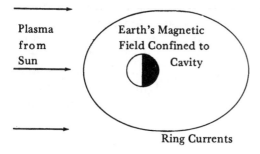

Figure 31. Chapman-Ferraro cavity. Ring currents set up around the earth were assumed to be the cause of magnetic field effects observed during magnetic storms.

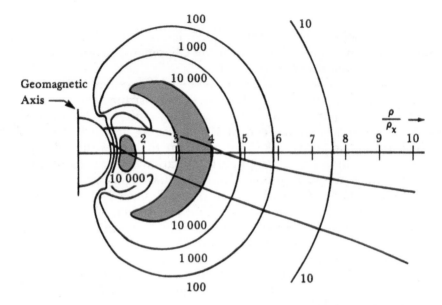

Figure 32. Radiation belts. Van Allen's picture of the inner and outer zones of the radiation belt made after Pioneer 3 *data returns. J. A. Van Allen and L. A. Frank, from* Nature 183 (1959): 430; copyright Macmillan Journals Ltd., 1959.

Meantime more information had been collected on the structure of the radiation belts. The inner zone was shown to be largely high-energy protons, many of which could be accounted for by the decay of neutrons splashed back from the atmosphere.[18] The neutrons were generated by cosmic rays colliding with nitrogen or oxygen nuclei of the air; being neutral, the neutrons could move upward unhindered by the magnetic field. But the neutrons decayed quickly and produced protons and electrons which, being charged, were trapped to form a part of the radiation belt. Detailed measurements revealed that both protons and electrons existed throughout the altitude range from the bottom of the so-called inner zone to the far edge of the outer zone. The apparent existence of two belts had been due to the insensitivity of some early instruments to lower-energy particles.

The boundary of the magnetosphere was first definitely located with instruments on *Explorer 10*, which was launched on 25 March 1961. The spacecraft was projected at an angle of roughly 130 degrees from the direction to the sun, that is, quartering away from the sun. Between the distances of 22 earth radii and the apogee of 47 earth radii, the satellite appeared to cross the boundary at least six times, suggesting that the boundary wavered in the wind. Inside the boundary the magnetic field was 20 to 30 gammas and steady, and there was no detectable plasma. Outside the boundary, however, the field weakened to between 10 and 15 gammas, and plasma was always observed. Data from *Explorer 12* in the direction of the sun showed a very sharp outer limit to the geomagnetic field, a limit that came to be called the *magnetopause*. Beyond the magnetopause was a region in which the magnetic fields were variable in direction and intensity, and the ambient radiation isotropic.[19]

Thus, by about the beginning of 1962, scientists began to envision a magnetosphere much as shown in figure 33. A continuous solar wind blowing against the earth's magnetic field was pictured as sweeping around the earth, confining the field to an immense cavity which extended to about 10 earth radii in the direction toward the sun, and to considerably more than this in the opposite direction. Inside the cavity lay the Van Allen Radiation Belt which showed considerable structure, with high intensities of energetic protons in the inner portions and large quantities of electrons in the outer reaches. Outside the magnetopause—that is, outside the boundary of the cavity—lay a region of turbulent magnetic fields and plasma. It was suggested that surrounding the turbulent region would be found a huge shock wave produced in the solar wind by the earth's magnetic field, which would act upon the high-speed plasma much as a blunt body would act upon a supersonic flow of gas in ordinary aerodynamics. By analogy with aerodynamics, estimates were made of where the bow shock might be found.

The bow shock was first detected by instruments in the Interplanetary Monitoring Platform, *Imp 1*, otherwise known as *Explorer 18*, which was launched in November 1963 into an orbit with an apogee at 30 earth radii.[20] In the course of its lifetime the spacecraft's instruments provided clearcut evidence that *Imp 1* had crossed the magnetopause and the bow shock many times. The data from a magnetometer installed by Norman Ness of the Goddard Space Flight Center were most convincing.[21] Figure 34 shows magnetic field data from orbit 11 of *Imp 1*. Inside 13.6 earth radii, a well-ordered field was noted, but from 13.6 to 20 earth radii the field was quite turbulent. Beyond 20 earth radii the field became quite steady at about 4 gammas, with some fluctuation in direction. The turbulent region from 13.6 radii to 20 earth radii was interpreted as a transition region between the shock wave in the solar wind and the magnetopause bounding the geomagnetic field. Plasma data from MIT and Ames Research Center instruments were consistent with this interpretation.[22] Beyond 20 earth radii the MIT instruments showed large fluxes in only one of six energy channels, presumably that due to the solar wind, whereas in the transition region the plasma probe indicated considerable turbulence, showing appreciable fluxes on all six channels of the instrument.

In December 1963 *Imp 1* found the interplanetary magnetic field, which was usually quite steady, to be disturbed, rising to about 10 gammas for a day or more. On the first day of this disturbance, 14 December, the moon was close to lying between the satellite and the sun. Ness originally

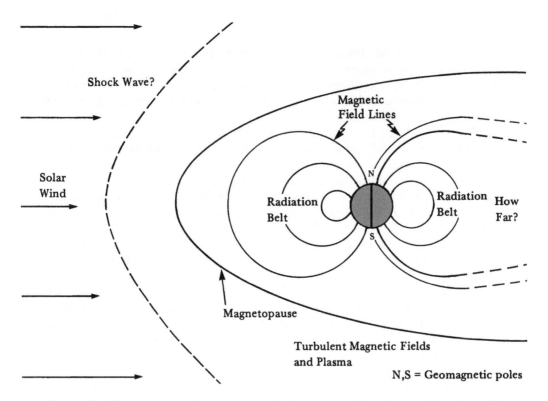

Figure 33. The magnetosphere as visualized early in 1962. Here and in figure 35, the lines emanating from earth represent magnetic field lines. Although the general structure was emerging, many features were still to be delineated.

attributed this unusual disturbance to a wake produced by the moon in the solar wind.[23] That the moon with almost no magnetic field should produce a wake detectable so close to the earth at once suggested that the much larger earth with a strong magnetic field would produce a similar wake reaching certainly to the orbit of the moon, and most likely well beyond. It began to appear that the earth's magnetospheric tail should extend to very large distances in the antisolar direction.

As investigation of the magnetosphere proceeded, it was clear that this region was intimately involved in many familiar phenomena, such as magnetic storms and auroral displays, serving in some way as a connecting link between the original solar radiations and the ultimate terrestrial effects. But the precise mechanisms involved eluded explanation. It was shown that both electrons and protons produced the auroras, with electrons of energies below 25 kiloelectron volts contributing most to the auroral emissions.[24] Størmer's theory that these particles came directly from the sun into the auroral regions of the earth had to be abandoned when both Soviet and U.S. deep-space probes showed that the fluxes of such particles in interplanetary space were insufficient. An alternate theory

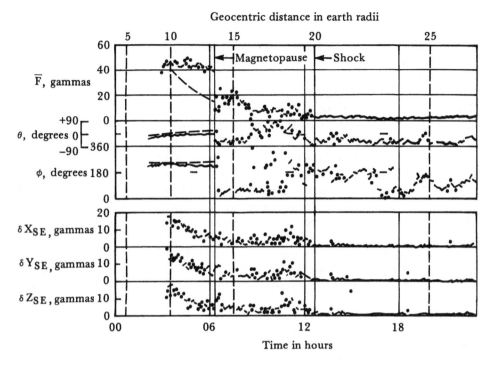

Figure 34. Magnetospheric bow shock as revealed by space-probe measurements. Magnetic field data from orbit 11 of Imp 1. *The magnetopause is at 13.6 earth radii. The second transition at 20 earth radii to an ordered field outside is the location of the bow shock wave. C. S. Scearce and J. B. Seek,* Journal of Geophysical Research *69 (1964): 3531–69; copyright American Geophysical Union, 1964.*

that the particles were accumulated in the trapping regions of the magnetosphere and then dumped or dribbled into the auroral zones to produce the auroras also ran into difficulties. Although both Soviet and U.S. measurements showed that the fluxes at the altitudes from which the particles could spiral along the field lines into the auroral regions were adequate to produce an aurora, the quantity of radiation was too low. The particles would be drained away in a few seconds, whereas auroras often lasted for hours.[25] Brian O'Brien observed, however, from instruments in Injun satellites of the State University of Iowa that trapped electrons in the radiation belt, electrons precipitated into the atmosphere of the auroral zone, and auroral light emissions all increased simultaneously.[26] One could conclude that the disturbances ultimately causing the auroras somehow also replenished the radiation belt, perhaps in this way making it possible to sustain a long-duration auroral display. Whether these additional electrons were inserted into the radiation belt from outside or came from lower energy electrons already existing within the belt and accelerated by some mechanism to the necessary higher energies was not known. Indeed, while many clearcut relationships between auroras and radiation belt activity had been established, at this stage the actual mechanism producing the auroras remained a mystery.

Also unexplained was the immediate cause of the main phase of magnetic storms. A ring current around the earth continued to be the most likely candidate, but how such a current was generated remained a puzzle. It could be shown that charged particles in the magnetosphere, in addition to spiraling around magnetic field lines bouncing back and forth between northern and southern reflection points, would also tend to drift longitudinally, the electrons drifting eastward and the protons westward.[27] Thus, these drift motions produced in effect a ring current, which S. Fred Singer suggested as the cause of the main phase of magnetic storms.[28] By the end of 1964, however, no spacecraft measurements had been able to locate the postulated ring current.

By the mid-1960s a very detailed, though by no means complete, picture of what the magnetosphere was like had evolved, as illustrated in figure 35. In the magnetospheric paradigm of 1964 the existence of the solar

Figure 35. The magnetosphere as visualized in the mid-1960s. Space-probe measurements have provided a wealth of detail. The principal research problems are shifting from describing the phenomenon to explaining the relationships and processes.

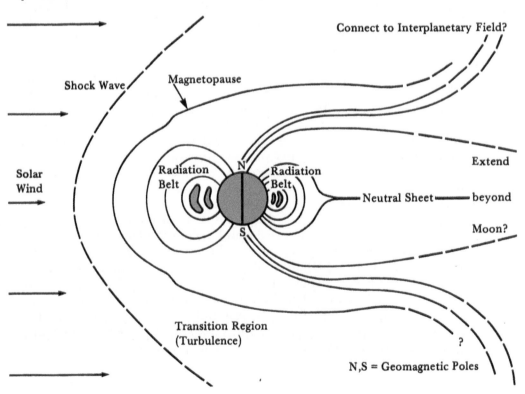

wind had been established. The wind consisted of protons mostly, with some alpha particles (helium nuclei), both of which had been observed. To be neutral the wind had to include equal numbers of electrons, but these had not been detected as yet. Embedded in the solar wind was an interplanetary magnetic field pulled out of the sun by the solar wind plasma. Near the earth the interplanetary field intensity was between five and six gammas. Blowing against the earth's magnetic field, the solar wind produced a huge shock wave sweeping around the earth much as an aerodynamic shock wave accompanies a supersonic airplane. But, whereas an aerodynamic shock wave is produced by compression of a gas consisting of air molecules all colliding with each other, the magnetospheric shock wave was set up by deflection of the individual plasma particles by the earth's magnetic field and was referred to as a collisionless shock wave.

Behind the shock was a region of turbulence. Here the magnetic fields became highly disordered; particle velocities, which in the solar wind were usually confined to a rather narrow range, suddenly varied widely. Closer to earth this transition region was bounded by the magnetopause enclosing the geomagnetic field now grossly distorted from the simple dipole configuration that would have existed in the absence of a solar wind. Some of the field lines that would otherwise have lain on the sunward side of the earth were swept backward in the antisolar direction and along with field lines on the night side were extended into a magnetospheric tail. The magnetic field lines that still enveloped closed regions near the earth contained the Van Allen Radiation Belt, which paradoxically appeared to be more limited in extent on the night side of the earth than on the daytime side, where the field was compressed by the solar wind. On the dayward side, toward the poles, where some of the field lines were swept out into the tail, appeared a cusp or dimple in the magnetopause. It was thought that where magnetic field lines of opposite direction came together near the equatorial plane of the tail, they might cancel each other producing a neutral sheet. Along this neutral sheet one could envision charged particles leaking from interplanetary space into the zones closer to earth, where they could then be steered by the field toward the poles.

In the steady state this magnetospheric configuration drifted slowly around the earth, always keeping the tail away from the sun as the earth revolved around the sun. The nose of the shock wave was about 14 earth radii from the center of the earth, and the nose of the magnetopause typically at about 10 earth radii. The extent of the magnetospheric tail was a matter of speculation, but it appeared certain to reach at least to lunar distances.

At times when the sun was disturbed, the magnetosphere and the radiation belts were affected. The spatial extent of the magnetosphere varied appreciably and trapped radiations were enhanced following solar storms.

There was a question as to whether during these disturbed conditions new particles were injected into the radiation belt or energy was transferred by hydromagnetic waves from the interplanetary plasma to particles already in the magnetosphere.

Many problems, of course, remained unsolved. An explanation of the auroras appeared tantalizingly close, yet elusive. The immediate cause of the main phase of magnetic storms was still to be found. How energy and particles were inserted from the interplanetary medium into the magnetospheric regions had yet to be explained. The existence of the neutral sheet had not been established, nor had its precise role in magnetospheric physics been described. How the field lines in the magnetospheric tail closed again also had yet to be described. Did they perhaps connect with magnetic field lines in interplanetary space, as some surmised? Related questions concerned the sun. How did the sun manage to eject the streams and clouds of highly energetic particles and magnetic fields that from time to time upset the normal conditions in the solar wind? There was reason to suppose that solar magnetic fields were the ultimate source of the energy conveyed to these clouds, but there was as yet no generally accepted explanation.

Most of the early research on the magnetosphere was directed toward describing it. As the subject became more familiar, more and more attention was devoted to achieving a coherent explanation of the magnetosphere and its relationship to the sun and interplanetary medium on the one hand, and to terrestrial phenomena on the other. By 1964 the major interest of the scientists lay in trying to understand the various processes in magnetospheric physics. There was, of course, still much to learn about *what* the magnetosphere and its most important phenomena were. But enough of the *what* had been learned that now investigators could profitably spend much of their time on the *how*, the immediate and ultimate causes of the auroras, magnetic storms, radiation belts, and the magnetospheric tail, and on the processes that related causes with effects. To understand these processes would be the principal objective of magnetospheric research in the years ahead.

SIGNIFICANCE

Clearly the discovery of the earth's radiation belt and the subsequent description developed for the magnetosphere constituted a major scientific achievement. It is natural, then, to ask what the significance of the achievement might be. Was magnetospheric physics really a new field of research, as some claimed? Did Van Allen's discovery set in motion a scientific revolution, or was the unveiling of the magnetosphere simply normal science? The attempt to answer these questions provides a good illustration of the difficulties in Kuhn's concepts of paradigm, normal science, and scientific revolution.

As to whether magnetospheric physics was a new field of research, certainly before the discovery of the radiation belt no one was consciously working on investigating the magnetosphere, since the existence of such a region was unknown. Following Van Allen's experiments, scores of researchers began to investigate the magnetosphere. One could then legitimately argue that here was indeed a new field of research, not being pursued before, now being pressed vigorously. But this seems too shallow a conclusion. For research on the earth's magnetic field, the auroras, sun-earth relationships, and cosmic rays had been of long standing when *Explorer 1* went aloft. From this, magnetospheric physics appears more as simply one aspect of those other fields—a remarkable and hitherto unforeseen aspect, to be sure, but integrally related.

Did, then, the unveiling of the magnetosphere constitute a scientific revolution in the related scientific fields? Certainly the magnetospheric paradigm that emerged from the first half-dozen years of satellite and space-probe research was new and unpredicted. One is tempted, then, to argue that the emergence of this entirely new paradigm was evidence of a scientific revolution. But again the quick answer may be too superficial. True, the trapped radiations and the magnetosphere as it was revealed were unpredicted. But that is not the criterion of a scientific revolution. One must ask instead whether the radiation belt and the magnetosphere were *unpredictable* from the existing paradigm in the sense of being fundamentally inconsistent with it. The answer to this question may well be no. In fact, the work of Størmer and others, based wholly on the existing paradigm, had provided an adequate basis for predicting the existence of trapped radiations in the earth's magnetic field. In this light the new magnetospheric paradigm appears as a straightforward extension of the previously existing paradigm, requiring no changes in *fundamental* principles or concepts. From this perspective, then, the magnetospheric research of the early 1960s was normal science—exciting, productive, important, yet normal science. But magnetospheric physicists are likely to consider the above perspective too broad. Norman Ness, one of the key figures in magnetospheric research, regards the progress made in the half-dozen years following the discovery of the radiation belts as revolutionary. In this assessment Ness considers the emergence of a new magnetospheric paradigm and the fact that no one predicted it as of primary significance.[29]

One major implication of the research on the earth's magnetosphere—which was immediately recognized—was that the way in which the interplanetary medium affects a planet depends strongly on whether the planet has a magnetic field. In a period when the idea of comparative planetology was being emphasized by the availability of spacecraft to carry scientific investigations to the different planets, scientists previously interested in sun-earth relations were beginning to talk about sun-planetary relations. It had already appeared as though the moon produced a detectable wake in the solar wind, although later measurements by *Explorer 35* would show that the lunar wake extends only a few lunar radii downstream, instead of to the vicinity of the earth as originally supposed.[30] The moon presented the case of a planetary body with very little magnetic field and no atmosphere. Solar wind particles might be expected, then, to strike the lunar surface directly. In the case of Venus, which also has little magnetic field but which has an atmosphere perhaps 100 times that of Earth, the solar wind would impinge on the top of the atmosphere but would not be able to reach the planet's surface. Mars would present the case of a planet with little magnetic field and an atmosphere about one percent that of Earth. Jupiter, on the other hand, with its very strong magnetic field would have a huge magnetosphere. If radio bursts that were observed to come from Jupiter were from trapped particles, the Jupiter radiation belt would prove much more intense than Earth's. At the end of 1964 these were principally ideas for future research. Knowledge of Earth's magnetosphere invested that future research with considerable promise.

SATELLITE GEODESY

Satellite geodesy also made a substantial contribution to the deepening perspective in which men could view their own planet. But the new perspective differed in an interesting way from that provided in magnetospheric physics. For the latter, rockets and satellites revealed a wide range of hitherto unknown phenomena. In contrast the subject matter and problems of geodesy were well known; it was increased precision, the ability to measure higher order effects, and the means for constructing a single global reference system that space methods helped to provide.

Geodesy may be divided into two areas: *geometrical geodesy* and *dynamical*, or *physical geodesy*. The former seeks by geometrical and astronomical measurements to determine the precise size and shape of the earth and to locate positions accurately on the earth's surface. The latter is the study of the gravitational field of the earth and its relationship to the solid structure of the planet. As will be seen, geometrical and physical geodesy are intimately related.

Geodesy offers many practical values. Accurate maps of the earth's surface depend on a knowledge of both size and shape. Into the 20th century the requirements for precision were rather modest. Individual countries could choose their own reference systems and control points and, using geodetic measurements made within their own territories, produce maps of sufficient accuracy for national purposes. The appreciable differences among the various geodetic systems did not appear to matter. As late as 1947, disagreements among Danish, Swedish, German, Norwegian, French, and English systems ranged from 95 meters to 250 meters, while in the absence of adequate surveys, errors between the various continents and ocean islands could be a kilometer or more.[31] For demands in the mid-20th century, the most obvious being those of air and marine navigation and missilery, such errors could at times appear enormous, and there was a growing agitation among geodesists to generate a world geodetic system that would use a common reference frame and tie all nets around the world into a single system. At this point the artificial satellite appeared on the scene and was able to provide some help. To understand how the satellite could contribute, a few basic concepts are needed.

The science of geodesy began when the Greek Eratosthenes (c. 276–c. 192 B.C.), believing the earth to be spherical, combined astronomical observation with land measurement to estimate the size of the globe (fig. 36).[32] He had learned (actually incorrectly) that at noon in midsummer the sun shown vertically down into a well in Syene (now Aswan). Observing that at the same time the sun as seen from Alexandria was 7.2° south of the zenith, he concluded that the arc along the earth connecting Syene and Alexandria had to subtend an angle of 7.2° at the earth's center. The arc accordingly had to be 7.2/360 or 1/50 of a total meridian circle. He was told that a camel caravan took 50 days to travel from Alexandria to Syene. Using a reasonable camel speed he deduced a length for the arc, which multiplied by 50 gave him a rough estimate (16% too large) of the length of a whole meridian circle. Such estimates of the earth's dimensions improved over the centuries as different persons used better measurements, and eventually better techniques.

Concerning techniques, the next major step in geodesy came when Tycho Brahe conceived of the method of triangulation, which was developed into a science by Willebrord Snell. In this technique (fig. 37) the points A and P, between which the distance is to be determined, are connected by a series of interlinking triangles. The length of one side of one of the triangles that is convenient to measure—say the side AB of the first triangle—is then measured to a high degree of accuracy. One then measures the angles of the first triangle, which can be done with precision much more easily than measuring length. Using the law of sines, the initial side of the next triangle down the chain can then be calculated. The process can be repeated to get the length of the initial side of the third triangle of the chain. Moving step by step from triangle to triangle, one finally gets to the last triangle, of which P is a vertex. With the lengths of all the sides of the triangles known, it is then possible to compute the distance between A and P along the terrestrial sphere. For great distances one must, of course, introduce appropriate corrections to take into account that the sum of the angles of a triangle on a sphere is greater than 180°. With care a high degree of accuracy can be achieved. By using nets of triangles one can proceed outward along one chain to the selected point P, and back along a different chain to calculate the measured baseline AB. If the calculated value of AB is sufficiently close to the originally measured

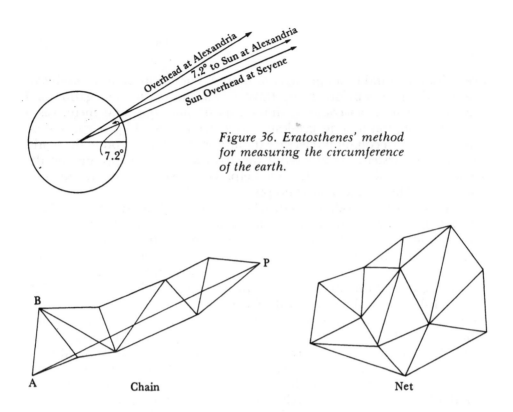

Figure 36. Eratosthenes' method for measuring the circumference of the earth.

B

P

A

Chain

Net

Figure 37. Triangulation. This technique provided a step-by-step method of accurately determining the distance between widely separated points on the earth's surface.

value, the confidence in the calculated value of AP can be high. Jean Picard (1620–1682) employed this technique in obtaining the value of the earth's radius that Isaac Newton used in deriving his law of gravitation.

The period from Eratosthenes to Picard has been referred to as the *spherical era* of geodesy. During that time it was assumed that the earth was a sphere. This made the geodetic problem quite simple, for one had only to determine the radius of the terrestrial sphere, and the rest came out of simple geometry (spherical trigonometry). But in the 17th century it became clear that the earth was not spherical. From this period on the earth was visualized as essentially an ellipsoid of revolution, with its major axis in the equatorial plane and minor axis along the earth's axis of rotation. The bulge in the equatorial plane could be explained as due to centrifugal forces from the earth's rotation. Thus, the 18th and 19th centuries could be thought of as the *ellipsoidal period* of geodesy, and a prime task was to find the ellipsoid of proper size and flattening to best represent the earth. By the mid-20th century the equatorial radius of the reference ellipsoid had been determined as 6 378 388 meters, while the flattening— that is, the ratio of the difference between the equatorial and polar radii to the equatorial radius—was put as 1/297.[33] The tasks of modern geodesy grew out of this historical background.

Those seeking a single geodetic net for the world had to agree on a suitable reference frame. It is natural to take this as a rectangular coordinate system with origin at the earth's center and three mutually perpendicular axes, one along the earth's rotational axis and the other two in the equatorial plane. Alternatively one could use spherical polar coordinates locating a point by its distance from the origin of coordinates, and its latitude and longitude. In principle all measurements and calculations could be made in terms of these coordinates without any intermediate reference. To visualize the geometry, however, a reference surface approximating the actual surface of the earth is helpful. The most useful reference surface should satisfy two important criteria. First, it should be of such a size and shape that all locations on the earth are close to the reference. Secondly, the surface should be one on which calculations of positions, angles, and distances are mathematically simple. A sphere would satisfy the second criterion very nicely, since one could use the ordinary spherical trigonometry of air and marine navigation. But for any chosen sphere, many locations on earth would be unacceptably far from the reference. By flattening the sphere at the poles, however, to produce an oblate ellipsoid of revolution, both criteria can be met. For calculations the methods of analytic geometry can be used, and an ellipsoid of the equatorial radius and flattening given in the preceding paragraph—6 378 388 meters and 1/297 respectively—provides a good first order approximation to the actual size and shape of the earth. This ellipsoid of revolution, with center at the origin of coordinates, was often used as reference ellipsoid before the advent of satellites. As will be seen later, satellite geodesy provided an improved estimate of the size and flattening of this reference ellipsoid.

By furnishing the means of accurately positioning different sites and features on the earth, geometrical geodesy provides essential data for map makers, the fixing of political boundaries, civil engineering, and military targeting. But, the data also raise numerous scientific questions, such as why various features are where they are and what forces cause observed irregularities in the shape of the earth. Dynamical geodesy addresses itself to such questions.

Among the factors affecting the shape of the earth are the distribution of matter in the crust and mantle, centrifugal forces due to the earth's rotation, and gravity. The dominant factor is gravity, and an investigation of the earth's gravitational field lies at the heart of dynamical geodesy. To understand why, the concepts of *geoid* and *spherical harmonics* will be helpful.

First the geoid. To start, consider a simplified case. Suppose the earth to be perfectly homogeneous, plastic, and nonrotating. Then the earth would assume the shape of a perfect sphere (fig. 38). More significantly, level surfaces around the earth would also be perfect spheres. By *level sur-face* is meant a surface to which the force of gravity is perpendicular every-where. At any point on the surface the bubble of a spirit level held tangent to the surface would be centered. A pool of water on a level surface would experience no sidewise, or "downhill," gravity forces urging the water to flow (and were it not for internal pressures in the water and adhesion to the material of the surface, the pool would stay where it was).

The level surface that coincides with the actual surface of the earth is called the *geoid*. In the idealized case treated here, the geoid is a perfect sphere.

Now suppose a homogeneous, plastic earth rotates around a fixed axis. In this case the centrifugal forces of the rotation combine with gravity lessening the gravity and causing a bulge at the equator and producing a flattening at the poles (fig. 39). The earth's figure becomes that of an oblate ellipsoid of revolution, the surface of which is level. If the surface were not level, sideways forces on the plastic material would keep the material flow-ing until those forces were reduced to zero. Thus, for this case, the geoid is the ellipsoid of revolution comprising the earth's surface.

Next, return to the nonrotating earth, but this time suppose that near the surface is a large mass of material much denser than the rest of the earth (fig. 40). In this case, near the dense mass the level surfaces are no longer spherical. For, if one imagines holding a spirit level near the in-truding mass, its extra gravitational pull draws the fluid of the level toward the mass thus forcing the bubble away. To counter this effect the end of the level nearer the mass must be tipped up to recenter the bubble. In other words, the level surface tips upward as one approaches the mass,

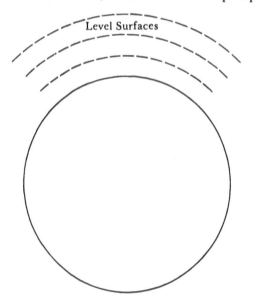

Figure 38. The geoid in the case of a homogeneous, plastic, nonrotating earth. For the idealized case depicted here, the geoid is a perfect sphere.

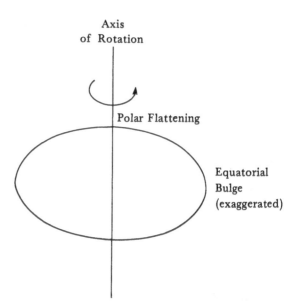

Axis
of Rotation

Polar Flattening

Equatorial
Bulge
(exaggerated)

*Figure 39. The geoid in the case
of a homogeneous, plastic, rotating earth. In this case, also an
idealized one, the geoid is a perfect oblate ellipsoid of revolution.*

thus forming a bulge. In a similar vein, near a mass deficiency, the level surface would show a depression.

For a rotating, nonhomogeneous earth, the same reasoning applies. Mass concentrations in the crust produce bulges in the geoid while mass deficiencies create depressions (fig. 41). Thus, the actual geoid, which for the real earth is defined as the level surface that over the oceans coincides with mean sea level, furnishes a good means of visualizing variations in the structure of the earth. Since mass and gravity go together, these struc-

*Figure 40. The geoid in the case of a nonhomogeneous, nonrotating earth. Varia-
tions in the earth's density generate bulges and depressions in the geoid.*

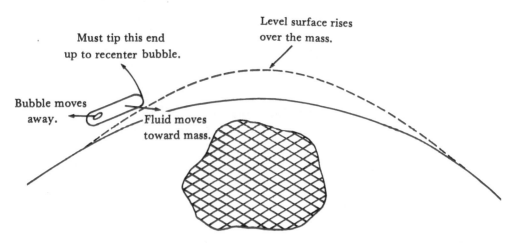

Level surface rises
over the mass.

Must tip this end
up to recenter bubble.

Bubble moves
away.

Fluid moves
toward mass.

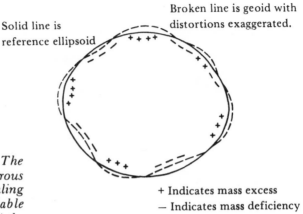

Solid line is reference ellipsoid

Broken line is geoid with distortions exaggerated.

+ Indicates mass excess
− Indicates mass deficiency

Figure 41. The actual geoid. The real earth's geoid has numerous bulges and depressions, revealing by their existence appreciable variations in the material of the planet.

tural features are revealed in their influence on the earth's gravitational field, intensifying the field near mass excesses and weakening it near mass deficiencies. These perturbations in the earth's field produce corresponding perturbations in the orbits of satellites which revolve under the influence of gravity. A precise analysis of these orbital perturbations can yield the features of the field. From these one finally gets back to the geoid and the earth's structure.

Mathematically the earth's gravitational field can be derived by calculus from what is called the geopotential function Ψ. Physically the surfaces over which Ψ is constant are the level surfaces discussed earlier in defining the geoid. Thus the makeup of the geoid and that of the earth's surface geopotential are identical.

The geopotential function can be expressed as the sum of an infinite number of terms (in general). Because these terms can be expressed in sines and cosines of latitude and longitude, they are referred to as spherical harmonics, by analogy with the harmonic analysis of a vibrating string where sines and cosines of the various multiples of the fundamental frequency of the string are used.* The amount of a specific harmonic in the expansion of the geopotential is given by a coefficient J_{nm}. The most general harmonics correspond to distortions of the geoid in both latitude and longitude. Some, called *sectorial harmonics*, reveal major distortions in longitude— for example, an ellipticity in the earth's equator. Of special importance are the zonal harmonics, which correspond to coefficients J_{nm} for which $m = 0$ and which depend only on latitude. The second zonal harmonic corresponds to the earth's equatorial bulge caused by the earth's rotation. The third zonal harmonic, if present in the expansion of the geopotential,

---

*More completely, like the sines and cosines, the basic functions out of which spherical harmonics are constructed form what mathematicians call an *orthogonal set.*

would add a pear-shaped component to the earth's figure, elevating the geoid at one pole and depressing it at the other.

It was in regard to the reference ellipsoid and the coefficients in the spherical harmonic expansion of the earth's potential that satellite measurements could aid the geodesist. The most straightforward contribution was to provide a sighting point in the sky that could be used to make direct connections between remotely separated points of the earth, supplementing the method of triangulation along the earth's surface. For this purpose simultaneous sightings of a satellite from two widely separated points were most useful. But once the orbit of a satellite had been accurately determined, simultaneous sightings were not required. One could relate separate sightings by computing the time and distance along the orbit from one sighting to the other, and again proceed to compute the distance between the two observing stations on the earth. By this latter method continental and transoceanic distances could be spanned, clearly a powerful aid in tying together different geodetic nets of the world.

The second major contribution that satellites could make was to help determine the different harmonic components of the earth's gravitational field, or alternately of the earth's gravitational potential. The orbit of a satellite is determined completely by its initial position and velocity and the forces operating on it. These forces include the gravitational influences of the earth, sun, and moon; atmospheric drag; solar radiation pressures; and self-generated disturbances such as those caused by gases escaping from the interior of the satellite. For a satellite near the earth yet well out of the appreciable atmosphere, the earth's gravity controls the orbit, the other effects amounting to corrections that have to be taken into account. As for the earth's field, Newton's inverse square law term constrains the satellite to an essentially elliptical orbit. But higher order terms also have their effects. The second order zonal harmonic or equatorial bulge causes the plane of the satellite's orbit and the line joining apogee and perigee to rotate in space. Still higher harmonics produce slight undulations in the satellite's orbit, which can be measured and analyzed to determine which harmonics, and how much of each, are producing the observed effects.

The application was simple in principle, but mathematically very complicated. Satellite orbits and their perturbations were directly related to the geoid, while the positions of the tracking stations and geodetic nets were tied to the reference ellipsoid, and a major objective was to improve the quantitative definitions of both geoid and ellipsoid. Because of the complexities, the modern computer was required to take advantage of the satellite opportunities. But with the computer the complexities and important results were quickly sorted out.

Some of the earliest came from the first Sputnik and Vanguard satellites. Using *Sputnik 2*, E. Buchar of Czechoslovakia was able to make an estimate of the earth's flattening. From the measured rate of precession of the satellite's orbit, which could be related mathematically to the flattening, Buchar obtained the value

$$f = 1/(297.90 \pm 0.18) \tag{12}$$

which is to be compared to the previously accepted value of 1/297. From a more extensive analysis of the *Vanguard 1* satellite, workers at the U.S. Army Map Service obtained

$$f = 1/(298.38 \pm 0.07) \tag{13}$$

while a U.S. Naval Research Laboratory group got virtually the same answer.[34]

While the measured flattening of the earth was smaller than that which had been in use, it was significantly greater than that which would exist in a plastic earth rotating at the present angular velocity of the earth, namely 1/300.[35] The implication was that the earth's mantle was not perfectly plastic. For a perfectly plastic earth, the flattening indicated by the satellite measurements would correspond to an earlier, faster angular velocity of earth. Instead, changes in the earth's equatorial bulge lag by a substantial period—tens of millions of years—behind the changes in the centrifugal forces producing the bulge.[36]

Vanguard 1 data also showed that the eccentricity of the satellite's orbit varied by 0.00042 ± 0.00003 with a period of 80 days. John O'Keefe and his colleagues concluded that this variation had to be caused mostly by the third harmonic in the earth's gravitational field. The distortion corresponding to this harmonic was very slight, amounting to only about 20 meters elevation of the geoid at the north pole and an equivalent depression at the south pole—widely described in the newspapers as the earth's "pear-shaped" component—but was significant in that it might imply a considerable strength in the earth's interior. O'Keefe and his colleagues estimated that a crustal load of 2×10^7 dynes/cm² (2×10^6 n/m²) was implied by their findings, producing stresses which they said "must be supported by a mechanical strength larger than that usually assumed for the interior of the earth or by large-scale convection currents in the mantle."[37]

It was possible by a detailed analysis of the orbital perturbations to derive a chart of the departures of the geoid above and below the reference ellipsoid, a chart which could suggest a great deal about the distribution of mass within the earth's crust. Using observations from five different satellites, William Kaula of Goddard Space Flight Center produced the chart shown in figure 42.[38] The positive numbers give elevations of the geoid in meters above the reference ellipsoid, which was taken to have the flattening of 1/298.24 indicated by satellite measurements. The negative numbers give depressions of the geoid below the reference ellipsoid. As an example of what one can read from such a chart, the elevations and depressions of the geoid shown in the equatorial belt strongly suggest that the earth's equator is not a circle, but an ellipse. This is consistent with an analysis by

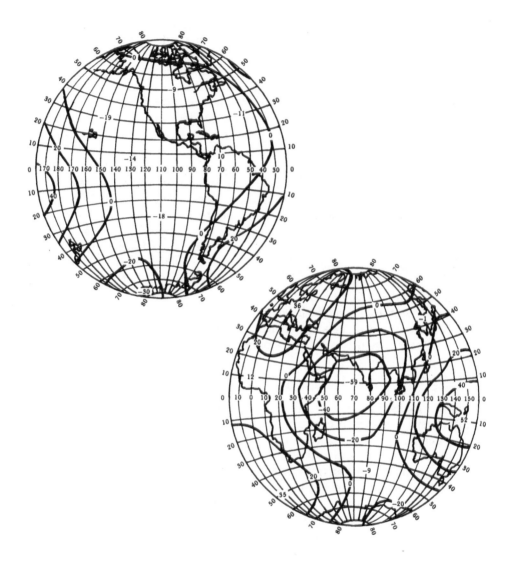

Figure 42. The geoid as revealed by satellite measurements. Geoid heights in meters referred to an ellipsoid of flattening 1/298.24, determined from observations of satellites 1959α₁, 1959η, 1960ι₂, 1961δ₁, and 1961αδ₁. W. M. Kaula in Satellite Geodesy, 1958–1964. *NASA SP-94 (1966), p. 33.*

C. A. Wagner of data from the communications satellite *Syncom 2* in synchronous orbit over the earth's equator.[39] Wagner found a difference of 130 ± 4 meters between the major and minor equatorial diameters, with one end of the major diameter at 19° ± 6° west of Greenwich.

While many workers were concerning themselves with physical, or dynamical, geodesy, others were working on the problems of mensuration and mapping. Using the large *Echo 1* satellite, French geodesists undertook to check the tie between the French geodetic net and that of North Africa.[40] With radio tracking techniques, the Applied Physics Laboratory of the Johns Hopkins University determined distances between stations at various points from the east coast to the west coast of the United States. From comparison of distances measured by ground surveys and those determined by satellite techniques, the APL workers concluded that, with a modest number of satellite passes and four or more observing stations, relative positions of ground stations separated by continental distances could be obtained with a confidence level of about 10 meters.[41] Using simultaneous photographic observations of the *Anna 1B* geodetic satellite taken from different stations, Air Force workers measured distances between stations separated by about 1000 km with good accuracy—better than 10 m. They concluded that their geodetic stellar camera system was "operationally capable of extending geodetic control to proportional accuracy of better than 1/100 000 when cameras in a network simultaneously observe a flashing satellite beacon."[42]

SIGNIFICANCE

The power of satellite geodesy was quickly evident, and significant results were forthcoming almost as soon as the satellites were available. Nevertheless, the first half-dozen years after Sputnik must be regarded more as a preparation period rather than as a full flowering of the new technique. One obtained a better estimate of the size and shape of the reference ellipsoid to use for a world geodetic system, and a good start on tying a number of existing networks together by using satellite triangulation. But most of the work of establishing a worldwide system had yet to be done.

Accuracies measured in meters were achieved in the early work, but these were far from the precision for which one could foresee a need and of which the satellite approach ought to be capable eventually. For some applications to oceanography, such as the determination of circulations and their causes, measurement of heights of the sea surface (which is not always at mean sea level) would be required to within centimeters, and correspondingly the altitudes of orbiting satellites to a comparable precision. To follow the motions of tectonic plates making up the earth's crust, and the drifting of continents—a few to perhaps 10 centimeters a year—comparable accuracy would be needed. The military services also had requirements for greater accuracy in geodetic measurements, not so much in the positioning of strategic locations around the world relative to each other as in the determination of higher order harmonics of the earth's gravitational field, which would affect the orbits of navigation and photographic satellites and the accuracy of long-range missiles. And the precise determination of the earth's field was of major scientific interest, being directly related to the structure and processes within the earth's crust and mantle, which in turn were probably connected with the causes of tectonic plate motions and continental drift. To obtain the great precision needed for such applications, many years of theoretical work and practical experience with the new geodetic techniques still lay ahead.

The power of satellite geodesy for gravitational field studies lay in the different avenue of approach it afforded from that of ground-based geodesy. For example, to locate the geoid from surface observations one might start with measurements of the deflection of the vertical from the normal to the reference ellipsoid. But the normal to the geoid surface (fig. 43), which is the direction of gravity at the point in question, usually differs from the direction of the perpendicular to the surface of the reference ellipsoid. The deviation is determined by astronomical measurements, using the stars as a fixed reference system in space. From the angle between the normals to the geoid and the reference ellipsoid at a chosen point, one can deduce for any direction along the ellipsoid's surface the slope of the geoid relative to the ellipsoid. By integrating this slope along various arcs the departure of the geoid above and below the ellipsoid of reference can be obtained. Tremendous quantities of data are needed to construct a chart of the geoid for the whole earth, or even for large regions. Furthermore, the technique cannot be applied over the oceans.

In the satellite orbit, nature has already done the integrating that the ground-based geodesist must do in depicting the geoid. For the motion of the satellite in its path is the result of the collection of forces operating on the spacecraft, the dominating one being that of the earth's gravity. Thus, the satellite geodesist has in the orbital data a worldwide picture of the geoid, his task being to sift out the different harmonics as an aid to visualizing the structure of the gravity field. The two approaches are in a sense from opposite directions. The ground-based geodesist starts from bits and pieces of the total picture and tries to reconstruct the total from those pieces. In contrast the satellite geodesist starts with an integrated effect and tries to discern from the total what the key building blocks are.

It had been extremely difficult to determine the higher harmonics in the earth's field from ground observations. But in a few years artificial satellites produced a respectable number of the more important harmonics. By the mid-1960s reasonably consistent values of the zonal harmonics J_2 ($= J_{2,0}$) through J_7 ($= J_{7,0}$) had been obtained by several different persons,

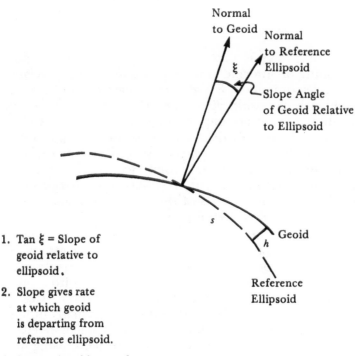

Normal
to Geoid

Normal
to Reference
Ellipsoid

ξ

Slope Angle
of Geoid Relative
to Ellipsoid

s

Geoid

h

Reference
Ellipsoid

1. Tan ξ = Slope of
geoid relative to
ellipsoid.

2. Slope gives rate
at which geoid
is departing from
reference ellipsoid.

3. Integrating this rate along arc s
gives actual departure h
of geoid from ellipsoid.

Figure 43. Determining the geoid from ground-based measurements. (1) Tan ξ is the slope of the geoid relative to the reference ellipsoid. (2) The slope gives the rate at which the geoid departs from the ellipsoid as a function of distance along the ellipsoid. (3) Integrating this rate along arc s gives the actual departure h of the geoid from the ellipsoid.

and a number of the more general coefficients—for example, J_{22}, J_{31}, J_{32}, J_{33}, J_{41}, J_{43}, J_{44}—had been computed.[43] In view of such successes it is natural to ask if satellite geodesy was destined to supplant ground-based geodesy. The answer to this question must be no. Just as sounding rockets continued to be essential for upper atmospheric research even after the satellite was available, so ground techniques, the power of which was greatly strengthened by the development of long-baseline interferometry using fixed sources in the sky like pulsars, remained essential to geodesy. In fact, over the years ahead geodesists would be discussing the necessity of combining ground-based and satellite methods to achieve required accuracies.

Nevertheless, satellite geodesy injected a remarkable stimulus into the field. By 1964 the world's astronomers were seriously considering incorporating the new results from satellite geodesy into the international system of astronomical constants.[44] As with the magnetosphere, one is led to ask if such actions, along with the impressive results flowing from geodetic research with satellites, did not indicate a scientific revolution in the field. Satellite geodesists could indeed think and speak of their work as "revolutionary."

As John O'Keefe observes, the satellite results had forced the abandonment of what had previously been accepted as the basic hypothesis of geodesy. In brief, the hypothesis held that there are no extensive areas on earth in which the difference between the earth's gravity and the gravitational field to be expected from a uniform ellipsoid is of a single algebraic sign. The basic hypothesis stemmed from the concept of *isostasy*, in which the earth was supposed to have an essentially plastic mantle in which stress differences would promptly be relieved. But satellite measurements revealed that there are substantial low-order harmonics in the earth's field, casting serious doubt on the idea of a plastic mantle, and imposing serious limits on the idea of isostasy. In O'Keefe's own words:

> The Vanguard results swept this whole philosophy into limbo. Stress differences clearly existed in the mantle. Isostasy was not true on a large scale, at least not to the extent demanded by the Basic Hypothesis of Geodesy.
> The meaning of this revolution in geodesy is not yet clear in terms of geophysics. We know that the hydrostatic assumption (isostasy) is not as good as we thought; but what takes its place? Hydrostatic equilibrium means that the earth behaves like a liquid at rest; that is not good enough; but is it because the earth is a liquid in motion (i.e., convecting) or is it because the earth can be thought of as a solid (i.e., having a non-zero ultimate strength)? . . . The ambiguity here has tended to obscure the significance of the revolution in geodesy. . . . The geodetic significance of the satellite revolution is clear; there *are* substantial low harmonics, and isostasy has very serious limits as a mode of describing the large-scale behavior of the earth. The Basic Hypothesis is false.[45]

Finally it must be noted that the expanded perspective afforded by satellite geodesy applied not only to the earth. The techniques that were proving so useful in the study of the earth would obviously be effective in investigating the moon and planets. Indeed, it was clear that a very long time would elapse before extensive ground-based geodesy could be practiced on another planet. For the foreseeable future satellites would have to be the principal tool of lunar and planetary geodesy. Fortunately circumstances had provided a considerable opportunity to practice with this new tool before lunar and planetary orbiters would be in operation. But there was little time to spare. By the end of 1964 the Lunar Orbiter was getting ready to go.

By the middle of the 1960s the sounding rocket and instrumented spacecraft had proved themselves to be revolutionary new tools for scientific research in the sense that they had made possible investigations and discoveries not possible at the ground. They had allowed men to look back on the planet from a new vantage point in space, enhancing the perspective in which one could view the earth. This perspective would continue to deepen as spacecraft and men reached other planets, making it possible to study the earth not as the only accessible planet but as one of many like objects. Broadly, these investigations might be regarded as normal science. But many scientists considered the impact of space science on both magnetospheric physics and geodesy as revolutionary, especially in the framework of the individual discipline. In any event, the degree to which space studies enriched those sciences was inspiration to the scientists and a satisfying return on the investments of the countries that supported the work.

Part IV

Love-Hate Relationships

Nec possum tecum vivere, nec sine te.

Martial

12

Who Decides?

Congress put the national space program in the hands of the civilian National Aeronautics and Space Administration without prescribing exactly what the program should be. Objectives of the space program as articulated in the NASA Act of 1958 left a tremendous leeway—in fact, a considerable responsibility—to NASA and the military to decide what should be done.[1] Not that NASA could decide such questions unilaterally, for as with all such complex programs there were many levels of decision, from within the agency to the Bureau of the Budget and the president, to the Congress, and back to the president. But assuming a reasonable amount of wisdom and attractiveness to what NASA proposed, the agency could expect approval of much of what it asked for, especially in the climate of competition with the Soviet Union that characterized the years immediately after launch of the first Sputnik.

While ultimate authority in such matters rests with Congress and the president, the initial stages in which a salable proposal is being developed may be all-important. The scientific community wanted to participate in the initial stages—not only in the conduct of experiments. Indeed, the firm determination to do so was a major force in the relations between NASA and scientists. While this desire made it easy to bring first-rate scientists into the planning of the program, it also generated tension when NASA undertook to make the decisions as to what to propose to the administration and Congress.

Many felt that the National Academy of Sciences as representative of the nation's scientists should call the shots in the national space science program. Immediately upon its creation in mid-1958, the Space Science Board solicited proposals and suggestions from the national scientific community for space science projects that should follow the accomplishment of the International Geophysical Year satellite program.[2] After assessing about 200 such proposals, the board sent recommendations to NASA shortly before NASA opened.[3] NASA managers were pleased to have these recommendations and incorporated them into program planning. But NASA was not willing to accept any implication that space science proposals should normally come to NASA through the Academy of Sciences, or that the Space Science Board should decide what experiments to conduct in the NASA program.

NASA's position was that operational responsibilities placed by law upon the agency could not be turned over to some other agency. Moreover, decisions concerning the space science program could not be made on purely scientific grounds. There were other factors to consider, such as funding, manpower, facilities, spacecraft, launch vehicles, and even the salability of projects in the existing climate at the White House and on Capitol Hill—factors that only NASA could properly assess.

So there followed a brief skirmish between the Academy and NASA as NASA insisted on deciding what space science it would include in its proposals to the administration and Congress.[4] Hugh Odishaw, executive director of the Space Science Board, pressed even further, urging that NASA rely entirely on the outside scientific community for its science program, and not create any NASA space science groups. The author—heading NASA's space science program—resisted and was supported by Silverstein and Dryden on the grounds that, aside from wanting to be involved in the scientific work, the agency had to have a scientific competence to work properly with the outside scientific community. Were NASA to limit itself only to engineering and technical staffs, day-to-day decisions in the preparation of satellites and space probes would have to be made without the insights into basic and sometimes subtle scientific needs that only working scientists could provide.

NASA created space science groups in a number of the centers, especially in the Goddard Space Flight Center and the Jet Propulsion Laboratory. A few years later John Simpson, physicist at the University of Chicago, confided to the author that he had been one of those who had opposed the idea of NASA's having in-house space science groups. He had, however, completely changed his mind after seeing how valuable it was for the outside scientists to have, as it were, full-time representation at the centers, and to have an understanding ear to turn to when problems arose. Simpson specifically cited the Interplanetary Monitoring Platform—an Explorer-class satellite conceived by Goddard Space Flight Center scientists for investigating cislunar space—as extremely valuable for space science, particularly for his own scientific interests. Yet he and his colleagues in the universities, working only part time on space research and without extensive engineering support, were unlikely to have created any such vehicles.

The first skirmish with the Academy and the outside scientists was not long lived, and NASA emerged firmly in control of space science as well as in other aspects of the space program. Nevertheless, NASA managers intended that the space science program be what the scientific community felt it should be. It was the firm conviction of NASA scientists that a high-quality science program could be attained only by supporting the research of top-notch scientists. The agency proceeded, therefore, to seek and to heed the best scientific advice it could get. In various ways NASA sought to bring scientists intimately into the planning as well as the conduct of the program. With NASA in the driver's seat, but the scientific community serving as navigator, so to speak, a tugging and hauling developed, with a mixture of tension and cooperation that is best described as a love-hate relationship.

SPACE SCIENCE BOARD

Certainly "love-hate" aptly describes relations between NASA and the Academy of Sciences. Given its role in bringing a space program into being, the Academy could claim the rights, if not of a full parent, at least of a godparent. After failing in its bid to prescribe the space science program, the Academy, through its Space Science Board, advised and served as watchdog for the scientific community.

On its part NASA strove to assimilate into its program the recommendations of the Space Science Board. That NASA and the Academy were setting the same course is plain from a comparison of the makeup of the space science program set forth in NASA's work papers of February 1959 and the book *Science in Space*, which the board sponsored and which set forth the areas of space research that board members considered promising.[5] But the scientists were impatient and more inclined to complain about deficiencies than to acknowledge what was acceptable in NASA's efforts. After the first year Berkner, as chairman of the Space Science Board, felt it necessary to direct his criticism to George Kistiakowsky, the president's science adviser (see pp. 124, 212). That criticism ranged over virtually all aspects of the space science program.[6] NASA people felt that Berkner had probably been moved by Hugh Odishaw, executive director of SSB, to complain to the president's science adviser instead of directly to the space agency. Odishaw had developed over the years a distrust of government and felt it incumbent upon himself to ensure that the Space Science Board properly discharged its watchdog function.

Berkner's missive elicited a response from Glennan, which the NASA administrator addressed to Kistiakowsky, agreeing in general with Berkner's objectives but taking exception to some of the allegations.[7] Nevertheless, space science managers were goaded into renewed efforts to shape the program to the satisfaction of the scientific community. The going was difficult, and criticism continued until in June 1960 the author felt impelled to put out a workpaper on the subject.[8] Specific criticism of NASA included officials' not visiting outside institutions enough, fear that the in-house publication policy of NACA would be followed rather than open publication in the scientific journals, inadequate involvement of the scientific community, too much emphasis on projects and not enough support to long-range university research, fear that NASA would release basic research data prematurely, desire that NASA provide engineering support to university scientists, a charge that NASA gave too much emphasis to vehicles (which in the jargon of the day included both launchers and spacecraft) and not enough to the experiments, concern that NASA scientists, much younger than their professional colleagues in the universities, were not sufficiently seeking and heeding the advice of the university scientists.

The author's paper outlined NASA's mode of working with the scientific community, a mode designed to foster broad participation by the scientific community. The intent was to work with the Space Science Panel of the President's Science Advisory Committee, then chaired by E. M. Purcell, and with the Space Science Board under Berkner. By this time the space scientists in NASA had become aware of the extensive use NACA had made of advisory committees, and by way of reassurance to outside scientists reference was made to this past practice. The paper referred to the creation of the Space Sciences Steering Committee in the Office of Space Flight Programs, with seven subcommittees containing outside consultants. It specifically referred to the list of suggested experiments sent to NASA by the Space Science Board in its first days; most of those proposals had been included in the space science program in one form or another. As to breadth of contact with the scientific community, through various channels—PSAC, the Space Science Board, and NASA's own committees—the agency had contact with about 200 scientists in a wide range of disciplines. Moreover, the author's paper stated it was NASA policy that no more than about 20% of the experiments in spaceflight missions be provided by NASA scientists, the remaining 80% to be provided by outside experimenters. It was felt that the recurring complaint that outside scientists did not know what NASA was planning stemmed less from an actual lack of communication than from disagreements over some of NASA's decisions.

If it did nothing else, the paper showed that NASA's space science managers were aware of the criticisms and were working to overcome them. In retrospect it can be seen that NASA's people did move steadily in the direction of making the space science program a creature of the scientific community. But it was a rocky road to travel and for a long time criticism outweighed approbation. Then success brought its own problems. As the program began to produce exciting scientific results, interest in the program grew, generating a new difficulty—the problem of the "ins" versus the "outs."

In the fall of 1961 when the Office of Space Sciences was formed at NASA Headquarters, the chairman of PSAC's Space Science Panel, Donald F. Hornig, wrote to Hugh Dryden expressing pleasure at the new organization, but at the same time referring to a "crisis of confidence between NASA and members of the scientific community who participate in the NASA program."[9] The author responded to Hornig's criticism pointing out that growing interest in the space science program had outrun NASA's ability to accommodate within the budget and the flight program all the good experiments that were being proposed and expressing the hope that the problem of limited flight space would soon be relieved with the appearance of observatory-class satellites.[10]

This difficulty was exacerbated by the fact that an experimenter in the NASA program usually had in mind an investigation, not just a single experiment. No sooner were the returns from one experiment in, than the experimenter was back with a follow-up proposal that was necessary to make the most of the experiment he had just completed. It made good scientific sense for the scientists on the advisory subcommittees to support such requests. In addition, there was a natural tendency for NASA to reappoint to these subcommittees those who had worked hard and had acquired a ready familiarity with the problems of planning and funding space science experiments. Thus, to those not yet in the program, the setup looked very much like a closed corporation.

It was in this climate that NASA asked the Space Science Board to conduct the first of what became a continuing series of summer studies of the NASA space program (app. G). The first study, at the State University of Iowa 17 June to 31 July 1962, essayed a comprehensive review of the entire NASA space science program, including some side glances at what the Department of Defense was doing or might do in space science.[11] The opportunity for the scientists to lay their various concerns not only before NASA officials but also before their scientific peers served to clear the air. When the smoke of battle settled, it appeared that the scientists approved of much of what NASA had been doing, but urged more attention to problems of a kind that continued to be a worry throughout the years. A few examples will illustrate.

NASA leadership, Abe Silverstein especially, had favored the development of large, observatory-class spacecraft. As Silverstein pointed out, the large-capacity spacecraft would permit a comparative study of many different quantities by measuring them simultaneously to seek relationships among them. Also, Silverstein thought that the larger spacecraft would probably give more science per dollar than the smaller ones (years later he expressed some doubt about this latter point).[12] But the scientists preferred small spacecraft (p. 149). Early in the summer study Herbert Friedman of the Naval Research Laboratory brought up this issue, stating that NASA's Orbiting Solar Observatory was more complicated than necessary for a number of scientific needs, such as the continuous monitoring of the sun. Also, more effort should go into providing cheaper, capable sounding rockets, which would be of great use in university research. Subsequently the Naval Research Laboratory developed and used to good advantage the Solrad, a smaller, simpler satellite than the Orbiting Solar Observatory. Also, with NASA support Van Allen's group at the State University of

Iowa built and used a small Explorer-class satellite, which they called Injun, for studies of the radiation belt and the aurora.

The astronomers supported Friedman in the bid for small satellites. But they also urged use of Orbiting Solar Observatories for many years to come and, looking beyond OSO, pointed to the future need for a more advanced observatory capable of obtaining resolutions of one arc second. The astronomers provided an interesting insight into the complex psychology that entered into relations between NASA and the scientists. While endorsing NASA's astronomy program, they nevertheless were uneasy about their own roles in the program. As Martin Schwarzschild, professor of astronomy at Princeton, confided to the author and some of his colleagues, the astronomers found it distasteful that NASA, not they, should be making the decisions. He added that the astronomers found it doubly infuriating—and infuriating was the word he used—that NASA managers appeared to be making the right decisions.

In his instructions to the summer study working groups, Berkner told the participants to concentrate on maximizing the science in the space program. He pointed out that the question of whether there should be a space program, or a space science program, was not an issue for them to debate— those questions had already been decided by the country. Yet the participants found it impossible to stay away from such matters, particularly when it came to manned spaceflight. Many expressed disapproval of the manned program, along with the wish that the monies going to Apollo might be diverted to space science. Some expressed concern that not only was Apollo going to proceed but that NASA would even seek to justify the program on the basis of science, and this the scientists strongly objected to.

In a lengthy and lively exchange, the author and his colleagues sought to direct the discussion into the channels indicated by Berkner. Study members were urged to recognize that the Apollo program would be carried out, that it concerned important national objectives other than science, a major one of which was the development of a strong national capability to operate with men in space. Since Apollo was going to be done, it behooved the scientists to take advantage of the opportunity before them and to help ensure that the science done in Apollo was the best possible. The Space Science Board had, after a lengthy discussion at its 10–11 February 1961 meeting, adopted a formal position supporting man in space, which position was communicated to the government on 31 March. Following President Kennedy's announcement of the Apollo program, the National Academy of Sciences had issued a release for 7 August 1961 in which it was stated that the Space Science Board had "recommended that scientific exploration of the Moon and planets should be clearly stated as the ultimate objective of the U.S. space program for the foreseeable future. . . . From a scientific standpoint there seems little room for dissent that man's participation in the exploration of the Moon and planets will be essential. . . ."

In keeping with this position, at the closing plenary session of the summer study, 31 July 1962, Berkner stated that man in space was a good thing and that exploration was science.[13]

But the debate went on many years thereafter, furnishing one of many examples that the scientific community is not of one mind, and that the Space Science Board did not necessarily speak for the community in some of its recommendations. Among others, Philip Abelson, distinguished chemist who during World War II had devised one of the methods for separating uranium isotopes, continued the battle against the Apollo program. Abelson urged that much more of value could be achieved by devoting to unmanned space science only a small fraction of the monies going into Apollo. As former editor of the *Journal of Geophysical Research* and editor of *Science,* Abelson had a ready outlet for his views. At one point he polled some 200 scientists, asserting that the results gave overwhelming support for his position.[14] The *Christian Science Monitor* in April 1965 devoted a page to the space program, in which Abelson attacked the manned program as not worth the cost and effort, while the author argued for a balanced program of both unmanned and manned missions.[15] The issue was, of course, not settled by argument, but by the final successful accomplishment of the Apollo missions.

Although the debate over Apollo was not ended at the summer study, some recommendations were made. Perhaps the most significant was that scientist-astronauts should be included in the program. The group also recommended that a scientist-astronaut be included on the first landing mission to the moon and that NASA create an institute for the training of scientist-astronauts to be administered by a university, or if not by a university by that part of NASA responsible for the space science program. The latter recommendations did not have the slightest chance of being accepted by NASA, but in time the agency did select scientist-astronauts.

In October 1964 a NASA press release announced the recruitment of scientist-astronauts for future manned flights. The more than 1000 applications received by NASA settled emphatically the question of whether any scientists were seriously interested in the manned spaceflight program. A preliminary screening reduced these to about 400 applications, which NASA then sent to the National Academy of Sciences. From these a special Academy committee chose, on the basis of scientific potential, 16 nominees to recommend to NASA. Of these, NASA selected 6. In the fall of 1966, NASA and the Academy of Sciences announced that more scientist-astronauts would be chosen. Following a process similar to that of the first selection, NASA chose 11 scientist-astronaut candidates from almost 1000 applicants.[16]

The new astronaut trainees started out with great optimism and hopes for the future of manned science in space. But they soon ran into difficulties that put another strain on NASA's ties to the scientific community: the

Johnson Space Center was not particularly enthusiastic about having scientist-astronauts in the program. The center certainly had not wanted the second batch, which overstaffed the center in scientist-astronauts, considering the probable number of manned space science missions. As the Apollo lunar landings approached and as plans were being developed for the Skylab space station missions, scientists increased their pressure on NASA to include scientist-astronauts on the missions. The Johnson Center resisted. Considering the newness and danger of the missions the center, out of a conviction that only astronauts with extensive test pilot training and experience could safely fly the spacecraft, was unwilling to consider the scientist-astronauts for any of the early missions. Even after the first successful landings on the moon, the scientists continued to have difficulty securing berths on flights. Discouraged and in protest, some resigned from the program. In a series of frank discussions with the author, these men described their frustrations, expressing the hope that something could be done to improve their lot in the program.[17] With continuing pressure from the Academy and with strong support from Deputy Administrator George Low, a few scientist-astronauts at long last did fly, geologist Harrison Schmitt on *Apollo 17* and one scientist-astronaut on each of the manned Skylab flights. Their experience in the Apollo and Skylab programs, however, emphasized the need for NASA managers to give careful thought to how manned space science would be accomplished in the 1980s with the Space Shuttle.[18]

The 1962 summer study surfaced a number of problems that recurred in one form or another over the years. One of these concerned space biology and medicine. Although there were recommendations for a life sciences program, interest was spotty, with considerable disbelief that much of real value for biology could be expected. Nevertheless, somewhat inconsistently, the life scientists made two recommendations that they continued to press for the next decade. One was that life sciences be elevated to a high level in the NASA organization. Scientists suggested that NASA might invite a respected person from the life sciences community to spend a quarter or a half year reviewing the setup within NASA and make recommendations. The hope was that this might lead to NASA's creating a life sciences directorship reporting to the administrator.

The second recommendation had to do with the selection of research for NASA to support. Accustomed to the peer review panels of the study sections of the National Institutes of Health, the life scientists recommended that NASA adopt such a procedure. The issue of how to work with the life sciences community and where to locate the program within the NASA organization burned for years. These topics are pursued in chapter 16.

Of prime concern to many of the summer study participants was NASA's relationship to the universities. James Van Allen, chairman of the summer study, had assembled an Ad Hoc Committee on NASA–University

Relationships, a draft report of which was presented during the Iowa City study.[19] "The Committee was unanimous in its favorable general impression of the NASA program. . . . It was . . . impressed by NASA's intention to perform its mission in such a manner as to strengthen existing universities. . . ." At the summer study the discussion ranged widely without always yielding specific answers to problems. NASA's Space Sciences Steering Committee and its subcommittees came in for a great deal of comment. Van Allen felt that the process of reviewing experiment proposals in the subcommittees, which required the experimenter to be more specific well in advance of performing his experiment than perhaps he could be, tended to erode the independent way in which the scientist worked. Others felt that the system had developed a group of ins and outs, although Van Allen didn't think so. In this connection the question arose again as to whether NASA centers should be participants in the actual science or only be service centers to the rest of the scientific community. In-house versus outside review and evaluation of proposals kept coming up, with the life scientists pushing for outside peer review groups. There resulted a rather confused recommendation to NASA to consider modifying its method of proposal review and experiment selection. Many people did not favor NASA postdoctoral fellowships, but both Fred Seitz, president of the Academy, and Berkner strongly supported them. Industry wanted more support for its space scientists, but the university scientists thought that this was a bad idea, since the higher industry salaries would draw researchers away from teaching posts.

All in all, on the university question (which is considered further in the next chapter) NASA came out in the best possible position. With a general agreement as to the soundness of NASA's approach and a diversity of views on many of the specifics, NASA could find ample support for a variety of courses the agency might wish to follow.

Once initiated to the ways of summer studies, NASA space science managers found them a useful device for examining many kinds of problems. Through the years NASA sponsored a considerable number of studies, some of them narrowly directed, others of broad scope. For many years the studies were concerned primarily with the content of the NASA program— what fields to support, which problems to attack, and sometimes which experimenters to support. The recommendations to NASA amounted to a list of good things to do, but when not all of them could be funded it was NASA's task to make the choices—as NASA had insisted in the first place.

But NASA people began to feel that it would be helpful if scientists would furnish additional advice as to priorities to observe in choosing among different researches when all were intrinsically desirable. In the summer study conducted by the Space Science Board at Woods Hole, Massachusetts, during July 1965, NASA spokesmen urged the participants to face up to the question of priorities, with little real success.[20] While scientists were willing to establish some order of preference within a single discipline, they shied away from doing anything of the sort for a mixture of disciplines.

Not until the summer study of 1970, also at Wood's Hole, which was devoted specifically to the question of priorities, did a genuine effort emerge on the part of the scientists to face up to the frustrations of making almost impossible choices.[21] The study group did an excellent job, but not without generating serious strains within the community. By choosing to ease off on magnetospheric and fields and particles research in favor of planetary research, it alienated the affections of the fields and particles workers. By emphasizing high-energy astronomy in preference to classical optical astronomy and solar physics, it created more dissidents. In the planetary field itself, which the group strongly supported, participants came close to reversing the support of earlier years given to the Viking project, because its costs were proving to be much greater than expected and were threatening other projects considered more desirable. NASA participants strove mightily during these discussions to bring home the disastrous consequences of withdrawing an endorsement of a project already well under way—largely because of their earlier endorsement—and on which a great deal of money had already been spent. NASA's concern was heightened by the fact that Congressman Karth himself was questioning Viking and showing signs of being willing to recommend canceling it. In the end the study participants agreed with NASA managers on this issue, but there can be little doubt that free of such concerns they would have scrapped Viking in favor of smaller missions such as Pioneers to Venus.

The association between NASA and the Space Science Board endured. Yet at times relations were precarious. The complacent assumption of the superiority of academic science, the presumption of a natural right to be supported in their researches, the instant readiness to criticize, and the disdain which many if not most of the scientists accorded the government manager, particularly the scientist manager, were hard to stomach at times. When Lloyd Berkner undertook in person to lay before NASA's first administrator some of the criticisms and demands of the Space Science Board, Glennan could not restrain an outburst of indignation at the arrogant presumptuousness of the scientists. His vexation was shared by Silverstein, who from time to time cautioned NASA's space scientists to guard against losing control of their destiny, a danger that Silverstein felt was being fostered by drawing outside scientists too intimately into the planning process.

Especially frustrating was the apparent unwillingness, or perhaps inability, of outside scientists to appreciate the problems with which NASA scientists had to wrestle. The complex array of emotions was best illustrated in Harold Urey, Nobel Laureate, enthusiastic supporter of the space program and severe critic of NASA. Periodically Urey would burst forth—in the Space Science Board, on the scientific platform, and in the press—with a sweeping polemic against the agency's handling of space science. Urey's most persistent complaint concerned NASA staffing. In May 1963 he wrote to the author to discuss remarks he had been making in the press about incompetence of NASA staffing in science, in particular lunar and planetary science.[22] Urey urged the author to drive out the second-raters from NASA and replace them with older, more experienced men who could give proper advice. He stated that he had talked about this matter with people from Washington, Pittsburgh, Chicago, Pasadena, and Los Angeles and regularly got the view that NASA people are second-rate by and large. Two years later, after taking violent exception to a paper presented at a space science symposium held by the Committee on Space Research at Mar del Plata, Argentina, Urey protested to the National Science Foundation and NASA. Since the objectionable paper had been given by a university scientist whose researches were supported by NASA, Urey wrote that "a serious consideration of personnel connected with the entire NASA program is in order."[23]

With regard to the outside scientists, whose research proposals had been reviewed and endorsed to NASA by experts in the field, Urey did not always seem willing to let the scientific process weed out those who were on the wrong track. As to NASA staffing, NASA people saw in the complaints of Urey and others a lack of understanding of what was involved in managing the space science program. Undeniably most of the managers in NASA Headquarters were not the top-notch scientists whom the critics said they would like to see there. But repeated efforts throughout the years to lure working scientists into NASA management only occasionally bore fruit. In spite of the enticement of top positions in the program, none of the senior "establishment" came. The administrative burden at headquarters was fearful, and the climate such as to devour whatever scientific and research competence an expert might bring with him, affording little opportunity for replenishment. Those experts most needed to help direct the evolving space science program were reluctant, especially in an era when university salaries were rapidly catching up with those of industry and government, to exchange the advantages of academia—their students and the independence to follow personal research interests—for a never-ending round of headaches plus an ambience that was bound in time to destroy the very competence for which they were sought out in the first place. To continue a scientific career in NASA one had to work in the centers.

Those scientists who did come to headquarters became resigned to a vicarious enjoyment of the research achievements of the program. Their personal satisfaction came from having contributed in an absolutely essential way to the program, and thus to the advancement of science. That, and the excitement of being at the center of action in one of the greatest of human dramas, was their reward.

The incessant criticism and insatiable appetites of the scientists put a severe strain on the tie between NASA and the Space Science Board. At times during the first years it seemed to the author as though, at the top management levels, only Hugh Dryden, Arnold Frutkin (head of the International Programs Office), and the author favored keeping the association. The rest of NASA seemed willing to cut the Space Science Board adrift, and to rely on NASA-sponsored committees for outside advice.

But the tie became stronger as time went by, particularly when the second chairman of the Space Science Board, Harry Hess, took over from Berkner. Hess, professor of geology at Princeton and originator of the revolutionary new concept of sea-floor spreading, brought with him from years in the Navy and working with the government a better appreciation of what agencies like NASA needed in the way of support from its advisers. Hess fostered a policy of not just tossing lists of recommendations at NASA and then leaving the agency to its own devices, but rather of assisting to realize the desired objectives. When Hess took over, the Executive Committee of SSB began to meet monthly with NASA representatives to provide more continuing assistance to the space science program. When Hess died in 1969, his successor, Nobel Laureate Charles Townes, continued the policy of working personally with NASA to accomplish SSB recommendations.

But in the early 1970s the Academy of Sciences began to show great concern over questions of conflict of interest and potential charges of being captive to those it advised. Thus, when a new chairman was needed for the Space Science Board, instead of consulting with NASA on possible choices as had been the custom, the Academy unilaterally—as it had every right to do—selected a candidate. James Fletcher, the fourth NASA administrator, had doubts about the choice—doubts that were shared by the author—since the proposed chairman had previously shown little evidence of giving thought to the negative effect that his outspoken criticism of various space science projects could have on NASA's efforts to defend its budget on the Hill. NASA objected to the choice; the Academy stood firm; and Fletcher gave serious thought to withdrawing NASA's financial support from the board and relying on NASA's own committees for advice. In the end NASA fortunately did not sever the relationship with the board, and the new chairman did an excellent job. Perhaps NASA's expressed concerns stimulated the Academy to special efforts to prove that NASA was wrong.

NASA's Advisory Committees

Next to being personally involved in space research, the best way of keeping close to the space science program was to serve on one of the NASA advisory committees. In fact, a prime motivation in the creation of in-house advisory groups, in addition to securing the advice of knowledgeable scientists, was to cement relations with the outside scientific community. After muddling along for a year with the several working groups established in early 1959, NASA put together the more systematized Space Sciences Steering Committee and subcommittees.[24]

In doing this the intention was not to undercut the role of the Space Science Board, but NASA managers felt the need for more frequent and intimate advice than could be expected from the board. Moreover, some operational tasks, such as assisting in the selection of experiments and experimenters, were not appropriate for a non-NASA group. Still, board members felt at first that NASA was weakening the ties to the Space Science Board and for a while questioned the need for the NASA subcommittees. To counter the disquiet, NASA management invited the board to name liaison representatives to attend and participate in the discussions of the subcommittees. Similarly by invitation from the Academy, NASA observers attended meetings of the board's committees, while Hugh Dryden and the author had a standing invitation to be with the board at its sessions.

Once under way the subcommittees began to develop a systematic approach to advising NASA on its planning, and particularly on the choice of experiments and experimenters for flight missions. For the flights, formal criteria were established and over the next few years refined from experience.[25] Through appropriate announcements, which later in the decade became quite formalized, NASA informed the scientific community of the existence of flight projects for which experiments were needed.[26] When proposals for experiments to go on these flights came in, they were reviewed by the appropriate subcommittees.

The NASA subcommittees sorted the proposals into four different categories. At the top went the proposals of outstanding merit, well conceived, addressing a critical problem of space science, and likely to yield significant new information. Proposals which were good, but not outstanding, were assigned to category 2. Category 4 experiments were those that the group advised NASA to reject as either unsuitable for spaceflight or incompetent. The third category was special, reserved for proposals that the subcommittees judged to be potentially of category 1 quality, but which needed a great deal of work before the experiments could be assigned space on a flight.

In rating the proposals the subcommittees were asked to consider a number of points:

- The originality and validity of the experiment;

- The importance of the problem addressed by the experiment;

- The suitability of the experiment for a space mission, with an eye to eliminating experiments that could better be done by other means;

- The competence of the experimenter and his group;

- The ability and willingness of the proposer's own organization—university, research institute, government laboratory, or industrial establishment—to provide the experimenter with the support he would need above what would be furnished by NASA.

Once the subcommittees had completed their ratings, the proposals were further reviewed by the NASA divisions and center project people to consider whether the spacecraft to be used could house the experiments and provide the necessary power, telemetering, orientation, or other special requirements. The ability of an experiment to fit into the spacecraft along with other experiments without undue interference had to be determined. After this engineering review, the division responsible sent its recommendations to the Space Sciences Steering Committee—later the Space Science and Applications Steering Committee—where the recommendations from the subcommittees and those of the division were compared. Then the steering committee sent its recommendations to the Associate Administrator for Space Science and Applications for final approval.

NASA customarily flew only category 1 experiments, a policy intended to maintain high quality in the space science program. Moreover, flying only experiments that respected members of the science community had judged to be outstanding blunted possible criticism. With an eye to the future, NASA often funded the research and development needed to raise a category 3 experiment to category 1.

NASA received a great deal of help from the discipline subcommittees, and the outside scientists seemed to appreciate the importance of what they were doing for space science. But, since much of the time of the meetings was taken up in evaluating proposals, there was a lot of drudgery, and little time was left for stepping back and viewing the whole program in perspective. Under the routine the consultants became restive and questioned how much they were influencing the overall planning of the space science program. In contrast, many not on the subcommittees felt that these groups were having too much influence. Since most subcommittee members were also participants in some of the projects on which the subcommittees made recommendations, it was felt that there was too much occasion for conflict of interest. NASA procedures for guarding against such conflicts of interest, such as asking a consultant to leave a meeting at which his proposals or those of a colleague were discussed, did not put the concerns to rest.

The growing dissatisfaction of NASA's advisers with their role, continuing concern over conflicts of interest, the increasing pace and scope of the space science program under Webb's administration, and the expanding involvement of the universities with NASA, led to the conviction that some changes were in order. Once more it appeared wise to secure outside advice, and in early January 1966 Administrator Webb wrote to Norman Ramsey, professor of physics at Harvard University, asking if he would chair an ad hoc advisory committee for NASA.[27] Among the questions on which NASA would appreciate having advice, Webb listed: how to organize major projects so that scientists and engineers could participate effectively; how to make it possible for academics to take part without damage to their academic careers (e.g., how an academic scientist could devote six to eight years helping to create an advanced biological laboratory or a large astronomical facility in space and still continue his academic career); what mechanisms to use for picking scientific investigations for the space science program; whether the orientation of some NASA centers should be changed; and how to improve the scientific staffing of the program.

Ramsey accepted, and the committee was formally established in February. Its task was different from that of former advisory groups, which had dealt primarily, though not entirely, with the content of the NASA program. This new committee was asked to advise not on *what* science to do, but on *how* to conduct the program. After numerous sessions both in Washington and elsewhere, in which the author and some of his colleagues had the benefit of hearing thorough discussions of Webb's questions and more, the committee submitted recommendations concerning advisory committees, NASA-sponsored research institutes, and relations with the universities and the scientific community.[28] Some of the recommendations NASA accepted, some not. Nevertheless, the agency felt that the value derived had been such that the committee, even though initially ad hoc, should be continued. Roger Heyns, chancellor of the University of California at Berkeley, succeeded Ramsey as chairman.

The most far-reaching of the recommendations that NASA did not accept was the creation of a general advisory committee—not purely scientific—for the administrator. Years before, the first administrator, T. Keith Glennan, had "strongly desired a broadly based General Advisory Committee—a consultative group analogous to a corporate Board of Directors in place of the Space Council chaired in those days by President Eisenhower, who did not want the Space Council to be active."[29] The ad hoc committee was convinced that NASA should have a general advisory committee and that such a committee would go a long way toward cementing relations between NASA and the outside community. But Webb was even more convinced that NASA should not set up a general advisory committee, which he averred would compromise the administrator's freedom of action. Such compromise could only be detrimental to the management of a hard-hitting, fast-paced program like NASA's. On the other hand, the continuation of the Ramsey committee under Heyns was a partial accommodation to the committee's views.

Among the recommendations that NASA did accept were two of considerable importance: modification of the agency's advisory structure and creation of a lunar science institute in Houston, adjacent to the Johnson Space Center. It is the former that is of concern here.

Even as the Ramsey committee deliberations were in progress, NASA was taking steps to create two broadly interdisciplinary advisory groups: the Lunar and Planetary Missions Board and the Astronomy Missions Board.[30] The strongest motivation in setting up the boards was to provide consultants with a forum in which they could view the NASA program in the perspective they had missed in the discipline subcommittees. Advice from such interdisciplinary boards was expected to help produce a more coherent, better integrated space science program. Because of the scope of each board's purview, panels or committees of specialists were expected to be set up under the boards. The Astronomy Missions Board, for example, would be considering a program including solar physics, optical astronomy, radio astronomy, x-ray astronomy, gamma-ray astronomy, and cosmology, for each of which a specialist group might be needed. Disciplinary groups would continue to review and recommend on specific experiment proposals, but by arranging suitable overlapping memberships with the boards and their committees, discipline committee members would be afforded an opportunity to take part in the broader programmatic discussions.

Characteristically, advisory groups want to report to the highest possible levels, in NASA preferably to the administrator himself. But the administrator was not in a position to assimilate all the recommendations that might be given to him by the highly technical groups or to appreciate the significance for the space science program of the more specialized recommendations. In contrast, the program offices, where the programs were formulated in the first place, could be expected to understand the nuances as well as the major thrusts of board recommendations. To make the boards as effective as possible, NASA managers conceived a double-pronged connection to the agency's management. The boards reported formally to the associate administrator, but worked with the space science program office, which also furnished the administrative and secretarial support for them. When desired, the boards could be heard at the administrator's level. But working with the program people they were continually feeding their ideas and recommendations into the agency at the working level, where those ideas could have greatest impact. Board discussions were lively, interesting, and productive, and for several years their reports fed into the NASA planning process a great deal of valuable advice, which for the most part was assimilated into ongoing program plans.[31]

Toward the end of the 1960s, however, the cycle of discontent repeated. The immediate cause was a mistake by the author and some of his colleagues in material for a report to President Nixon recommending directions for the future of the space program. After taking office, Nixon had established in February 1969 a Space Task Group, consisting of the vice president as chairman, the secretary of defense, the acting administrator of NASA, and the science adviser to the president, to provide him with a "definitive recommendation on the direction which the U.S. space program should take in the post-Apollo period."[32] The Department of Defense and NASA provided extensive material to go into the report, which came out in September.[33] As the deadline approached for the completion of the report, often only hours were available for making hasty revisions requested by the report staff. In the course of one of these quick changes, part of the planetary program was modified. NASA people supposed that the change was in keeping with the desires of the Lunar and Planetary Missions Board— but it wasn't.

The board reacted strongly, and for a while there was talk of the members' resigning en masse. Actually the NASA error was not in itself enough cause for such a strong reaction on the part of the advisers. The problem had been growing for some time. In a period when the entire NASA program was under scrutiny, the board no longer felt that it had the necessary perspective to make proper recommendations. In fact, the chairman, John Findlay, confided to the author that if the Lunar and Planetary Missions Board had known of all the program possibilities that were being considered in the Space Task Group planning—in other areas as well as for the moon and planets—some of the board's recommendations would have been quite different.

As a first order of business, the Lunar and Planetary Missions Board was pursuaded not to resign, and NASA managers committed themselves to working out some better arrangements for the advisory structure. After much discussion within the agency and with consultants, NASA decided to create a Space Program Advisory Council.[34] The council was asked to advise on the entire space program—science, technology, and engineering, manned and unmanned. Under the council were four interdisciplinary committees: physical sciences, life sciences, applications, and space systems. The chairmen of these interdisciplinary committees made up about half the membership of the council, the rest consisting of the council chairman and members at large. It was the author's intention that the committees would themselves have specialist panels working with and reporting to them, so that the committees and their panels would be analogous to the previous missions boards and their committees. The new element was the council, which was supposed to provide the across-the-board perspective that the missions boards had lacked. Once more, to make the advisory structure as effective as possible, a two-pronged connection with the agency was established. The council reported to the deputy administrator of NASA, but was expected to work with the program offices, the Office of Space Science and Applications providing administrative and secretarial support. The committees reported to the associate administrator and were expected to work directly with the appropriate offices—the Physical Sciences Committee with space science divisions, the Applications Committee with the applications groups, Life Sciences with space biology and space medicine people, and the Space Systems Committee largely with the Office of Manned Space Flight.

Although the council had grown out of discontent with the previous advisory structure, and although it had been designed especially to provide consultants with a deeper insight into NASA programs and planning, it was not as effective as the missions boards had been. The arrangement was unwieldly and required a tremendous amount of attention from NASA personnel just to provide the necessary secretarial and administrative services. But most important, the council and its committees lost touch for a while with the divisions in the program offices. Program managers and program scientists did not understand the arrangement. Sitting at the top of an imposing hierarchy, the council appeared too much as an arm of the Administrator's Office, remote and not easily accessible to program planners. The same was true of the interdisciplinary committees, though to a smaller degree. As a consequence, when program divisions needed specialized advice, they created their own working groups—like the highly successful Planetary Sciences Planning Committee set up by the lunar and planetary people in the Office of Space Science and Applications. The existence of these proprietary working groups further separated the top-level advisory groups from the lower ones. Thus, while the council might have a grand perspective, it was in danger of losing touch with the realities of detailed program planning. A great deal of management time was required to keep these centrifugal forces under control.

The effectiveness of the Space Program Advisory Council and its committees was improved with the passage of time. But the unwieldiness was intrinsic and constantly invited reconsideration of the advisory structure.

SPACE SCIENCE PANEL

After the first few years in which the Space Science Panel of the President's Science Advisory Committee saw NASA get under way, advised on such matters as orbiting astronomical observatories, sounding rockets, and universities, and supported NASA in the battle over classification of geodetic satellite data, interest among panel members appeared to wane. For some years attention to what NASA was doing was somewhat desultory. In the latter half of the 1960s, however, when NASA's bumbling over what to do in post-Apollo years helped to precipitate a crisis of confidence over the agency's planning, interest reawakened.

After studying the matter of manned space stations at some length, the panel issued a report endorsing the Apollo Workshop, or Skylab, program as a one-time-only experiment in determining what man could do in a space station, but withholding support from any continuing space station work until Skylab results were available and more mature consideration could be given to the matter. When the author testified on the Hill shortly after the report came out, he found that Congressman Karth had been giving a great deal of attention to what the panel had to say; and his copy of the report, amply marked up either by the staff or Karth himself, was a source for a great many questions on the space science program.[35]

The panel's negative attitude toward permanent space stations persisted throughout the work of preparing a draft report for the Space Task Group. In this the panel was opposed to Administrator Paine, who wanted to proceed at once with a permanent space station, as the next natural step in the space program. It was the panel's view that before proceeding with a space station, a more economical and versatile means of transportation to and from the station should be developed. In this respect—although with some ambivalence and quite tentatively—it supported a Space Shuttle project as the next major manned spaceflight effort for the nation. How much influence the panel had in securing ultimate support for the Shuttle program is moot. But at any rate in this last great issue to come before the panel before its demise in January 1973 at the hands of President Nixon, the members were pointing in the direction NASA came to regard as the right one for the country.

THE SCIENTISTS DECIDE

By virtue of science's being very much what scientists do, the space science program, if it was to be a good one, had to be what space scientists made of it. Recognizing this, NASA built its space science program on advice from the best scientific minds it could get to think about the program. Over the years June Merker, assistant to the author, kept a running record of recommendations made to NASA by the many advisory bodies with which the space science office had to deal. For each recommendation she put down what NASA's response had been. A simple perusal is enough to convince one that NASA did pay careful attention to what the scientists were telling the agency.[36]

This accommodation to the scientific community did not come about without much stress and strain. Scientists are a contentious lot, habituated to open debate and free expression of views, and the tremendous opportunities of the space program inspired them to more intense dispute than usual. One reviewer of this manuscript raised the question of why so much attention should be paid to the quarrelsomeness of the space scientists.[37] Others expressed the view that even more attention should be given the subject. In view of their special role and position in the program, a certain noblesse oblige fell on the space scientists.[38] Nevertheless, much of the tension in the program stemmed from the scientists' presumption of special privilege, which at times Congress found irritating. Many scientists however—like Harry Hess, Charles Townes, John Simpson, Eugene Parker, Fred Seitz, John Findlay, and Gerard Kuiper—were invariably courteous and helpful.

But it should not be supposed that all the stresses and strains were between NASA and those outside. There were plenty of internal problems, and the space science program had its share, some of which are discussed in chapters 14–16.

13

The Universities: Allies and Rivals to NASA

For several reasons the universities were important to NASA, particularly to the space science program. First, much of the research embraced by space science—such as astronomy, relativity and cosmology, atmospheric studies, and lunar and planetary science—was done in or in conjunction with universities. As a consequence the best informed and most competent researchers important to space science were to be found on campus. While many of the investigators would have to spend long hard hours learning to use the new rocket and spacecraft tools, their years of working with the problems to be solved would give them a substantial head start.

Second, the university was the only institution devoted extensively to the training of new talent. As the space program was getting under way, various groups outside of NASA expressed concern that the new endeavor would lure scientific and technical expertise away from other areas of more immediate national concern. NASA managers argued that many researchers entering the space program would continue their ongoing research, except that now they could apply powerful space techniques to their investigations. In space science the argument was easy to make. Astronomers would continue to do astronomy, and solar physicists would continue to study the sun, but with the inestimable advantage of having their instruments above the atmosphere, which hitherto had hidden most of the wavelength spectrum from the observer on the ground. Atmospheric and ionospheric researchers would continue their investigations, but having their instruments in the very regions of study would shorten considerably the long chains of reasoning previously needed to go from ground-based observations to conclusions. And sending instruments to the moon and planets would furnish new data, the lack of which had for decades stymied efforts to understand these bodies.

But in applications and technology, the argument was not as persuasive. While one might grant that satellites should contribute to the observation and forecasting of weather and to the improvement of long-distance communications, still there was the usual feeling that conventional approaches needed the more immediate attention. As to the usefulness of space technology, the connection was even less direct and the value of diverting manpower to space technology research more doubtful.

A significant effect of the Soviet Union's precedence in space was to set aside such arguments for a number of years. But those arguments were bound to recur unless steps were taken to counter any imbalances the space program might generate through the absorption of highly trained manpower from other activities. As a remedy, NASA undertook to support the universities in training substantial numbers of graduates in science and engineering, and even in aspects of law and economics related to space.

In providing support to the universities for research and the training of graduate students, NASA created a staunch ally. For space science especially, as the agency sought to bring university experts into planning the program as well as into the research, relations became quite intimate. But by simultaneously establishing space science groups of its own at NASA centers, NASA generated a substantial strain on the growing tie with the universities. For it was inevitable that the NASA space science groups would appear to have the inside track to funding and space on NASA's rockets and spacecraft.

Although in time NASA space science groups came to be seen by outside scientists as important points of contact, university researchers continued to worry that, in the face of budget cuts, the continuity of NASA space science teams would be ensured while university groups would be in jeopardy, and that university projects would be more likely to suffer from whims of NASA administrators than would those in the centers. Thus, while the alliance between NASA and the universities strengthened as the program unfolded, the element of rivalry was also there, a rivalry that at times displayed hues of outright antagonism when hard decisions had to be made—like the cancellation of the Advanced Orbiting Solar Observatory, which terminated important university research projects. It was a classic example of a love-hate relationship in which mutual interests and respect conflicted with a natural competition for support and position. For space science, at least, this element of ally and rival must be kept in mind as an important feature of the NASA university program.

The program itself got off to a slow start. NASA inherited little in the way of a university program from the National Advisory Committee for Aeronautics. Oriented primarily toward in-house research, NASA's predecessor supported only a limited amount of university research.[1] At first NASA's relations with the university community assumed an administrative complexion and during Glennan's years the group responsible for handling university matters remained on the administrative side of the house. When Administrator James E. Webb took over, the Office of Grants and Research Contracts, which had prime responsibility for NASA's university affairs, was still under Albert Siepert, NASA's director of administration.[2]

Only gradually did the idea of a university program as such emerge. From the provisions of Public Law 85-934, which went into effect in the fall of 1958, NASA acquired the authority to make grants in support of research pertinent to the NASA mission.[3] But for a time NASA did not have the authority to provide for building research facilities on campuses. In May 1959, when Glenn Seaborg, Edward Teller, and some of their colleagues from the University of California at Berkeley met with Hugh Dryden and the author seeking funds to construct a building to house a space institute, Dryden had to tell them that NASA lacked authority to provide such support. The agency was, however, seeking to remedy this situation in the authorization request then before the Congress.[4] But, not until the summer of 1961 did the agency gain the legal basis for making facilities grants to universities.[5] In spite of its slowness, NASA in its first two years laid the basis for what might be called a conventional program to support space research on university campuses. Webb, the second administrator of NASA, added some decidedly unconventional elements to the program.

STEPPING UP THE PACE

To meet NASA's own needs, and under prodding from Lloyd Berkner and the Space Science Board, NASA space science managers during 1959 and 1960 gradually evolved a program for support of space science in the universities. By the fall of 1960, a policy for the program had taken shape. In November 1960 the author set forth some elements of policy to be followed with universities and nonprofit organizations. NASA would support basic research in these institutions for the purpose of developing space science, but could not support science in general. NASA would use multiyear funding and would seek to provide continuity of support to academic research groups.[6] With these thoughts the shape of the conventional part of the university program was beginning to emerge. It remained to match the size of the program to the need.

During the spring of 1961 the determination grew to strengthen NASA's association with the universities. Meeting with his staff on 22 June 1961, Webb decided that NASA must encourage university participation in the space program and, moreover, must share in the necessary funding to make it possible for universities to take part. Webb assigned the author the task of organizing an intra-NASA study of how to proceed and to assemble an outside group of consultants. The very next day the author and his associates began to develop a list of topics to take up in the proposed studies, such as support of research, the differing requirements of laboratory research versus spaceflight research, the development of graduate education, the development of schools, the use of grants as opposed to contracts, fellowships, and the construction of facilities on campuses.[7] Simultaneously a panel of university presidents, deans, and department heads (app. H) was lined up to meet on 14 August 1961 on the questions NASA posed and other questions that they themselves might bring forward.

On 30 June the author chaired a meeting of representatives from interested offices in the agency to review plans for the meeting.[8] A number of the university consultants came to NASA Headquarters in early July for a preliminary look at the questions to be taken up in the August session. Thus, by the time of the meeting, considerable thought had already been devoted to the problems of concern.

At the August meeting, to set the stage for the discussion, Richard Bolt of the National Science Foundation presented some statistics on basic research in universities. Total basic research in the United States, he said, amounted to $1.8 billion annually, of which half was spent in universities. The government provided two-thirds of the university share—$600 million. According to Bolt, universities direly needed money to build new facilities, with an immediate requirement of $500 million and a total over the next 10 years of $2.8 billion.[9] It was certainly not NASA's responsibility to provide these huge sums, but any funding would help to relieve the total problem.

Predictably the advisory committee recommended that NASA enhance its university program, providing money for research, graduate training, and construction of new laboratories. With the committee's welcome endorsement, NASA stepped up the pace of its program. Much of the research supported in universities was funded by the various program offices. Although in the long run such funding proved to be more stable than that from the university office itself, university officials saw serious shortcomings in this procedure. NASA program managers naturally tied their dollar support to specific projects with prescribed objectives, and with firm deadlines for flight experiments. For researchers engaged in such projects, the funding was essential and the imposed requirements unexceptionable. But the question remained of how advanced work, the exploratory research that was needed before an investigator could propose a flight experiment, would be supported.

Out of the need to do advance work grew the concept of the *sustaining university program*. This terminology was never much liked either in NASA or on the Hill, since it seemed to denote a program to sustain the universities, which was not NASA's legitimate business. But no better language was devised to describe that part of the university program that was designed to make possible university participation in spaceflight research. For example, the sustaining university program would provide long-term funding of a rather broad nature that would permit the university to build up and maintain a continuing research group and to pursue the ground-based research prerequisite to spaceflight investigations. The broad-based support was achieved by funding research in very general areas pertinent to space sciences, leaving the choice of specific research problems and the setting of schedules largely to the investigator himself.

Continuity of support was achieved by step funding.[10] At the initiation of a grant, two years' funding was committed, one full year's worth to be applied to the first year, two-thirds of a year's worth assigned to the second year, and the remaining one-third of a year's funding to be used in the third year (fig. 44). Each year thereafter adding a full year's funding would continue the arrangement for the next three years. Thus, if at any time NASA had to, or chose to, discontinue the grant, enough money and time would remain so that the work could be phased out in an orderly and rational way—or perhaps support found from some other source. The program offices were also encouraged to use step funding whenever they could see their way clear to do so.

Mindful of the criticism that NASA was using persons critically needed elsewhere, the agency put together a program to fund the training of graduate students in areas related to the space program. During 1962, as NASA managers were trying to determine how large a training program to support, the so-called Gilliland report on the nation's need for scientific and technical people was in preparation. Although the report did not appear until December 1962, many of its conclusions were widely known well before that time and had a decided impact on NASA's planning. The report stated that to meet the nation's needs, the country would have to double by 1970 its production of engineers, mathematicians, and phys-icists.[11] This equated to adding about 4000 persons at the doctorate level to the work force each year. Comparing its total budget with those of sister agencies like the Department of Defense, the National Science Foundation, and the National Institutes of Health, NASA arbitrarily adopted one-fourth of the goal as its fair share. Finding itself unable to meet this goal, the university office ultimately aimed at a steady-state level of 1000 trainees in its program.[12] Before many years had passed NASA trainees could be found in universities and colleges of almost every state in the Union. This geographical spread in the program earned the agency considerable praise from members of Congress, although one or two legislators held that

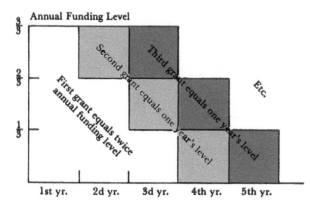

Figure 44. Step funding. At the start, two years' funding was assigned, of which one year's worth was allocated to the first year, two-thirds of a year's worth to the second year, and one-third of a year's worth to the third year. At each annual renewal, one year's funding allocated in thirds to the next three years restored the original funding status for another three years.

NASA had exceeded its authority in entering upon any such training program. By leaving the administration of the training program, including the selection of trainees and their research projects, to the universities—NASA's prime requirement was that the research be clearly related to space—the agency also earned the appreciation of the universities.

A most important aspect of the sustaining university program was building laboratories for universities interested in the space program. The magnitude of the national need for new university construction had been indicated in Bolt's brief résumé for NASA's ad hoc advisory committee. From all over the country university administrators came to see NASA officials about the possibility of obtaining funds to construct new buildings. As indicated earlier the pilgrimages had begun even before NASA had the necessary authority to help. Always the story was the same. University interest in doing space research was running high, but facilities were already overloaded by other research and by teaching requirements. To take advantage of the opportunities presented by the space program and to help NASA conduct the space science and other space programs, the university required additional facilities and equipment. Out of this need grew the facilities portion of NASA's sustaining university program.

By the end of Webb's first year and a half in office, NASA's university program had begun to take the shape it would display throughout the 1960s: a component supported by the technical program offices and the sustaining university program supported by the Office of Grants and Research Contracts. The former supported research closely connected with specific programs and projects of the agency, while the sustaining program provided funding for graduate training, the construction of facilities, and continuing research in rather broad areas. As the program was expanded, the underlying policy was also firmed up. To the points listed in the author's memorandum of November 1960, Webb added an important guideline: NASA was to work with universities in such a way as to strengthen them while at the same time getting NASA's job done. This particular policy of Webb's, often repeated in conversation and writing, evoked approbation from the Space Science Board's Ad Hoc Committee on NASA-University Relationships in the spring and summer of 1962.

Once launched on a path of renewed growth, the university program increased steadily to more than $100 million a year (fig. 45). The sustaining university program flourished for a number of years before running into peculiar problems that markedly altered its character and greatly reduced its size. To run the program Thomas K. L. Smull had taken over from Lloyd Wood, its initial mentor. Smull, formerly of the NACA, had a broad acquaintance with university administrators and a keen sense not only of the capabilities of universities but also of their needs. He was not an easy conversationalist, and his writing tended to be labored, but these shortcomings were overcome by his imaginativeness and the soundness of

Excludes obligations to or through federally funded research and development centers (JPL, Lincoln Lab, SREL).

Change in status from "university" to "nonprofit" for three organizations accounts for $10.1 million of the difference between FY 1973 and FY 1974 (Draper Lab., ERIM, Dudley Observatory). Office of University Affairs includes 1962-1971 sustaining university program.

Omitted obligations: transition period (July–Sept. 1976), $27.7 million; FY 1959, $3.6 million.

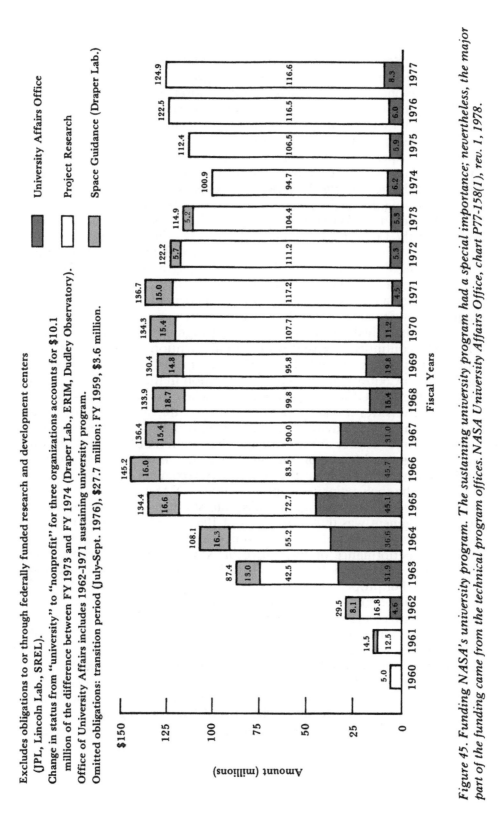

Figure 45. Funding NASA's university program. The sustaining university program had a special importance; nevertheless, the major part of the funding came from the technical program offices. NASA University Affairs Office, chart P77-158(1), rev. 1, 1978.

his thinking. Because of his obvious interest in the welfare of the universities with which he dealt, a welfare which he put on a par with that of the National Aeronautics and Space Administration which he represented, Smull gained a solid acceptance in the university community.

To assist in the management of the program Smull brought in a number of key persons. John T. Holloway, a physicist of notably sharp intellect and equally cutting tongue, brought years of experience with universities from the Office of Naval Research and the Office of Defense Research and Engineering. Holloway was effective in promoting all parts of the sustaining university program, and in time became deputy to Smull. Frank Hansing from Agriculture, Donald Holmes from Defense, and John Craig from the Central Intelligence Agency were other new recruits to the office. Almost single-handedly Hansing managed the training grants program, earning the great respect of both colleagues and outsiders. Holmes wrestled in respectable fashion with the more tricky facilities grants, where he encountered a number of vexations not experienced in other parts of the program. Craig took responsibility for the research grants.

Having launched the NASA university program on a career of expansion, Webb continued to give it his personal attention. As part of the reorganization of NASA in the fall of 1961, Webb moved the Office of Grants and Research Contracts from its obscure location in the Office of Administration to the new Office of Space Sciences, where an intimate association with the universities was an important feature of the operating program.[13] Following the ad hoc advisory meeting of August 1961, the administrator engaged John C. Honey of the Carnegie Corporation of New York to continue to review and advise on the agency's program with the universities. While Honey added little to the substantive recommendations of the ad hoc group, he took pains to emphasize that if NASA was serious about elevating the university program, adequate staffing had to be provided. Honey's judgment was that at the start of 1962 NASA was grossly understaffed for its projected plans in the university area, and in particular to match the performance of the Office of Naval Research, the National Science Foundation, and the National Institutes of Health in working with universities. For proper effectiveness, Honey advised one manager for each $4 million of program.[14] By Honey's standards, NASA's university program was permanently understaffed.

As the program unfolded NASA made a practice of seeking outside evaluations. It was difficult to assess and assimilate such evaluations, for they tended to be highly flavored by the personal views and bents of the evaluators. In June 1964, Sidney G. Roth of New York University turned out a report on NASA–university relations which seemed to show greater concern about how to enhance the benefits of NASA's program to the universities than about how NASA might get what it needed from the program. It gave much attention to the mechanics of operating the university program. Roth recommended that NASA establish discipline divisions—a biology division, a physics division, etc.—for dealing with universities and make use of corresponding evaluation panels in deciding on grant awards.[15] NASA could not use such a recommendation, which failed to take into account the agency's need to organize along project lines. There just wasn't a definite sum set aside to go into university research in biology, and another sum for university physics, and so on. Rather, NASA's monies were earmarked for projects in lunar exploration, satellite astronomy, space communications, and the like. Money was directed into the academic disciplines as they appeared directly or indirectly to support the agency's assigned projects.

A year later, in a similar study for NASA, D. J. Montgomery of Michigan State University found almost no desire in his widespread discussions to have NASA change its methods of evaluating research proposals.[16] A key problem cited by Montgomery was that of communicating adequately to the university community NASA's intentions and the opportunities NASA could offer for university research.

In 1965 the NASA university program was in full swing.[17] The Office of Space Science and Applications was devoting about $30 million a year to the support of university research related to the space science and applications programs, and other program offices were also pouring sizable sums into the universities. In the sustaining university program, the training grants, which now consumed about $25 million a year, had attracted high-caliber students who appeared to be doing good research on important space problems. Twenty-seven research facilities grants had been awarded, and these with the broad research grants were enabling many of the major universities—the Massachusetts Institute of Technology, Princeton, the University of Wisconsin, and various campuses of the University of California, for example—to establish interdisciplinary space research activities. The sweep of the program and the widespread university interest was brought out in a NASA–university conference held in Kansas City 1–3 March 1965.[18] The conference was held to inform the universities of NASA's plans and to hear university reports of progress in their projects. Those attending comprised a veritable Who's Who of the university community.[19] The meeting evoked both praise and criticism of NASA's program. Illustrating the praise was a letter from Professor Martin Summerfield of Princeton University.[20] Summerfield wrote to compliment NASA both on the conference and on the substance of the NASA university program. He said that he found the same enthusiasm in his talks with colleagues at his university. Most appreciated was NASA's policy of supporting a university in what the institution found to be in its own self-interest.

But the glow of success blinded one to some serious defects. In the sustaining university program were problems that would soon destroy the program as originally conceived, replacing it with one of quite different thrust.

EXPERIMENTAL PROGRAM:
FACILITIES GRANTS AND MEMOS OF UNDERSTANDING

When James E. Webb began in the spring of 1961 to encourage NASA managers to expand and deepen the agency's association with the nation's universities, they naturally thought in terms of programs like those of the Office of Naval Research or the National Science Foundation, programs which have been characterized here as conventional. But Webb, out of an interest born of his long experience in government as director of the budget and as under secretary of state and many years of association with the Frontiers of Science Foundation in Oklahoma and Educational Services, Inc., in Massachusetts, had more in mind. He wanted to experiment, to create a closer, more fruitful government-university relationship than had existed before.

No sooner had the word gone out that NASA now possessed the authority to support the construction of university facilities and would be receptive to proposals suitably related to the space effort than the agency was deluged with requests for support of laboratories and institutes. The month of October 1961 illustrates the kind of interest that had been stirred up. Lloyd Berkner of the Southwest Research Institute in Dallas obtained from NASA a commitment to support the Institute at the rate of $500 thousand a year on a step-funded basis.[21] On 20 October 1961 Nobel Laureate Willard Libby, representing the University of California at Los Angeles, discussed with Webb and the author the possibility of getting money from NASA to erect a building to be devoted to research in the earth sciences. As a site for the building, the university was interested in acquiring title to some neighboring land belonging to the Veterans Administration. The land was understood to be surplus to current government needs, and Libby wondered if NASA might assist in obtaining the real estate for the university.[22]

Two days later, Governor Kerner of Illinois was in Webb's office to explore Illinois's interest in the NASA university program. Immediately on the heels of the discussions with Governor Kerner, James S. MacDonnell, president of MacDonnell Aircraft Company and a trustee of Washington University at Saint Louis, was inquiring as to how Washington University might be related to the space program.[23] On 26 October 1961 Professor Gordon MacDonald of UCLA followed up Libby's earlier visit seeking an earth-sciences laboratory for the university.[24] In these exploratory discussions, Webb's questioning began to reveal the germ of a new idea, that of getting universities to develop stronger university-community relations.

By 30 October, when Professor Samuel Silver of the University of California visited NASA Headquarters to solicit support for a space science center at Berkeley, Webb's idea had begun to take shape. Silver needed not only money for space research, but also funds to erect a laboratory to house the space-science center. Webb asked if the proposed center might take on two economists who, working closely with the physicists and engineers, would study the values of science and technology, their feedback into the economy, and how a university can help to solve local problems.[25] To Webb the fact that a laboratory provided by NASA would be devoted to space research, while an essential requirement, would not be adequate justification. There had to be more, and during the first half of 1962 the desired quid pro quo was worked out. Following the administrator's lead, on 3 July 1962 Donald Holmes set down a few notes on the policy that would be followed by NASA in making construction grants to universities. Holmes noted that in accordance with Public Law 87-98, and when the university had met criteria established by NASA, it would be NASA's intention to vest title in the grantee to the facilities acquired under the facilities grant program.[26] Among the criteria would be Webb's special requirement, of which, in connection with a proposed facility grant to the University of California at Berkeley, Webb wrote on 25 July 1962: "One of the conditions of the facility grant will be to require that each university devote appropriate effort toward finding ways and means to assist its service area or region in utilizing for its own progress the knowledge, processes, or specific applications arising from the space program." He further stated that a memorandum of understanding signed by senior officials of the university and NASA would be used to establish the conditions of the facility grant.[27]

These additional conditions for obtaining a facility grant from NASA may or may not have been necessary to justify the grants to Congress, but for Webb they were entirely in character. He repeatedly said that he liked to accomplish several things at once with any action he took. He saw in the universities not only a source of support for the scientific and technical research of NASA, but also the possibility of meeting a much broader need of the administration. In the last half of the 20th century, political, economic, and social problems had become so complex as to place them beyond the comprehension of any single individual or group. As never before in the history of man, statesmen needed advice and counsel based on the expertise, experience, and insight of many diverse talents. Where better to look for this than on the university campus where all kinds of talents and interests exist together, engaged in study and thought at the very frontier of knowledge and understanding? The task was to bring all this talent together in such a way as to derive from it practical and timely advice to administrators and lawmakers.

So, as NASA people sought specific help from the universities in their individual projects and programs, Webb sought to give this developing university program a broader and deeper character. He would support the training of large numbers of graduate students and the construction of limited numbers of buildings for space science and engineering and the aeronautical sciences if university administrations in return would commit themselves to developing new and better ways of working with local governments and industry to solve common problems and advance the general welfare. Webb was especially interested in seeing what could be done to develop readily tappable centers of advice for local, state, and national government.

The universities were quite ready to sign agreements along the lines that Webb desired, but actually showed little understanding of what Webb was talking about. Most university administrators seemed to feel that the agreements were purely cosmetic—showpieces that could be used in Congress to justify the construction grants and other subventions to the universities. A few produced some results, but nothing approaching what Webb had hoped for.

Webb's dream was a desirable objective, but may have been impossible of achievement in the university environment. The independence of the individual researcher, which academic tradition guarantees, fosters the expertise and specialized knowledge that Webb wished to tap. To place such expertise and knowledge on ready call to be applied on command to problems of someone else's choosing—that is, on demand from the government seeking advice, or the university administration seeking to serve the government—would destroy the very independence that generated the unique expertise in the first place. This meant that one would have to rely on voluntary contributions to the activity by individual professors, which left the university administrators in a position of attempting to persuade their professors to join an undertaking the administrators themselves did not understand well enough to describe in very persuasive terms.

To add to the dilemma, university researchers often feel that their best personal contributions to society are to be made through their personal research, which is the thing that they do best. Thus, when Webb asked individual department members if they didn't feel an obligation to their university administration to help carry out a memorandum of agreement like those with NASA, the answer was no. Such an answer, which was regarded as natural and proper by the university researcher, seemed outrageously callous and irresponsible to Webb.

That NASA could apply only a few tens of millions of dollars in the university area afforded Webb very little leverage. As Richard Bolt of the

Science Foundation had pointed out, university needs nationwide for buildings, equipment, and other capital investments were variously estimated in the vicinity of several billions of dollars, against which NASA's few millions made little showing.

The fortunes of the sustaining university program rode the wave of Webb's interest in drawing the universities into the broader role in political, economic, and social matters to which he felt they could contribute so much. One may argue over whether Webb's objectives were achievable at all; but they could hardly have been realized in the few years that he allowed for their accomplishment. In 1965, when the university program appeared to be riding high, Webb, instead of taking satisfaction in its accomplishments, began to show disappointment in its shortcomings. On 19 February 1965 he wrote to the author that "no university, even under the impetus of the facilities grant accompanied by a Memorandum of Understanding, had found a way to do research or experiment with how the total resources of the university could be applied to specific research projects insofar as they are applicable."[28]

Webb met frequently with university heads to press them for reports of progress. He asked for independent reviews of the program. One of these, conducted by Chancellor Hermann Wells of the University of Indiana, included an extensive tour of the universities owning buildings paid for by NASA. The report did not give Webb the encouragement he sought. When the president of one of the universities Webb felt most likely to produce good results stated that most of the universities believed Webb had introduced the memo of understanding purely to satisfy Congress and that he really wasn't serious about requiring performance under the agreement, the administrator's disenchantment was complete. As 1966 rolled around it became clear to his associates that Administrator Webb was planning to wind down the sustaining university program.

In an effort to forestall any such curtailment, the author wrote a 13-page memorandum to the administrator pointing out the importance of the universities, the substantial accomplishments already achieved in the NASA university program, the highly successful training-grant program which was already bringing many competent young recruits into the space program, and the increasing flow of results from the research grants.[29] The author argued for a strong, continuing program, emphasizing that current accomplishments were the results of steps taken many years before and that a successful program of the kind NASA now had was the best possible basis from which to try to achieve the special objectives Webb had in mind. It was too late; events had overtaken the program. Added to Webb's disappointment with lack of performance on the memoranda of understanding was an emergent suspicion that the Gilliland Committee report might have grossly overestimated the need for new technical people in the nation's work force. Physicists and engineers, especially in the aerospace field, were beginning to have difficulty in landing jobs, and it was just possible that NASA's sizable graduate-training program might be exacerbating a serious national problem. Simultaneously President Johnson, disturbed by unrest and violence on the campus and smarting from what he regarded as gross ingratitude for all that his administration had done to help students pursue their education, was disinclined to provide any further assistance. (That the dissidents were associated with departments other than the scientific and technical ones with which NASA was concerned was obscured by the emotions of the period.) As Webb later told the author, he had been instructed by the president—in a memorable meeting—to wind down the training program. In the existing climate, Webb proceeded to phase out the facilities grant program also. This dropped the sustaining university program to about one-quarter its previous level by FY 1968, for the time being consisting principally of the broad area-research grants. Numerous congressmen, like Joseph Karth of Minnesota, who had found the sustaining university program to their liking, expressed disapproval when the new budget requests showed how much it was being curtailed. Nevertheless, the cuts stood.

Ultimately Smull and Holloway became casualties of Webb's disillusionment over the NASA university program. To Smull and Holloway—and to the author also—the basic university program was amply justified by the important, often essential, contributions made to the prime NASA objectives in space science and technology. Webb's desire for a broader government-university relationship, while understandable and laudable, seemed best regarded as a hope for an additional benefit that might or might not be attained.

But Webb didn't see it that way. To him the broader objectives were the most significant contribution that the university program, or at any rate the sustaining university program, could make. Without that contribution the program forfeited his endorsement. He came to feel that Smull and Holloway favored the conventional program too much and did not put enough effort into achieving the newer relationships he sought. From accompanying Smull on numerous visits to universities and hearing him urge on university people Webb's desire for performance under the memos of understanding, the author knows that the administrator was wrong in this estimate. But the lack of mutual understanding grew, exacerbated by Holloway's sharp tongue and Smull's failure to display to the universities the image of NASA that Webb desired. Finally Holloway left to take a position in another agency. Smull moved to another office in NASA.

Francis Smith, an electronics engineer from the Langley Research Center who had achieved considerable success in conducting various investigations and planning activities for NASA, was put in charge. Phoenixlike, out of the ashes the Office of Grants and Research Contracts rose again in form of an Office of University Affairs, for a short while reporting directly to the administrator and then for a number of years to the associate administrator. Honest, witty, and bedeviled by Webb's assignments to duties he really didn't care for, Smith nevertheless displayed a willingness to experiment that put him in great favor with the administrator. But Smith did not long stay at the post, leaving NASA to go to the University of Houston. Thereafter Frank Hansing took over and proceeded to mold the university program to the needs of NASA as perceived by top management.

EXPERIMENTAL PROGRAM: RESEARCH INSTITUTES

NASA's evident willingness to experiment with new relationships and management devices antedated Webb's administration. In this a prime mover with regard to academic ties was Robert Jastrow, a physicist who had come to NASA from the Naval Research Laboratory in November 1958. An imaginative theorist, Jastrow over his years with NASA interested himself in atmospheric and magnetospheric physics, meteorology and atmospheric predictability, the origin of the moon and planets, and astrophysics and cosmology. He was a superb speaker, able to hold both lay audiences and professional colleagues spellbound with his descriptions of space science topics, an ability that served NASA well when Jastrow appeared before congressional committees in defense of the agency's space science budget request. He produced numerous books and articles of both technical and popular level.[30] On television he was a frequent exponent of the many benefits mankind was receiving from the space program.

Immediately upon joining NASA Jastrow busied himself with promoting space science. He joined forces with Harold Urey to agitate for an early start of a lunar program. But Jastrow was also convinced that the best minds could be attracted into the space program only if the agency could establish the right atmosphere in dealing with university researchers. He set about trying to establish such an atmosphere.

In December 1958 Jastrow suggested to Administrator Glennan the establishment of a NASA fellowships program to be administered by the National Research Council of the Academy of Sciences.[31] Jastrow urged that the fellowship provide a large enough stipend that a post-doctoral researcher could afford to take advantage of it. The fellow would come to NASA to work on a problem of his own choosing, NASA's only requirement being that the problem be pertinent to space. Having the program operated by the National Research Council might free it, in the minds of prospective fellows, from the taint of bureaucratic bias and parochialism. The suggestion was approved, and a formal announcement of the program appeared the following March.[32]

The program attracted national and international interest and brought many first-rate researchers to NASA. From the Goddard Space Flight Center, where it started, the fellowship program spread to other NASA centers including the Jet Propulsion Laboratory.[33] In this way NASA developed an association with hundreds of competent scientists throughout the United States and the rest of the world, and these scientists became personally interested in space research.

Jastrow soon came to feel that continuous attention to the theoretical basis for space investigations was essential to a sound and productive program. He therefore joined the newly formed Goddard Space Flight Center, where he took on the task of assembling what eventually became the Theoretical Division of the center. Still not satisfied with this setup, because it lacked the drawing power to attract the best minds, Jastrow then proposed that a small study group be in a location more easily accessible to visiting scientists. He chose a set of offices in the Mazor Building in Silver Spring, Maryland, which was not too difficult to get to from downtown Washington or the National Airport. Here, supported with contracted computing capabilities, he initiated what was to become one of the most interesting experiments in government relations with the scientific community and academia. Visiting researchers were welcomed to work on space science problems, and such luminaries as Gordon MacDonald, a leading geophysicist, and Harold Urey, lunar and planetary expert, came. MacDonald remained for more than a year. Leading experts from around the world were invited to frequent work sessions on important space science topics, like particles and fields in space, the solar system, and cosmology. Presentations treated the most advanced aspects of their fields and were thoroughly discussed by the attendees. These sessions and the ongoing work of Jastrow's group were the source of numerous ideas for space science experiments.

An important element of Jastrow's concept was close working relations with local universities, for teaching and working with doctoral students was considered one of the best ways to keep a researcher on his toes and was one of the best stimuli imaginable for generating research ideas. In this respect Jastrow found the Washington area deficient. Although good relations were established with the University of Maryland, Catholic University, and others, still the quality of their contributions was not up to what Jastrow sought. In October 1960 Jastrow wrote to Abe Silverstein—head of the Office of Space Flight Programs, which housed the space science office—proposing that NASA create a center for theoretical research. By 13 December the proposal had evolved into one to establish an Institute for Space Studies in New York City, where close relations could be developed with leading universities like Columbia, New York University, and Princeton. With support from both Silverstein and the author, Glennan quickly approved the proposal.[34]

Whether the Institute should become an independent center or remain part of the Goddard Space Flight Center was seriously discussed. In the end, the tremendous obstacles that would stand in the way of creating another new NASA center, and the uncertainty that in the face of political jockeying NASA could sustain the choice of New York City for its location, led to abandoning the notion of a separate center. The Institute for Space Studies was set up in New York, in rented quarters, as an arm of the Goddard Space Flight Center, but with considerable autonomy over the choice of its research activities.[35]

The permanent staff was intentionally small, a half-dozen key researchers plus secretarial and administrative help. Most of the researchers on site were to be visiting experts who would spend from a few weeks to as much as a year at a time at the institute working on space science problems and joining in the discussions of the frequent work sessions. A large computer was rented with programming staff, and later purchased. As time went on the computing capability was enlarged and improved, giving the institute one of its most attractive features.

Among those who came to the institute for extended stays were the ubiquitous Harold Urey, who seemed to turn up wherever exciting space topics were being pursued; H. C. van de Hulst, astronomer, solar physicist, and first president of the international Committee on Space Research; and W. Priester, pioneer worker in high atmospheric structure, who did much to determine seasonal and other variations in upper air densities.

In New York it was possible to arrange the kinds of university faculty appointments needed to give the institute the desired academic ties. Visiting professors lectured at the institute. Institute members taught at Columbia and other universities and became faculty advisers to doctoral candidates working on space science topics. With contracts the institute gave several hundred thousand dollars worth of funding support annually to university research of mutual interest. Through these associations the institute became a unique experiment in government-university relationships.

Among the early areas of interest at the Goddard Institute for Space Studies were lunar and planetary research, the origin of the solar system, and astrophysics and cosmology. Studies of energy balance in the earth's atmosphere occupied a great deal of attention, and later considerable work was done on predictability in the earth's atmosphere, a topic central to making long-term forecasts of weather and climate. When the exciting possibilities of infrared astronomy became apparent, the institute, although predominantly devoted to theoretical research, set up a small experimental activity alongside the theoretical work.[36]

Papers flowed into the journals. Many of the work sessions gave rise to books on frontier topics, like Jastrow's *Origin of the Solar System*.[37] A. W. Cameron published prodigiously on theoretical investigations into the origins of the solar system, stars, and other celestial objects.

The Goddard Institute gave NASA a firm connection with a number of important universities and with a broad spectrum of working scientists;

but key members of what one often referred to as the scientific establishment remained aloof, apparently not hostile so much as indifferent. So Jastrow proposed still another experiment, a meeting of top NASA people with foremost leaders of the scientific community. On 20-21 June 1963, at Airlie House near Warrenton, Virginia, James Webb, Hugh Dryden, Harry Goett, and the author listened as Jastrow, Gordon MacDonald, and others presented to the elite of physics in the United States (app. I) an exciting review of the kinds of problems that could now be attacked with rockets and spacecraft. An interest was aroused, and the group agreed to meet periodically to keep in touch with the space program. Formally designated as the Physics Committee, the group operated more as a colloquium than as the usual advisory committee. Robert Dicke of Princeton, expert on relativity and cosmology, became its first chairman.[38] Some of the most exciting experiments for the NASA space science program—in such areas as x-ray astronomy, relativity, and cosmology—were on the bill of fare. As time went on ideas from the many discussions found their way into the flight program of the agency. One may cite as examples Bruno Rossi's work on high-energy astronomy, the work of Stanford University on the relativistic precession of accurate gyroscopes in orbit, and the corner reflectors implanted on the moon by the Apollo astronauts to support precise geodetic measurements.

With the NASA fellowship program, the establishment of the Goddard Institute for Space Studies, and the formation of the NASA Physics Committee, Jastrow had contributed immeasurably to providing NASA with a well rounded tie to the university community, particularly in physics and cosmology. Many came to hope that the Goddard Institute could serve as a pattern for other space institutes—for example, in lunar research, planetary studies, and astronomy, as the Ramsey Committee seemed to favor (p. 217). But in the late 1960s conditions were different from those prevailing when the Goddard Institute had been established. Setting aside the question of how much the institute's success owed to Jastrow's leadership, special difficulties were encountered. Budgets were rising in the early 1960s, falling in the latter half. The pattern of associations with the newly formed NASA still had to be developed in the early years of the agency, while in the late 1960s working patterns and vested interests had already been established which outside scientists would be loath to disturb. Nevertheless, following the report of the Ramsey Committee, Webb wished to experiment once more, this time introducing a new element, that of the university consortium.

Recognizing that there would be a vast store of lunar samples and other lunar data housed at the Johnson Space Center and that the center would have facilities and equipment needed to analyze and study these data, Administrator Webb desired to evolve some mechanism for facilitating the use of those resources by outside scientists, particularly university researchers. The success of the Goddard Institute for Space Studies suggested that a lunar institute might be set up as an arm of the Johnson Space Center. But the image that the center had acquired of not understanding the needs of science or being particularly interested in science made such an arrangement unattractive to many outside scientists—and also to the Office of Space Science and Applications in NASA Headquarters.

Instead of an institute managed by the center, Webb turned to the possibility that an institute might be managed by a university or a group of universities. Fred Seitz, president of the National Academy of Sciences, showed an interest. An existing consortium, University Research Associates, considered setting up and managing an institute for NASA. The possibility that Rice Institute might either by itself or as one of a number of universities provide the desired link between academia and the resources of the Johnson Space Center was also weighed. In the end a group of universities on 12 March 1969 formed a new consortium called the University Space Research Association and took over management of the Lunar Science Institute, which in its impatience NASA had already set up with the aid of the Academy of Sciences.[39] The new institute was housed in a mansion adjacent to the Johnson Space Center, provided by Rice Institute and refurbished by the government. At once the Lunar Science Institute began to hold scientific meetings, invite visitors to use its facilities, and foster lunar research.

The pattern of activities at the Lunar Science Institute was, at least on the face of things, similar to that that had proved so successful with the Goddard Institute for Space Studies. But the LSI at the end of the 1960s faced a number of vicissitudes that the Goddard Institute had not encountered. For example, after 10 years of working with NASA, some academic scientists had already managed to overcome the previously mentioned difficulties to establish personal ties with the Johnson Space Center and did not wish to see a new organization interposed. In contrast foreign scientists who did not have such close associations with the space center found the LSI a boon.

On its part, the Johnson Space Center was ambivalent about LSI. Such an institute could be useful in working with the scientific community, serving as a buffer when difficult issues had to be wrestled with. But when the institute's managers pressed for an independent research program plus rather free access to such resources as Apollo lunar samples and various lunar data, there was trouble, which occasional personality clashes enhanced.

Most fundamental, however, was the decline of NASA's budgets in the late 1960s and early 1970s, and a number of times NASA's space science managers considered withdrawing financial support from the Lunar Science Institute. The Goddard Institute for Space Studies was also beset by similar financial pressures, but its established position in the NASA family made it easier to weather these storms than it was for the Lunar Science Institute, which still had not had enough time to prove itself.

Thus, while the Lunar Science Institute could not be called a failure, its success in the severe climate in which it was launched, was an uneasy one. There could be little question that when time came to consider establishing an astronomy institute in support of an orbiting astronomical facility, or a planetary institute in support of more intensive exploration of the solar system, such propositions would receive long and searching scrutiny before being implemented.

A SLOWER PACE

Following the closeout of the facility grants program and the phasing down of the training grants, the sustaining university program became a low-key operation. It was used to stimulate advanced research in areas important to space applications and to provide seed grants to a large number of minority institutions. There was some experimenting—for example, with the development of new engineering curricula in the universities to meet modern needs—but the earlier flair was gone. Always the largest dollar component of the university program, the project grants of the technical program offices became the main thrust of NASA's university program. But the cutback on the sustaining university program had its impact on the project grants. In space science, for example, more money than before had to be devoted to support of the more advanced research to lay the groundwork for spaceflight experiments, much of which had come out of the graduate research projects of NASA space science trainees. The effect was not easy to measure, but there were tangible signs. Program managers found it more difficult to provide step funding than before, and earlier step funding was often allowed to lapse to gain a year's funding and thereby ease the current squeeze on the budget. Thus, although the total university program remained in the vicinity of $100 million per year, the more liberal flavor that had ensured a considerable continuity of support and had afforded the universities the ability to plan future staffing and research projects in a rational manner, was gone.

14

Programs, Projects, and Headaches

As with its predecessor, the National Advisory Committee for Aeronautics, NASA's principal technical strength lay in the field centers. At the time of the metamorphosis into an aeronautics and space agency, NACA had three principal centers: the Langley Aeronautical Laboratory near Hampton, Virginia; the Ames Aeronautical Laboratory at Moffett Field, California; and the Lewis Flight Propulsion Laboratory in Cleveland. In addition there was a High Speed Flight Station at Edwards Air Force Base in California and a small rocket test facility on the Virginia coast at Wallops Island.[1] The first four of these became under NASA the Langley, Ames, Lewis, and Flight Research Centers, the research orientation of which Deputy Administrator Hugh Dryden was so desirous of protecting. Wallops Station was assigned primarily to the space science program.

To the former NACA installations, NASA added six more: the Goddard Space Flight Center in Greenbelt, Maryland; the Jet Propulsion Laboratory in Pasadena; the John F. Kennedy Space Center at Merritt Island, Florida; the George C. Marshall Space Flight Center in Huntsville, Alabama; the Lyndon B. Johnson Space Center (which for many years was known as the Manned Spacecraft Center) in Houston; and, briefly, an Electronics Research Center in Cambridge, Massachusetts, which was transferred to the Department of Transportation.[2] A sizable facility for testing large rocket engines was established in Mississippi not far from New Orleans and placed administratively under Marshall, which had prime responsibility for the Saturn launch vehicles used in the Apollo and Skylab programs.[3] The Jet Propulsion Laboratory and Marshall were transferred to NASA from the Army; the others were created by NASA. As its original name suggests, Johnson was in charge of the Mercury, Gemini, and Apollo spacecraft and most of the research and development was related to those programs.[4] Kennedy, originally the Launch Operations Directorate of Marshall, provided launch support services for both manned and unmanned programs, but the former required by far the greater capital investment and manpower.[5] Both Goddard and the Jet Propulsion Laboratory were principal centers for the space science program, the former for scientific satellites, the latter for planetary probes.

Management at headquarters guided the space program, directed the overall planning, developed and defended the budget for the agency, and fostered the kinds of external relations and general support that the space program needed. In a very real sense headquarters people labored at the center of action where the political decisions were made that permitted the space program to proceed. Yet the story of headquarters activity is mostly one of context, of background—essential, indispensable, but background nevertheless—against which the actual space program was conducted. Research, the essence of the space science program, was done by scientists at NASA centers, in universities, and at private and industrial laboratories.

It follows that the mainstream of space science must be traced through the activities of these institutions. The important role of the universities was the subject of the preceding chapter. With occasional exceptions, like the upper atmospheric research of the Geophysical Research Corporation of America and the pioneering work of American Science and Engineering in x-ray astronomy,[6] the contribution of industry was more to the development and flight of space hardware than to conducting scientific research. It remains, then, to take a look at the part played by the NASA centers.

The principal space science centers were the Goddard Space Flight Center and the Jet Propulsion Laboratory (JPL being operated by California Institute of Technology under contract to NASA). Wallops Island, which for a time was placed administratively under Goddard, provided essential support to the sounding rocket and Scout launch vehicle programs.[7] But not all NASA space science was done at these centers. The Ames Research Center managed the Pioneer interplanetary probes and took the lead in space biology and exobiology—a term coined to denote the search for and investigation of extraterrestrial life or life-related processes. Langley had responsibility for the Lunar Orbiter and later the Viking Mars probe. Most notable was the lunar research fostered by Johnson in the early 1970s with the samples of the moon and other Apollo lunar data, which for a time made Houston a veritable Mecca for lunar scientists.[8] But Apollo lunar science was an exception generated by the special nature of the manned lunar exploration program; and, generally, Dryden's policy stood in the way of more than a limited participation of the research centers in space projects.

Over the years the NASA centers built up an enviable reputation of success on all fronts, in manned spaceflight, space applications, and space science. In the last mentioned, by 1970 Goddard had flown more than 1000 sounding rockets, more than 40 Explorer satellites, 6 solar observatories, 6 geophysical observatories, and 3 astronomical observatories, most of them successfully. In applications Goddard enjoyed comparable or better success rates with weather and communications satellites. The experience of the

Jet Propulsion Laboratory was similar. By the end of the 1960s JPL had sent 3 Rangers and 5 Surveyors on successful missions to the moon and dispatched 5 Mariners to Mars and Venus.[9] These achievements are bound to be recounted repeatedly and will rightfully be judged as success stories. Success, however, was not bought without a price of some mistakes, temporary failures, and occasionally severe personal conflict, which form an instructive part of the total history. In reviewing the struggles and problems that preceded the achievements, a proper sense of perspective is important, for troubles often tend to magnify themselves in the eye of the beholder. The difficulties were, after all, overcome in the ultimate successes that were achieved. Still, as part of the total story, perhaps as illustrating the natural and usual course of human undertakings, those difficulties are important to the historian. They should also be instructive to later managers. Thus, without at all deprecating their splendid achievements, it is appropriate to delve briefly in this and the next chapter into some of the trials endured by the Goddard Space Flight Center and the Jet Propulsion Laboratory.

THE CHARACTER OF THE FIELD CENTERS

The different centers in NASA had distinctive personalities that one could sense in dealing with them. As might be expected the former NACA laboratories kept as NASA centers many of the characteristics they had acquired in their previous incarnation. One trait was the fierce organizational loyalty that had been displayed as part of NACA. Thus, while officials at those centers were convinced that the real power of the agency lay in the centers and felt very strongly that they should have some voice in formulating orders, and also that once given an assignment they should be left alone to carry it out, they also recognized that the ultimate authority lay in headquarters. Given marching orders they would march much as ordered.

The new centers in NASA had their difficulties in this regard, to varying degrees. The Marshall center reflected the background and personality of its leader, Wernher von Braun, and his team of German rocket experts. Bold, with a bulldog determination, undaunted by the sheer magnitude of a project like Saturn, they could hardly be deterred by request or by command from their plotted course. The effort to superimpose the Juno space science launchings and the Centaur launch vehicle development on the Marshall team, when Saturn represented its real aspiration, simply did not work out. The Juno launchings had to be canceled after a string of dismal failures, which space science managers at headquarters felt were caused by lack of sufficient attention on the part of the center.[10] Centaur, in the midst of congressional investigation into poor progress, was reassigned to the Lewis Research Center.[11] The Manned Spacecraft Center developed an arrogance born of unbounded self-confidence and possession of a leading role in the nation's number-one space project, Apollo. A combination of self-assurance, the need to be meticulously careful in the development and operation of hardware for manned spaceflight, plus a general disinterest in the objectives of space science as the scientists saw them, led to extreme difficulties in working with the scientific community. But the art of being difficult was not confined to the manned spaceflight centers. In this both the Goddard Space Flight Center and the Jet Propulsion Laboratory were worthy competitors. So, too, was headquarters, for that matter.

The Goddard Space Flight Center's collective personality stemmed from its space science origins. As the first new laboratory to be established by NASA, Goddard inherited most of the programs and activities of the International Geophysical Year, like the Vanguard satellite program and the Minitrack tracking and telemetering network. Also, many of the scientists and engineers of the Rocket and Satellite Research Panel and the IGY sounding rocket and scientific satellite programs joined Goddard to make up, along with the Vanguard team, the nucleus out of which the center developed. These origins indelibly stamped Goddard as a space science center, even though science accounted for only about one-third of the laboratory's work (and by the nature of things, most of that effort went into the development, testing, and operation of sounding rockets, spacecraft, and space launch vehicles required for the scientific research). In actuality only a small fraction of the Goddard Space Flight Center's personnel was engaged in space science research. Nevertheless, the presence of those persons in key positions, which they came to fill as charter members of the laboratory, imparted to the center a character that accounted simultaneously for its success in space science and for many of the difficulties experienced with upper levels of management.

As professional scientists, these persons were by training and experience accustomed to deciding for themselves what ought to be done in their researches. While subjecting themselves to a rigorous self-discipline required to accomplish their investigations, they nevertheless approached their work in a highly individualistic manner. They questioned everything, including orders from above. While they could and did work effectively as groups, their cooperation included a great deal of debate and free-wheeling exchange on what was best to do at each stage. To trained engineers in NASA—for whom a smoothly functioning team, accepting orders from the team leader as a matter of course, was the professional way of going about things—the seemingly casual approach of the Goddard scientists looked too undisciplined to work.

The Goddard scientists had also been accustomed to determining their own objectives and pacing themselves as they thought best. The accomplishment of an experiment that produced significant new information was what counted; costs and schedules were secondary. That a project took longer to carry out than had originally been estimated was of little consequence so long as the project succeeded, particularly if the additional time was put to good use improving an experiment and ensuring success. This peculiarly science-related sociology of the space scientists at Goddard reinforced the tensions that naturally come into play between a headquarters and the field in large organizations, and led to a major confrontation in the mid-1960s.

FIELD VERSUS HEADQUARTERS

Headquarters and field in any effective and productive organization support each other, working as a team in the pursuit of common goals—those of the organization. Yet many aspects in even the most normal of headquarters-field relationships serve to pit one against the other at times. When circumstances exacerbate those normal centrifugal tendencies, serious trouble can arise. To understand the nature of the problem, a few words about the difference in headquarters and center jobs in a technical organization like NASA are in order.

At the heart of the difference is the matter of programs and projects. The raison d'être of an agency is reflected in its various programs, where the term *program* is used to mean a long-term, continuing endeavor to achieve an accepted set of goals and objectives. NASA's overall program in space included the exploration of the moon and the planets, scientific investigations by means of rockets and spacecraft, and the development of ways of applying space methods to the solution of important practical problems. Each of these programs could be, and when convenient was, thought of as a complex of subprograms, such as a program to develop and put into use satellite meteorology, a program to improve communications by means of artificial satellites, or a program to investigate the nature of the cosmos. Barring an arbitrary decision to call a halt, one could foresee no reason why these programs, incuding the subprograms, should not continue indefinitely. Certainly, if past experience is a good indicator, the effort to understand the universe must continue to turn up new fundamental questions as fast as old ones are answered. As for exploration, the vastness of space, even of that relatively tiny portion of the universe occupied by the solar system, is so great that generations could visit planets and satellites and still leave most of the job undone. And it would be a long while before diminishing returns would call for an end to applications programs.

Unlike a program, a *project* was thought of as of limited duration and scope, as, for example, the *Explorer 11* project to measure gamma rays from the galaxy and intergalactic space. A program was carried out by a continuing series of projects, and at any given time the agency would be conducting a collection of projects designed to move the agency a number of steps toward the agency's programmatic goals and objectives. The *Explorer 11 project* contributed to the *programmatic* objective of understanding the universe by determining an upper limit to the rate of production of gamma rays in intergalactic space, which eliminated one candidate version of the continuous creation theory of the universe.

A project like a sounding rocket experiment might be aimed at only a single specific objective, last only a few months or a year, and cost but a few tens of thousands of dollars. Or a project could require a series of space launchings, many tens or even hundreds of millions of dollars, and take years to accomplish. The Lunar Orbiter, with five separate launchings to the moon, and the Mariner-Mars project that sent two spacecraft to Mars in 1971 were examples. Some projects were huge in every aspect, as was Apollo. In fact, because of its size and scope, Apollo was more often than not referred to as a program, although more properly Apollo should be thought of as a mammoth project which served several programs, among them the continuing development of a national manned spaceflight capability, the exploration of space, and the scientific investigation of the moon.

With these definitions of program and project in mind, one can describe rather simply the difference between headquarters and center jobs. Headquarters was concerned primarily with the programmatic aspects of what NASA was up to, whereas the task of the centers was mainly to carry out the many projects that furthered the agency's programs. The distinction is a valid but not a rigid one. Occasionally headquarters people participated in project work, but this was an exception to the general rule. The most notable exception was Apollo, the size and scope of which were such as to make the administrator feel that the uppermost levels of management for the project should be kept in Washington. Nevertheless, the prime task of headquarters, working with the centers and numerous outside advisers, was to put together the NASA program, to decide on the projects best designed at the moment to carry out the program and assign them to the appropriate centers for execution, and to foster the external relationships that would generate the necessary support for the programs and projects. As an essential concomitant to programming, much time was occupied in preparing budgets, selling them to the administration, and defending them before Congress.

Also, each center, while project-oriented, had its center programs toward which the center directed its own short- and long-range planning. Thus, the research centers conducted programs of advancing aeronautical and space technology. In addition to a program of space science, the Goddard Space Flight Center pursued extensive programs of space applications and space tracking and data acquisition, with tracking and acquisition occupying almost 40 percent of the center's manpower. Unmanned investigation of the solar system was the Jet Propulsion Laboratory's principal program.

Although the qualifications should be kept in mind to have the correct picture, nevertheless the main distinction between the responsibilities of headquarters and those of the centers is clear. Center personnel members were primarily occupied with project work, while headquarters people spent—or should have spent—their time on program matters. That is where difficulties arose, for numerous pressures drove headquarters managers to get involved in project or project-related work. Such actions could only be regarded by a center as undue interference from above.

Naturally, NASA space science managers were vitally interested in what was happening in the various space science projects. They were responsible for proper oversight. But there was more to it than that; project work was where the action was. That was where interesting problems were being attacked and where exciting results were being obtained. Alongside project work, programmatic planning often seemed like onerous drudgery. As a consequence oversight tended to degenerate into meddling, to the distress of project managers and center directors. Even when headquarters managers took pains to couch their thoughts in the form of mere suggestions, their positions in headquarters made suggestions look more like orders. That program chiefs in headquarters occupied staff, not line, positions often was lost sight of in the shuffle, and some headquarters managers became adept at wielding what amounted in practice to line authority.

To this natural tendency to get into the act were added the pressures of the job. As the NASA program grew in size, scope, and expense, upper levels of management demanded more and more detail on schedules, costs, and technical problems. Nor was the demand for information confined to NASA management. Becoming increasingly familiar with the programs and their projects, the legislators also demanded what seemed an impossible amount of detail, either to provide while still getting the job done or for the congressmen to assimilate. On the science side, members of the authorizing subcommittee in the House, under Chairman Joseph Karth of Minnesota, frequently concerned themselves with the details of engineering design decisions and were not loath to second-guess space project engineers on matters that seemed to NASA people to lie beyond the competence of the legislators to judge. An example of this searching interest was furnished by the investigation of the Centaur liquid-oxygen and liquid-hydrogen fueled rocket stage which Karth's subcommittee undertook in 1962. NASA and contract engineers found it difficult to defend the propellant feed system which they had chosen and which could be shown to be most efficient for a rocket the size of Centaur, against a different system for which the congressmen expressed a preference and which admittedly would likely have more growth potential.[12]

Because of this increasing demand for information of various kinds, headquarters in turn demanded of the centers the detailed reporting that centers felt was appropriate for project managers but went far beyond what headquarters really needed. While program managers were willing to concede that the information they were calling for was more than they ought to need, yet they were caught in the middle; to do their jobs as circumstances were shaping them, they did need the data. They were forced, therefore, to insist, and the extensive reporting required, with its implied involvement of headquarters with what were strictly center responsibilities, remained as a continuing source of irritation.

The irritation transferred to headquarters when centers were late or deficient in their reporting, especially when a center simply refused, sometimes through foot dragging, sometimes in open defiance, to supply the information requested. A center might be reluctant to respond when it felt that the request was premature, that the data were not yet properly developed, and that the center might later be called to task if the information supplied prematurely turned out to be incorrect.

A related source of irritation arose in connection with the center's management process. At almost any time throughout the year a program manager might be called upon to furnish information about projects in his program. It was essential, therefore, for him to be continuously aware of the status of projects on which he might have to report. For this it was not enough to rely on written reports which came only so often. In addition, space science program managers kept in close touch with the project managers and attended many of the meetings held by the project managers with their staffs and with contractors' representatives. This practice came to be a particularly sore point with the management of Goddard Space Flight Center.

STRAINS ON THE FAMILY TIE

The Goddard Space Flight Center and NASA Headquarters, only half an hour's drive apart, were connected by close ties. Between the two staffs, many personal associations dated from the days of the Rocket and Satellite Research Panel and the sounding rocket and satellite programs of the International Geophysical Year. An easy relationship existed from the very start of the center. John Townsend—who served as acting director of the center until the permanent director, Harry Goett, formerly of NACA's Ames Aeronautical Laboratory, took over—had been associated with John Clark and the author at the Naval Research Laboratory. For many years Townsend had been the author's deputy in the NRL's Rocket Sonde Research Branch. Harry Goett and Eugene Wasielewski, whom Goett brought into Goddard as associate director, had long been acquainted with Abe Silverstein from the days of the National Advisory Committee for Aeronautics. These friendships served to mitigate the divisive forces between headquarters and field, but were not enough to avert an ultimate break.

Harry Goett assumed the directorship of Goddard in September 1959. As was his nature he quickly entered personally into every aspect of the center's work. From his first day until he left, he kept in close touch with every project. As an untiring battler for the center and his people, Goett endeared himself to his coworkers. He was a warm, emotional person who showed a deep interest in the men and women working for him, and on both sides a deep affection developed.

In the first weeks and months of NASA's planning for its program, many center people had been drawn into headquarters working groups to help get things under way. But as center project work grew, these assignments, which tended to persist, began to interfere with center duties. Finding Goddard people still working on headquarters tasks a year after NASA's start, Harry Goett began to protest that his personnel should be relieved as fast as possible of these additional duties. On the other hand, center people's taking part in headquarters planning was advantageous to the center. Both organizations tried to keep center participation within reasonable bounds.

As Goett, Townsend, and their people built up Goddard and launched their initial projects, program managers were developing their own methods of keeping themselves and their superiors informed. Simultaneously the Congress was increasing its demand for detailed information, which it was incumbent on headquarters to supply. As the requirements for reporting increased, project managers complained that they were spending too much time with program managers and in preparing reports, time that would be better spent in getting on with the projects. In mounting crescendo Goett complained to the author and his deputy in the headquarters space science office, Edgar M. Cortright, that headquarters managers were getting in the way of center management. Goett urged that headquarters people keep their hands off project management.

While agreeing in principle with the Goddard director, Cortright and the author strove to get him to see that in the existing climate of continuing congressional scrutiny, keeping informed was an important part of headquarters work. That, space science management insisted, was an absolutely essential part of the program manager's job, but not to usurp the project manager's duties or to interfere with his work. Cortright and the author urged upon their people great care in working with the project managers to avoid any kinds of action that would undercut, or appear to undercut, the project manager's responsibilities and authority. It was no advantage to the program for any project manager to feel that his responsibilities had been in any way lifted from his shoulders.

Headquarters was far from Simon pure in these matters, unfortunately, and there was considerable justice in Goett's complaints. The natural urge to meddle plus the incessant pressure to keep informed led many program managers to get into the project business. Sometimes this led to strong adversary relations between program and project managers; at other times to close "buddy-buddy" relations. Both situations caused problems for center management and called for continuing attention.

By the fall of 1962, Goett found the situation so disturbing that he felt impelled to complain openly at a NASA management meeting held at the Langley Research Center that headquarters got too much into projects and should stick to program management. His barbs were aimed not only at space science managers, but also at those responsible for applications programs and for tracking and data acquisition. He felt that there was not enough contact between the center director and the associate administrator.[13] Goett also felt he did not have enough contact with the author. The last complaint stemmed from the mode of management the author had adopted, about which a few words are in order.

Being a scientist, the author felt it wise to name as deputy an engineer whose training and experience would complement his own. Edgar M. Cortright, an aeronautical engineer with considerable research experience in the National Advisory Committee for Aeronautics, filled the bill very nicely. An implication of this philosophy of organization was that the deputy should be more than an understudy, more than just someone to sit in when the principal was away. Rather, the deputy should take responsibility for important aspects of the top management job that came within his sphere of expertise. This was the arrangement agreed on between Cortright and the author. Cortright would handle engineering matters, which meant oversight of much of the project work, dealing with contractors, and a great deal of the relations with the space science centers. The author would work on program planning, advisory committees, and most of the space science program's external relations including those with the Academy of Sciences, the scientific community, and the universities. Such an arrangement had worked well at the Naval Research Laboratory, where John Townsend's engineering and experimental bent had complemented the author's theoretical background. Moreover, in addition to providing the top level of management in the office with talents and experience complementing those of the director, it was an effective way of providing a deputy with substantive work and to continue his professional growth. A deputy with nothing more to do than to wait around for the principal to be away must find life deadly dull, unrewarding, and stultifying.

Under this arrangement, problems of the kind Goett was wrestling with would normally have been taken up by Cortright. But Goett was not willing to deal with a deputy. As director of the Goddard Center—even though the author was meticulously careful to support agreements Cortright worked out—Goett felt that he should deal directly with the principal in the office for which the center was working. Under the circumstances the author took special pains to make it clear that he was available to Goett at any time, yet expressed the hope that Goett would work with Cortright in the normal course of day-to-day matters.

The strain caused by the project-management versus program-management conflict took increasing amounts of time and attention. A great deal of the time spent with Goett was devoted to this problem. John Townsend, Goett's man for space science matters, pointed out that if a program manager had only one project under way in his program, then it became very difficult to draw a line between program and project, and the pressure on the program manager to get into project management was overwhelming. Townsend recommended that programs be put together in such a way that a program manager would have several projects to deal with. Under such an arrangement a program manager could no longer give the single-minded attention required by a project, and should find it much easier to confine himself to program matters. Cortright and the author agreed and tried to avoid single-project programs.

Goett pointed out that it was not just the cases in which program and project managers were at odds that gave trouble. When the two got along well together, often they would team up to promote their project over other projects which the center management—taking into account existing constraints on dollars, manpower, and facilities—might judge to be more appropriate. Thus, program and project managers working hand in glove for their own projects—perhaps to enlarge them or to extend them beyond existing commitments—were not always working for the best interests of the center.

Goett was most disturbed to have program managers, in the name of keeping in touch, attend meetings with outside contractors. Even if the headquarters people came with the determination to keep their mouths shut, contractors' representatives had a penchant for tossing questions to the headquarters representatives, with the implication that that was where the final word would lie. And when headquarters people did volunteer comments, their comments tended to take on more weight than the word of the project manager. These difficulties became even worse when the headquarters man was technically more competent than the project manager—which Goett didn't feel could happen very often. In that case the project manager tended to defer to the headquarters person for decisions and recommendations that the project manager should make himself, and the contractors were easily confused as to who was calling the shots.

Goett's solution to these problems would have been to keep program managers away from project management meetings, and especially away from meeting with contractors. Considering the program manager's basic responsibility to see to the health of his program and the corresponding need to keep informed—a need that was enhanced by the growing amount of attention given by congressional committees to NASA's programs and projects—Goett's solution was not acceptable. Cortright and the author spent a great deal of time trying to get Goett to appreciate headquarters' needs and to agree to some middle-of-the-road way out of the dilemma. A written description was prepared of the distinction between program management and project management, and the author committed himself to ensuring that his program people understood the bounds of their authorities and responsibilities. But the author also insisted that the way be kept open for headquarters people to keep adequately informed. Goett was not satisfied. In a letter to Associate Administrator Robert C. Seamans 5 July 1963, he outlined some of the problems as he saw them.[14] Shortly thereafter, on 26 July 1963, the Office of Space Science and Applications proposed a revision of NASA Management Instruction 37-1-1.[15] In Appendix A were specific definitions of *program* and *project*. The instruction made the point that the headquarters job concerned itself with program matters primarily, while project managers normally were at field centers. On 5 November 1963 the author wrote Harry Goett on the subject of headquarters-center relations.[16] The letter outlined agreements that it was hoped had been reached to keep headquarters people properly informed, without undercutting the center's position with contractors. But matters continued to deteriorate.

Complaints were not confined to the center side. In a talk given to a number of managers of space science and applications projects, at Airlie House near Warrenton, Virginia, the author spoke on relations between program managers in headquarters and project managers in the centers.[17] By giving what was viewed by headquarters people as too much emphasis to the rights and prerogatives of project managers, the author drew forth some howls from the former. On 30 December 1963 the staff of the Office of Space Science and Applications met to discuss relations with the Goddard Space Flight Center.[18] Program people complained that Goddard seemed to be waging a war to keep headquarters at arm's length. It was difficult to find out about contractor meetings in time to attend. Although Goett had stated that headquarters should keep itself informed by means of the reports it received, still Goddard habitually did not turn in reports on time. The center was being too independent in formulating its plans for supporting research—i.e., the general background research of the kind all centers undertook in support of their project work. Program chiefs felt a need to specify reporting requirements for this supporting research, since most of the money for such research came from portions of the budget for which the program chiefs had responsibility. Another complaint concerned Requests for Proposals, documents which centers sent to potential contractors asking for bids on work that the center wanted done. Program people were required to follow the progress of such RFPs through the headquarters paper mill and to assist in expediting their progress. It was important, therefore, for them to keep in close touch with the formulation of the work statements that would go into the Requests for Proposals. Yet the center appeared to be making it difficult for the program managers to keep in touch. The Interplanetary Monitoring Platform project was considered an illustration of the center's intentions in this regard. Since a decision that program people would attend "working group" meetings of projects, Paul Butler, manager of the IMP project, had ceased to hold working group meetings. Instead he held what he called "coordination meetings" with his staff, which headquarters people were explicitly told they were expected not to attend.

While the managers in the Office of Space Science and Applications were most intimately involved in the day-to-day relations with the center, the problems also had the continuing attention of Webb, Dryden, and Seamans. Concerned about overruns and schedule slips in NASA projects, the Administrator's Office noted that many of the bad examples were Goddard's.[19] As general manager of the agency, Associate Administrator Seamans maintained pressure on the Office of Space Science and Applications to correct the deficiencies. Although Seamans had known and worked with Harry Goett since 1948 and admired him very much, Seamans could not accept Goett's insistence that headquarters leave Goddard to its own devices. As Seamans wrote years later:

> . . . it was essential if NASA was to continue to receive Congressional support, that we tighten the management of our projects in order to keep costs and schedules closer to plan. We could not, in the public interest, take it on faith that Harry Goett was doing all that could be done to manage these projects properly. It was necessary for NASA Headquarters to have direct access to a variety of management data as was the case with other NASA centers. I kept Dr. Dryden and Mr. Webb fully informed of the Headquarters/ Goddard relationships and of important issues.[20]

But the problems did not end. Discussions with Goddard management seemed to elicit too much explanation of why it was in the nature of things for schedules to slip and not enough desire to change matters. The Goddard scientists especially could not see why there should be any urgency about adhering to a schedule if additional work would produce a better experiment. As for the experiments, usually there was no reason why they should be done now rather than later, unless, of course, they had to be timed to coincide with some natural event. But NASA's record of doing what it said it would do on time and within cost was important to those who had to fight for the agency's appropriations. Schedules and costs were most visible to a carefully watchful Congress, and for years NASA continued to feel that it had to sell itself. Besides, it was just plain good management to estimate costs and schedules correctly and then keep to those estimates.

Whatever opinion the Administrator's Office might have had as to who was the more to blame for the strains caused by projects versus programs, the apparent unresponsiveness of the center on tightening up project management overshadowed the other concerns. Both Associate Administrator Robert Seamans and his deputy, Earl Hilburn, pressed continually for better performance. But when, in a stressful meeting with Seamans, Goett took such a rigid position that he left no maneuvering room for headquarters, the associate administrator decided that Goett had to go. With the concurrence of both Webb and Dryden, on 22 July 1965 Seamans removed Goett from the directorship and replaced him with Dr. John F. Clark, who had been chief scientist in the Office of Space Science and Applications.[21]

It was a traumatic experience for Harry Goett and for others. The author found it a most unpleasant duty to go out to the Goddard Space Flight Center to meet with key managers and inform them that their director was being replaced. Goett was beloved of his people; he had been a conscientious, hard-working, imaginative director, under whose regime the center had achieved most of the space accomplishments of NASA's first few years. Goett himself had played a key role in establishing a productive relationship with the academic community. Those accomplishments were, of course, the real story of the Goddard Space Flight Center, not the struggles over how to manage. It was tragic that Goett's obsession over one concept of headquarters-field relationships—born perhaps of his past experience in the NACA—made him unable to appreciate the new climate in which NASA had to operate. It was unfortunate that the author was unable to work out some accommodation that would have kept Goett at the Goddard helm. Harry Goett's departure was a distinct loss to NASA.

Not having Goett's flair for the controversial, John Clark projected a more pedestrian image for the center. Yet under his administration, Goddard continued its record of successful space science and applications flights. The problems remained, and both center and headquarters had to work continuously to keep them under control. But both sides approached the problems with a better understanding of each other's needs. In short order Clark was telling headquarters where to head in, and headquarters was pressing him to get on with the job of better resource and schedule management.

The difficulties experienced by the Office of Space Science and Applications with the Goddard Space Flight Center occurred in various forms and varying degrees with all the other centers. The task of finding ways for headquarters and field to work together harmoniously and effectively is never ending. Nor is it to be expected that tension between headquarters and field will ever disappear. Should this happen, one or the other will probably not be doing its best job.

15

Jet Propulsion Laboratory: Outsider or Insider?

In the summer of 1958, before NASA had begun to operate, the author flew to California to visit the Jet Propulsion Laboratory in Pasadena. The purpose of the visit was to talk with the director, William Pickering, and his key staff members about the possibility that a group from the Rocket Sonde Research Branch of the Naval Research Laboratory in Washington might transfer to JPL. Discussions within the Department of Defense that had accompanied the congressional debate on the nation's space program during the first half of 1958 had made clear that the Navy, in spite of its pioneering contributions in the rocket exploration of the upper atmosphere and in developing the Aerobee, Viking, and Vanguard rockets, would probably not have a key role in space research and development. Some members of the Navy's high-altitude rocket research group were, therefore, casting about for a more promising situation for pursuing their research in the years to come.

There was good reason for the NRL researchers to consider the Jet Propulsion Laboratory as a possibility. Since the 1930s it had been at the forefront of rocket research and development in the United States. During the pioneering years of the 1940s and 1950s, the laboratory had furnished strong leadership to the country in rocket propulsion, making numerous contributions to the development of solid propellants and of rockets like the Army's Corporal and WAC-Corporal. Moreover, JPL had furnished the Explorer satellite that rode the Army Ballistic Missile Agency's Jupiter C rocket in the country's first successful response to the Sputnik challenge.[1] It seemed logical that the Jet Propulsion Laboratory would be deeply involved in rockets and space research as it had been in the past.

The laboratory staff expected to play a role, but Pickering and his associates were not sure just what role. The summer of 1958 was primarily a time to wait and see, and anyone who joined the laboratory would have to recognize the uncertainties and take his chances along with the rest of JPL.

Back in Washington the author reported to his NRL colleagues that JPL would probably have much to do with the space program, including space science, but that there was no assurance that the space science at JPL would be the atmospheric and solar research that the Naval Research Laboratory investigators had worked on for the past decade. Moreover, the real center of action on space would doubtless lie with the new National Aeronautics and Space Administration itself. As a consequence the thought of joining JPL was shelved, and the author and his colleagues pursued the idea of going to NASA, where over the next half year many of them found positions either in headquarters or in the newly formed Goddard Space Flight Center.

The Jet Propulsion Laboratory also joined the NASA family, transferred by presidential order on 3 December 1958.[2] Once fully under way, having cleared the initial hurdles of switching from largely ground-based research to primarily spaceflight projects, the laboratory proceeded during the 1960s and early 1970s to add luster to its already enviable reputation. Although there were mistakes and various kinds of problems to overcome, in time these minuses were greatly overshadowed by the pluses of spectacular achievements with Rangers and Surveyors to the moon; Mariners to Mars, Venus, and Mercury; and amazing feats in space communications using the JPL deep-space tracking network. The network included ground-based radar sounding of the planets. Most of what JPL did during NASA's first decade and a half concerned space science—the scientific investigation of the moon and planets with unmanned spacecraft—a natural extension of the laboratory's work in the 1950s, when its director was a member of the Rocket and Satellite Research Panel.[3]

A detailed review of these activities is beyond the planned scope of this book. Here only one issue will be treated, that of developing an effective working relationship between NASA and the Jet Propulsion Laboratory. The complex and frequently emotional matter consumed a great deal of time on the part of NASA space science managers on the one side and people of JPL and the California Institute of Technology on the other. The subject is important in illustrating how nontechnical issues can often make the accomplishment of technical objectives far from straightforward.

Singling out one topic from a rich and varied story like that of the Jet Propulsion Laboratory could distort the overall picture by undue emphasis on the one aspect. The reader should remember in what follows that even as the participants wrestled with knotty issues in human relations, the Jet Propulsion Laboratory's engineers and scientists were laying the groundwork for the phenomenal successes that were later achieved in investigating the moon and planets. While the very human strife between NASA Headquarters and the laboratory in the first half of the 1960s loomed large at the time in the minds and emotions of those involved, it was a passing phenomenon. The real and permanent image of the laboratory was to be seen in the utter dedication and superlative competence of its people and in their achievements.

THE QUESTION OF RESPONSIVENESS

Part of the problem was rooted in the unique status of the Jet Propulsion Laboratory in the NASA family. While the laboratory grounds, buildings, and equipment belonged to the government, the laboratory itself as an organization, a working team, was a creature of the California Institute of Technology. Within NASA a frequent question was whether the laboratory should be regarded as another center in the NASA complex—that is, as an insider—or be treated purely as a contractor—that is, as an outsider. For its part, JPL took great pride in its connection with Cal Tech, tenuous and neglected as this connection was. The association gave JPL a special access to the academic world. Also, in true academic fashion, Cal Tech accorded the laboratory a great deal of independence to plan and carry out its own research programs, although, as JPL Director Pickering later complained, Cal Tech's desire to have space science done on campus rather than at JPL sometimes stood in the way of JPL's developing the kind of program that NASA wanted.[4] It was an independence that the Army had accommodated and to which the JPL staff had become thoroughly accustomed.

In taking possession from the Army, NASA kept the arrangement under which Cal Tech would continue to exercise administrative oversight over the laboratory—for a substantial fee, "which in the early years of the association [with NASA] Cal Tech did very little to earn," as the first administrator, Glennan, put it.[5] But the space program would have an entirely different dimension from that of the projects previously engaging the attention of JPL, and NASA would request many things that the laboratory had previously shunned. The question quickly arose as to whether the Jet Propulsion Laboratory would accept program direction from NASA Headquarters or would negotiate a mutually acceptable program with NASA. Space agency managers like Abe Silverstein assumed without question that it would be the former, while the laboratory's management was determined that it be the latter. In fact, JPL people thought there should be no question about it, since the contract just signed with NASA actually did contain a mutuality of interest clause that called for NASA and JPL both to agree on programs and projects assigned to the laboratory.[6]

For years, until it was finally eliminated, this mutuality clause in the NASA–Cal Tech contract was a source of disagreement. From the very first, NASA Administrator Glennan, Deputy Administrator Dryden, and Associate Administrator Richard Horner were faced with a showing of independence and what headquarters viewed as a lack of responsiveness by JPL. These administrators had to spend what they regarded as an inordinate amount of time on questions of prerogative, time that would have been better spent on getting ahead with the space program. As Glennan would write years later:

> I think that JPL was the beneficiary of tolerance by NASA peers, was not really thought of as a responsibility by Cal Tech. I suppose that the payoff of success is the final answer—but did it need to cost so much in dollars, in tolerance and accommodation by Newell and others?[7]

As will be seen, the problem was not quickly resolved and if anything was even more intense when the second administrator, James E. Webb, took over in January 1961. Hugh Dryden and Robert Seamans, who had succeeded Richard Horner as associate administrator, continued to strive for a resolution of the problem.

But to the Jet Propulsion Laboratory, the mutuality clause was essential to preserve a cherished way of life that the laboratory viewed as a right, not only inherited from the past but also earned by competence and achievement. Moreover, JPL personnel could hardly be chided if from time to time they told themselves that it was circumstance rather than any previous history of leadership in rocketry that had put so many employees of the National Advisory Committe for Aeronautics in the driver's seat in the space program. To NASA managers, however, being in charge imposed responsibilities upon the agency. Were the Jet Propulsion Laboratory a Civil Service center, there would be no question about the authority of NASA Headquarters to decide on project assignments to the center. As a contractor the laboratory should be no less responsive to NASA direction.

Thus, while NASA and the Jet Propulsion Laboratory began their association with enthusiasm and great expectations, they also started with an arrangement that was interesting, to say the least. Add to this the principal players in the drama that was about to unfold, and conflict became a virtual certainty. Abe Silverstein, self-assured and customarily certain about what was the right way to go, would run a taut ship. He would welcome ideas and suggestions, but, once the decision was made—by NASA—he would expect his team to fall in line.

William Pickering was as stubborn as Silverstein was domineering. He had worked in cosmic ray physics at the California Institute of Technology, had been a charter member of the Rocket and Satellite Research Panel, and had shared in the pioneering of rocket instrumentation.[8] In 1954 he became director of the Jet Propulsion Laboratory. More than almost anyone else in NASA, except perhaps Wernher von Braun, he had a keen sense of his role as champion of his team, and he was not about to relinquish any of the laboratory's traditional independence without a fight.

When James E. Webb became administrator of NASA, the potential for conflict between NASA and the California Institute of Technology was substantially increased. Webb saw in the unique setup with JPL an opportunity to pursue within the NASA sphere itself the kinds of objectives he sought with individual universities in the memoranda of understanding he later attached to NASA's facility grants (pp. 232–35). Webb expected Cal Tech, with the Jet Propulsion Laboratory as a powerful drawing card, to foster and facilitate in the university community—particularly in California—interest and participation in space research. In this Webb would be pressing his hopes upon Lee DuBridge, president of the California Institute of Technology.

DuBridge, with an illustrious career in physics to point to and the successful management of the Radiation Laboratory at the Massachusetts Institute of Technology during World War II on his record, had no doubts about his ability and that of Cal Tech to run the Jet Propulsion Laboratory properly. An extremely sensitive person, DuBridge found any expressed or implied criticism of his institute or its laboratory distressing, and not always understandable. But he also found it difficult to satisfy Webb, or even to understand what the administrator wanted.

So the stage was set, and the story began to unfold in the fall of 1958.

MOON AND PLANETS

From the outset most assumed that the Jet Propulsion Laboratory would concentrate on the investigation of the solar system. This was much to the laboratory's liking, but the real interest was in the planets, not in the moon. In this, JPL immediately came into conflict with Administrator Glennan's desire to tackle the moon first. Just as the earth sciences had come before the moon and planets in the orderly and moderately paced development of space science, so in Glennan's view the moon should come before the planets. The JPL managers were, however, convinced that the Soviet Union, with the great lead it had already gained in space exploits, would quickly move ahead in the investigation of the moon also. America's only chance of recapturing the lead, they felt, would be to proceed at once to the planets.

This and other differences of view came out in a series of meetings of Abe Silverstein, the author, and other NASA representatives with William Pickering and his associates. The meetings at JPL in mid-January 1959 were devoted to a discussion of plans and policies,[9] the hope being to found a close working partnership between NASA and the laboratory.

Pickering made it clear that JPL would like to do nothing in 1959 that did not contribute to deep space probes. In particular he urged the development of a spacecraft fully stabilized in three axes, which would be a most effective vehicle for investigating deep space and the planets. The laboratory would do the engineering itself, using outside firms as subcontractors. The laboratory's past experience lay on the experimental side, and JPL wished to continue being the doer, keeping the supervision of other NASA programs to a minimum.

In turn, Silverstein emphasized the rugged job that lay ahead of NASA in monitoring the national space program and the hope that JPL would consider itself a part of NASA, not an outsider. As a member of the NASA family the laboratory would have to bear its share of monitoring outside contracts. Pickering responded that the laboratory would be glad to participate in headquarters committees, analyses, planning, and the like, but would refuse to undertake the detailed technical supervision of contracts. In that reply can be seen the underlying insistence on negotiating mutually acceptable work assignments that would be a central issue for the next several years.

In spite of the differences, the laboratory moved out on its assigned work and during the next two years well into the development of the Ranger lunar spacecraft and the planetary Mariner, largely in-house with assistance from outside subcontractors. For its part, NASA supplied the resources for expanding the laboratory's facilities and equipment and for increasing the staffing. NASA also undertook to reestablish the military channels previously open to JPL when it had worked for the Army—for example, to the Army Ballistic Missile Agency in Huntsville and to Cape Canaveral. In addition NASA continued to press JPL to expand its productivity through outside contracting. When work was begun on a Surveyor spacecraft to be soft-landed on the moon, a contract was given to Hughes Aircraft to do the job under the supervision of the Jet Propulsion Laboratory.[10]

With JPL, as with the rest of NASA, the first year produced both progress and wasted motion. It was a period of learning. At the request of the JPL leaders, the Vega upper stage intended for deep-space missions was assigned to the laboratory in the first months, only to be canceled within the year in favor of the Centaur stage.[11]

By the end of 1959, NASA management found it necessary to restrain its centers from diversifying their activities too broadly. Centers naturally tended toward self-sufficiency. An interesting line of research was often followed beyond the initial goal, even when this led a center into an area in which some other center was already competent. NASA management decided, therefore, that centers should be required to specialize more than they appeared to be doing and to avoid gross duplications. To this end Associate Administrator Richard Horner sent out letters assigning roles and missions to each center. The letter that went to William Pickering on 16 December 1959 confirmed that the Jet Propulsion Laboratory would have responsibility for lunar and planetary missions. On 21 December Abe Silverstein wrote Pickering, giving guidance on lunar and planetary missions for the immediate future. A week later the author with several of his colleagues from headquarters visited JPL to discuss the guidelines.[12]

Pickering quickly pointed out that the Jet Propulsion Laboratory had recommended emphasizing planetary investigations, whereas Silverstein's guidance seemed to start with a great deal of lunar work. Much to the displeasure of the JPL people, the NASA representatives made clear that the agency indeed was stressing the lunar work initially. In a lengthy discussion of policy for the space science program, it was agreed that NASA Headquarters would make tentative selections of experiments and experimenters for JPL missions, with the collaboration of the laboratory. The scientists would then develop prototype models or experiment designs and deliver them to the laboratory for evaluation. Final selection would be made on the basis of the JPL evaluation. Although this procedure was followed for awhile, actually it assigned to the laboratory more authority in allocating space on NASA payloads than was eventually permitted in NASA policy.

In this discussion the ever recurring issue of how to work with university and other outside scientists came up. Here the problem was how to meet both the needs of the project engineers who wanted to pin specifications down and fix schedules as early as possible and those of the scientists who wished to polish their experiments until the very last minute. NASA people sensed an inflexibility in this matter on the part of JPL engineers that boded trouble for the future.

Another topic that would recur many times over the years was how to attract the scientific community into the program. For the lunar and planetary areas JPL proposed to set up a committee along lines Pickering had suggested in an earlier letter to Silverstein,[13] but the NASA representatives indicated that headquarters would do this. After some debate it finally emerged that the laboratory was afraid that NASA would use the committee already established under Robert Jastrow for this purpose. The Jet Propulsion Laboratory would find it anomalous and disturbing to have a man from another center—the Goddard Space Flight Center—chairing a committee in a field that had been assigned to JPL. Once they appreciated what was disturbing the JPL members, the NASA people agreed to find a headquarters person to chair the committee.

That, however, was not the end of the matter. On 22 March 1960 Pickering returned to the subject in a letter opposing the idea of scientific discipline subcommittees to the Space Science Steering Committee in headquarters.[14] Pickering recommended that NASA get its advice on experiment proposals directly from the centers. JPL felt that the centers could, through their contacts with the scientific community, adequately represent the interests of that community. Pickering's proposal failed to recognize that NASA centers would also be competitors with outside scientists in seeking space on NASA spacecraft, and that there was a need to shield the centers from charges of conflict of interest—or even theft of ideas as was alleged on a few occasions—by having headquarters groups ultimately responsible for the selection of experiments and experimenters.

As work progressed, trouble continued to brew. NASA managers came to feel that the Jet Propulsion Laboratory's traditional matrix organization, which might have been fine for general research and smaller projects, was totally inadequate for large-scale projects with pressing deadlines. NASA also found the laboratory's record keeping, contract administration and supervision, and reporting inadequate. As a result NASA began a campaign to get Pickering to tighten up the organization and to improve the administrative side of the house. Since Pickering spent a great deal of time on outside matters—for example, with the American Institute of Aeronautics and Astronautics, in whose establishment he had played a leading role, and with the International Astronautical Federation and the International Academy of Astronautics—headquarters at first urged and later demanded that Pickering appoint a deputy to give continuous attention to the internal running of the laboratory. This last suggestion was especially disturbing to Pickering, who, despite NASA management's doubts about the quality of his leadership,[15] felt keenly his role as defender of his people. The question of a deputy for the laboratory remained a bone of contention for a long time, and even when one was appointed NASA felt that Pickering did not make proper use of the position.

The laboratory had its own complaints. At the NASA management meeting at the Langley Research Center in October 1962, at which Harry Goett had lashed out at headquarters for meddling too much in center affairs, Brian Sparks of JPL ran through an almost identical list of charges, showing that headquarters looked pretty much the same to the different centers. Sparks said that the laboratory felt headquarters took on too much project as opposed to program responsibility. For example: JPL did not have any real say on the matter of launch vehicles to be used; headquarters program chiefs dealt personally with individual project personnel instead of going through the project manager; the program office inserted itself into contracting matters and even asked contractors to quote prices for additional units on contracts managed by JPL; and headquarters insisted on approving the use of assigned construction funds. Additional complaints were that the Office of Space Sciences insisted on passing on the acceptability of every project the laboratory undertook, including study contracts, while the Office of Advanced Research and Technology similarly insisted on approving all advanced research before any funds could be released. JPL found it particularly irritating that other centers had been

encouraged to compete with JPL for planetary projects, expecially when
the planetary area had long since been assigned to the laboratory. Wernher
von Braun echoed Sparks on behalf of the Marshall Space Flight Center.[16]

In the debate that ensued headquarters people undertook to rationalize
their actions, but the important point had been made that headquarters had
to set its own house in order even as it pressed the Jet Propulsion Laboratory to make improvements. Each side worked hard, and with sincerity, on
these problems. But accommodation was difficult since two different philosophies were involved. The laboratory continued to insist on its inde-
dependence and fell back on the mutuality clause in the contract with
NASA to sustain its position. NASA insisted that JPL was a member of the
NASA team with the same responsibilities to headquarters that other NASA
centers had.

As time passed, technical problems—not unexpectedly—arose in JPL
projects, piling additional stress on that caused by the philosophical differences. The successful flight of *Mariner 2* to Venus in 1962 was encouraging, but the momentary elation was muted by a series of failures in the
Ranger project.[17] JPL might take some consolation in that it was the
launch vehicle, not the JPL spacecraft, that was the culprit in the first
several Ranger failures, but could hardly evade overall responsibility for
the missions. Both launch vehicle and spacecraft had to work to achieve a
successful mission, and until that happened both the laboratory and NASA
were on the spot.

To add to the difficulties JPL was also getting a reputation among
scientists of being intolerably difficult to work with. A subtle issue was the
construction of flight equipment to go on JPL spacecraft. JPL usually
insisted on taking prototype instruments developed by the scientists and having the flight hardware made itself.[18] The logic of this procedure was
obvious, but the potential impact on the scientific experiments was serious,
and experimenters usually objected. Many of the instruments were new,
developed specifically for the experiments to be performed. Only the experimenter and original designer of the instrument, who thoroughly understood the principles and details of the experiments to be performed, could
sufficiently appreciate the idiosyncrasies of the equipment to ensure suitable calibration. It was essential, therefore, that experimenters participate
in the preparation not only of prototypes but of flight hardware as well.
Indeed, in recognition of these points NASA policy was that experimenters
be held responsible for the proper functioning of their equipment. The
JPL approach kept the experimenters at arm's length and tended to frus-
trate their attempts to discharge their responsibilities.

Illustrating the difficulty of working with JPL as the scientists saw it
were complaints that Herb Bridge of the Massachusetts Institute of Technology and John Simpson of the University of Chicago aired in the fall of

1963.[19] Both scientists had similar stories of extreme difficulties in trying to work with JPL: no focal point for getting timely decisions; too many people in the loop; delays at JPL in meeting requirements of the scientists making the laboratory for all practical purposes the selector of experiments to go on a payload, rather than the Space Science Steering Committee in NASA Headquarters; intolerable delays in getting contracts out and money flowing to the experimenters so that they could get their work done on time; correspondence unanswered; a mixture of arrogance and rigidity, as, for example, when JPL considered itself sufficiently competent to try new instruments and techniques but would not allow the experimenters to do so. Van Allen of the State University of Iowa told of the frustration of having Iowa-built equipment pass all the tests that had been prescribed by the Jet Propulsion Laboratory only to have JPL people open up the equipment and then reject it because the construction techniques used were not those employed by JPL. If the tests prescribed by the laboratory were valid, then equipment that passed the tests, Van Allen insisted, should be accepted for flight.

JPL difficulties with the university community were of special concern to Administrator Webb. As pointed out earlier, he expected the connection with the California Institute of Technology to enable the laboratory to do a superior job in dealing with the university scientists and thus in making the opportunities to do space science more readily available to academic institutions than might perhaps be possible in the government centers. In fact, the administrator made much of this expectation in justifying to Congress NASA's paying Cal Tech more than $2 million a year to manage the JPL contract. Webb spent a great deal of time with Cal Tech President Lee DuBridge trying to get him to appreciate the importance of producing more for the annual fee than the mere routine administration of a contract, which NASA could have done for itself more cheaply.

As 1963 drew to a close, NASA stepped up its efforts to get the management of the Jet Propulsion Laboratory to improve performance and to strengthen the organization for managing big projects. Earl Hilburn, deputy to Associate Administrator Robert Seamans, was assigned the task of working with Cal Tech to resolve some of the fundamental differences. Hilburn, a technical man himself and a hard-nosed businessman to boot, insisted, with the weight of the Administrator's Office, that the laboratory find a suitable general manager. On 24 December 1963, Pickering informed the author by phone that he was in the process of setting up a new position of assistant laboratory director for technical divisions. Brian Sparks would be the new assistant director.[20]

This was progress, but in NASA's view fell far short of what was needed. An assistant director would not be the equivalent of a deputy director responsible for the internal management of the laboratory and empowered to make decisions binding on the director. But for the moment this appeared to be about as far as Pickering would budge. It took the dramatic failure of *Ranger 6* to break the logjam.

ACCOMMODATION

When *Ranger 6* separated from its launch vehicle on 30 January 1964 and slid onto a perfect trajectory toward its intended target on the moon, spirits ran high. As the telemetry record continued to show that the spacecraft was operating properly, success at long last appeared to be at hand. Three days later project people, NASA and JPL managers, contractors, experimenters, congressmen, and numerous visitors followed the progress of *Ranger 6* as it approached the moon; and when in the last seconds of the flight the signal was sent to turn on the television cameras, all were prepared to heave a sigh of relief. But then the unbelievable happened. The cameras didn't work![21]

The dejection of JPL and NASA personnel was complete. Although Congressman Miller, chairman of NASA's authorizing committee in the House of Representatives, expressed confidence in the Ranger program and congratulated NASA and JPL on hitting the target aimed at, there was no avoiding a thorough review of the project by the Congress. The author, at NASA headquarters, forwarded Congressman Miller's letter to Pickering with a note assuring JPL that NASA would work vigorously alongside the laboratory and expressing confidence that Ranger would succeed.[22] To determine what had gone wrong and what was needed to fix the spacecraft, NASA set up a review board under the chairmanship of Earl Hilburn.[23] On the Hill, Joseph Karth, chairman of the Subcommittee on Space Science and Applications in the House, got the job of probing the Ranger failure. From 27 April to 4 May 1964 the author and his colleagues in NASA, Jet Propulsion Laboratory managers, and JPL contractors—particularly the Radio Corporation of America, which had been responsible for the television equipment—were on the carpet.[24]

Although the congressmen were deeply interested in the technical side of the story and delved deeply into what had gone wrong, they gave their most serious attention to management matters. Karth, well aware of the mutuality clause in the NASA contract with the laboratory, appeared to feel that laboratory unresponsiveness to NASA direction might be the underlying cause of the trouble. Moreover, he wondered what, if anything, the government was getting in return for the large management fee paid to the California Institute of Technology. Although these were the very questions that NASA continually debated with Cal Tech and JPL, during the congressional inquiry NASA and the laboratory closed ranks in mutual defense. During the hearings the author tried to make the point that at the heart of the so-called unresponsiveness of the Jet Propulsion Laboratory lay the sort of individual competence and self-reliance that NASA was seeking to use in the space program. Testimony also pointed out that the kinds of problems that NASA was having with JPL at the moment stemmed from the very difficult undertakings being attempted, and in the nature of things the agency had the same kinds of difficulties with its Civil Service centers. When the chairman observed that if relations with NASA's centers were as bad as with JPL, then perhaps the investigation ought to be broadened to include the management of all NASA centers, the author replied that the proper point of view was that management relations with JPL were basically as good as with the other centers. But it had become clear that that line of defense was one to abandon as quickly as possible.

As a result of the investigation, the congressmen made clear to NASA that they were unhappy with the management arrangements between NASA and the Jet Propulsion Laboratory. They expected NASA to tighten up the government's control and to get rid of the mutuality clause in the contract with Cal Tech and JPL. Moreover, they were not at all convinced that the government was getting its money's worth for the $2 million annual fee to the California Institute of Technology. Considerable pressure was put on NASA to eliminate that fee, one way of doing which would be to convert the laboratory to Civil Service.

The pressure to remove the fee was excruciating to Lee DuBridge, for over the years Cal Tech had built the fee into its funding structure so that now it formed about 10 percent of the university's basic support. Sudden withdrawal of that sum would cause considerable difficulty. This possibility, and the publicity generated by the Ranger investigation, finally drew the attention of the Cal Tech Board of Trustees, who pledged themselves to help find a solution to the problem.[25]

Fortunately for Cal Tech, Webb was not in favor of pulling out. As mentioned earlier, the NASA administrator saw in the Cal Tech–JPL arrangement great possibilities for the kind of university-government relationships he was hoping to develop in the broader aspects of the agency's university program. Webb, therefore, stood firm against the outside pressure to change the management arrangement and renewed his efforts to wrest from DuBridge and Cal Tech the benefits he sought. As long as DuBridge was at the helm at Cal Tech, Webb strove in vain, for if ever two people spoke the same language with different meaning, those two were Webb and DuBridge. In whatever he said, Webb had in mind the broad, sweeping contribution that he thought a university should be able to make to government in expertise and wise counsel, while DuBridge never relinquished his dedication to the traditional independence of academic institutions and of the individuals within those institutions.

In his hopes Webb was repeatedly disappointed. Cal Tech showed little interest in broadening the use of the JPL capability by other universities, which Webb very much hoped to bring about in the national interest.

To make matters worse, JPL proved to be pretty good at antagonizing outside experimenters assigned by NASA to JPL spacecraft, by keeping them at arm's length and imposing unreasonable schedules and what seemed to the experimenters to be arbitrary and unnecessary construction and test requirements for their instruments.

Having got through the congressional inquiry, the Ranger managers bore down again on preparations for another flight. Oran Nicks, head of the Lunar and Planetary Program Office within NASA Headquarters, and his people—who through all that had happened had remained unshakeable in their faith in and esteem for the laboratory—redoubled their efforts to assist JPL in whatever ways they could. Walter Jakobowski, Ranger program manager, did what he could to facilitate the work. Benjamin Milwitsky, program manager for Surveyor, which was having plenty of troubles of its own, worked assiduously to keep Surveyor from repeating the *Ranger 6* fiasco. But it was the JPL engineers and their contractors who pulled it off. They left no possibility for trouble unprobed, no component, no subsystem unchecked, no test undone, to ensure that the next flight succeeded.

The going was not easy. Troubles continued to turn up in ground-based tests, so that in June 1964 the Office of Space Science and Applications set up its own review board separate from that chaired by Hilburn, its purpose to leave no stone unturned in the effort to make Ranger succeed.[26]

Simultaneously pressure continued for JPL to tighten up its management and to be more responsive to NASA direction. After all that had happened following the *Ranger 6* failure, after all that had been said about the need to tighten up management and improve responsiveness, one would have thought that JPL had the message. The author and his deputy, Edgar Cortright, were shocked, therefore, to learn in a conversation with Pickering in early July of 1964 that JPL considered Surveyor a low-key project which could be kept on the back burner, with the contractor left pretty much to his own devices. Cortright and the author disagreed on the spot, and on 13 July a letter went out to Pickering underlining that Surveyor was considered one of the highest priority projects in the space science program and that the project had to have proper management attention. The letter asked that Pickering be certain "that JPL is properly staffed and organized, the Hughes contract is adequately monitored, and NASA Headquarters appropriately informed of Surveyor needs, to insure the earliest and fullest possible success of the Surveyor program."[27]

The following day a second letter to Pickering dealt with management problems. It requested that JPL develop a more formalized discipline in both business and project management. In particular NASA requested that the rather loose matrix organization that JPL had favored be tightened into a more direct project organization. The letter expressed concern that space science had a fuzzy sort of place in the laboratory structure and asked that it be given a firmer, more independent status. NASA asked that JPL work on improving relations with experimenters. The following September the author repeated these requests to Lee DuBridge, president of Cal Tech and accordingly Pickering's boss.[28]

The continuing lack of response to NASA's requests led NASA management to give serious consideration to insisting that Cal Tech remove Pickering as director of the Jet Propulsion Laboratory. But Pickering had too much to offer to make this a palatable move. Another option seriously considered was that of converting the laboratory to Civil Service as some congressmen had favored. But again the administrator considered this too drastic. Setting aside the question of whether the necessary personnel authorizations could be obtained from an administration that was trying to reduce the total number of government employees—and ignoring the dislocations that would be generated in adjusting to Civil Service salaries, retirement plans, and fringe benefits—there was still the question of how many of the employees would stay. The fierce pride that JPL people took in their heritage as part of the Cal Tech family left grave doubts as to whether the laboratory could be converted without seriously disrupting the ongoing program.

At any rate, none of these unsavory options was adopted. Instead the contract with the California Institute of Technology was revamped.[29] The mutuality clause was removed, and JPL was required to be responsive to NASA direction. Specific organizational and management arrangements were required, including the strengthening of contract administration and provision for adequate accounting, record keeping, and reporting. On Webb's insistence the new contract called for NASA managers to evaluate semiannually the performance of Cal Tech and JPL, with the total fee to Cal Tech depending on the rating received in the evaluation. Of all the provisions in the new contract, the one requiring the institute and the laboratory to undergo periodic evaluation—an indignity that DuBridge pointed out was not imposed on other NASA centers—rankled the most. Sweetening the pill, however, NASA agreed to provide a small fund (a few hundred thousand dollars annually) for the director of the laboratory to use at his own discretion to support research he deemed especially important.

The new contract provided no magic solution. Much still had to be done to settle the dust of battle and to establish a smooth working pattern. That occupied an appreciable amount of management time during the next several years. But the road had been cleared and it was a matter of bending to the task. Moreover, with the Ranger hurdles behind, successes became the rule, failures the exceptions, on JPL missions. In the light of these successes the earlier troubles faded farther and farther into the background. On 28 July 1964 *Ranger 7* took off from Cape Kennedy for the moon, matching *Ranger 6* in the flawlessness of its flight. But this time the television worked perfectly. The cameras returned superb pictures of a lunar mare—later designated Mare Cognitum, or "Known Sea," by the International Astronomical Union. Those pictures taken just before the spacecraft hit the moon were a thousandfold more detailed than any that could be obtained through ground-based telescopes. On 31 July, three days after the launching and immediately following the completion of the mission, Dr. Pickering and a beaming JPL team held a happy press conference in which some of the Ranger pictures were shown and their scientific value discussed. Then Pickering and the author flew to Washington to brief President Johnson, who expressed his great pleasure in the achievement. On 11 August Congressman Karth, who half a year earlier had dug so grimly into the Ranger troubles, inserted into the record of the House of Representatives a paper by the U.S. Information Agency describing the worldwide admiration that *Ranger 7* had evoked.[30]

Ranger 8 (20 February 1965) and *9* (24 March 1965) were equally successful and more visible, since they were covered on live television. Then, after excruciatingly troubled years of development and testing, the very first Surveyor landed gently on the moon's surface on 2 June 1966 and began to send pictures and other lunar data back to earth.[31] Not a vestige of doubt remained that the Jet Propulsion Laboratory could match technical performance with the best that the country had to offer.

Not that the laboratory itself or those in NASA's lunar and planetary office had ever doubted that they could do it. Oran Nicks and his people would frequently say that they were working with the most competent team in the space science program. In the end, results were eminently satisfying.

At the division level much effort had been invested in trying to understand each other's needs and aspirations. NASA representatives had spent a great deal of time at JPL keeping in touch with what was going on. In return JPL members had been invited to spend tours of duty at NASA Headquarters to become familiar with the problems on the Washington end. Without doubt this was helpful. On returning to JPL, Gregg Mamikunian wrote the author in May 1966 expressing appreciation for the opportunity to work at NASA Headquarters for a while. He expressed his "painful realization and awareness that decisions in regards to projects or missions at headquarters are not arbitrarily or whimsically arrived at (as is the . . . consensus at the centers and universities) but with . . . regard . . . to the objectives of the scientific community at large and of the nation."[32]

Webb's new contract requirement for a periodic evaluation of the laboratory was intended to generate at the upper management levels the kind of familiarity with each other's views that those at the working level had already achieved to some extent. In this the device was successful. A pattern developed in which, before the actual evaluation, NASA and the laboratory agreed on the items to be rated, on both the technical and administrative sides. Then a preliminary written evaluation was drawn up from suggestions from the various NASA managers. Cal Tech and JPL were given an opportunity to review the preliminary evaluation and prepare for a face-to-face meeting with NASA, where JPL and Cal Tech could take exception to ratings they deemed unfair. Following the meeting the Office of Space Science and Applications revised ratings as appropriate and submitted the resulting evaluation to the administrator for approval.

Fortunately, by the time of the first evaluation in June 1965 the Jet Propulsion Laboratory had a number of items on which it could be given a rating of outstanding, including recent Ranger successes.[33] But it was quite a while before many outstanding ratings could be handed out for the administrative side. Nevertheless, as time went on the ratings improved.[34] The process forced a continuing attention to the many administrative problems that had dissatisfied NASA in the past, and the ratings provided JPL and Cal Tech with a measure of how well they were meeting the NASA requirements. Thus, as the 1960s drew to a close and JPL was preparing for the spectacularly successful flights of Mariner to Mars in 1969, administrative relations between the center and headquarters were on an even keel. Not that all problems were solved, but the most significant matters were now the technical ones, as one would want.

In retrospect, given the Jet Propulsion Laboratory's former style of in-house engineering and distaste for much that was required in contracting with industry for projects, given also the laboratory's priority over the National Advisory Committee for Aeronautics in rocket research, and considering the strong personalities involved, an intense struggle between JPL and its new bosses was predictable. No doubt, in time some sort of accommodation would have been worked out by degrees. But the *Ranger 6* failure did not permit the gradual course. To preserve the arrangement that Administrator Webb wished to exploit in the university community, NASA had to tighten up management and insist on a visible improvement in performance. A revamped contract provided the basis for working out a solution. Strong efforts by men of good will on both sides made it work.

16

Life Sciences: No Place in the Sun

Throughout the 1960s the life sciences were something of an enigma to the highest levels of NASA management. Partially this was because no individual near the top of the hierarchy had training in any of the life science disciplines. But there was more to it than that. One could sense an ambivalence in the life science community concerning the space program, a fascination with its novelty and challenge mixed with skepticism on the part of most that space had much to offer for their disciplines.

Not that NASA wasn't concerned with life sciences in a variety of ways. The list of NASA interests was a long one: medical support to manned spaceflight, environmental control and life-support systems for manned spacecraft, spacesuits and other protective systems, nutrition, aviation medicine, man-machine relationships, space biology (the study of terrestrial life forms exposed to conditions in space), exobiology (the search for and study of extraterrestrial life and life processes), plus occupational medicine and employee health programs. But much of this interest was incidental to other, primary objectives of the agency. Aviation medicine and man-machine relationships supported the development of aeronautical instrumentation and techniques. Although an extensive amount of work was required, nevertheless spaceflight medicine, environmental control, life support systems, spacesuits, etc., were narrowly constrained to the minimum needed to ensure the attainment of the Gemini, Apollo, and other manned spaceflight objectives. Only space biology and exobiology could be regarded as pure science, and these fell into the space science program.

NASA's philosophy concerning the life sciences was simple: where science was the objective, make the most of space techniques to advance the disciplines; in other areas do only what was essential to meet the need. A natural outcome of this philosophy was to disperse the different life science activities throughout the agency, placing each in the organizational entity it served. Thus, except for the brief period from March 1960 to November 1961 when the agency had an Office of Life Sciences Programs in headquarters,[1] space biology and exobiology were placed with the other space science groups; aviation medicine and related activities were in the Office of Advanced Research and Technology, which had responsibility for NASA's aeronautical program; and space medicine was placed under the direction of the Office of Manned Space Flight. The single life sciences office had not worked, doubtless for a variety of reasons; but one reason that suggested itself was the separation of the life sciences activities from the other activities with which they were most naturally associated in the NASA program. The Office of Space Sciences, for example, already had a group producing and launching sounding rockets and unmanned space-craft for space research. Rather than duplicate such a group in another office it seemed to make sense to place space biology and exobiology close to their tools in the Office of Space Sciences.

While the dispersion of the life sciences throughout the organization made sense to NASA managers, and the arrangement appeared to function more effectively than had the temporarily integrated one, the setup was not to the liking of the outside life sciences community. Dissatisfaction with the way NASA handled its life sciences program endured throughout the 1960s. Since it was principally the researchers who were most vocal in ex-pressing their displeasure, NASA space science managers came in for a great deal of the flak directed at the agency.

Although space medicine, which in the NASA setup formed a part of the manned spaceflight organization, achieved extensive results, space biology and exobiology produced only modest returns during the 1960s.[2] Even though some interested experimenters had used sounding rockets in the pre-NASA period to expose seeds, mice, and other biological specimens to the rigors of rocket flight and high-altitude radiations,[3] nevertheless when NASA came on the scene the life scientists were not ready to keep pace with the astronomers and physicists in the space science program. Whereas the latter could bring space instrumentation directly to bear upon fundamental problems already engaging their attention—earth and plane-tary atmospheres, solar activity and sun-earth relationships, stellar spectra, cosmic rays, and cosmology, to mention some—the same was not true for the life scientists. During the 1950s and 1960s a revolution was in progress in the life sciences for which the center of action was the ground-based laboratory. There researches in areas like molecular biology, the genetic code, immunology, and information storage and transfer in biological sys-tems held the attention of the best investigators. It was not clear in what way space research could make more fundamental contributions than these ground-based studies.

A number of experimenters however, wanted to try their hand at space research. Catering to this interest, a small but determined group within NASA worked hard to promote the field of space life sciences.

The first biologists in NASA Headquarters included Richard S. Young, who in 1958 and 1959 had flown sea-urchin eggs in recoverable Jupiter nose cones launched by the Army Ballistic Missile Agency. In February 1960 Young went to the Ames Research Center to start NASA's first life sciences laboratory.

In the space science group, which initially was almost entirely preoccupied with the physical sciences, the author persuaded Freeman Quimby, a biologist from the Office of Naval Research in San Francisco, to come east and work with NASA to make a place in the space science program for space biology and exobiology. Later, at the time John Holloway and Donald Holmes joined the NASA university program staff, their boss—Orr Reynolds, head of research in the Office of Defense Research and Engineering—also came to NASA to take charge of the biology division in the new Office of Space Sciences. A physiologist, Reynolds was skilled in the ways of government programs and how to make them work. Appreciating the opportunities for biological research afforded by rockets and spacecraft, yet at the same time recognizing the factors that would militate against any widespread interest, he set about trying to acquaint his colleagues with what might be done in space. He was remarkably successful, and under his guidance interest in space biology grew.

In fact, many important questions could be examined with space experiments. What effects, for example, might prolonged weightlessness have upon living organisms? What would happen to plants grown in the absence of gravity? How would frog's eggs and sea-urchin eggs fertilized in space develop in a weightless environment? What light could space experiments cast upon the role and importance of gravity in the development of such eggs on the ground? How would a frog's otiliths—the tiny stones in the ear that sense the direction of gravity—function in the absence of gravity? What might exposure to radiations in space do to biological specimens, particularly in the production of mutations? Although the physicists insisted that order of magnitude considerations showed that there could be no significant effect, still some biologists wondered if exposure to radiations under weightlessness might produce different effects from those observed at one g on the ground. Then a whole class of intriguing questions concerned the rhythms that organisms exhibit in the environment existing at the earth's surface. Many of these rhythms are linked in some way to external periodicities such as the day-to-night variation in sunlight or the lunar month. In orbit a new set of periodicities would exist, those associated with the spacecraft's period of revolution in its orbit. How would these influence the circadian—i.e., nearly daily—and other rhythms of plants and animals in orbit? How would these rhythms respond to flight on an escape trajectory from the earth on which there would be no orbital periodicities?

To enable experimenters to study some of these questions, NASA flew a number of sounding rockets and several recoverable satellites named Biosatellite.* Of the three Biosatellites placed in orbit, two were recovered for further studies of the specimens after flight.[4]

Although the experimenters themselves were enthusiastic about the opportunity to experiment in space, many scientists considered NASA premature in the Biosatellite project. There were two different philosophies. The satellite experimenters were willing to conduct exploratory investigations, to learn what they could from their initial experiments, but—more important—to use the early research for obtaining an insight into just how rockets and spacecraft could contribute in future experiments. To others such suggestive experiments were not enough. More in keeping with life science tradition, they would hold off from experimenting in space until laboratory research had made it virtually certain that definitive experiments could be performed. To these persons the results from the first successful Biosatellite (*Biosatellite 2*, 7-9 September 1967, in which plants, flies, and other living organisms were flown to determine the effect of space conditions on living organisms) were perhaps interesting but not particularly significant. When in *Biosatellite 3* (29 June-7 July 1969) prolonged weightlessness appeared to generate critical fluid imbalances in an instrumented monkey, who actually died from the stresses produced, that was considered significant but not definitive, since the experiment was marred by incomplete preparatory research and inadequate ground and other controls.

In the light of these thoughts, the Space Science Board, NASA advisory committees, and various members of the life sciences community continually advised NASA to support a great deal of advanced research on the ground to establish an adequate basis for experimenting in space. NASA did support such research, but the program was considered inadequate. Indeed, many life scientists would have preferred to see all the money that went into satellite work devoted to laboratory research until a better basis could be laid for going into space.

In this respect those interested in exobiology were in better shape. For one thing, hardly anyone would disagree that the discovery of life on some other planet would be an exciting event with tremendous philosophical and scientific implications. The study of such life in comparison with earth life would be of fundamental importance. Even if no such extraterrestrial life were to be found in the solar system, the opportunity to investigate other planets in pristine condition and to study the prebiological

---

*The NASA program drew extensively upon Air Force technology; and Air Force interest in aerospace medicine, which antedated the creation of NASA, continued after NASA was formed.

chemistry of these bodies would be a valid line of investigation for the life sciences. In the United States a number of competent investigators—Nobel Laureate Joshua Lederberg, Wolf Vishniac, and Norman Horowitz among them—were attracted by the intriguing possibilities.[5] Worldwide interest in the subject stimulated much discussion in the Committee on Space Research and other scientific circles, and led to international agreements on planetary quarantine (pp. 303-05). Since at NASA's inception it would still be many years before an automatic laboratory might be landed on Mars (the primary target of the United States) or on Venus (where the Russians made their first successful landings), time was ample for the sort of preparatory work that NASA's advisers urged. More than a decade of such advanced research preceded the launching of Viking in 1975, which was instrumented to probe the Martian surface for evidence of microbial life.[6]

In the intervening years, scientists did their best with photographs and spectrograms of the planets to glean any hints on the possibility of extra-terrestrial life. They searched for signs of water or other molecules that might be associated with life. The pictures of Mars obtained from a Mariner spacecraft in 1964, which showed a moon-like surface that appeared to be perfectly dry, held out little encouragement for exobiolo-gists. But when *Mariner 6* and *7* in 1969 and *Mariner 9* in 1971 obtained closeup pictures of the planet revealing features that looked like ancient water channels and alluvial fans, and a considerable amount of water ice in the polar caps, hopes ran high once more and experimenters bent more vigorously to the task of preparing for the Viking lander flights to come.[7]

Nevertheless, in the 1960s criticism of NASA's life sciences program remained high. During the decade almost every advisory committee meeting on the subject deplored some aspect of the program. The critics were impartial in bestowing their criticism. While the space science program was called to task for not supporting enough preparatory research and not including adequate controls in the space experimenting that did take place, manned spaceflight was berated for not doing enough background research to ensure the safety of the astronauts. This latter criticism increased following the monkey's demise in *Biosatellite 3*. Even though the experiment was thought to have been carried out poorly, the results were alarming to many who felt that similar disasters might befall astronauts unless proper steps were taken to forestall the difficulties. To do this would require understanding thoroughly what was going on, and research was needed to get that understanding.

As the scientists criticized the substance of NASA's program during the 1960s, they also had much to say about its organization. The dispersal of life sciences activities throughout the agency was the main target of their displeasure; indeed, this criticism often seemed more intense than their dissatisfaction with the program. In fact, the two criticisms were related. The various disciplines in the life sciences, including the applications of research results to medicine, were interrelated, the critics pointed out, and an effectual total program could be achieved only if all parts were properly integrated into the total. This could be done only by someone trained and competent in the life sciences who had the authority to pull it all together. Such a person would have to be in top management so that he could bring adequate weight to bear on planning, budgeting, and the use of funds.

Such were the thoughts of the biologists and medical researchers at the space science summer study in Iowa City in 1962, when they urged NASA to reverse its recent action in dispersing the different life science activities throughout the agency.[8] Much was made of the fact that all of the technical people in NASA's top management were trained in the physical sciences or engineering. Along with this recommendation went a related one, that NASA use the peer review system used by the National Institutes of Health to decide which research proposals to support.

Although Associate Administrator Robert Seamans did engage a physiologist, Nello Pace of the University of California at Berkeley, to review and recommend on NASA's life sciences organization, still when the time came to make a decision Seamans was not prepared to accept the summer study's recommendation. For one thing, NASA's life sciences effort was relatively so small that a separate office for it would be incongruous alongside the other much larger program offices. More important, except for space biology and exobiology, NASA was not conducting life sciences research for its own sake. As pointed out before, most of NASA's work was directed toward other ends. Finally, with regard to using a peer system for reviewing research proposals, NASA did not have large sums of money set aside specifically for university research. As with the physical sciences and other areas, the program offices distributed their monies where they would best support the flight objectives of the agency.

The life scientists, however, were serious and constant in their recommendations. In spite of their paradoxical general disinterest in space life sciences, the community continued throughout the 1960s to send the same recommendations to the agency. And for the reasons that Seamans had cited originally, the agency continued to hold back until in 1969 and 1970 two different committees once again urged on NASA the importance of strengthening its setup in the life sciences.

In November 1969 the President's Science Advisory Committee released a biomedical report from what was called the Stead Committee of PSAC's Space Science and Technology Panel.[9] The committee noted that manned spaceflight afforded a good opportunity for biomedical research and urged that the necessary ground-based research be done to develop the cadre of people needed to take best advantage of this opportunity. The recommendation derived from the long-standing complaint that NASA tailored its biomedical program too closely to the operational needs of manned spaceflight and that hence a great deal of potentially valuable research was being left undone.

Half a year later the Academy of Sciences conducted a space science summer study at the University of California at Santa Cruz, under the chairmanship of Kenneth Thimann. The subject was space biology. The study presented its report to the Space Science Board on 13 January 1970.[10] The report was critical of NASA's space biology program, strongly recommending that more preparatory work be done on the ground. NASA people who had audited the summer study discussions were already aware of what was coming. In fact, Wolf Vishniac—research biologist from the University of Rochester, experimenter in the NASA program, and a member of the Space Science Board—complained to the author and John Naugle, then head of space sciences, that the board was rigging the study of space biology in such a way as to kill the program, by choosing participants who could be expected to return a negative report.[11]

Because of the continuing displeasure being expressed, which had seemed to increase in intensity in the last year or so, the author wrote Philip Handler, president of the Academy, asking that the Academy conduct still another study on life sciences in NASA.[12] The letter was discussed at the same Space Science Board meeting at which the Thimann Committee report was reviewed. There was some reluctance to make another study in the wake of so many previous ones. It was pointed out that NASA already knew the scientific community's views on the subject and could, if the agency so wished, even now take the advice it had been receiving for a decade. But many of the more recent studies had specialized in only one aspect of the life sciences, such as space medicine or space biology, whereas NASA wanted an up-to-date look at the entire program, including questions of organization and management. The Academy agreed to do it.

It was an illustrious group that met at Woods Hole in the summer of 1970 under the chairmanship of renowned biologist Bentley Glass to go over the NASA life sciences program once more. After weeks of thorough review and discussion, the committee prepared its report. As predicted, the recommendations updated those that NASA had been receiving for the past 10 years: strengthen the ground-based research program; use various devices to attract better researchers into the program, such as NASA life sciences fellowships much along the lines of the resident research associateships that Robert Jastrow had instituted 11 years earlier; pull all life sciences in NASA together into a single office of equivalent status to the other program offices; and provide a more effective arrangement for getting advice from the life sciences community.[13]

This time, at the author's urging, NASA decided to accept the main thrust of the summer study's recommendations.[14] At the time, declining budgets and the political climate made it unwise to create a whole new office of life sciences. Nevertheless, the agency decided to do the following:

- Place responsibility for all NASA life science activities in the hands of a single director.

- Put much of the life sciences staffing under the new director; however, to maintain certain natural working relations—for example, between those working on man-machine interactions and the aeronautical research groups—a few life science elements would still be placed elsewhere in the NASA organization.

- Require the new director to review and approve all life science budgets, so that a properly integrated total life science program could be developed.

- Make the associate administrator the point of contact, within the Office of the Administrator, for the life sciences director.

- Arrange for frequent meetings of the administrator and deputy administrator with the director of life sciences to discuss progress and problems.

- Create a Committee on Life Sciences under NASA's Space Program Advisory Council.

Also, the agency would support a number of life sciences fellowships along the lines recommended by the summer study. Since most of the life sciences budget went into the biomedical program associated with manned spaceflight, the new office was placed administratively under the associate administrator for manned spaceflight. In mid-November the author called Bentley Glass, chairman of the summer study, to inform him of NASA's plans relative to his committee's recommendations.[15] Although NASA's plans did not go as far as the committee had asked, Glass was pleased with the agency's positive response. NASA's failure to put the program office for life sciences at the same level as the other program offices was a disappointment, but in the circumstances understandable. Placing all life sciences under a single director was the improvement most sought by the scientists.

The Academy of Sciences made a long list of potential candidates for the new job available to NASA, and several Space Science Board members offered their assistance in trying to get one of these to take the job. But here again, as Administrator Webb had found years before in searching for a chief scientist for NASA, it was not possible to lure first-rate researchers away from their academic posts to take on the bureaucratic headaches of administering a program that had yet to sell itself. So, after considerable search for someone from outside, Dale Myers, head of the manned spaceflight office, appointed a NASA man, Dr. Charles A. Berry, a clinical M.D.

who had achieved phenomenal success in dealing with the needs of the medical program for Gemini and Apollo. NASA's advisers were worried about two aspects of this appointment: first, Berry was not a research man; second, he came from the Johnson Space Center, which had consistently frustrated efforts of the community to get NASA to expand the research component of the biomedical program. But having failed to come through with anyone from the outside research commuity to take the job, the scientists were in a rather weak position to complain.

Under Berry the new arrangement made sluggish progress toward the objective of a properly unified life sciences program. When Berry left in 1974 to assume the presidency of the University of Texas Health Sciences Center at Houston, Dr. David Winter from the Ames Research Center was named to replace him. Winter, a research man, was closer to the sort of person the life sciences community had hoped to see as director of NASA's life sciences program. In November 1975, after the close of the Skylab project, the life sciences office was transferred from manned spaceflight—now renamed the Office of Space Flight—to the Office of Space Sciences. Although this still left life sciences lower down in the organization than the scientists would like, nevertheless the new location afforded the research atmosphere they desired.

Thus, in the 1970s NASA was in a better position than before to work closely with members of the life sciences community in putting space techniques to use for medical and biological research. Inasmuch as the 1970s were to be a period of transition from the use of expendable rockets to the use of the Space Shuttle—which appeared to hold particular promise for life science research in space—it was doubly satisfying that NASA had found a way of accommodating itself more closely to an important group of its clients.

17

Leadership and Changing Times

Of all the responsibilities placed on the newly created National Aeronautics and Space Administration, perhaps the most obvious yet the most difficult to define was that of leadership. Glennan's leadership, embracing an enthusiasm for space research and exploration tempered by a willingness to build slowly and solidly, was ideal for getting the nation's space program under way. Space science managers were able to put together a wide-ranging program of earth and planetary sciences, solar physics and astronomy, and some space life sciences. Of equal importance, they were able to establish with the scientific community the kind of relationship that would draw researchers of high quality into the program.

NASA's Administrators

Like Glennan before him, the second administrator, James E. Webb, strongly supported a balanced program of science, technology, application, and exploration. His policies assured each of the areas a place in the overall program. On the space science side relations with the scientific community continued to follow the patterns established during Glennan's tenure. The principal changes were those brought about by the expansion of the program that took place under Webb, in which Gemini and Apollo were undertaken, the university program was increased, and the pace of the space science program was stepped up.

All in all, the course of leadership during Glennan's time and in the first years of Webb's tenure was relatively smooth. Reasonably well-thought-out projects were relatively easy to sell. With rapidly increasing budgets it was not too difficult to maintain a respectable balance among the various areas, even though different interests might quarrel with the relative emphases NASA gave to the different parts of the program.

The problems facing the agency were those having to do with getting on with the program.[1] Manned spaceflight people had to decide on the mission mode for Apollo: whether to use direct ascent, which Abe Silverstein favored; or to go first into a near-earth parking orbit and then on to the moon, which the President's Science Advisory Committee strongly urged; or to go into a lunar parking orbit from which to land on the moon, which the agency finally chose. Applications managers had to work out relations with industrial users of space technology and with other government agencies like the U.S. Weather Bureau and the Department of Defense. Decisions were to be made on the kinds of weather and communications satellites to develop and who would operate them. On the space science side, it was necessary to determine what balance to maintain between observatory-class spacecraft, which Abe Silverstein favored, and the smaller, cheaper ones that the scientific community preferred. Experiments and experimenters had to be selected for the missions to be flown. How much ground-based work should be funded as preparation for later flight experiments had to be decided. Much management time was devoted to resolving conflicts between the manned flight and space science programs—for which purpose George Mueller, associate administrator for manned spaceflight, and the author, associate administrator for space science and applications, finally agreed on the creation of a special manned space science division. It was headed by Willis Foster, one of the scientists who had come to NASA from the Office of Defense Research and Engineering in the Pentagon. Contrary to one of the cardinal principles of organization and management, Foster was to have two bosses—Mueller and the author—an arrangement that was intended to give his division equal access to both the Office of Manned Space Flight and the Office of Space Science and Applications.[2] Foster's was an extremely difficult role to play, for the manned spaceflight office tended to view science as something that might support the achievement of the Apollo missions, whereas the space science managers wanted the agency to view manned spaceflight as a technique that could serve pure science and other primary objectives of the agency.

Yet, difficult though they were, these problems, including those of Foster's division, were relatively straightforward. In a climate of positive support to the space program, they were part of the price to pay for accomplishing established goals. But in the late 1960s, demands on leadership changed severely in character. Under the best of circumstances the Apollo 204 fire on 27 January 1967 would have been difficult to live down.[3] But coming at a time when the country was becoming more concerned about a variety of problems other than whether the United States was or was not ahead of the Soviets in space, the impact of the accident upon the agency was immeasurably increased. A great deal of Administrator Webb's time was taken up in recouping for NASA the respect it had been building up in the Mercury, Gemini, and other programs, and in regaining the confidence of the Congress. That in Apollo the United States was on trial, as it were, before the whole world had much to do with the program's continuing to receive support. But in the aftermath of the congressional hearings and internal NASA reviews, Webb began to sense a slackening of support for the space program.

After peaking in 1966, NASA's annual expenditures began to decline sharply as spending on the building of the Apollo hardware passed its peak. Normally one might have expected at this stage to begin a small amount of advanced work on some new project to replace Apollo after it had been completed. And after the considerable effort put into selling Apollo as a project to develop a national capability to explore and investigate space, it was natural for NASA managers to think of putting the Apollo and Saturn equipment to use. NASA planners began to talk of an Apollo Extension System.[4] But when the idea of extending the Apollo project did not go over too well, a new concept was introduced: the Apollo Applications Program.[5] The name was meant to emphasize "applying" the Saturn and Apollo capability to other research, thereby capitalizing on the very large investments the country had made to bring that capability into being.

During the muddy period of planning for an Apollo Applications Program that was not going to sell, Webb often stated to his colleagues in NASA that he did not sense on the Hill or in the administration the support that would be needed to undertake another large space project. When NASA managers wanted to come to grips with the problem, to decide on some desirable project like a space station or a manned base on the moon and then work to sell the idea, Webb preferred to hold back and listen to what the country might want to tell the agency. It was his wish to get a national debate started on what the future of the space program ought to be, with the hope that out of such a debate NASA might derive a new mandate for its future beyond Apollo. But no such debate ensued. In a country preoccupied with Vietnam and other issues, the space program no longer commanded much attention. If any leadership was to be provided, NASA would have to do it, since that vague "they" out there were not going to.

In this climate the administrator became increasingly concerned about the timing of the decision to send astronauts off on their first flight to the moon. Added to the Apollo fire, a disaster out in space in which astronauts were killed in full view of the world might well destroy not only the Apollo project, but NASA itself. In the summer of 1968, as the Manned Spacecraft Center people were coming down the final stretch in their preparations for a circumlunar flight, Webb was in Vienna attending the international symposium on space applications sponsored by the United Nations Committee on the Peaceful Uses of Outer Space (p. 300). Thomas O. Paine, who had been appointed deputy administrator when Robert Seamans decided to leave the agency,[6] was at home in Washington minding the shop, and it fell to him to guide the agency toward the first manned lunar flight. When Webb resigned in October,[7] the final go-ahead came from Paine as

acting administrator. While mindful of the hazards, still it was clear to
Paine that the flight had to be attempted some time, and if the Apollo
team was ready it should be now. When *Apollo 8* came through with flying colors, the decision was fully justified and NASA recaptured for the
time being the admiring attention of the world.[8]

Webb's resignation had anticipated the change in administrations that
would bring a searching reappraisal of the space program. Although prepared to reap political harvest from each Apollo success, incoming President Richard Nixon was committed to an all-out attack on inflation that
would call for some painful belt tightening. To those who chose to read
the signals, it was clear that the Republican administration was not about
to let the space budget climb again to its mid-1960 levels. The big question
in the minds of space planners was how low Nixon would let the budget
drop.

As an early step in assessing the space program, on 3 December 1968
President-elect Nixon asked for recommendations from a group of outside
consultants under the chairmanship of Nobel Laureate Charles Townes,
who was chairman of both the Space Science Board and NASA's Space
Technology Advisory Committee. Nixon received the report of the task
force on 8 January 1969, but did not at the time choose to release the
document.[9] The report recommended continuation of a $6-billion-per-year
space effort, with one-third of the funding for the Department of Defense
and two-thirds for NASA. The task force disapproved of any commitment
to a large, orbiting, manned space station, but supported the development
of a space shuttle. The scientists urged a strong program of unmanned
planetary probes. Of major importance would be a reorientation of the
NASA organization away from the manned-unmanned dichotomy that had
existed throughout the 1960s. The report strongly recommended that, in
any mission, NASA plan to use whatever mode—manned or unmanned—
would be most effective in achieving the objectives sought. To this end
NASA should stop flying men just to fly them, and should focus on a
search for the most appropriate role for human beings in the system.

With the recommendations of the outside scientists in hand, the president then called for a governmental study of future possibilities for the
space program. On 13 February Nixon sent a note to the vice president, the
secretary of defense, the acting administrator of NASA, and the president's
science adviser, asking them to meet as a task group and to provide "in the
near future definitive recommendation on the direction which the U.S.
space program should take in the post-Apollo period."[10] The president
said that he would like to receive a coordinated proposal by 1 September
1969.

At the president's request, Vice President Spiro Agnew acted as chairman. The secretary of defense appointed Robert C. Seamans, secretary of
the Air Force and formerly deputy administrator of NASA, to represent the

Department of Defense on the Space Task Group. Invited observers were U. Alexis Johnson, under secretary of state for political affairs; Glenn T. Seaborg, chairman of the Atomic Energy Commission; and Robert P. Mayo, director of the Bureau of the Budget. The group immediately arranged for their respective staffs to conduct the necessary background studies. The science adviser, Lee DuBridge, with personnel from the Office of Science and Technology, served as coordinator of the staff studies.

Both Paine, whom the president appointed in March to the post of NASA administrator, and the vice president favored an expanded space program, Agnew speaking out a number of times for sending men to the planets. Paine felt the country could well afford many times what it was spending on space and pressed for a program that would include large manned space stations, lunar bases, and the development and use of a reusable space transportation system to replace the older, expendable boosters used during the 1960s. In these views Paine came into conflict with those of the Townes committee, the President's Science Advisory Committee, and Secretary Seamans. In spite of his former NASA connection, Seamans was strongly opposed to an expansion of the space program in times that called for fiscal conservatism. He would not support a large space station, and the shuttle could have his endorsement only if it could be shown that it would indeed generate the economies claimed for it.[11] The President's Science Advisory Committee called for a program of lower costs that would focus on using space capabilities for benefiting the nation and the world. The committee placed great emphasis on expanding the use of unmanned, as opposed to manned, techniques in space research and application. It also recommended studying, "with a view to early development, a reusable space transportation system with an early goal of replacing all existing launch vehicles larger than Scout with a system permitting satellite recovery and orbital assembly and ultimately radical reduction in unit cost of space transportation."[12]

During this period Thomas Paine worked continuously to revive national interest in a bold and imaginative space program. He described the large space station in near-earth orbit as "the next logical step" in the development of space. A lunar base would continue man's exploration of his corner of the universe, provide the means for doing much valuable science, and capitalize on the extensive investments already made in Apollo. A reusable space transportation system, consisting of a shuttle and various auxiliary stages for orbital and deep-space operations, would tie all the endeavors together and make space stations, lunar bases, and other advanced space missions economically attractive. Seeking additional support, Paine traveled to Europe pressing for international cooperation in the development and use of a space shuttle system.[13]

As contributions to the staff studies for the Space Task Group, the President's Science Advisory Committee, the Department of Defense, and

NASA prepared reports of their own.[14] Within NASA the study staff drew on planning material from the agency's Planning Steering Group (p. 378). When the output from that activity, reflecting a judgment by the planners that only a modest program had any chance of selling, proved to be too conservative for Paine, the administrator asked the author to include among the NASA options a program that would rise to $8 billion a year by the mid-1970s.[15] While this option was conveyed to the president in the Space Task Group's report, it received no serious consideration from the administration. Indeed, NASA's lowest option, which would rise to above $5 billion a year by 1976, was more than the White House planners were ready to bargain for.[16] All in all the Space Task Group's report did not show the conservatism the White House desired and was not adopted as the president's blueprint for the future in space.

NASA accordingly continued to seek some sort of guidelines from the president under which to plan for the future. After a period of negotiation the sought-after guidelines appeared in the form of a statement from President Nixon on 7 March 1970 (app. J).[17] Pointing to the many critical problems on our own planet that needed attention and resources, he nevertheless stated that the space program should not be allowed to stagnate. The nation's approach to space should continue to be bold, but balanced, and the country should not try to do everything at once. The general purposes of the space program should be exploration, the acquisition of scientific knowledge, and practical applications to benefit life on earth. In support of these general purposes he set forth six specific objectives: lunar exploration, planetary exploration including eventually sending men to Mars, reduction in the cost of space operations, extension of man's capability to live and work in space, expansion of practical applications of space technology, and encouragement of greater international cooperation in space.

By now the administration's conservatism as far as space was concerned was patent. It was to be seen in the qualifying language of the president's space message. Yet Administrator Paine chose to focus on the president's call to be bold, rather than on his admonition to proceed at a measured pace. Paine likened the space program to the great voyages sponsored by Prince Henry the Navigator, and encouraged his people to swashbuckle, as he put it (although years later Paine would question the appropriateness of that term).[18] The responsibilities of leadership, he felt, required him to get approval for as large a space program as the traffic would bear, and to this end he pressed for a wide variety of new starts with budgets that would quickly mount up in the years ahead to levels exceeding those of the Apollo era. To raise NASA planning out of the conservatism to which it had been depressed by the political climate, in June 1970 Paine assembled NASA center directors, program directors from headquarters, and other key persons for a five-day meeting at Wallops Island, Virginia, to consider NASA's future. Arthur Clarke, whose book *The Exploration of Space* had been in the 1950s a kind of blueprint for the future, was invited to be the keynote speaker in the hopes of starting the discussion on a sufficiently imaginative level.[19]

But NASA talking to itself this way had little effect, certainly none in raising budgets that continued their downward plunge. It was not in the cards to escalate the space program at that time. NASA was outvoted at every turn. The administration was absolutely dedicated to cost cutting. Industry was dubious about the value of increased expenditures in space and communicated its doubts to the White House. The Department of Defense, potentially NASA's strongest ally, was having budget troubles of its own and would not encourage a large competitive drain on national resources. The scientific community, not about to endorse another large, manned spaceflight project, preferred to phase out manned spaceflight— save only a possible shuttle program—in favor of more automated missions. There was much sympathy for Van Allen's call for a severely reduced space budget, $2 billion or less annually, devoted primarily to applications and science. Although Van Allen and Thomas Gold (the latter noted for his role in propounding the theory of continuous creation of matter) were opposed to the shuttle,[20] other scientists would support a shuttle if it was really to be developed and used as a tool to improve space operations and reduce their costs.[21]

There is a difference of opinion as to whether Paine's attempts to force the space budget far above the levels the administration wanted to see kept it from falling lower than it did, or were counterproductive. At any rate, after Paine resigned in September 1970,[22] Acting Administrator George Low made a conscious and visible effort to accommodate to the administration's desires to keep spending down. The new administrator, James C. Fletcher, not only continued Low's policy, but moved toward a constant-level budget, which made the process of getting White House approval much easier. One of the great difficulties NASA had been experiencing in introducing new projects was the shape of the funding curve in the years ahead. While the initial funding for a new project might fit into the current year's budget, increasing costs in future years often called for the total budget to rise again. If the budget rise was not approved, then projects recently started would have to be canceled—a painfully difficult thing to do. By eliminating this future bow wave in the funding curve, Administrator Fletcher was in a much stronger position than Paine had been to ask for assurances that NASA would be able to follow through on new projects that the agency started.

This was important in selling the Space Shuttle. In the cost-conscious climate of the Republican administration, the Space Shuttle became the only salable manned spaceflight project. After the Skylab flights in 1973 and the Apollo-Soyuz mission in 1975, it would not be possible to gain support for more of the very expensive missions, any of which would drive the NASA budget skyward again. In contrast, the Space Shuttle costs as finally approved would fit into a budget profile for the 1970s which, when computed in 1971 dollars, would allow only a slight rise in the first half of the decade.

MANNED SPACE SCIENCE

Leadership at the top provided the template, as it were, for the leadership exercised by NASA managers lower down. As stated earlier the straightforward, though frantically busy, years of Glennan's tenure and Webb's first years as administrator afforded an ideal climate for program people to establish their working relations both inside and outside the agency. As for space science, at times NASA was pulling and the scientific community reluctantly following, as with observatory spacecraft, the manned spaceflight program, and later Viking. At other times NASA was being pushed by an impatient clientele, as was illustrated by the scientists' desire for more sounding rockets, for individually assigned Explorer-class satellites, more Pioneer-class probes to Venus and other planets, and the utter dissatisfaction with NASA's organizational arrangements in the life sciences.

For space science one of the most difficult problems of leadership, both inside and outside NASA, concerned the manned spaceflight program. Underlying the prevailing discontent in the scientific community regarding this program was a rather general conviction that virtually everything that men could do in the investigation of space, including the moon and planets, automated spacecraft could also do and at much lower cost. This conviction was reinforced by the Apollo program's being primarily engineering in character. Indeed, until after the success of *Apollo 11*, science was the least of Apollo engineers' concerns. Further, the manned project appeared to devour huge sums, only small fractions of which could have greatly enhanced the unmanned space science program. It has been seen how such concerns colored the proceedings of the space science summer study in Iowa City in the summer of 1962 and led to Philip Abelson's campaign against the manned spaceflight program (p. 209).

The science program managers in NASA rallied in support of the agency's manned spaceflight projects, but they had their difficulties internally. As the nation's top priority space project, Apollo enjoyed a commanding position when it came to funds and requests for support from other parts of NASA. With regard to the latter, Ranger and Lunar Orbiter pictures of the moon and Surveyor data on properties of the lunar surface were, to Apollo people, a source of engineering information that had come too late to be used in the original design of lunar spacecraft and were none too soon for planning the Apollo missions.[23] Apollo's need for lunar data tended to constrain the planning of unmanned investigations of the moon. Apollo engineers sought from the unmanned program specific discrete items of information, such as the bearing strength of the lunar soil and the distribution of craters and rubble on the surface. But space scientists insisted the desired information could be had from an investigation of the moon that would provide an understanding of the basic processes that had gone into the creation of the moon and its surface features. Moreover, such an understanding would make it possible to answer specific questions not now foreseen that might come up later; concentrating too narrowly on unrelated individual measurements could be self-defeating in the long run. The engineers were not convinced and this insistence of the scientists on a thorough scientific investigation appeared like an unwillingness to be helpful, or worse, a self-centered desire to have it one's own way. Moreover, the Apollo people pointed out, the manned missions would make possible all that the unmanned spacecraft could do and more, and the scientists ought to wait for Apollo to provide the means for making the definitive studies they desired.

The troubles between the space scientists and the manned spaceflight engineers were enhanced by a decision of the associate administrator, Robert Seamans, that the Office of Space Science and Applications would assume responsibility for all space science in the NASA program, including that done on manned missions, but that the monies for manned space science projects would be put in the manned spaceflight budget, where they would be less likely to be cut in the congressional review process.[24] The Office of Space Science and Applications understood that this was simply a budgeting device and that after NASA's appropriations had been secured the manned space science monies needed for advanced research and the design and prototype work on manned space science experiments would be transferred to space science. But George Mueller did not do this. Instead he undertook to review and pass on the intended space science work before releasing money from his budget. In this way Mueller exercised the control over the manned space science program that had supposedly been assigned to the Office of Space Science and Applications. It is, in fact, a cardinal principle of management that the one who has the money has the control.[25] Thus Seamans had given the space science managers a responsibility for which they didn't have the necessary clout.

The author met periodically with Mueller in an effort to develop a satisfactory working relationship. The manned space science division under Willis Foster was one of the devices agreed on to bring the two offices closer together (p. 284). But, while Mueller appeared to the scientists to be fairly lavish in allocating funds to the engineering aspects of the manned spaceflight program, he suddenly became very cost conscious when it came to supporting science. In this climate the scientists were unable to discharge properly the responsibility that Seamans had assigned to them, and they quite naturally tended to direct their attention to the unmanned space science program. Noting this, manned spaceflight personnel accused the space scientists of neglect. Ignoring that their office was withholding the monies that had been slated for support of manned space science, they asked why the Office of Space Science and Applications wouldn't put some of its own funds into the important area of manned space science. To the science managers who were already having enough difficulties meeting the needs and demands of the scientific community, this question appeared infuriatingly obtuse.

As for the scientific community, to cut back on the unmanned program to fund a manned space science program would have generated a major crisis. Supporting an adequate unmanned program could keep the periodic attacks of the scientists on the manned program within bounds. Those scientists who did participate in the manned flights found the exercise much more difficult than working in automated spacecraft. Schedules were tighter and oriented toward engineering and operational requirements, rather than toward science. Documentation and test requirements were an order of magnitude greater than those for unmanned missions, where the life of an astronaut was not in the balance.

The frustrations felt by the scientists were illustrated by those expressed by Eugene Shoemaker, a geologist from the U.S. Geological Survey who early went to work with the Manned Spacecraft Center in preparing for the Apollo missions to the moon. Shoemaker's participation in the NASA program was in keeping with an arrangement between NASA and the Geological Survey that Thomas Nolan, director of the Survey, and the author had agreed on. Nolan committed the support of USGS scientists, while NASA (the author) agreed to use this support and not to build up within the agency another little Geological Survey. The agreement was informal, arrived at over lunch at NASA Headquarters, and never went to the Administrator's Office for his blessing. In spite of the informality, however, the agreement had a major effect on the shape of the geology portion of the space science program.

Under the aegis of this agreement—perhaps without ever being aware of its existence—Gene Shoemaker worked long and hard with the Manned Spacecraft Center and the astronauts to plan a lunar exploration program, to develop cameras and instruments for photography and measurements of the moon, and to help train the astronauts in the geological sciences and in the techniques of field work. Shoemaker was instrumental in arousing and maintaining the interest of the earth sciences community in lunar science. He and his colleagues contributed much to the success of the Apollo astronauts in their geological exploration of the moon.

It was a shock, then, to manned spaceflight personnel members when, after the resounding success of *Apollo 11*, their colleague and former mentor began to blast them for alleged shortcomings in Apollo. Shoemaker contended that in the name of engineering and safety requirements serious scientific shortcomings had been designed and built into the Apollo hardware—unnecessarily. Shoemaker contended that this was due primarily to utter insensitivity to the needs and interests of science and that if properly designed, Apollo would have been able, without danger or compromise to operational requirements, to contribute far more to science than it was going to. Shoemaker was particularly incensed over the canceling of several planned Apollo missions. In a talk before the American Association for the Advancement of Science in December 1969, he likened the first Apollo landing to the first exploratory trip of John Wesley Powell down the Colorado River and through the Grand Canyon. Both were courageous and fruitful ventures. But Powell's first trip was followed by years of intensive study of the geology of the Grand Canyon, which provided the real scientific return and practical benefit of those explorations, whereas Apollo was going to be cut off too soon after the initial landing to capitalize properly on the tremendous investments already made.[26]

Shoemaker seized every opportunity to take up the issue and to castigate NASA. NASA people felt the affront deeply. Shoemaker had come into the space program an unknown, just beginning his scientific career. Sizable sums of money had been devoted to support his research, and NASA had in effect financed much of his career. Why couldn't Shoemaker criticize in private and praise in public? And, anyway, what good was all the criticism going to do? NASA lacked the funds to continue Apollo landings much longer. Moreover, voices on the Hill were asking why the agency didn't just stop all further lunar missions, since each new flight exposed NASA and the country to a possible catastrophe and the loss of much of the good that had already been achieved—a possibility that was not lost upon those conducting the missions.

Little could be done in the way of backtracking and redoing the Apollo hardware, but some steps could be taken with regard to how the existing hardware was used. Yielding to strong pressure from the scientific community, supported by scientists within NASA, the Johnson Space Center* inaugurated a new era in relations with the scientific community. Lines of communication between the experimenters and astronauts and engineers were strengthened, during both preparation and flight periods. The experimenters' feeling of effectiveness increased steadily with each new Apollo mission until with *Apollo 17*, which carried geologist Harrison Schmitt, the scientists were positively ecstatic.

Teams assembled from the outside scientific community by the Johnson Space Center, to help plan for the lunar exploration and to advise on the allocation and analysis of lunar samples, did yeoman service, and their advice was heeded. Their efforts received considerable praise from the

---

*The Manned Spacecraft Center was renamed the Johnson Space Center 17 February 1973.

scientific community, although naturally there were always those who were dissatisfied with some of the recommendations made for allocating lunar samples. With the advice of these groups NASA supported the outfitting of laboratories and the preparatory work necessary to get ready to analyze samples when they should become available, thus building up a team of hundreds of scientists around the country to take part in this unique project. The result was a revitalization of lunar science, and more importantly, a development of new and improved instruments and techniques for geochemical, geophysical, and mineralogical analysis. Following the return of the first lunar samples, it was not long before more than 700 researchers around the world were thoroughly involved in their analysis and study. The Johnson Center sponsored annual meetings in Houston to make reports and discuss results. Attended by hundreds of scientists from various fields and many countries, these attracted considerable attention from the press.

This success gave rise to another of the debates in which NASA so often found itself embroiled. After the first rush of analysis and study of the lunar samples was over, the time arrived for more work on integrating disparate results into a connected and understandable whole, in particular to attempt to discern what the new information meant with regard to the origin and development of the moon, the solar system, and the earth. At this stage there was no longer need for so many hundreds of investigators, and it made eminent sense for NASA to plan to fund only a part of those previously supported by the agency. Such steps were unequivocally recommended by NASA's advisers, who accordingly shared in playing the role of the villain in the reductions that followed.

FINDING THE WAY OUT

The very success that James Webb had had in selling Saturn and Apollo as projects to develop a powerful national capability to operate in and use space as the country might decide in the national interest set the stage for the dismal lack of success in the first attempts to plan for a follow-on to Apollo. For it had not been foreseen that Apollo and Saturn hardware would have to be regarded as "first generation," highly experimental, and much too costly to maintain as the basis of a continuing national space capability. Naturally the manned spaceflight people wanted to stay in business, and Webb's reluctance to let go of his original dream fostered planning to continue to use Apollo hardware. Paine's desire to swashbuckle reinforced the efforts to keep the Saturn and Apollo manufacturing lines open. That scientists openly opposed any continuation of the Apollo kind of operation was ascribed to their usual idiosyncrasies. That the scientists kept insisting they could not come up with any real requirements for science in a space station was overlooked, and planning went ahead. First Mueller proposed a "wet workshop," in which the spent S-IVB stage of Saturn would be dried out in orbit and outfitted there for use as a temporary space station. Later that was replaced with the "dry workshop," in which the Saturn stage would be outfitted on the ground and then launched into orbit for the same purposes. In a vague way the workshop was thought of as a transition to a more permanent program of using Saturn and Apollo hardware for the continuing exploration and investigation of space.

Not until NASA finally recognized that Saturn and Apollo had to go, could a way to the future be plotted. Then the Skylab workshop could emerge as a limited project that, with the Apollo-Soyuz mission, would wind up the Apollo era. Once the decision had been taken to close out Apollo and to concentrate on reducing the costs of operating in space, the Space Shuttle could fall into place as the keystone of the future.

During NASA's first 10 years—when the agency had led with a reasonable and acceptable, yet aggressive, program—NASA had enjoyed a strong followership. But in the late 1960s when NASA had attempted, in a totally unsuitable climate, to continue to use the costly Apollo hardware, that followership was almost lost. In a country at the moment only peripherally interested in space, James Webb had found it impossible to generate a national debate that might furnish some guidance for the agency. Thomas O. Paine's efforts in 1969 and 1970 to gain approval for a very large, very expensive program including space stations, lunar bases, and shuttles had gained neither administration nor grass roots support. But the fourth administrator, James C. Fletcher, found the country willing to support an imaginative program as long as costs could be kept down. In a program dedicated to economy and usefulness, Fletcher was able to include the development of a Space Shuttle which would put manned spaceflight to use in serving the agency's scientific and applications objectives. NASA then recaptured the leadership that for a brief time had faltered.

Thus, with a program dedicated to service and economy, NASA emerged from the confusion and uncertainties of the late 1960s with a renewed commitment to a strong U.S. presence in space. Starting with an uncertain lease on life, with each passing year the Shuttle strengthened its position in NASA's future. In the light of the Shuttle's advancing development, the 1970s became a period of transition from the pioneering era of the 1950s and 1960s to the 1980s, when many expected space operations to become routine.

Part V

The Scope of Space Science

I am the owner of the sphere,
Of the seven stars and the solar year,
Of Caesar's hand, and Plato's brain,
Of Lord Christ's heart, and Shakespeare's strain.

   Ralph Waldo Emerson, "The Absorbing Soul"

18

International Ties

Only a few years after Sputnik, the scope of space science reached beyond America's shores to many other parts of the globe. The predisposing cause was doubtless the political interest of U.S. leaders in recapturing leadership in space while projecting an image of peaceful purpose and cooperativeness in the world. In this regard science could help to broaden the base of the space program.

A fundamental assumption underlying the practice of science is that the laws of nature hold throughout the universe. Nature itself makes science international in character and provides a strong basis for international cooperation in science. The history of science bears out the point, for the insights that gave rise to great new advances in science have come from all over the globe.

It was, therefore, to be expected that NASA would quickly be involved in international activities in space science. In fact, with roots in the International Geophysical Year, which had already generated a lively interest in the potential of satellites for scientific research, one might argue that the appearance of an international component in the NASA space science program was inevitable. The moment NASA took over responsibility for the Vanguard program from the Naval Research Laboratory, the agency acquired a number of international commitments, like those of the satellite geodesy program that proved so touchy for a while.

POLITICAL CONTEXT

Sputnik 1 got the attention of an entire world. In various ways political and scientific organizations made their interests in space research and applications known. It seemed natural, for example, for the staff of the North Atlantic Treaty Organization to suggest cooperative efforts in space under the aegis of NATO. In spite of their superficial reasonableness, these overtures were not supported by Department of State or NASA managers, primarily because a cooperative program under NATO would reintroduce those military overtones Congress had already rejected in not assigning the

U.S. space program to the Pentagon. Thus, in spite of his long association with NATO's Advisory Group for Advanced Research and Development, Dryden, with guidance from the State Department, turned down these suggestions from the NATO staff, pointing out that individual NATO countries could cooperate with NASA on their own initiative without invoking the NATO name.

The United Nations was another matter. Here among a large number of the world's nations, a deep interest was to be expected in activities that would fly rockets and spacecraft over the sovereign territories of U.N. members. On 19 November 1958, the United States and 19 other countries jointly introduced a resolution into the General Assembly of the United Nations calling for the creation of an ad hoc committee on the peaceful uses of outer space.[1] The committee was established in December and met from 6 May to 25 June 1959 at U.N. Headquarters in New York City to discuss a variety of subjects related to international interest in space matters.[2] It was soon realized that the United Nations was in no position to assume operational responsibilities in a space program—although for a brief period there was some discussion of such things as launching sites run by an organ of the United Nations. International competence in science resided in the International Council of Scientific Unions and its unions, while many aspects of practical applications of space would apparently fall under already existing U.N. organizations such as the World Meteorological Organization and the International Telecommunications Union. As a consequence the ad hoc committee recommended against the creation of either a new agency or any sort of central control for space activities. Instead it was suggested that there be a focal point—in the nature of an international secretariat—to facilitate international cooperation in the peaceful uses of space.[3]

Differences of view between the United States and the Soviet Union tended to dominate the discussions in the committee for the first two years or so. The Soviet Union wished to establish at the outset a set of general principles to guide space activities, while the United States preferred to develop an international policy by practice, moving step by step with individual, limited agreements. After extensive exploratory discussion, the way was clear to move ahead on a firmer basis, and General Assembly resolution 1721 created a permanent Committee on the Peaceful Uses of Outer Space with 28 members.[4] The resolution became a basic document on space, among other things commending to member states that international law apply to outer space and celestial bodies, that both space and celestial bodies be free to all and not subject to national appropriation, and that member states should report space launchings to the U.N. for registration.[5]

The permanent committee provided for two subcommittees, one legal, the other scientific and technical. Although the deliberations of the Legal Subcommittee occasionally touched upon the interests of the scientists, the other subcommittee usually provided the forum for space science matters. The function of the committee and its subcommittees was regarded as one of aiding and encouraging members rather than one of getting into operational programs. In this vein, at the meeting of the Scientific and Technical Subcommittee in Geneva in May and June 1962, the subcommittee gave special attention to helping the less developed countries to pursue some of their interests in space. Much discussion was devoted to training and education for scientists and engineers of the smaller countries, and various means of meeting this need were recommended. The subcommittee recommended publication of information on national space programs and of technical information needed by nations just beginning space research. A major recommendation asked for United Nations sponsorship of sounding rocket ranges that met prescribed conditions, including openness and accessibility to all member states.[6]

By the next year India, with assistance from a number of countries in supplying launchers, tracking equipment, computers, and aircraft, was well along in construction of a sounding rocket range on the geomagnetic equator at Thumba. Since the range was to be operated in keeping with the principles laid down by the United Nations, U.N. sponsorship was accorded the range, under which aegis India hosted a great many launchings by other nations.[7]

When, half a decade later, in August of 1968, the committee sponsored a symposium on the peaceful uses of outer space, space had become big business; and almost fourscore countries participated in one way or another. Administrator Webb considered the symposium of sufficient importance to attend in person. The United States, the Soviet Union, and numerous other countries could report on a wide variety of space science results. As a prelude to the imminent American manned flights to the moon, both the U.S. and the USSR reviewed results from their unmanned lunar spacecraft. On the science side problems were minimal; but some knotty questions were raised in space applications, such as international cooperation in commercial space communications systems and the delicate subject of space photography for earth-resource surveys. In earth-resource photography, the prospect of substantial benefits to themselves led the countries to acquiesce, at least as far as accepting research satellites. Operational satellites were a question that could be resolved later.[8] These questions, bearing on international relations in space applications, are beyond the scope of this book.

The natural arena for international cooperation in space science was that of the International Council of Scientific Unions, which had sponsored the International Geophysical Year. As might have been expected, IGY spawned a number of continuing activities, for which special committees were formed, such as the Special Committee for Antarctic Research and the Special Committee for Oceanographic Research. Among them was the Committee on Space Research (COSPAR).[9]

For a brief period substantial difficulties loomed. Miffed at the high proportion of Western representation on the committee, the Soviet Union chose to introduce political considerations into sessions of the nonpolitical COSPAR. At the March 1959 meeting in The Hague, Prof. E. K. Federov, a tough, hard-line negotiator, showed up instead of Anatoly Blagonravov to represent the USSR. Federov's insistence that not only Soviet bloc countries, but also the Ukraine and Byelorussia, should be admitted to COSPAR as independent members evoked a general consternation. American attendees pointed out that this was like asking that a couple of states like Texas and New York be members in addition to the United States. The committee would not go along, and Federov read what was apparently a prepared statement that under the circumstances the USSR would not be able to participate in the Committee on Space Research.[10]

It looked as though COSPAR might have to proceed without the participation of one of the two major launching nations. But the U.S. delegate, Richard Porter, put forth a counterproposal that any nation interested in and engaged in some way in space activities could be a member of the committee. Porter's motion was adopted, paving the way for admitting Soviet bloc countries. Also the committee agreed to accept on its Executive Committee a Soviet vice president and a U.S. vice president, thus assuring both countries of permanent positions on the executive body of COSPAR. With these compromises, the Soviets did not pull out, and for future meetings Blagonravov returned as the Soviet representative.

There were two kinds of membership in COSPAR—representation from a number of interested scientific unions, like the Unions of Geodesy and Geophysics, Scientific Radio, Astronomy, and Pure and Applied Physics; and national members. The former provided the ties with the international scientific organizations. But the ultimate strength of COSPAR lay in the national memberships, for, as with the International Geophysical Year, the individual countries would pay for and conduct research. When at the same March 1959 meeting attended by Federov the United States offered to assist COSPAR members in launching scientific experiments and satellites, the future of COSPAR seemed assured.[11]

COSPAR met annually, varying the place of meeting to give different countries the opportunity to act as host. The sessions consisted normally of two parts, a scientific symposium on recent space science results or on some topic of importance to space science,[12] and discussions of plans and problems. To facilitate the latter, working groups were established with appropriate representation from interested countries and unions. Perhaps the most notable, and controversial, of these was the group set up to look into undesirable side effects of space activities. Because of concerns in the scientific community over possible compromise of other scientific activities by space research—for example, interference of radio signals from satellites with ground-based radio astronomy—the International Council of Scientific Unions passed a resolution in 1961 calling on COSPAR to examine proposed experiments that might have potentially undesirable effects on scientific activities and observations, and to make careful analyses and quantitative studies available to scientists and governments.[13] COSPAR responded with resolution 1 (1962) setting up a Consultative Group on Potentially Harmful Effects of Space Experiments, under the chairmanship of Vikram Sarabhai, physicist and later head of India's atomic energy agency.[14] There were representatives from the two major launching countries, Russia and America, and several "neutral" members. The president of COSPAR, H. C. van de Hulst, felt the consultative group's task sufficiently important that he himself should also serve. The group plunged into a study of such matters as the effects of rocket exhausts on the atmosphere and of high-altitude nuclear explosions on the earth's radiation belts.

The purpose of the consultative group was initially scientific but the subject was bound in time to bring in political considerations. At the May 1963 meetings of the Scientific and Technical Subcommittee of the Committee on the Peaceful Uses of Outer Space, Soviet Delegate Blagonravov kept bringing up the matter of U.S. high-altitude nuclear tests and the West Ford experiment, which placed clouds of tiny copper needles in orbit to test their usefulness for reflecting radio signals from one point on the ground to another.[15] On 20 May 1963 he delivered a blast against the United States, accusing it of fostering war and ignoring the welfare of the world and world science. His remarks were utterly cynical in that the Soviet nuclear tests had put much more radiation into the lower atmosphere than had U.S. experiments. Also, careful analyses of West Ford had shown that the metallic dipoles would not adversely affect ground-based radio astronomy, a conclusion that the Soviet Union did not refute. Continuing to press the matter, on 21 May 1963 Blagonravov submitted a paper on contamination of outer space, urging that the U.N. committee ask the Committee on Space Research to study the harmful effects of such experiments in outer space.[16] Again the cynicism of the Soviet delegates was apparent in that they consistently showed little interest in the COSPAR consultative group and had a very poor attendance record at its meetings. Their greater concern with the political issues than with the scientific aspects of the subject was apparent.

The most persistent task of the consultative group had to do with the protection of the moon and planets from biological contamination. If there was life, or evidence of past life, or evidence of how the chemistry of a planet evolves toward the formation of life, on any other planetary body, life scientists wanted to preserve the opportunity to study that evidence uncompromised by any form of terrestrial contamination.

In this matter the consultative group by no means had to start from scratch. As far back as 1956 the International Astronautical Federation had begun to worry about interplanetary contamination. *Sputnik 1* called forth similar concerns in the U.S. Academy of Sciences, and on 8 February 1958 the academy passed a resolution urging that "scientists plan lunar and planetary studies with great care and deep concern so that initial operations do not compromise and make impossible forever after critical scientific experiments." Lloyd Berkner, president of the International Council of Scientific Unions, carried the resolution to ICSU, which in March of 1958 established an ad hoc Committee on Contamination by Extraterrestrial Exploration, with Marcel Florkin, Belgian biologist, as president. The committee developed a code of conduct for space missions and continued for a number of years to work and advise on problems. The COSPAR consultative group inherited the mantle along with the work and thinking of the ad hoc committee.

The aims were simple, but the problems were exceedingly complex and difficult to resolve. One could not ask for 100 percent sterility in planetary and interplanetary spacecraft. To seek such an unachievable goal would be prohibitively expensive. It was necessary, therefore, to deal with probabilities, and to seek to keep at an acceptably low figure the probability that planets of interest might be contaminated. The scientists had to work out a compromise between asking for so low a probability that the costs of engineering spacecraft to prescribed standards would be forbiddingly high and allowing so high a probability that the chances of compromising scientific research were too great. Unfortunately, as with many questions dealing with probabilities, there were a great many opinions as to what probabilities were reasonable and as to how to go about the engineering. The interminable discussions of the scientists were a vexation to the engineers who had to translate prescribed standards into engineering criteria. All in all, it took a decade to agree on an international set of objectives. At its 12th plenary session in Prague, 11-24 May 1969, the Committee on Space Research reaffirmed the basic objective of keeping the probability of contaminating Mars and other planets at or below one part in a thousand for an anticipated period of biological exploration. The period was taken to be 20 years, extending through 1988, during which period it was estimated that approximately 100 missions would be flown.[17]

Requirements were more easily stated than met. Unsterilized space probes meant to fly by planets in the period concerned had to be so aimed that their combined probabilities of hitting and contaminating the planets should remain within the stated limit. Also, spacecraft to land on a planet had to be so designed and treated—by exposure to lethal radiations, chemical cleansing, or heating—that again the combined probabilities of producing contamination for all spacecraft landed during the period should remain within the established limit.

Opinions differed considerably as to how the evolving requirements should be translated into engineering criteria for the construction and processing of spacecraft. The United States tried to facilitate the discussion of this complex subject at COSPAR meetings by describing in detail the processing of its spacecraft. The Soviet Union, on the other hand, consistently refused to give details, saying only that it would decontaminate its spacecraft. This refusal to participate openly generated considerable uneasiness among scientists from the other countries, for proper protection of the planets from contamination could be achieved only by the full cooperation of all. Laxness on the part of only one country could vitiate the efforts of others to preserve this scientific opportunity, which once lost could never be recovered.

The United States spent many millions of dollars developing materials, components, and techniques for producing planetary spacecraft that were as close to sterile as possible. One had to accept on faith that the USSR was doing the same sort of thing. But when it came to the moon, after some initial attempts to develop sterile or nearly sterile spacecraft, a revolt set in. The extremely low probability of finding any lunar life, or of propagating any terrestrial life deposited on the moon, led NASA, vigorously supported by the physicists, to insist that "cleanliness" as opposed to sterility was enough. Although the life scientists objected, this policy prevailed for the moon.

Such were the problems taken up in the various COSPAR working groups, although most problems did not have the drama associated with them that those of the consultative group displayed. Members developed plans for cooperative meteorological programs, solar studies including eclipse expeditions, geodetic observations, and the like. But, as with the International Geophysical Year, it would be the member nations that would carry out the planned programs. Accordingly most actual cooperative projects took place between pairs or small groups of nations.

NASA INTERNATIONAL PROGRAM

That was how most of NASA's international program developed. As has been seen, NASA quickly became involved with the United Nations and the Committee on Space Research. But those served more as a backdrop than as the arena for NASA's international activities. By far the most frequent arrangement was a bilateral one between NASA and a counterpart agency in another country, sometimes with a covering government-to-government agreement. The State Department provided guidance and a considerable amount of assistance and in the dealings with the United Nations took the lead. But except for U.N. matters, NASA, while keeping contact with the State Department, was pretty much on its own.

The variety of the program was remarkable. By 1962, 55 nations plus the European Preparatory Commission for Space Research were engaged with NASA in various space activities.[18] Twenty-four were helping with operational support to NASA missions through the Minitrack, Mercury, and Deep Space tracking networks; the optical tracking network inherited from the International Geophysical Year; the volunteer program of satellite observations called Moonwatch; and data acquisition. Assistance varied all the way from simply providing the real estate on which to erect and operate ground stations, to assuming a substantial responsibility for their staffing and operation.

Thirty-four nations were working with NASA in cooperative projects using satellites, sounding rockets, and ground-based work in meteorology and communications. As in the United Nations space committee, many nations expressed a great need for scientific and technical training related to space. By 1962 13 foreign Resident Research Associates were at NASA centers, 5 foreign students were being trained in American universities under NASA sponsorship, and 13 engineers or technicians were training at NASA centers or ground stations. Visitors from 42 countries plus the European Preparatory Commission had come to explore their interests in the space program. By the 1970s 94 countries or international organizations were cooperating in some form with NASA.[19]

By the time Deputy Administrator Hugh Dryden, Arnold Frutkin (who had become head of NASA's Office of International Programs), and the author journeyed to Aachen in September 1959 to attend meetings of the NATO Advisory Group for Advanced Research and Development, the guidelines for NASA's international activities were pretty much in mind. They were referred to time and again in discussions with dozens of scientists from the different countries who sought out the NASA people to explore ways of participating in the space program. In any cooperative project that might develop, the guidelines called for:

- Designation by each participating government of a central agency for the negotiation and supervision of joint activities;

- Agreement on specific projects rather than generalized programs;

- Acceptance of financial responsibility by each participating country for its own contributions to joint projects;

- Projects of mutual scientific interest;

- General publication of scientific results.

A decade later virtually the same guidelines were still in force.[20]

Generally the guidelines were readily accepted. Only the third, calling for no exchange of funds, occasioned some expressions of dismay. Accustomed to being funded by the U.S. for a variety of things, some had hoped that they might be supported in space research by American dollars. But Dryden and Frutkin pointed out that a project in which a country was willing to invest some of its own money was more likely to be of genuine interest and value than one that was undertaken simply because someone else was willing to pay for it. In time the policy came to be accepted as natural and proper, and in fact one could sense—possibly because one wished to—a greater feeling of satisfaction and pride on the part of those who were paying their own way.

There was a decided difference between the East and the West in space cooperation. By far the greater part of NASA's international program was between NASA and Western countries. While some cooperative projects were agreed upon between Blagonravov and Dryden, which were confirmed by the two governments, these were of very limited scope. Moreover, in Soviet-American cooperation during the 1960s it proved generally impossible to achieve the kind of openness and freedom necessary for more than arm's length relations.

COOPERATION WITH WESTERN COUNTRIES

A sizable portion of NASA's *cooperative* programs in the first decade—that is, programs with other countries in pursuit of common objectives, in contrast to activities like the operation of satellite tracking stations that were purely in support of NASA's own program—was devoted to space science. The tacit recognition that this would be true can be seen in the guidelines cited earlier, which were oriented toward scientific projects. Under those guidelines, which were sufficiently flexible to admit of a broad range of endeavors, many different kinds of cooperative projects sprang up. A few examples will illustrate.

Sounding Rockets

Sounding rockets were a popular medium for entering into space science. They were inexpensive, handling and launching them was simple compared to the large launch vehicles, and they afforded the means for accomplishing some significant research. As more than a decade of research in the United States and the USSR had shown, sounding rockets could be used to attack problems of the atmosphere and ionosphere, the magnetosphere, solar physics and astronomy, cosmic rays and interplanetary physics, and biology. Supplementing the U.S. and Soviet ranges, including that at Fort Churchill in Canada, throughout the 1960s new sounding rocket ranges appeared around the world—at Woomera, Australia; Sardinia in the Mediterranean; Andöya, Norway; Jokkmokk, Kronogård, and Kiruna, Sweden; Chamical, Argentina; Natal, Brazil; Hammaguir, Algeria; Kourou, French Guiana; Thumba, India; Sonmiani Beach, Pakistan; and Huelva, Spain.

At the same time NASA joined with other countries in a variety of cooperative rocket soundings, some from U.S. ranges, others from ranges overseas. For example, in Australia in the fall of 1961 ultraviolet-astronomy experiments used the British Skylark rocket to obtain data on the southern skies to compare with northern hemisphere data. During 1961 and 1962 NASA and the Italian Space Commission cooperated on a series of rocket firings in Sardinia to measure upper atmospheric winds by tracking glowing clouds of sodium vapor released at altitude from the rockets. Throughout the 1960s and into the 1970s launchings at Andöya studied the aurora and also the ionosphere within and near the auroral zone. From time to time there were special expeditions, like that to Cassino, Brazil, in November 1966, in which 17 sounding rockets were fired to investigate solar x-rays and the effect of the solar eclipse of 12 November 1966 on the earth's upper atmosphere. Occasionally ships were used, as with the solar eclipse expedition to a spot near Koroni, Greece, in May 1966. All in all, by the early 1970s some 19 countries spanning the globe had engaged with NASA in a productive program of sounding rocket research, much of which required the special geographic locations afforded by the different ranges.[21]

Ariel

Throughout the 1960s the United Kingdom also cooperated with the United States in conducting sounding rocket experiments at various locations around the world. But the scientific satellite exerted an even greater attraction than the sounding rocket, and the countries that could afford it quickly approached NASA with ideas for cooperative satellite projects. The United Kingdom was among the first to seek such cooperation, and the U.K. satellite *Ariel 1* was the first international satellite that NASA put into orbit—on 26 April 1962.[22] Every few years thereafter additional Ariels followed, the fourth going into orbit from the Western Test Range in California on 11 December 1971. In addition, British experimenters were successful in competing for space on U.S. satellites. By 1973, 13 British scientists had put experiments on Explorers; on solar, geophysical, and astronomical observatories; and on Nimbus weather satellites.[23]

NASA's association with the United Kingdom was typical of many of the international cooperative programs in that by and large the British experiments were primarily of interest to British scientists. The mutuality of interest was there, of course, and NASA considered the U.K. experiments to be a valuable supplement to U.S. space science. But the ionospheric research conducted in the Canadian satellites Alouette and Isis was more intimately related to the NASA program.

Alouette and Isis

For a time a race was on between Britain's Ariel and Canada's Alouette to see which would be the first in orbit. Ariel won, and *Alouette 1* followed half a year later, going into a nearly polar orbit so that its revolution about the earth would bring it repeatedly in range of Canadian ground stations.[24] Once in orbit Alouette proceeded to establish a record (for its time) of 10 years of successful operation on orbit.

The major purpose of the Alouette experiments was to investigate the ionosphere, particularly by sounding the ionosphere from above. Following the experiments of Breit and Tuve in 1926,[25] ionospheric soundings had been made from the ground by sending pulsed signals upward and recording the returned signal on film as a function of time after the initial transmission. The effect of the ionosphere was to spread the reflected signal out in time, and from the shape of the returned signal one could estimate ionospheric heights and ionization intensities. But the uppermost regions of the ionosphere could not be sounded from the ground. Also, the complexities of the ionosphere often produced confusing signals, and experimenters hoped that soundings from above would help to resolve some of the ambiguities.

This important research lay at the heart of NASA's plans for studying the ionosphere. Since the Canadians proposed to do it, NASA scientists proceeded to build their own ionospheric program around that of Canada. The Canadian work thus not merely supplemented, but actually supplanted research that NASA scientists would otherwise have done. In monetary terms the Canadian contribution to the ionospheric program freed some tens of millions of dollars that could be used on other projects. Moreover, the Canadian researchers working on Alouette added to the total competence of the ionospheric team.

Following a second Alouette in November of 1965, Canada and the United States moved on to Isis—International Satellite for Ionospheric Studies—an improved satellite that would carry the topside ionospheric sounders plus 8 to 10 additional experiments furnished by both Canadian and American scientists. The first Isis went aloft on 30 January 1969 into a polar orbit, the second on 31 March 1971 into a nearly polar orbit.

Pencil and Its Descendants

Cooperation with Japan was of an entirely different character from that with European nations. Although there were many meetings and much discussion about cooperation, it gradually became apparent that the Japanese were firmly committed, emotionally as well as politically, to developing a space capability for themselves. But the realization by NASA scientists that this commitment existed came only after extensive exchanges on various possibilities of working together. The first indication of possible interest in cooperative projects came from some exploratory discussions of William Nordberg and William Stroud of the Goddard Space Flight Center with Japanese scientists. The NASA scientists, who in their previous positions at the Army's Signal Engineering Laboratories in New Jersey had pioneered the use of grenade explosions at high altitude to measure upper-atmosphere temperatures, had aroused the interest of Japanese scientists in the possibility of a joint program of atmospheric research using sounding rockets carrying grenades. But they had also left the impression that NASA might be willing to furnish funds to the Japanese for the work. NASA Headquarters was somewhat embarrassed, since the policy not to exchange funds on cooperative projects had already been established.

Since, however, there seemed to be considerable Japanese interest— above and beyond the grenade-rocket work—in the possibility of cooperating on space projects, the author visited Japan in May 1960 for a series of conversations with scientists and administrators. Professor Hideo Itokawa of the University of Tokyo showed some of the progress that had been made in developing Japanese sounding rockets. Starting with Pencil, a tiny rocket about 30 cm long and 2.5 cm in diameter, and a miniature launching range no bigger than some American back yards, the engineers had conducted horizontal firings, testing small-scale launchers, multistage combinations of Pencil, and techniques for separating stages during flight. From these exploratory tests engineers had gone on to larger rockets, which were being flown from a launching pad at Akita on the western coast facing the Sea of Japan. In step by step fashion, they planned to work up to the multistage Kappa, Lambda, and Mu rockets, some of which would eventually be capable of putting satellites into orbit. The first objective, however, was to produce a reliable high-altitude sounding rocket.[26]

On the political side were extensive conversations with members of the Science and Technics Agency. Minister Nakasone, head of the agency, was most desirous of working out some kind of cooperative agreement, and it seemed as though a great deal of progress were being made.[27] Subsequent conversations with Professor Hatanaka, astronomer at the University of Tokyo, and other scientists, however, revealed that Japan was torn by internal strife between the Science and Technics Agency and the university community, which accorded its allegiance to the Ministry of Education.[28] It suddenly became apparent that NASA could easily find itself in the middle, and one wondered what could come of the talks with Nakasone and his people. As it happened, within weeks after the author had returned to Washington Prime Minister Kishi's government fell, Nakasone was out, and negotiations with Japan were set back momentarily.

The talks, however, had cleared the air and NASA was in a much better position to understand the Japanese situation when delegations soon thereafter came to Washington. The groundwork had been laid in Tokyo for cooperating on some sounding rocket firings to compare results from American and Japanese ionospheric instruments, and an agreement was soon completed. The first firings took place at NASA's Wallops Island facility in Virginia in April and May 1962, and continued in the autumns of 1963 and 1964.[29] Every few years thereafter the United States and Japan cooperated on rocket soundings, joint firings in 1968 and 1969 taking place from the Indian range at Thumba.

Although Japan might have gained something from a joint satellite project with the United States, Japan preferred to go it alone. Work continued on a Japanese launch vehicle, and on 11 February 1971 a Lambda multistage rocket successfully placed the *Ohsumi* satellite in orbit, making Japan the fourth nation—France had been the third—to orbit a satellite with its own launch vehicle.[30] The U.S. assisted in tracking *Ohsumi* and other Japanese satellites that followed in 1971 and 1972.[31]

San Marco

Italy also wished to launch its own satellites, but was willing to use an American launch vehicle for the purpose. Out of this desire Project San Marco was born. The project was conducted in three phases. In the first phase an Italian team, under the supervision of NASA engineers, became familiar with the Scout rocket by conducting suborbital launches from Wallops Island in April and August 1963. In the second phase the Italian team launched the satellite *San Marco 1* on a Scout 15 December 1964, also from Wallops Island. For the third phase, the project moved to the coast of Kenya, where the Italians had constructed a launching pad on a towable platform of the kind used in drilling for oil beneath the ocean. Located on the equator, the San Marco platform was anchored off shore in the Indian Ocean. Here *San Marco 2* went aloft 26 April 1967, and *San Marco 3* 24 April 1971, to investigate the atmosphere and ionosphere above the earth's equator.[32]

The San Marco platform had special value in making it possible to launch satellites directly into orbits above the earth's equator. For this reason NASA requested use of the platform for launching a number of U.S. satellites. With costs reimbursed by NASA, *Explorers 42, 45,* and *48* were sent up from the platform in 1970, 1971, and 1972. The first of these, named *Uhuru,* the Swahili word for freedom, produced exciting data on celestial x-ray sources.[33]

Ground-Based Projects

Not all cooperative programs in space science involved direct participation in sounding rockets or spacecraft launches. Many countries cooperated in ground-based projects. A number of nations cooperated in ionospheric research by making observations from the ground, to be coordinated with satellite experiments. Twenty-seven foreign stations in 13 countries photographed GEOS and PAGEOS geodetic satellites to help improve the accuracy of geodetic results. France participated with NASA in analyzing data obtained by tracking French and U.S. geodetic satellites with lasers. Such ground-based cooperation was even more extensive in the applications area, where many countries undertook ground-based observations or the analysis and use of satellite data in meteorology, communications, and earth-resource surveys.[34]

On the scientific side the outstanding example of this kind of cooperation has to be that associated with the analysis of samples of the moon obtained by the Apollo astronauts. In the United States hundreds of scientists turned their attention to deciphering from lunar samples and other Apollo data what could be learned about the moon and its origin and by inference about the earth. They were joined in these efforts by 89 principal investigators and more than 260 foreign coinvestigators from 19 different countries. The coinvestigators were associated with both foreign and American principal investigators. From 24 countries 97 foreign scientists took part in the lunar science conference held in Houston in January 1970; 90 from 16 countries attended the 1971 conference; and 108 from 15 countries plus the European Space Research Organization came to the third lunar science conference in 1972.[35]

For a while the study of the lunar samples made the investigation of the moon and planets appear like the hottest field in science. It was a far cry from the 1950s, when a graduate student in astronomy who confessed to an interest in studying the planets was inviting disdain. The big change, of course, was the abundance of new data. Some idea of the extent of the change can be obtained by looking at the almost overwhelming mass of results published in the proceedings of the lunar science conferences.[36]

THE SOVIET UNION

An entirely different climate surrounded the efforts of NASA to cooperate with the Soviet Union. In this case competition, born of the Cold War, went far beyond mere rivalry and militated against the free and open cooperation that was readily possible with Western countries. Moved by an inherent idealism, U.S. scientists thought of cooperation in space science as a good means for reducing tension between the two countries, whereas the more realistic Soviet scientists every so often would have to point out to their U.S. colleagues that it was the other way round. Intimate or large-scale cooperation would have to await the resolution of political difficulties.

Nevertheless, it was an American trait to cling to the idealistic approach, and scientists made persistent efforts to encourage exchanges with the Soviet Union.[37] None of these overtures, however, bore any fruit until in February and March 1962 President Kennedy provided a basis—not a very firm one, but a basis nevertheless—for exploring more intimately ways in which to cooperate. In an exchange of letters with Nikita Khrushchev, Chairman of the Council of Ministers of the USSR, Kennedy expressed the hope that representatives of the two countries might meet at an early date to discuss ideas "for immediate projects of common action."[38]

Working mainly with the State Department, NASA assembled a long list of possibilities for cooperation. From that list four main proposals—

cooperation in satellite meteorology, spacecraft tracking, studies of the earth's magnetic field, and communications satellites—were selected and dispatched along with a number of other ideas to Khrushchev 7 March 1962. With uncharacteristic speed for the Russians, he replied within two weeks, furnishing a list of proposals and agreeing to a meeting of appropriate representatives to discuss the matter.[39] President Kennedy named Hugh Dryden, deputy administrator of NASA, as the U.S. delegate. The USSR was represented by Anatoly A. Blagonravov. Between 27 and 29 March 1962, exploratory talks were held in New York City. More definitive talks in Geneva—which went on in May and June simultaneously with, but outside of, meetings of the Scientific and Technical Subcommittee of the United Nations Committee on the Peaceful Uses of Outer Space—led to a draft agreement 8 June 1962.[40] After a period of review, both governments approved the agreement, and James Webb, administrator of NASA, and President Keldysh of the Soviet Academy of Sciences exchanged letters putting the agreement into effect on an agency-to-agency level. The initial agreement called for working together on three separate projects: (1) exchange of satellite weather data over a communications link to be set up between Washington and Moscow; (2) each country to launch a satellite instrumented with magnetometers to study the earth's magnetic field during the International Years of the Quiet Sun beginning in 1965, coordinating their orbits and exchanging magnetic field data including those obtained from ground-based instruments; and (3) cooperative communications experiments using the next U.S. Echo satellite to be launched.[41]

Responding to a Soviet initiative in May 1964, a fourth project was added—to publish a book, prepared jointly, reviewing past Soviet and U.S. work in space biology and medicine, also giving some attention to future problems.[42] There was, however, no Soviet response to a suggestion Kennedy had put forth in a 20 September 1963 speech at the United Nations that the two countries consider joining forces to put a man on the moon. In this case the Soviet negativism was matched by that of the U.S. Congress, which quickly made known its distaste for the idea.

To implement the program agreed to in 1962, Dryden and Blagonravov continued to meet, in Rome during March 1963 and in Geneva the following May. Appropriate working groups were established. But it should be emphasized that the joint efforts were not *integrated* projects; they did not require putting together joint teams for preparing hardware, conducting launchings, analyzing data, or any such arrangement that might adversely affect one program if the other country failed to perform. Instead the projects were *coordinated;* the two national programs proceeded separately, but were to be conducted in such a way as to facilitate cooperative tests, as with Echo, or the exchange of data and information, as with the meteorology and magnetic field projects.[43] The performance of the Soviet participants on these projects for many years is best described as indifferent.

The primary interest to space science lay in the magnetic field studies and the book on space medicine and biology. In the magnetic field project a difficulty arose that was typical. American scientists considered it essential to the program to exchange data on the position of the satellites when measurements were taken. Otherwise, the field data could not be properly interpreted. But the Soviet Union consistently refused, shying away from providing any data that might reveal the capabilities of its electronic tracking equipment. The USSR had not, in fact, accepted another U.S. suggestion—to cooperate in the tracking of spacecraft. This difficulty also appeared to affect the communications project, in that finally the Soviet participants would agree only to receiving signals reflected to them from Echo, refusing to transmit any. Work on the book progressed exceedingly slowly, publication finally being achieved in 1975.

The difficulties that lay in the way of working with the Soviet Union in anything approaching the fashion of the cooperative projects with Western nations were formidable. Repeated frustrations led Arnold Frutkin in 1963 to prepare a set of internal NASA notes on Soviet deficiencies in their dealings with the Committee on Space Research. Frutkin listed eight ways in which the Soviet members appeared to be not forthcoming in their participation in COSPAR international activities: a lack of promised information on the Soviet sounding-rocket program; at one COSPAR meeting no papers on the sterilization of planetary probes, even though the Soviet Union had itself proposed that there be such a discussion; failure of Soviet members to attend the first two meetings of the COSPAR Consultative Group on Potentially Harmful Effects of Space Experiments; lack of any specific information on the Soviet Cosmos satellites; Soviet attempts to introduce political issues into COSPAR deliberations, e.g., nuclear testing; failure to provide information on radio tracking stations; and bypassing screening arrangements for papers to be presented at COSPAR. All of this Frutkin felt added up to a "retrogression in Soviet attitudes toward, and participation in, COSPAR."[44]

Two years later, in his book *International Cooperation in Space*, Frutkin moderated his assessment somewhat, noting modest progress and urging imaginative, aggressive efforts—tempered with a proper sense of realism—to "widen and deepen the cooperation which has already been won in the space field."[45] In spite of this commendable positivism, Frutkin's book brings out the stark contrast between the U.S. and Soviet space programs in openness and willingness to share with others. Frutkin once observed that he had written his book too early, a remark occasioned by the U.S.-Soviet cooperation of the 1970s, which included the joint docking mission of the Apollo-Soyuz Test Project in 1975. Certainly a book written in the late 1970s on U.S.-USSR cooperation on space projects would have many more positive elements to present than a book published in 1965. But there would also be a risk that the rockiness of the soil that had first to be tilled might be overlooked.

ESCALATION

Cooperation in space projects with the Western nations, while diverse and forthcoming, was nevertheless limited in scale. If it was true that a powerful benefit of the space program was to be gained from the management and conduct of large, complex projects, as Webb and French journalist J. J. Servan-Schreiber thought, these were benefits that other countries were not going to get from cooperation on individual experiments or even in the preparation of Explorer-class satellites. In December 1965 on the occasion of West German Chancellor Ludwig Erhard's visit to Washington, President Johnson, drawing on suggestions from NASA, invited European countries to pool their resources in a major spacecraft project as an advanced technological exercise of considerable scientific merit.[46] Following up on Johnson's suggestion, Frutkin and the author went to Europe in February 1966 to begin discussions with European countries and the European Space Research Organization, exploring the possibility that these nations might find it to their advantage to step up their space research to larger, more complex projects.

NASA suggested a spacecraft to send probes into the Jupiter atmosphere as the kind of project that was sufficiently advanced to task both management and industry and was bound to advance European technology in important ways. At the same time NASA emphasized that the Jupiter probe suggestion was only illustrative, and that other projects would serve the same purpose. Another possibility might be a solar probe to go very close to the sun to investigate magnetic fields and the interplanetary environment in the vicinity of our star. The NASA delegation spoke with groups from West Germany, France, the Netherlands, Italy, the United Kingdom, and the European Space Research Organization, beginning with scientists and government officials in Germany.[47] The reaction was surprising.

Most of those spoken to found the projects fascinating, but showed skepticism about the ultimate usefulness of such projects for advancing technology. Representatives in England were disbelieving and quite cool to the idea. At a time when the nation was having great economic troubles, leaders could not bring themselves to recommend investing in projects so far removed from immediate needs of the country. Throughout Europe one encountered the feeling that it would be better to invest directly in applications satellite projects that would have clearly foreseeable benefits. In fact, the NASA delegates encountered more than mere skepticism: Europeans believed that NASA was seeking additional financing for large-scale projects that Congress was no longer eager to support. While admitting that the financial aspect was an important consideration, the NASA representatives stated that both the U.S. and Europe would realize an important return on an investment in the kind of project proposed. But there was also suspicion that America was dangling the Jupiter probe in front of Europe to divert attention toward science and away from more practical projects like communications satellites.

More basic was European concern about dependence upon American technology. Both the European Space Research Organization and the European Launcher Development Organization had been formally established in March of 1964 after two years of intensive debate over the need of Europe to master the technology of space.[48] The principal purpose of these two organizations was to foster the development of technical know-how, ELDO especially to develop a sufficient launch capability to make Europe independent of the United States for a good number of its space missions. Europeans were, for example, convinced that the United States would not launch applications satellites for European countries if those satellites appeared to compete undesirably with U.S. industry—as communications satellites might do.

Only West Germany was interested in an expanded program with the United States, and out of these discussions came several cooperative projects, one of which was the solar probe Helios, intended to make magnetic field and other measurements within the orbit of Mercury. Costing Germany more than $100 million for the satellites, Helios was a sizable project, certainly well beyond the Explorer class in technological difficulty. As its share the United States provided the two launchings required and furnished some of the experiments. The first Helios probe was launched toward the sun in December 1974.[49] Other than the German projects, little came of the 1966 overtures to Europe. The proposals had, however, started a serious train of thought toward larger, more demanding programs, so that when the third administrator of NASA, Thomas Paine, began to press for some sort of cooperation in the Space Shuttle project that was being debated in the United States, a more receptive climate prevailed.

The same questions had to be faced again that had arisen earlier, and those concerning communications satellites had acquired an even greater force because of intensified airing of differences in the communications satellite consortium, where European members felt that the United States was dominating the consortium to the disadvantage of Europe. But cooperation on a Space Shuttle project was of a different character from joining in a scientific project like sending a probe to Jupiter. The Shuttle offered the opportunity to join in the development of a whole new technology, which in the view of the promoters would completely revolutionize space operations of the future, outdating and supplanting most of the expendable boosters used in the 1960s and 1970s.

After a long-drawn-out, careful assessment of values and costs, European countries in the European Space Research Organization, soon to give way to a new organization called the European Space Agency, agreed in September 1973 to develop a manned laboratory—Spacelab, originally called a sortie module in the United States—to be carried aboard the Space Shuttle.[50] In this fashion the increased cooperation with Western countries initially sought in 1966 came about. While the kind of cooperation on space experiments and satellite research that had gone on before would continue, it would be colored during the 1970s by Space Shuttle and Spacelab developments and was slated to be fundamentally modified when the new vehicles came into operational use in the 1980s.

On the Soviet side escalation came about in a different manner. In international circles the openness of the U.S. space program and America's readiness to enter into a variety of cooperative endeavors came in for a good deal of favorable comment. NASA people could sense a strong pressure on the Soviet scientists to do the same, a pressure that at times the Soviet delegates to international meetings seemed to find uncomfortable. Still, very little changed, except possibly some of the Eastern bloc countries found it a little easier to get assignments to support the Soviet program with ground-based observations. Also, in 1967 France, under de Gaulle's anti-U.S. leadership, managed to enter into a cooperation with the Soviet Union that went on for a number of years.[51] But for the United States to accomplish more, once again a change in the political climate was a prerequisite. In the move toward detente, political overtures on the part of the Nixon administration set the stage for new agreements in the space field.

In April 1970, Administrator Paine talked in New York with Anatoly Blagonravov about the possibility of combined docking operations in space. The idea was picked up by President Handler of the U.S. Academy of Sciences and discussed in Moscow in June with Mstislav Keldysh, president of the Soviet Academy of Sciences. In a letter to Keldysh, 31 July 1970, Paine made the first formal proposal for exploration of the subject.[52] Discussions were held in Moscow in October, and agreement was reached to work together to design compatible equipment for rendezvous and docking in space. Work got under way at once and, although the first plans did not specifically include actual missions, the Apollo-Soyuz Test Project to carry out a docking in space eventually emerged.[53]

While the Apollo-Soyuz Test Project, which was carried out in 1975, did include some scientific experiments, the project goes beyond the planned scope of this book. But in the climate established by the discussions on rendezvous and docking, it was possible to broaden the cooperative agreements arrived at between Dryden and Blagonravov a decade before. During January 1971, George Low, acting administrator of NASA after Paine resigned, met with Keldysh in Moscow to discuss further possi-

BEYOND THE ATMOSPHERE

bilities for cooperation. They agreed to exchange lunar surface samples and agreed on procedures for expanding earlier cooperative activities.[54] These Low-Keldysh agreements, as they came to be called, established a basis for increased cooperation between the two countries in both space science and applications. It remained to be seen whether the agreements would lead to further integrated undertakings, such as Apollo-Soyuz, or would continue to produce coordinated programs like the lunar sample exchanges.

19

Space Science and Practical Applications

In many ways space science contributed to the realization of important space applications—which may be defined as the use of space knowledge and techniques to attain practical objectives. Indeed, at the start of the program numerous potential applications required much advance research, including some space science, before their development could begin. Moreover, to many persons the development of applications appeared as the ultimate payoff of investments in the space program. Although the scientists would probably not have put it so strongly, nevertheless they could appreciate that point of view. As a consequence space scientists often pointed to potential applications of their work as one of the justifications for giving strong support to science in the space program.

Yet, in pointing to ultimate applications as one of the benefits to expect from their research, the scientists encountered a strange paradox. Although not appreciated for most of the 1960s, it finally became clear that in many respects applications—the "bread-and-butter work" of the space program—found it more difficult to gain support, especially on the executive side of government, than did space science.

Most space applications depend on or are affected in some way by properties of the atmosphere or conditions of space, which are subjects of the investigations of space science. For example, weather forecasting and the prediction of climatic trends depend on a knowledge of atmospheric behavior. The atmosphere is an exceedingly complex mechanism, a heat engine that receives solar heat which it reradiates into space. In the interval between receiving the energy and returning it to space, the atmosphere displays a bewildering variety of phenomena. The energy is converted into mechanical energy of winds and giant circulations that transport the excess energy received at the equator toward the polar regions. Clouds form and dissipate, storms are generated, water is taken up into the atmosphere from oceans, lakes, and rivers and released again in some form of precipitation. Interactions between the atmosphere and the land and oceans account for much of the complexity of weather phenomena. Weather forecasting consists of deducing from current data on the state of the atmosphere, and an imperfect knowledge of how the atmosphere behaves, the state of the atmosphere at a chosen time in the future. To do this requires knowing how long certain circulation patterns may be expected to persist, the ways in which energy exchanges are likely to occur within the atmosphere and between the atmosphere and the land and sea, and how all these are influenced by the continuous input of energy from the sun.

As a consequence meteorology assumes a dual aspect, the practical one of forecasting weather and climate and the scientific aspect of research on the atmosphere. Thus, when meteorological satellites were sent aloft to obtain pictures and other atmospheric data from around the globe—filling in tremendous gaps that had previously existed in weather data—the purpose was both practical and scientific. Because of its importance to both civilian and military needs, the practical aspect naturally stood out, and much progress in this phase of meteorology was achieved during the 1960s.

But exceedingly difficult scientific problems remained. The ground-based studies of decades had not unraveled the complexities of the long-term predictability of large-scale atmospheric circulations, of severe storm phenomena, of the puzzles of tropical meteorology, or of the causes of climatic change. It was hoped—expected—that space science and ground-based research together could move faster than ground-based studies alone.

When in the 1970s detailed study of other planets became possible, atmospheric scientists sought from the planetary atmospheres new insights into the difficult problems of the terrestrial atmosphere with which they were wrestling.[1]

Navigation satellites have great military and economic importance.[2] The principle of operation is quite simple. The artificial satellite substitutes as a reference point for the moon, sun, or stars; but since the satellite can be tracked by radio day or night, in fair weather or cloudy, it is available to the navigator whenever it is above the horizon. As in using the natural celestial bodies, if the navigator knows accurately the position of the artificial satellite, radio sightings of it permit him to locate his position on the earth. But, just as the celestial navigator has a problem with refraction of light by the atmosphere, for which he has to make corrections, so the satellite navigator must worry about refraction of radio signals. For him, the ionosphere produces the major effects, which are large enough to render the navigation system useless were it not possible to make correction. Here is where the ionospheric physicist's knowledge of the spatial and temporal variations of both the normal and disturbed ionosphere are essential. Again, the tie between space science and an important practical application is close.

Sometimes the connection between space science and a particular application was too close for comfort. The use of satellites for geodesy is a case in point.[3] For the scientist, more accurate geodetic measurements would provide more information on the size and shape of the earth and could give clues to the distribution of mass in the earth's crust and stresses in the mantle. With a system of sufficient precision, the very slow motions of continents relative to each other could be measured. More accurate mapping of the earth's surface could provide a better basis for plotting important information, like geological data, geographic locations, crops, forests, water resources, and land-use patterns. But the very measurements that made geodesy important to the scientists were also invaluable to the military.

And there was the rub. The military applications, which are fairly obvious, seemed to call for classifying the satellite data and restricting their distribution. This gave rise to controversy between the scientific community and the military (pp. 117–19), in which NASA was caught in the middle, appreciating the needs of the military but wanting to meet the demands of the scientists. The President's Science Advisory Committee and his science adviser were drawn into the debate, as was the Space Science Board. Congressman Karth and the Space Science and Applications Subcommittee of the House Committee on Science and Astronautics took up the cudgels on behalf of the scientists. These various counterpressures eventually forced an accommodation in which the distribution and use of data obtained in the geodetic programs supported by the military would be controlled by the military, while data obtained in NASA's open space-science program would be made available to the scientific community.

Just as intimate was the relation of science to the use of satellites for surveying and monitoring earth's resources. Here the geophysicist's study of the earth from space would furnish much of the basis for putting satellite observations of forests, agriculture, glaciers, oceans, geological formations, and mankind's use of land for cities, roads, farming, water storage, etc., to practical use.[4] The scientist's knowledge and the user's need would be brought together in a system that would convert satellite data into information required by the forest manager, the civil engineer, the irrigation planner, or the crop expert.

These close relationships between space science and applications led Administrator Webb to speak often to the author and others of the value of having the two together in a single Office of Space Science and Applications. In many ways this association was a good one, from which both programs benefited.

Such, too, were the reasons why the Space Science Board took a strong interest in space applications from the start. It often criticized NASA for doing too little scientific research to support the development and use of applications systems.[5] The criticism was especially strong in connection with the earth-resources survey program, a new field opened by the availability of observational satellites.[6] Because of its newness the field was highly scientific in character at the start, and there was concern in NASA that attempts to press these applications too rapidly before an adequate scientific basis had been laid, might prove abortive and seriously damage the ultimate prospects of what appeared to be a most promising area for practical returns. The Space Science Board's interest in space applications and the importance of space science for supporting those applications persisted. When the Academy of Engineering finally set up a Space Applications Board—an analog to the Academy of Sciences Space Science Board—in 1973,[7] the SSB immediately took steps to arrange for an exchange of liaison representation between the two boards.

Although in general experimenters personally were concerned only with the fundamental science they were doing, many scientists were genuinely interested in practical applications of their work. The members of the Upper Atmosphere Rocket Research Panel derived considerable satisfaction from the fact that properties of the atmosphere obtained from sounding rocket measurements contributed to the refinement and extension of the International Standard Atmosphere of the International Civil Aviation Organization,[8] used in designing aircraft and calibrating aeronautical instruments. It was also regarded as something of a triumph when ionospheric experimenters learned that their data on the properties and temporal variations of the ionosphere were proving useful to radio operators in scheduling and conducting shortwave, long-distance radio communications.

But, aside from personal interests, the possibility of deriving important applications was used to justify many parts of the space program, including space science. Potential military uses accounted for the numerous studies on the launching and use of artificial satellites conducted by the various services during the 1940s and 1950s[9] and for the military support of the sounding rocket program of the Rocket and Satellite Research Panel. Panel members became quite adept over the years at pointing out practical returns the services might derive from their investment in high-altitude rocket research (pp. 41–42). Equally adept were members of the Space Science Board and other scientific committees advising NASA. The agency's science program managers devoted much time to providing Congress with examples of how space science results had produced or might produce practical benefits.[10]

In this attempt to relate science to ultimate practical returns, scientists were heeding what was considered an obvious lesson of history. The power of the products of science and technology in prosecuting World War II was apparent. Following the war Congress was disposed to listen to the scientists and to give strong support to scientific research.* Scientific leaders took the opportunity to explain the nature and importance of science.[11] Paradoxically, the support for science was more assured than was support for many specific applications that scientists continued to invoke as justification for their own researches.[12]

For the first decade of NASA's existence those in the space science program had a rather straightforward view of how space applications fit into the picture—and, for that matter, so did those managing the applications program. In the belief that an attractive idea for the practical use of some space technology would sell itself, the scientists were accustomed to presenting in broad outline possible applications that might come from their research, and let it go at that.[13] But while space scientists were pointing to the support that they could give to applications as one justification of their own research, those in the applications program were experiencing strange difficulties selling their wares, especially in the latter half of the 1960s. Indeed, it often seemed that space science was easier to sell for its own sake than space applications were for their practical worth.

It took many years for this paradox to be appreciated, even though indications of the fundamental problems faced by those seeking support for applications programs had appeared in NASA's first few years when meteorological and communications satellites were being developed. The Advanced Research Projects Agency had begun the work on meteorological satellites. Once NASA was operating, the work was transferred to the new agency, primarily because of potential civilian benefits and because of the long-standing tradition of the government's providing weather services to the public through a civilian agency, the U.S. Weather Bureau of the Department of Commerce.

Among NASA's earliest successes was Tiros—Television Infrared Observational Satellite—which formed the basis for the country's first operational weather satellite system.[14] Tiros satellites were successful not merely because they worked technically, but equally because the Weather Bureau—later the Environmental Science Services Administration—could afford them.

---

*The salesmanship of the scientific community seems to have been successful, and a deep-seated conviction was established in Congress that a certain amount of scientific research—including pure science—was vital to the nation's interest. This conviction appears to have persisted even in the years following the Korean War when support to science began to decline and legislators refused to give the scientists all they wanted. The key point is that Congress was unwilling to let the size of the science budget in the U.S. increase indefinitely year after year, or to rise above some "reasonable" level. For some reason, the acceptable level appeared to be about 10 percent of the total U.S. budget for research and development, with the other 90 percent going to work on military and other practical systems.

Not so with Nimbus, the proposed successor to Tiros, Nimbus was a large, observatory-class satellite intended to provide a wide range of meteorological data, worldwide, day and night. The military, the Weather Bureau, and NASA had agreed on Nimbus as the next logical step beyond the more primitive Tiros satellites. NASA managers were shocked, therefore, when the Department of Commerce suddenly withdrew its support from Nimbus, precipitating a crisis of confidence in NASA in its congressional committees. But to Commerce the problem was simple. The projected price tag of $40 million or more per satellite was far beyond what even future meteorology budgets would be able to accommodate. Of more immediate concern, schedules were slipping and cost overruns would exceed available funds. Then, too, there was the question of how soon NASA would release control of Nimbus satellites to the Weather Bureau, a matter of prime concern if Nimbus were to be a part of an operational weather service.

Following rejection of Nimbus by the Department of Commerce, NASA agreed to upgrade Tiros satellites in a series of steps to improve observational capabilities while keeping costs down. Nimbus would be retained, with NASA paying the costs, as a research platform for testing new instruments and for trying new operational procedures. The success of Nimbus, and the operational use made of Nimbus data by the Department of Commerce, attested to the technical soundness of the satellite. It was, however, not economically viable; at any rate, it was not acceptable to the principal intended user.

Although vexing to NASA managers who had considered Nimbus a particularly fine example of a valuable space application, the Nimbus case was relatively uncomplicated. More complex was the range of difficulties encountered in developing a communications satellite system.[15] Whereas it had become traditional for the government to supply weather data to the public, industry provided most communications services—for a fee. The profit motive was a prime consideration and vested interests abounded. These complications were enhanced by international desires to share in the profits as well as in the technological benefits. The issue of how best to proceed was further beclouded by the military need for reliable and secure communications at its command. Congress, which consistently pressed heavily on NASA to push space applications, was torn between the desire to bring the benefits of satellite communications quickly to the country and its conviction that industry, if it was going to make a profit from providing satellite communications services, should bear its share of the development costs.

How the administration and Congress resolved these issues goes well beyond the subject of this book. But it is important to note that there was clearly going to be more to bringing a space application into being than simply demonstrating its technological feasibility. For applications, the harsh realities of the market place had a controlling influence.

Nowhere was this more evident than in the field of earth-resource surveys. Here vested interests were to be encountered at every turn. Also, in many cases the new satellite approach came immediately into competition with previously established ways of operating. Land-use surveys had been made with the aid of aerial photography, and a small industry had grown up around this technique. Estimates of grain production were compiled from aerial photos and thousands of individual reports made by farmers to county agents. Although NASA was convinced that satellite photography would be not only more effective for many uses, but also far more economical than traditional ground and aircraft surveys, many disagreed, even to the extent of wanting to discourage research to test the point. A particularly vigorous stand was taken by those who favored the use of aircraft. Even when numerous technical studies by NASA contractors began to show that the satellite approach would be quicker, more economical, and as accurate, the agency's troubles were not over.

Many of these questions were being debated during the Nixon administration, which regarded inflation as the major problem to solve. In this climate the administration was not inclined to encourage investment in expensive new systems—even if they were better—when the old systems were adequate. Moreover, there was concern that the old ways would not be supplanted, but merely supplemented by the new, piling additional costs upon the old for gains that were not essential, valuable though they might be. A member of the Office of Management and Budget in the Executive Office of the White House shocked NASA managers into a realization of how serious these questions were considered to be by conjecturing that, if the meteorological satellite program had still lain entirely ahead, it might not have been possible in the Nixon era to get approval for proceeding. The administration managers were dead serious about this, even in the face of the eminent success and value of the existing Tiros program.

A little later, when the Arab oil embargo and the energy crisis weighed heavily on the nation, NASA managers fully expected to be called on for extensive research and development on problems related to the emergency, and were prepared to forgo some of their space research to help. But, while NASA had developed an image of success and great technological capability in connection with its Apollo and other programs, there was great doubt as to how well the agency could cope with the practical problems the nation then faced.

In Apollo, it was pointed out, NASA had enjoyed a green light all the way. NASA was both the developer and user of its hardware and systems. To a large extent the agency set its own technological and operational objectives, and established its own criteria for success. In the commercial, social, and political world, matters would be different. NASA might develop elegant systems for energy, transportation, health care, or what have you; but NASA would not be the ultimate user of these systems, and hence not the judge of whether they were acceptable. It would not be enough to establish the technical feasibility of an idea. There would still remain the necessity to match it to the way the user chose to carry on his business, and to make it economical. In the Office of Management and Budget there was serious doubt as to whether NASA could adapt to these realities, a doubt that was fostered by John Young, the division chief who handled various technical budgets, including NASA's. For many years Young had been a key figure on the administrative side of the NASA organization. The familiarity he had acquired of NASA's methods, plus numerous scars from vigorous encounters with Administrator Webb, had left Young with the conviction that NASA did not understand the very difficult problems in pushing applications from the laboratory to the market.[16] He felt strongly that NASA was not the agency to put to work extensively on the nation's energy and resource problems, in spite of the widely prevailing, opposite view in Congress and elsewhere. Young expressed these views in no uncertain terms to the author during extended discussions between the Office of Management and Budget and NASA on the subject. It is not at all clear that Young and OMB were right in their assessment of NASA, but probably largely because of their opinions NASA was called on at the time for only a limited amount of help. Instead the agency was encouraged to pursue its work in space and aeronautics.

It is not within the scope of this book to probe into the problems faced by those responsible for developing space applications. Such matters are very complex and require a careful analysis to set them in their proper perspective. The subject does, however, bring out how the simplistic view of the scientists—both inside and outside of NASA—as to how their researches might lead to practical uses was extremely naive. For all the trouble scientists took to justify their work in terms of practical benefits, it can be seen in retrospect that, as far as science was concerned, Congress was prepared to take the long view. How else can one explain the sizable budgets approved for astronomical satellites, relativity studies, interplanetary investigations, and lunar and planetary exploration, the ultimate practical benefits of which surely had to lie in the very dim future? If the space scientists had appreciated the strength of their position, they might have felt more secure in letting space science, with its long-term implications, speak for itself.

20

Continuing Harvest: The
Broadening Field of Space Science

As the decade of the 1960s neared its end, space science had become a firmly established activity. While the past had been immensely productive, the future promised much more and thousands of scientists around the world bent to the tasks that lay ahead. A steady stream of results poured into the literature; universities illustrated courses in the earth sciences, physics, and astronomy with examples and problems from space research, and a few offered courses devoted entirely to space science. For their dissertations graduate students worked with their professors on challenging space science problems. With the loss of that air of novelty and the spectacular that had originally diverted attention from the purposefulness of the researchers, the field had achieved a routineness that equated to respectability among scientists.

Maturity underlay the field's hard-earned respectability. Starting about 1964, in addition to the individual research articles published in the scientific journals, more comprehensive professional treatments of the kind that characterizes an established, active field of research began to appear.[1] It is interesting, for example, to compare the book *Science in Space* published in 1960 with the second edition of *Introduction to Space Science* issued in 1968.[2] The matter-of-fact tone of the latter, which discussed what space science had already done and was doing for numerous disciplines, contrasts with the promotional tone of the former, which could only treat the potential of space science, what rockets and spacecraft might do for various scientific disciplines.

SPACE SCIENCE AS INTEGRATING FORCE

The breadth of the field as it evolved was impressive. Among the disciplines to which space techniques were making important contributions were geodesy, meteorology, atmospheric and ionospheric physics, magnetospheric research, lunar and planetary science, solar studies, galactic astronomy, relativity and cosmology, and a number of the life sciences.

The assured role of space science in so many disciplines in the late 1960s was a source of considerable satisfaction to those who had pioneered the field, an ample justification of their early expectations.

But more significant was the strong coherence that had begun to develop among certain groups of space science disciplines. Perhaps the most profound impact of space science in its first decade was that exerted upon the earth sciences. Sounding rockets made it possible to measure atmospheric parameters and incident solar radiations at hitherto inaccessible altitudes and thus to solve problems of the atmosphere and ionosphere not previously tractable. Satellites added a perspective and a precision to geodesy not attainable with purely ground-based techniques. The improved precision laid a foundation for establishing a single worldwide geodetic network essential to cartographers who wished to position different geographic features accurately relative to each other. The new perspective gave clearer insights into the structure and gravitational field of the earth. These examples illustrate one of several ways in which space science was affecting the earth sciences; that is, making it possible to solve a number of previously insoluble problems.

Following James Van Allen's discovery of the earth's radiation belts and the growing realization over the ensuing years that these were but one aspect of a tremendously complex magnetosphere surrounding the earth, magnetospheric research blossomed into a vigorous new phase of geophysical research. This was a second way in which space science contributed to the earth sciences, opening up new areas of research.

But probably the most significant impact of space methods on geoscience was to exert a powerful integrating influence by breaking the field loose from a preoccupation with a single planet. When spacecraft made it possible to explore and investigate the moon and planets close at hand, among the most applicable techniques were those of the earth sciences, particularly those of geology, geophysics, and geochemistry on the one hand and of meteorology and upper atmospheric research on the other. No longer restricted to only one body of the solar system, scientists could begin to develop comparative planetology. Insights acquired from centuries of terrestrial research could be brought to bear on the investigation of the moon and planets, while new insights acquired from the study of the other planets could be turned back on the earth. Delving more deeply into the subject, one could hope to discern how the evolution of the planets and their satellites from the original solar nebula—it being generally accepted that the bodies of the solar system did originate in the cloud of gas and dust left over from the formation of the sun—could account for their similarities and differences.

The wide range of problems served to draw together workers from a number of disciplines. Astronomers found themselves working with geoscientists who came to dominate the field of planetary studies that had once been the sole purview of the astronomers. Physicists found in the interplanetary medium and planetary magnetospheres a tremendous natural laboratory in which they could study magnetohydrodynamics free from the constraints encountered in the ground-based laboratory. Also known as hydromagnetics, this field was an extension of the discipline of hydrodynamics to fluids that were electrically charged (plasmas), particularly their interactions with embedded and external magnetic fields. The scientific importance of the field stemmed from the realization that immeasurably more of the matter in the universe was in the plasma state than in the solid, liquid, and gaseous states of our everyday experience. An outstanding practical value lay in the fact that magnetohydrodynamics was central to all schemes to develop nuclear fusion as a power source. Physicists also found the opportunity to conduct experiments on the scale of the solar system attractive for the study of relativity, and many of them began to devise definitive tests of the esoteric theories that were in existence. It is safe to say that this interdisciplinary partnership was a valuable stimulation to science in general.

The expanding perspective derived from space science was, in the author's view, the most important contribution of space methods to science in the first decade and a half of NASA's existence. While it was natural for individual scientists to concentrate attention on their individual problems, to those who took the time to assess progress across the board, the growing perspective was clearly evident even in the early years of the program. In a talk before the American Physical Society in April 1965, the author addressed himself to the growing impact of space on geophysics, which even then appeared much as described above.[3] NASA managers in their presentations to the Congress began to emphasize the important perspectives afforded by space science. As a case in point, the spring 1967 defense of the NASA authorization request for fiscal 1968 described space science as embracing (1) exploration of the solar system and (2) investigation of the universe.[4] Gathering the different space science disciplines into these two areas was not simply a matter of convenience. Rather it reflected a growing recognition of the broadening perspective of the subject, a point that was further developed by Leonard Jaffe and the author in a paper published in *Science* the following July.[5] At the time it was much easier to treat of the impact of space science on the earth sciences, which already offered many examples. While it would probably take a number of decades to achieve a thorough development of the field of comparative planetology, with an appreciable number of missions to the moon and planets behind and more in prospect, the powerful new perspectives available to the geoscientists were quite clear.

As for astronomy—the investigation of the universe—the deeper significance of the impact of space science on the discipline appeared to be unfolding more slowly. To be sure, the most obvious benefit—that of making it possible for the astronomer to observe all wavelengths that reached the top of the atmosphere, instead of being limited to only those that could reach the ground—began to accrue with the earliest sounding rockets that photographed the sun's spectrum in the hitherto hidden ultraviolet wavelengths. This benefit grew steadily with each additional sounding rocket or satellite providing observations of the sun and galaxy in ultraviolet, x-ray, and gamma-ray wavelengths. The value of these previously unobtainable data was inestimable. But in the long run, a deeper, more significant impact of space methods on astronomy could be expected, as Prof. Leo Goldberg and others pointed out: the advent of a much more powerful means of working between theory and experiment than had ever existed before.

At one time the author tried to persuade the House Subcommittee on Space Science and Applications that, as far as the origin and evolution of natural objects were concerned, the scientist knew more about the stars than about the earth. The statement was intentionally phrased in a provocative fashion to get attention, which it did. The Congressmen reacted immediately in disbelief, and it took quite a bit of discussion to develop the point, which went as follows.

Certainly men living on the earth, as they do, had been able to amass volumes and volumes of data on the earth's atmosphere, oceans, rocks, and minerals of a kind and in a detail that could not be assembled for a remote star. But, when it came to the question of just when, where, and how the earth formed and began to evolve many billions of years ago, the scientist was limited to a study of just one planet—the earth itself. From an investigation of that one body and whatever he could decipher of its origin and evolution, he had to try to discern the general processes that entered into the birth and evolution of planets in general. Only in such a broad context could the scientist feel satisfied that he really understood any individual case. Having only the earth to study, he was greatly hampered.

For the stars, however, the astronomer had the galaxy containing 100 billion stars to observe, and billions of other galaxies of comparable size. In that vast array the astronomer could find, for any object he might want to study, examples at any stage of evolution from birth to demise. With such a display before him in the heavens, the astronomer could proceed to develop a theory of stellar formation and evolution and test the theory against what he observed. In such an interplay between theory and observation the theorists did develop a remarkable explanation of the birth, evolution, and demise of stars.[6] So, in this sense, the astronomer could claim to understand more about the stars than the earth scientist did about the earth.

But there was a shortcoming in this theoretical process. The theory was based on observations of those wavelengths that could reach the ground—mostly visible, with a little in the ultraviolet and infrared, and after World War II critical observations in some radio wavelengths. Yet that very theory predicted that vitally important stellar phenomena would be manifested in the emission of wavelengths that the astronomer could not yet see. The early formation of a star from a cloud of gas and dust that was beginning to aggregate into a ball would be revealed primarily in the infrared as gravitational pressures caused the material to heat up. At the other end of the spectrum very hot stars would be emitting mostly in wavelengths shorter than the visible, presumably mostly in the ultraviolet. Little attention was paid to x-rays or gamma rays, yet among the most important discoveries of space science have been x-ray emissions from the sun and more than a hundred stellar sources.[7] Here is where the most profound impact of space science upon astronomy could be expected in the decades ahead. Just as the new-found ability to study other planetary bodies than the earth immeasurably broadened the perspectives of the earth sciences, so the ability of the astronomer to observe in all the wavelengths that reached the vicinity of the earth could be expected to strengthen the interplay between theory and experiment in the field of astronomy. By the 1970s the process had already begun, but the full power would doubtless have to wait until astronomers had the benefit of a variety of satellites more powerful than the solar and astronomical satellites of the first decade. In addition to large, precise optical telescopes, which one naturally thought of in the 1950s and early 1960s, there would also have to be specially instrumented spacecraft to pursue the new field of "high-energy astronomy" which leapt into prominence with the early discovery of x-ray sources. One would also need both infrared and radio telescopes in orbit. In short, to make the most of the opportunity that had burst upon the astronomical community, there would have to be established in orbit a rather complete facility consisting not just of a single instrument, but of a set of instruments ranging across the whole observable spectrum.

As for the life sciences, space appeared able to contribute in a variety of ways. One could expose biological specimens—including the crews of manned spacecraft—to the environment of space and observe what happened. But the biologists agreed that the most significant contribution of space science to their discipline could well be in exobiology—the study of extraterrestrial life and the chemical evolution of planets.[8] This subject was subsumed under the study of the solar system, since the evolutional histories of the planets, the kinds of conditions they developed on their surfaces and in their atmospheres, would have much to do with whether life formed on the planet, or with how far a lifeless planet moved toward the formation of life.

Thus, as one moved into the 1970s, although space scientists could take much satisfaction in the wide variety of individual disciplines to which they had been able to contribute, it was the new perspective that brought groups of disciplines together in a common endeavor that was most important.

EXPLORATION OF THE SOLAR SYSTEM

By the end of the 1960s study of the solar system extended from Earth to the nearest planets, Mars and Venus; and men had landed on the moon. In the next two years the Apollo astronauts made a searching exploration of the moon, after which Skylab crews turned attention toward Earth and the sun. During the early 1970s unmanned spacecraft also added Mercury, the asteroids, and Jupiter with some of its satellites to the list of objects in the solar system that scientists' instruments had been able to reach and observe, and had begun the long trek to Saturn and the outer planets.

Geodesy

Around Earth, satellite geodesy continued to advance steadily. As shown in chapter 11, the first half-dozen years of geodetic research using space techniques were largely years of preparation for the ultimate coup, that of establishing a worldwide geodetic net referred to a common reference ellipsoid. Year by year geodesists moved steadily toward that goal. By 1970 such a reference system was in use, at least among the principal experts, and positions relative to the common reference could be given to 15 meters or better. As for Earth's gravitational field, good estimates had been obtained for coefficients for the various harmonics in the spherical harmonic expansion of the geopotential up to at least the 19th degree. Geodesists were quick to point out that, in a little more than a decade, scientists had increased the quantitative knowledge of global positioning and of the size and shape of Earth by 10 times, and knowledge of the gravity field by 100 times—the same order of improvement as had been achieved in the previous 200 years.[9]

By the mid-1970s another order of magnitude had been realized in positioning techniques, and one could begin to zero in on accuracies sufficient to match variations in mean sea-level height (centimeters) and the very slow movements of tectonic plate motions and continental drift (centimeters per year). Using a combination of satellite techniques, observations on quasars and pulsars with a method called very long baseline interferometry, and extremely accurate clocks (1 part in 10^{16}) that might be developed with superconducting cavities, one could aspire to positional accuracies of several centimeters relative to the reference ellipsoid.

But, as anticipated, geodesy did not remain Earthbound, although the first extension to another planetary body was more by chance than otherwise. Tracking of Lunar Orbiters, five of which were put into orbit of the moon between 10 August 1966 and 1 August 1967, showed a peculiar perturbation of the orbital motion over a number of the circular maria of the moon. Analysis soon indicated that these disturbances were probably caused by unusual concentrations of mass in the maria basins. These mascons, as they came to be called, became one of the many puzzles in connection with the evolution of the moon that scientists had to try to explain. Some suggested that they were caused by heavy metallic material from slow-moving iron meteorites that had gouged out the basins where the mascons are now found. Or they might be due to lavas of different densities that filled the basins with the material now observed in the maria. They could be plugs of denser material from a lunar mantle that was shoved upward after the basins had been formed by impacting meteorites. Those making the above suggestions considered the mascons as strong evidence that the moon was quite rigid. But there were other opinions, as John O'Keefe points out:

> This [a rigid moon] is more or less how Urey saw it. What most of the rest of us saw was that the moon was in imperfect isostatic equilibrium like the earth. Apart from the mascons, there was isostatic equilibrium. Kaula pointed out that the earth, like the moon, has important deviations from isostasy, like the Hawaiian mass. Urey . . . overestimated the significance of the "mascons" as indicating a difference between the earth and the moon. He maintained that the moon was more rigid than the earth right up to the time when actual measurements showed that in fact it is less rigid.[10]

Whatever the mascons might turn out to be, however, they were an exciting and clearly significant discovery of the first extension of geodesy into the rest of the solar system.

Atmospheric and Ionospheric Studies

Like geodesy, upper atmospheric research and ionospheric physics continued to build upon the groundwork established earlier (chapter 6). Indeed, to the nonexpert the research and results could easily appear to be more of the same. But to the expert progress in the field was nothing short of phenomenal. Questions that had been uppermost in the experimenter's mind a decade before were quickly answered. It was shown that above 200 kilometers the neutral atmosphere became essentially isothermal, varying with the solar cycle from around 600 kelvins at sunspot minimum to somewhat more than 2000 K at times of very high solar activity. As had been expected the lighter gases, helium and hydrogen were found to predominate above 600 or 700 km, the outermost portions being largely hydrogen. Above 1000 km the positive ions He^+ and H^+ of helium and atomic hydrogen exceeded the concentrations of atomic oxygen O^+, which was the major ionic constituent of the F_2 region from 150 km to above 800 km.[11] At considerably lower altitudes, the D region of the ionosphere was found to

be surprisingly complex. In 1965 experimenters found that D-region ionization below 80 km consisted primarily of hydrated protons. Some years later
it was shown that negative ions in the D region tended to form complex
hydrated clusters.[12] Most important, the various solar radiations responsible for the ionization of the different layers were completely mapped. As
Herbert Friedman of the Naval Research Laboratory put it in 1974:

> The solar spectrum is now known with high resolution over the full elec-
> tromagnetic range. Spectroheliograms in all wavelengths have revealed the
> spatial structure over the disk [of the sun] and in the higher levels of the
> corona. We may conclude that the major features of the electromagnetic
> inputs to and interactions with the ionosphere are well understood.[13]

One of the most significant findings was that the upper atmosphere
and ionosphere could not be considered to be in or even near equilibrium,
previously a common assumption. For example, temperatures of the
electrons in the upper ionosphere were found to be appreciably higher
than the temperature of the ambient neutral gas, while the ion temperatures were intermediate between the two.[14] A continuing round of dynamic
processes characterized these regions, as solar radiations and the day-night
cycle set in motion a complex chain of chemical and physical reactions.
Also, in the light of space discoveries, the ionosphere was seen to be a part
of the much greater magnetosphere within which one could identify a
plasmapause, inside of which the hot plasma composing the ionosphere
rotated with Earth and outside of which particle radiations, no longer co-
rotating with Earth, exhibited convective motions induced by the solar
wind.[15]

But while such detail may be of interest as answering questions that a
decade before still puzzled researchers, their true significance lay in the fact
that by the 1970s all known major problems of the high atmosphere and
ionosphere had a satisfactory explanation based on sound observational
data. From then on research in the upper atmosphere and ionosphere
could be regarded as largely a mopping-up operation, the investigation of
the finer details of what was going on.

Meteorology

As for meteorology, the story was somewhat different. Sounding
rockets provided the means for making measurements at all levels within
the atmosphere, and satellites furnished worldwide imaging of cloud systems plus observations of atmospheric radiations and temperature profiles.
These data amplified by orders of magnitude the amount of information
available to the meteorologist, filling in enormous gaps that had existed
over the oceans and uninhabited land areas. But most of the impact of
these data was on the forecasting of weather and climatic trends, where
their contribution was of inestimable value. In a decade and a half of giant

strides in practical meteorology brought about in part by space methods, nothing of a revolutionary nature was contributed to the science of the lower atmosphere. In the mid-1970s atmospheric scientists still had to admit that there had been no breakthroughs attributable to space observations, although a wealth of new information was available and was under continuing intensive study. Most researchers were, however, optimistic that in the years ahead space data would share with ground-based, balloon, and aircraft measurements in leading ultimately to the breakthroughs in that understanding of the atmosphere needed to provide long-term forecasts of both weather and climate and to predict accurately the place and time of occurrence of severe storms.

If there was no general breakthrough, there were several intriguing contributions from space research. For one thing, as with other areas of the earth sciences, the perspective afforded by satellite imaging was a great stimulus to research. The ability to see and assimilate with ease the distribution and kinds of clouds, the location and nature of weather disturbances, the distribution of vorticity, etc., gave the researcher a new handle on his subject. As one result, tropical meteorology, once regarded as a dead-end field, sprang to life; and scientists began to develop new insights into the relations between the tropics and mid-latitudes.

Most of the sun's radiation lies in the visible wavelengths, which, along with some infrared and ultraviolet, control Earth's weather. This portion of the sun's spectrum is remarkably constant over time, although the question of just how constant remains open, and is one of the problems that space methods may help to solve. The short wavelengths, on the other hand, are extremely variable, their changes causing enormous variations within the high atmosphere. A natural question, then, was whether these upper atmospheric and ionospheric changes might not be related to meteorological changes. But, although sudden warmings of the stratosphere appeared to be associated with solar ultraviolet radiations, the general view was that this portion of the solar spectrum, containing less than one one-millionth of the energy carried in the visible wavelengths, could hardly have any significant effect. Yet, after more than a decade of space research, intriguing hints of relationships between upper atmospheric and meteorological activity began to appear.[16] For example, particles-and-fields research had shown that the interplanetary medium around the sun was divided into sectors in some of which magnetic fields were directed away from the sun, while in others the magnetic fields were pointed generally toward the sun (fig. 46). The two kinds of sectors alternated with each other in going around the sun.[17] Quite remarkably the boundaries between these sectors appeared to be associated with changes in atmospheric vorticity. Here was a phenomenon that could have a profound significance and the existence of which lent substance to the question of magnetospheric and upper atmospheric relationships to meteorology.

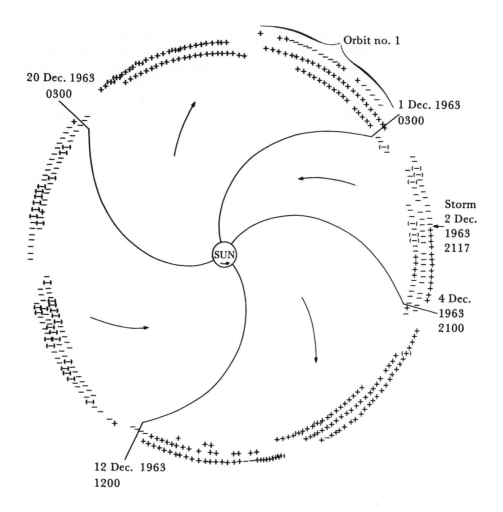

Figure 46. Sector structure of the solar magnetic field. The data are for December 1963 to February 1964. The direction of the average experimental interplanetary magnetic field during three-hour intervals is denoted by plus signs (away from the sun) and minus signs (toward the sun). Parentheses around a plus or minus sign indicate that the field direction fluctuated significantly. The solid lines represent magnetic field lines at sector boundaries. Alois W. Schardt and Albert G. Opp in Significant Achievements in Space Science, 1965, *NASA SP-136 (1967), p. 42.*

As space scientists were getting a firm grip on the physics and chemistry of Earth's upper atmosphere, their attention was simultaneously being drawn toward the planets. What was known about planetary atmospheres had come from the efforts of a small, select group of scientists, mostly astronomers.[18] Even more remote from the astronomers than Earth's upper atmosphere had been from the geophysicists, the atmospheres of Mars, Venus, and the other planets taxed the investigators' ingenuity. Gross uncertainties often existed in their estimates of atmospheric properties. As with Earth's upper atmosphere, measurements from space probes promised to eliminate or reduce many of the uncertainties.

The Soviet Venera spacecraft in 1970 and 1972 removed any doubt that the ground-level pressure of Venus's atmosphere was about 100 times that of Earth, and the Jet Propulsion Laboratory's Mariners showed that Mars's atmosphere was roughly one percent that of Earth. The atmospheres of both Venus and Mars were established as primarily carbon dioxide, as had been concluded from ground-based observations. As radio astronomers had already shown, the surface of Venus was confirmed to be in the vicinity of 700 K and fairly steady, while that of Mars was somewhat colder than Earth's and varied appreciably with the seasons. Whereas both Earth and Mars rotated rapidly with approximately the same periods, ground-based radar measurements showed Venus to turn very slowly—once every 243 days—in the opposite direction to that of the other planets.[19]

From the point of view of comparative planetology, the relations between Venus, Earth, and Mars were ideal. Earth was clearly intermediate between the two others in many respects, and many scientists felt that a detailed study of all three should be of special benefit in understanding Earth. An example of the kind of interplay that was possible was furnished by the study of the role of halogens in the atmosphere of Venus. The investigations led to the suspicion that chlorine produced in Earth's stratosphere from the exhausts of Space Shuttle launches or from freon used at the ground in aerosol sprays might dangerously deplete the ozone layer, which was known to shield Earth's surface from lethal ultraviolet rays of the sun. In a similar manner, when Venus's atmosphere was found to exhibit a single circulation cell, global in extent,[20] it was recognized that a careful study of this special example could yield important insights into the terrestrial atmosphere, where numerous circulation cells interact on a rapidly rotating globe. The clouds on Venus had long been a mystery, in which stratospheric aerosols now appeared to play a key role. The unraveling of the precise role of aerosols in the Venus atmosphere would certainly benefit studies of chemical contamination of Earth's atmosphere. At the other end of the scale, the role of dust storms in the thin Martian atmosphere could lend an important additional perspective to the role of dust in modifying Earth's climate. On a much grander scale, as Pioneer spacecraft passed by Jupiter in 1973 and 1974 it was learned that the famous red spot[21] was a huge hurricane large enough to engulf three Earths. What might be learned from the Jupiter hurricane about atmospheric dynamics that could be applied to the case of Earth remained to be seen.

The first planetary ionosphere other than Earth's to be detected experimentally was that of Mars. By observing the influence of the planet's atmosphere on radio signals from *Mariner 4* as those signals traversed the planet's atmosphere just before the spacecraft was occulted by the planet, it was possible to obtain an estimate of electron density as a function of altitude. The ionosphere, which was observed at a period of minimum sunspot activity, was somewhat less developed than expected from terrestrial analogies. Later, at times of greater activity, *Mariner 6, 7,* and *9* revealed a slightly more intense ionosphere, showing a noticeable dependence on the solar cycle. The same sort of occultation experiment on *Mariner 5* (1967) gave electron density profiles for Venus, both dayside and nightside. The nightside ionosphere was almost two orders of magnitude less intense than the daytime ionosphere, which showed a distinctly higher electron density than that of Mars.[22] Jupiter was shown to have a well-developed ionosphere.[23]

Thus, during the 1960s, while satisfactory answers were being obtained for all the known, major problems of Earth's high atmosphere, a good start was made on the investigation of the atmospheres and ionospheres of other planets.

Magnetospheric Physics

A genuine product of the space age, magnetospheric research also moved on apace. During the first half-dozen years the principal task was for the experimenters to produce an accurate description of the magnetosphere, although once the existence of the radiation belts, a terrestrial magnetosphere, and the solar wind had been revealed by actual observation in space, a host of theorists vied with each other to devise explanations. Indeed, as was pointed out in chapter 11, Eugene Parker's seminal paper on the solar wind antedated the detection of the wind and met with a great deal of flak until his critics were silenced by the space observations. The situation was typical for science; often many competing theories produce a continuing argument which no one can win until specific measurements become available to weed out those theories that don't fit the data. After the initial years of discovery and survey, however, the main action shifted to the theorists,[24] although the experimentalists continued to amass additional data from both satellites and space probes.[25]

At the end of the 1960s the theorists could explain many features of sun-earth relations, the interplanetary medium, and the magnetosphere, but a large number of fundamental questions remained to be resolved. To the layman a schematic picture of the magnetosphere drawn at the end of the decade might look much like that of figure 35 (chapter 11), produced a half-dozen years earlier. But the expert would read into that diagram a new collection of rather subtle questions that still had to be answered before one could claim to have a thorough understanding of magnetospheric physics.[26] The initial reaction to the discovery that Earth's magnetic field generated a huge bow shock in the rapidly moving solar wind was to apply hydrodynamical theory. The general shape and position of the bow shock and the magnetopause could be understood from magnetohydrodynamical principles. Also, it appeared that one might explain the sudden commencement and initial phase of magnetic storms in a straightforward way. But the theory could not explain why the solar wind appeared to apply a surface drag to Earth's field, pulling some of the field lines out into the long geomagnetic tail that was a spectacular aspect of the magnetosphere. Instead one would expect the magnetospheric cavity to close off in a teardrop shape.

To resolve some of these difficulties attention turned to the idea that the solar wind was a collisionless plasma, and the bow shock a collisionless phenomenon. Since under this assumption particle-to-particle collisions would be negligible, one had to seek the cause of the bow shock in cooperative field effects, such as interactions between electrostatic fields of the charged particles and magnetic field components perpendicular to the direction in which the gas velocity changed. As the 1970s opened, a great deal of study was going into collisionless shocks, particularly turbulent shocks, which observations showed Earth's bow shock to be.

Other problems that required attention were the wide range of geomagnetic activity, the acceleration of particles within the magnetosphere, the production of the auroras, and the formation of the geomagnetic tail. In connection with these matters, the idea that parallel and opposite magnetic fields might merge and annihilate each other aroused stormy debate. One could point to cases in which such a process might be important. For example, if the field in the solar wind had a southward component when it struck Earth's magnetic field at the nose of the magnetosphere, where the terrestrial field would be northward, merging and annihilation might take place. Or merging could occur when field lines in the interplanetary medium happened to be essentially parallel to field lines in the magnetospheric tail.

With such problems the magnetospheric physicists had an agenda that would keep them amply occupied during the 1970s. Moreover, early in that decade they got their first look at another planetary magnetosphere—that of Jupiter.[27] As the fascinating complexity of Earth's magnetosphere and its important role in sun-earth relationships had unfolded, physicists had immediately thought of the possibility of other planetary magnetospheres. It was known from observation of Jovian radiations that Jupiter had a very strong magnetic field.[28] As a consequence no one doubted that the first spacecraft to reach the giant of the solar system would encounter a well defined magnetosphere.

Until instruments could be sent to the nearer planets, the question remained open as to whether they also had magnetospheres. The first Soviet Luniks showed that the moon had very little magnetic field, an observation that was confirmed repeatedly in later U.S. missions.[29] Although some magnetism was found in the lunar crust, it became quite clear that the moon did not have a poloidal field such as the dipolar field of Earth. Accordingly there was no lunar magnetosphere analogous to that of Earth. Similarly the Mariner that flew by Venus in 1962 could detect no magnetic field, nor could *Mariner 4* when it reached Mars in 1965.[30] But, surprisingly, Mercury when reached in 1974 turned out to have an appreciable field.[31]

These circumstances provided the space scientists an opportunity for comparative studies in sun-planetary relations, an opportunity that in the early 1970s was still largely unexploited. As has already been shown, Earth provided the case of a planet with a sizable atmosphere and a strong magnetosphere, both of which were involved in intricate ways in the processes by which the sun exerted its influence on the planet. At the other extreme the moon provided the case of a body with neither magnetosphere nor atmosphere, so that solar radiations impinged directly upon the lunar surface. For Venus the sun's particle radiations struck the atmosphere directly, unmodified by a magnetosphere; but the extremely dense atmosphere shielded the planet's surface completely from these radiations. Mars, with its thin atmosphere but also without a magnetosphere was intermediate between the moon and Venus. Mercury, on the other hand, provided an example of a planet with a magnetosphere but no atmosphere. How such a magnetosphere would differ from Earth's, which was continually interacting with the planet's atmosphere and ionosphere, was an interesting subject for investigation, one that doubtless would be explored over the years ahead.[32]

In December 1973 *Pioneer 10* reached Jupiter, followed a year later by *Pioneer 11*. Their instruments revealed a huge magnetosphere reaching 7.6 million kilometers into space. Within the magnetosphere were radiation belts ten thousand to a million times as intense as those of Earth. Jupiter's magnetosphere reached well beyond the orbits of its four largest satellites, those observed first by Galileo centuries ago. The innermost satellite, Io, appeared to interact strongly with the magnetospheric radiations.[33] But to pursue these fascinating investigations any further would go beyond the scope of this book.[34]

Planetology

The final topic of this section concerns the planetary bodies themselves. While investigation of Earth's atmosphere, ionosphere, and magnetosphere—and related solar studies—were naturally the first areas of research in the space science program, they had only limited appeal to the layman. As exciting as these challenges were to the researchers, the average person could hardly relate to himself the magnetohydrodynamic concepts, terrestrial ring currents, or complex photochemical reactions in the ionosphere. But the investigation of the moon and planets was different. Here in pictures one could see landscapes and clouds—often strange, to be sure, but landscapes and clouds nevertheless. One could envision spacecraft orbiting a planet or landing on its surface and could identify personally with astronauts stepping onto the bleak and desolate moon. As a consequence NASA had little difficulty in capturing and holding a widespread interest in this aspect of the space science program.

The exploration of the moon and planets began with the Soviet Luna flights in 1959. From that time on, every year at least one mission to the moon or a planet was attempted by the United States or the Soviet Union. The American assault on the moon began with Pioneers, followed by Rangers, then the soft-landing Surveyors. In the summer of 1966 the first of five Lunar Orbiters began the task of mapping almost the entire surface of the moon. Even an Explorer was injected into lunar orbit to study the space environment around the moon. The climax was reached when the Apollo missions began manned exploration of the moon with the orbital flight of *Apollo 8* in December 1968 and the first manned landing in July 1969. While Apollo was in progress the Soviet Union conducted a series of sophisticated unmanned lunar missions that included circumlunar flights of Zond spacecraft, which were successfully recovered with pictures they had taken of the moon. More advanced Luna spacecraft soft-landed on the moon, carrying roving vehicles to investigate the lunar surface in situ and radio the information to Soviet stations on Earth, and in some cases to send samples of lunar soil to Earth for investigation in the laboratory. The success of the Soviet unmanned rovers and sampling missions sparked an intense debate between the scientists and NASA, many of the scientists feeling that the unmanned approach to the study of the moon was the wiser, and by far the more economical.

Criticism was blunted, however, by the tremendous success of Apollo. The astronauts brought back hundreds of kilograms of lunar rocks and soil from six different locations, the analysis and study of which quickly engaged the attention of hundreds of scientists throughout the United States and around the world. In addition to collecting lunar samples, the astronauts also set up nuclear-powered geophysical laboratories instrumented with seismometers, magnetometers, plasma and pressure gauges, instruments to measure the flow of heat from the moon's interior, and laser corner reflectors for geodetic measurements from Earth. The geophysical stations operated for many years after the astronauts had left, radioing back volumes of information on the moon's environment and its seismicity. Twice satellites were left behind in lunar orbit for lunar geodesy and to make extended chemical analyses of the lunar surface material from observations of the short-wavelength radiations of the moon.

The United States claimed the first success in planetary exploration when *Mariner 2*, launched in the summer of 1962, passed by Venus the following December, probing the clouds, estimating planetary temperatures, measuring the charged particle environment of the planet, and looking for a magnetic field.[35] In 1965 *Mariner 4* flew by Mars to take 21 pictures, covering about one percent of the planet's surface. Then *Mariner 5* visited Venus, this time getting substantially more data on the atmosphere, including estimates of the ionosphere. It was back to Mars in 1969 with *Mariner 6* and *7*, which returned some 200 pictures of the surface along with a variety of other measurements. *Mariner 9* went into orbit around the Red Planet in November 1971 at a time when the planet was almost completely obscured by a global dust storm. During the next month and a half the spacecraft monitored the clearing of the dust storm, which itself provided much interesting information about the planet. After that the spacecraft's cameras were devoted to the first complete mapping ever achieved of another planet—*Mariner 9* returned 7329 pictures of Mars and its two satellites, which permitted drawing up complete topographical maps showing the true nature of the markings that had so long puzzled astronomers.[36]

During the 1960s the Soviet Union had also set its sights on the planets. In fact, its attempts appreciably outnumbered those of the United States, since the USSR seized virtually every favorable opportunity to try a launching. But early Soviet planetary endeavors were about as dismal as America's early lunar tries. Not until a Venera spacecraft in 1970 succeeded in penetrating the atmosphere of Venus, returning data on the composition and structure of the atmosphere, did fortune smile on these planetary attempts. In December 1970 *Venera 7* landed and returned data from the surface of Venus; *Venera 8* followed suit in July 1972, for 50 minutes sending back surface data and analyses of the soil of Venus. Less fortunate, however, were the two Soviet Mars landers which, like *Mariner 9*, also arrived at the planet during the great dust storm of 1971. The storm that provided the Mariner with an unexpected opportunity to observe the dynamics of the planet's atmosphere may have been the cause of the Soviet spacecraft's failure to land successfully.[37]

In November 1973, *Mariner 10* left on a journey that would take it first by Venus and then on to Mercury, where the spacecraft arrived in March 1974 taking pictures and making a variety of other measurements. Having completed its first Mercury mission, *Mariner 10* was redirected by briefly firing its rockets so that the spacecraft would visit Mercury again in September of 1974. By visiting Mercury several times, *Mariner 10* provided the scientists with the equivalent of several planetary missions for little more than the price of one.[38] Also, with the visit to Mercury, scientists at long last had close looks at all of the inner planets of the solar system, including the two satellites of Mars.

The result of all these space probe missions was the accumulation of volumes of data on the moon and near planets, illuminated with thousands of highly detailed pictures. The photo resolutions exceeded by orders of magnitude what had been possible through telescopes. When Rangers crashed into the moon the closeup pictures sent back just before the impact were a thousandfold more detailed than the best telescopic pictures previously available (fig. 47). After landing on the surface with its television cameras, Surveyor afforded another thousandfold increase in resolution, revealing the granular structure of the lunar soil and a considerable amount of information on the texture of lunar rocks (fig. 48). In the laboratory Apollo samples put the moon's surface under the microscope, as it were (fig. 49). As for the planets no detail at all had been available before on the surface characteristics of distant Mercury or clouded Venus. Some *Mariner 10* pictures afforded better resolution for Mercury than Earth-based telescopes had previously given for the moon (figs. 50–51). Ultraviolet photos of Venus from passing spacecraft showed a great deal of structure in the atmospheric circulation that was hitherto unobservable (fig. 52), while radar measurements from Earth penetrated the clouds to reveal a rough, cratered topography.[39] For Mars, the indistinct markings observable from Earth were replaced with sufficient detail to show craters, volcanoes, rifts, flow channels, apparently alluvial deposits, sand dunes, and structure in the ice caps (figs. 53–60). Added to such pictures, data on planetary radiations in the infrared and ultraviolet; surface temperatures; atmospheric temperatures, pressures, and composition (when there was an atmosphere); and charge densities in the ionosphere (when there was an ionosphere)—this wealth of information completely revitalized the field of planetary studies, which had long been quiescent for lack of new data. By the early 1970s comparative planetology was well under way, although one must hasten to add that the task ahead of understanding the origin and evolution of the planets was one of decades, not merely months or years.

Nevertheless, progress was rapid. Much was learned about the mineralogy and petrology of the moon, and by extrapolation probably about the other terrestrial planets. Radioactive dating of lunar specimens led to the conclusion that the moon is probably some 4.6 billion years old, an age consistent with the ages of meteorites and the presumed age of Earth. The moon was found to be highly differentiated; that is, the lunar materials, through total or partial melting, had separated into different collections of minerals and rock types. The maria were mainly basalt, similar to but significantly different from the rocks of the ocean basins on Earth. In contrast the lunar highlands were rich in anorthosite, a rock consisting mainly of the feldspar calcium aluminum silicate. Both maria and highlands were much cratered—as could already be seen from Earth—and one could now see that the crater sizes extended down to the very small, showing that the moon had been bombarded by very small particles as well as by very large objects. The entire surface was covered with fine fragments and soil, broken rocks and rubble—crustal material chopped up by the cratering process. A considerable amount of glass was found, some of it in coatings splashed onto other rocks, much of it in the form of tiny glass beads of a variety of colors dispersed through the soil.

Reading through the record imprinted on the lunar surface one could bit by bit piece together the course of the moon's evolution. Contrary to expectations of many, before the unmanned and manned space missions showed differently, the present lunar surface was not a virginal record of conditions that existed at the time of the moon's formation. Instead one could discern a continual, sometimes violent process of evolution. Either at the time of formation or very shortly thereafter the moon's crust was molten to a considerable depth. Presumably during this phase the differentiation of lunar materials took place, producing the light-colored, feldspar-rich highlands. After solidifying about 4.4 billion years ago, the highlands were bombarded for hundreds of millions of years, probably by material left over from the formation of the moon and planets. This process produced the cratered topography of the highlands still visible today. Between 4.1 and 3.9 billion years ago the bombardment of the moon became cataclysmically violent, with asteroid-sized meteorites gouging out the great basins, hundreds and even thousands of kilometers across (fig. 61). Then, as radioactive dating of lunar materials showed, between 3.8 and 3.1 billion years ago a series of eruptions of basaltic lavas filled the basins to form the maria, or dark regions of the moon clearly visible from Earth. By 3 billion years ago the violent evolution of the moon had come to an end. Cratering and "gardening" of the lunar surface continued, but at about the rates observed today, most of the impacting particles being micrometeor and grain sized, occasionally cobble sized, with the very large impacts exceedingly rare.[40]

The moon observed by Apollo instruments was very quiet. The lack of any substantial organized magnetic field suggested that there was no molten core, but the presence of magnetism in lunar rocks indicated that there might have been a core at some time in the past—on the assumption that a liquid iron core was required to generate the lunar field that magnetized the lunar rocks. Seismometer data showed that the moon was at least partly molten below 800 km and suggested a small, possibly iron, core a few hundred km in radius. The energy released by moonquakes detected by the Apollo seismometers was nine orders of magnitude—a billion times—less than that released by earthquakes over a similar period of time, which, however, seemed consistent with the relative sizes of Earth and moon. The seismic data yielded the picture of a lunar crust 50 to 60 km thick, four times the average thickness of Earth's crust, underlain by a mantle solid at the top but partially melted toward the bottom (fig. 62). The very thick crust could account for the lack of any recent volcanic activity.[41]

Before these investigations it was common to think of the moon as a small body and to suppose that small bodies would remain cold and essen-

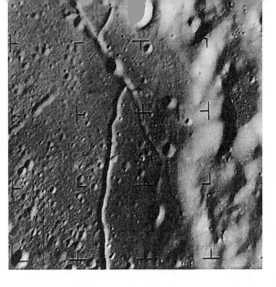

Figure 47. Ranger photos of the moon. The closeup picture, above, of the crater Alphonsus floor taken by Ranger 9 *in 1965, shows detail not available in telescopic pictures taken from Earth.* Ranger VIII and IX, JPL Tech. Rpt. 32-800, pt. 2 (15 March 1966), p. 353, fig. 7.

Figure 48. Surveyor photos of the moon, top right and right. Sitting on the moon's surface, Surveyor 1 *in 1966 provided a millionfold increase in resolution over that of Earth-based pictures.* Surveyor Program Results, NASA SP-184 (1969), p. 259, fig. 7-42.

Figure 49. Thin section of Apollo 16 lunar sample 67075. Samples of lunar material brought to Earth made it possible to examine minerals of the moon under a microscope. The area of the section photographed is 2.3 mm long.

tially unchanged throughout their histories. Now it was clear that bodies the size of the moon, and even smaller ones, whether formed in the molten state or melted after formation, undergo a substantial evolution. This conclusion was borne out further by data from the other planets. Mercury appeared even more cratered than the moon. There was widespread evidence of lava flows on the planet. Large cracks and long scarps were visible in the *Mariner 10* pictures. There was no doubt that Mercury underwent a great deal of evolution after its formation.[42]

The evidence of activity on Mars was even more striking. After a period of discouragement for scientists when the *Mariner 4* pictures appeared to show a dead, moonlike planet, the pictures of *Mariner 6* and *7* revived interest, and those of *Mariner 9* aroused excitement. Those pictures showed huge volcanoes, one of them—Olympus Mons—twice the diameter of the largest known volcanic structure on Earth, namely, the big island of Hawaii. A great deal of the Martian surface was cratered, indicating an ordeal of bombardment like that experienced by the moon and Mercury. Some of the surface was smooth and featureless, indicating a process of filling in as with blowing, drifting sands. A huge rift, comparable in length to the width of the United States, indicated considerable tectonic activity (fig. 56). Numerous channels hundreds of kilometers long looked as though they might have been produced by flowing water (fig. 57). Consistent with this observation were the frequent formations that looked like alluvial fans produced by the deltas of terrestrial rivers or sedimentary deposits of meandering streams (fig. 58). While the variable frost cover observed in the north and south polar caps was shown to be solid carbon dioxide, a substantial base of frozen water was also found. Various estimates suggested that a considerable amount of water must have outgassed from the planet over time, and if past conditions on the planet were just right there could have been ponds and rivers. But the question of the role of water in the evolution of the planet remained unsolved.[43]

It was very clear that Mars is an active planet, by no means dead, as some had prematurely concluded. Some investigators thought they could detect in the marked differences between the cratered highlands of the planet and the volcanic provinces the suggestion of an incipient separation into individual tectonic plates as on Earth, a picture not generally accepted by students of Mars. Paul Lowman, however, was led to conclude that the evidence was piling up that earthlike planetary bodies would follow similar courses of evolution.[44] In the larger bodies like Earth, the rates of evolution would be faster and the duration longer than for the smaller planets. Of the inner planets, Earth, still vigorously active, was most advanced in its evolutionary course (fig. 63). Venus might be in a comparable stage, but no life evolved there to convert the carbon dioxide atmosphere to one with large amounts of oxygen. Mars was following the course

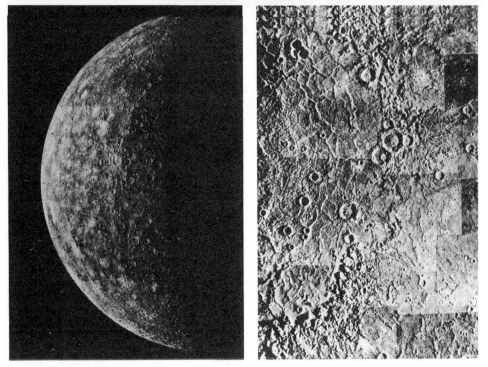

Figures 50–51. Photomosaics of Mercury. Fig. 50, left: Eighteen pictures, taken at 42-second intervals, were enhanced by computer at the Jet Propulsion Laboratory and combined into this mosaic. The pictures were taken from Mariner 10 during 13 minutes when the spacecraft was 200 000 kilometers and 6 hours away from Mercury on its approach to the planet, 29 March 1974. About two-thirds of the portion of Mercury seen in this mosaic is in the southern hemisphere. Largest of the craters are about 200 kilometers in diameter. Illumination is from the left.

Fig. 51, right: The semicircle of cratered mountains in the left half of the mosaic forms the boundary for the largest basin on Mercury seen by Mariner 10. The basin is near a subsolar point when the planet is at perihelion, leading investigators to suggest the name Caloris for it. The ring of mountains is 1300 kilometers in diameter and up to 2 kilometers high. The basin floor consists of severely fractured and ridged plains.

Figure 52. Cloud structure on Venus. The structure reveals the pattern of atmospheric circulation on the planet. The picture was taken in ultraviolet light by Mariner 10 cameras 6 February 1974.

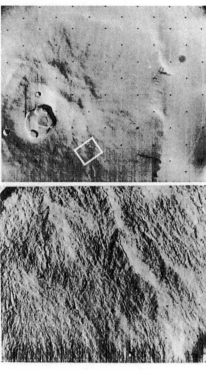

Figure 53. Craters on Mars. The slightly darker region to the right is Hellespontus; the lighter region to the left, Hellas. Apparently an escarpment forms the boundary between the two regions. The area of the photo, taken by Mariner 7 in 1969, is 720 by 960 kilometers.

Figures 54–55. Volcano on Mars. Olympus Mons was photographed 7 January 1972 by Mariner 9. The top photo shows an area 435 by 655 kilometers. The lower, high-resolution photo, corresponding to the inscribed rectangle at the top, shows details of lava flow down the mountain side.

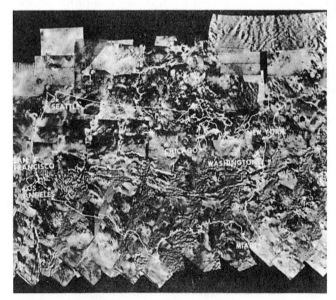

Figure 56. Rifting on Mars. The superimposed outline of the United States emphasizes how the great canyon on Mars dwarfs the Grand Canyon on Earth. The photomosaic was made from Mariner 9 pictures obtained in several weeks of photographic mapping of the planet. The area covered reaches from –30° to +30° latitude and from 18° to 140° longitude.

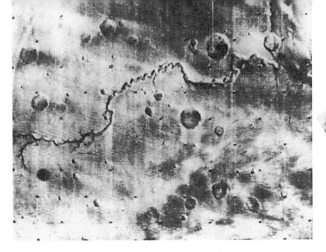

Figure 57. Flow channel on Mars. This valley, some 400 kilometers long, resembles a giant version of an arroyo on Earth.

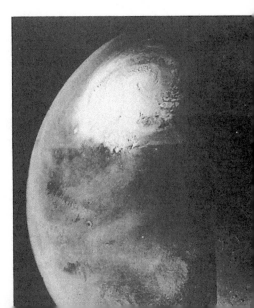

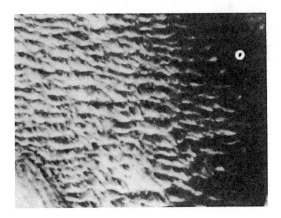

Figure 58. Evidence of water flow on Mars, at right. Braided channels at -6° latitude, associated with Vallis Mangala in the Amazonis region of Mars near longitude 150°. Such features are common to sediments deposited during meandering stream flow. Each of the two frames in this composite covers about 30 by 40 kilometers.

Figure 59. Sand dunes on Mars, above. The presence of sand and dust on Mars was dramatically emphasized by the great dust cloud that enveloped the planet as Mariner 9 approached. After the cloud of dust had settled, deposits of sand were observed to shift about from photos of the same region taken at different times.

Figure 60. Ice cap on Mars, below. The photo shows the northern hemisphere of Mars from the polar cap to a few degrees south of the equator. At the stage shown, the ice cap is shrinking during late Martian spring.

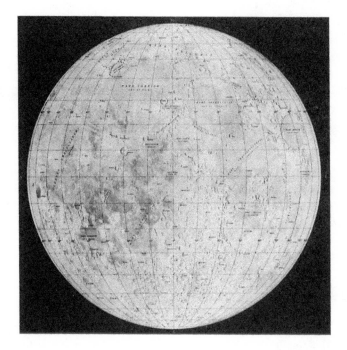

Figure 61. The moon. The great maria, like Mare Imbrium in the upper hemisphere, were probably gouged out by huge meteorites, and subsequently filled by extensive flows of dark basaltic lava.

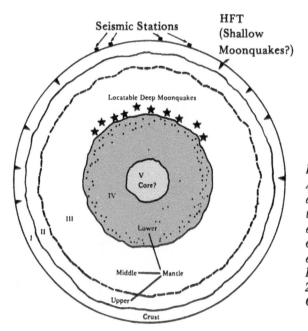

Figure 62. Structure of the moon. As on Earth, seismic data reveal a great deal about internal structure. Y. Nakamura, G. Latham, et al., in Proceedings of the Seventh Lunar Science Conference, *ed. R. B. Merrill (New York: Pergamon Press, 1976), pp. 3113–21; reproduced by permission of Gary V. Latham.*

HFT = High-frequency teleseismic event.

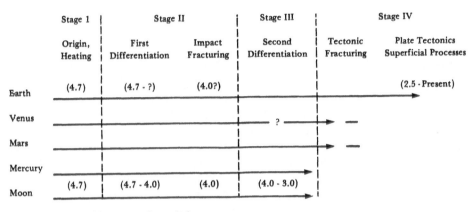

	Stage 1	Stage II		Stage III		Stage IV	
	Origin, Heating	First Differentiation	Impact Fracturing	Second Differentiation	Tectonic Fracturing	Plate Tectonics Superficial Processes	
Earth	(4.7)	(4.7 - ?)	(4.0?)			(2.5 - Present)	
Venus				?			
Mars							
Mercury							
Moon	(4.7)	(4.7 - 4.0)	(4.0)	(4.0 - 3.0)			

Ages ( ) in billions of years before present;
refer to peak of events; events may overlap.

Figure 63. Crustal evolution in silicate planets. Evidence suggests that earthlike planets all follow similar courses of evolution. Paul D. Lowman, Jr., in Journal of Geology 84 (Jan. 1976): 2, fig. 1; reproduced courtesy of Dr. Lowman.

taken by Earth, but was well behind, only now approaching the tectonic plate stage. The moon and Mercury had long since run the course of their evolution, which terminated well before a tectonic plate stage.

This picture, although consistent with much of the data, could hardly be regarded as more than tentative. It would have to pass the test of further observations and measurement, and stiff debate. But one satisfying feature was the emphasis the theory gave to the kinship of the planets with each other. As theorists had pointed out, if the planets did form from the material of a solar nebula left over after the creation of the sun, then their individual characteristics should depend to a considerable extent on their distances from the sun (fig. 64). Near the sun, where the nebular material would be heated to rather high temperatures by the sun's radiations, one could expect to find planets composed primarily of materials that condense at high temperatures, the silicates and other rock-forming minerals. Moreover, the densities of the planets could be estimated according to distance from the sun by considering what compounds were likely to form at the temperature to be expected at Mercury's distance, which at the distance of Venus, which in the vicinity of Earth, etc. From what was known of the inner planets, they did indeed fit such a picture.

As for the outer planets, one would expect them to consist of large quantities of the lighter substances—hydrogen, helium, ammonia, methane—which could condense out of the solar cloud only at the low temperatures that would exist so far from the sun. Qualitatively, the outer planets also fitted this picture, but quantitatively there were discrepancies. To develop the true state of affairs in proper perspective, an intensive investigation of the outer planets was called for, and was on the agenda for the 1970s and 1980s. The investigation got off to an exciting start with the visit of *Pioneer 10* to Jupiter in 1973. It was clear that an exciting period in planetary exploration lay ahead as scientists began to amass data on the atmospheres, ionospheres, and magnetospheres of these strange worlds. While these planets themselves would be quite different from the terrestrial planets, their satellites could be expected to resemble the latter in many ways. Moreover, as many persons pointed out, the satellite systems of Jupiter and Saturn might turn out to be very much like miniature solar systems, particularly the satellites that formed along with the parent planet rather than being captured later. Supporting this view was the early discovery from the Pioneer observations that the four regular satellites of Jupiter decreased in density with increasing distance from the planet, as though they had formed from a cloud of gas and dust that was hotter near the planet than it was farther away.[45] The opportunities for important research seemed endless.

Finally there was the question of extraterrestrial life. Space research on fundamental biology was early divided into two areas: (1) the study of terrestrial life forms under the conditions of space and spaceflight, and (2) exobiology. The former was discussed briefly in chapter 16. The latter came to mean the search for extraterrestrial life and its study in comparison

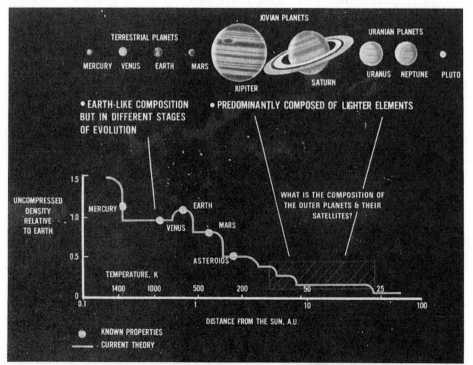

Figure 64. Origin of solar system planets. The higher temperature materials, like silicates, condense nearer the sun; the more volatile substances, farther away.

with Earth life. Most scientists considered the chance of finding life elsewhere in the solar system to be minute, but it was universally agreed that the discovery of such life would be a tremendously important event. Thus, while recognizing the unlikelihood of finding extraterrestrial life, many considered that the potential implications offset the small chance of finding any, and accordingly devoted considerable time to studying in the laboratory the chemical and biological processes that seemed most likely to have been part of the formation of life. They sought out Earth forms that could live under extremely harsh conditions—like arid deserts, the brines of the Great Salt Lake, or the bitter cold Antarctic—and paid special attention to them. And they devised experiments to probe the Martian soil for the kinds of life forms deemed most likely to be there.

But, after a decade and a half, the problem of life on other planets remained open. No life was found on the moon, nor was there any evidence that life had ever existed there. A careful search was made for carbon, since Earth life is carbon-based. In lunar samples a few hundred parts per million were found, but most of this carbon was brought in by the solar wind. Of the few tens of parts per million that were native to the moon, none appeared to derive from life processes.[46]

Nor was any life found on Mars, even though the two Viking spacecraft with their samplers and automated laboratories were set down in 1976 in areas where once there might have been quite a bit of water.[47] Still, the subject could hardly be called closed. If life had been found, that would have settled the question. But that life was not found in two tiny spots on Mars did not prove that there was no life on the planet. So, although the first attempts were disappointing, it could be assumed that future missions to Mars would pursue the question further. Nor would it be likely that exobiological research would be confined to the Red Planet. At the very least, one would expect that, as scientists studied the chemical evolution of the planets and their satellites, they would keep the question of the formation and evolution of life in mind.

While the foregoing has touched upon but a few of the results accruing from the exploration of the solar system, still the reader should be able to derive some insight into the impact that space science was having upon the earth and planetary sciences. For one thing the study of the solar system was revitalized after a long period of relative inactivity. Second, lunar and planetary science became an important aspect of geoscience, attracting large numbers of researchers. Third, the new perspective afforded by space observations gave an immeasurable boost to comparative planetology, a field that made great strides during its first 15 years. Nevertheless, no one doubted that in the mid-1970s comparative planetology still looked forward to its most productive years.

All of which leads to the usual question. If space science was having such a profound impact on Earth and planetary sciences, was space science producing a scientific revolution in the field? In the broad sense, no. But, viewed strictly from within the discipline there were indeed numerous revolutionary changes. Much new information was accumulated, permitting the theorist to deal in a realistic way with topics about which one could only speculate before. Many had to relinquish pet ideas about the nature of the lunar surface or the markings on Mars. Proponents of a cold moon were faced with incontrovertible evidence of extensive lunar melting. No pristine lunar surface was to be found; instead a substantial evolution had marked the moon's first one and a half billion years. Far from being an inert planet, Mars turned out to be highly active. Of course, in different aspects of the subject many investigators had been on the right track. Noted astronomers R. B. Baldwin and E. J. Öpik had correctly anticipated that many of the features of Mars were due to craters.[48] Gerard Kuiper had been sure that volcanism was important on the moon, as he explained many times to the author. Thomas Gold had been certain that the lunar surface would contain a great deal of fine dust. Yet, no one had succeeded in putting the separate pieces together in satisfactory fashion. Thus, the most revolutionary aspect of space science contributions to the earth and planetary sciences was probably in helping to develop an integrated picture of the moon and near planets. This was an enormous expansion of horizons, an expansion that could be expected to continue with each new planetary mission.

Investigation of the Universe

As space scientists were busily altering the complexion of solar system research, space methods were also profoundly affecting the investigation of the universe. Here space science could contribute in a number of ways to solar physics, galactic and metagalactic astronomy, and cosmology, including a search for gravitational waves, observations to determine whether the strength of gravity was changing with time, and studies of the nature of relativity. But the contribution of space techniques to these areas was qualitatively different from those in the planetary sciences. Whereas rockets and spacecraft could carry instruments and sometimes the observers themselves to the moon and planets to observe the phenomena of interest close at hand, this was not possible in astronomy. The stars and galaxies would remain as remote as before, and even the sun would continue to be a distant object extremely difficult to approach even with automated spacecraft because of the tremendous heat and destructive radiation.

The connection between the scientist and the objects of study would continue to be the various radiations coming from the observed to the observer. But rockets and satellites would increase the variety of radiations that the scientist could study by lifting telescopes and other instruments above Earth's atmosphere, which was transparent only in the visible and some of the radio wavelengths. This extension of the observable spectrum proved to be as fruitful to the prober of the universe as were the lunar and planetary probes to the student of the solar system.

As stated in chapter 6, rocket astronomy began in 1946 when sounding rockets were outfitted with spectrographs to record the spectrum of the sun in hitherto hidden ultraviolet wavelengths. In 1948 x-ray fluxes were detected in the upper atmosphere, after which rocket investigations of the sun ranged over both ultraviolet and x-ray wavelengths. Inevitably experimenters turned their instruments on the skies, and when they did various ultraviolet sources were found. In 1956 the Naval Research Laboratory group found some celestial fluxes that might have been x-rays, but the real significance of x-rays for astronomy had to await more sensitive instruments that did not become available until the early 1960s.

In the meantime sounding rocket research on the sun's radiations moved on apace. Investigators from a number of institutions continued to amass detail on the sun's spectrum in the near and far ultraviolet, which was important in understanding the quiet sun and normal sun-earth relationships. But the real excitement proved to be with the x-rays. It was these, rather than the ultraviolet wavelengths, that came into prominence with high solar activity. When satellites came into being, they were put to use in making long-term, detailed measurements of the sun's spectrum in all wavelengths. On 7 March 1962 the first of NASA's Orbiting Solar Observatories went into orbit, to be followed by a series with steadily improving instrumentation. The Naval Research Laboratory built and launched a series of Solrad satellites, intentionally less complex than the OSOs, to provide a continuous monitoring of the sun in key wavelengths. But, while satellites came into prominence in the 1960s, sounding rockets, some of them launched at times of solar eclipse, continued to yield important results. In fact, some scientists felt that the most significant work on the sun came from sounding rockets rather than from the far more expensive satellites.

NASA's Orbiting Solar Observatories continued into the 1970s, the first one of the decade being *OSO 7*, launched on 29 September 1971. An important event for solar research was the launching of Skylab in 1973. In this space laboratory astronauts studied the sun intensively using a special telescope mount built for the purpose. Although the high cost of Skylab's solar mission in dollars and time to prepare and conduct the experiments was distressing to many of the scientists, nevertheless the results were extremely important for solar physics, some of them providing solutions to long unsolved problems.

Sounding rocket experiments were also fruitful in stellar astronomy. Perhaps the most significant event in rocket and satellite astronomy occurred when American Science and Engineering experimenters, with an Aerobee rocket flown on 12 June 1962, discovered the first x-ray sources outside the solar system to be clearly identified as such. As will be seen later, this discovery proved to be of profound significance to modern astronomy. During the 1960s sounding rockets continued to search for and gather information on these strange sources, but progress was slow. Long-term observations with more precise instruments were needed, a need NASA was much too slow in supplying. The breakthrough came with the launching of NASA's first Small Astronomy Satellite, *Explorer 42*, on 12 December 1970.

During the 1960s the course of ultraviolet astronomy from satellites also proceeded slowly. The principal satellite designed for such studies, NASA's Orbiting Astronomical Observatory, proved difficult to bring into being, and it wasn't until the end of the decade that *OAO 2* (7 December 1968), the first successful astronomical observatory, went into orbit. It took another four years to get *OAO 3* aloft (21 August 1972). From OAO observations the Smithsonian Astrophysical Observatory compiled the first complete ultraviolet map of the sky, issuing the results in the form of a catalog for use by astronomers.[49] With *OAO 3*, which had been named *Copernicus* in honor of that dauntless pioneer in scientific thought, Princeton University experimenters obtained a number of significant results. They showed that, while hydrogen in the interstellar medium was almost entirely in atomic form, most interstellar clouds had an abundance of neutral hydrogen molecules, the relative abundance being consistent with a balance between the catalytic formation of H_2 on grains of material and the competing dissociation of the gas by absorption of light. Much of the galactic disk was found to be occupied by a hot coronal gas at half a million kelvins, with a hydrogen density of one particle per liter. The Princeton workers also observed that the relative abundance of the different chemical elements in the interstellar gas was what would be expected if, starting with a mixture of elements in the ratios found in the overall cosmic abundance, the materials of high-condensation temperatures had already condensed out to form small solid particles or dust grains. Finally, flowing from most very hot stars were stellar winds of some thousands of kilometers per second.

Other experiments were also on the OAOs, some of them concerning x-rays and gamma rays, and the Orbiting Astronomical Observatories were clearly proving very fruitful. Yet one could detect the feeling that OAO was a bit out of step. The satellite had been sufficiently difficult to construct that it had delayed satellite optical (visible plus ultraviolet) astronomy for about a decade, whereas a series of cheaper, simpler satellites could have kept research moving while work on a larger instrument proceeded. Also, now that it had come, OAO was well behind both existing telescope technology and current needs. For most of the problems of greatest concern to the optical astronomers, a larger aperture (2.5 to 3 m), more precise telescope was required. As agitation developed for the construction of a

Space Shuttle it was quickly realized that one of the things that the Shuttle could do ideally was to launch such an instrument into orbit and service it throughout the years. The launching of such a telescope became one of the prime scientific missions for the Shuttle.

But in the 1970s the circumstances surrounding astronomy had changed. Whereas in 1959 and 1960 the most important tasks for satellites had appeared to be in ultraviolet astronomy, in the late 1960s and early 1970s both ground-based and space research had changed the picture. Now the high-energy end of the spectrum—x-rays and gamma rays—was the center of attention for many astronomers, particularly for the large number of physicists who had moved into the field of astronomy. As a result a number of NASA's sounding rockets and small astronomy satellites were devoted to this area of research. In addition the agency began to plan for outfitting and launching a series of multiton satellites for x-ray and gamma-ray astronomy—to be called High Energy Astronomical Observatories. The importance of this work in the eyes of scientists was shown by the fact that Britain, the Netherlands, and the European Space Agency all instrumented satellites of their own for high-energy astronomy work.

The result of all the research with sounding rockets and satellites was an outpouring of data, not obtainable from the ground, at a time when ground-based astronomy was each year turning up new, exciting, often unexplainable discoveries. The quasars had extremely large red shifts in their spectra, suggesting that they were among the most remote of objects observed, yet if they were as remote as indicated, then they would have to be emitting energy at rates that defied explanation. Strange galaxies appeared to have violent nuclei, emitting unexplainable quantities of energy. The discovery of pulsars introduced the neutron star to the scene. Radio galaxies gave evidence of cataclysmic explosions in their centers. The rocket and satellite could not have appeared at a more propitious time.

As with the exploration of the solar system, the flood of new data and information was far beyond what could be covered in a brief summary like the present one. To keep within bounds, it is necessary to illustrate the impact of space science upon the field by means of selected examples. Two will be given: x-ray astronomy and some of the contributions of space science to solar physics.

X-ray Astronomy

Once stars are born the major part of their evolution can be followed in the optical wavelengths—that is, the visible and ultraviolet. For this reason most attention was directed at launching space astronomy in the direction of ultraviolet studies. Very little thought was given to the higher energy wavelengths, although these were proving to be extremely informative about the sun, particularly about solar activity. But there were a few who thought that one ought to look for celestial x-ray sources. Perhaps the most insistent was Bruno Rossi, professor of physics at the Massachusetts Institute of Technology. As Rossi later said, while he had not been in a position to predict specific phenomena like the x-ray sources that were eventually discovered, he had had "a subconscious trust . . . in the inexhaustible wealth of nature, a wealth that goes far beyond the imagination of man."[50] Moreover, there was a very compelling reason to try to look at the universe in the very short wavelengths. Much has been made of the fact that sounding rockets and satellites gave experimenters their first opportunity to look at all the wavelengths that reached the top of the atmosphere. But, like the atmosphere, interstellar space also had its windows and opacities, and not all wavelengths emitted in the depths of space could reach the vicinity of Earth. While the interstellar medium was quite transparent to wavelengths all the way from radio waves through much of the near ultraviolet, at and below 1216 Å absorption by hydrogen, the most abundant gas in space, cut off radiations from distant objects. Farther into the ultraviolet, absorption decreased again, rising once more when the absorption lines of helium, also abundant in space, were encountered in the far ultraviolet. Not until the x-ray region could one again see (with instruments, not with the eyes) deep into the galaxy. The existence of that window in the spectrum was an important reason for sounding out the possibilities of x-ray astronomy. One could study the universe only in the wavelengths that were available for observation, and all known windows ought to be exploited.

Rossi urged his ideas upon Martin Annis and Riccardo Giacconi of American Science and Engineering, who were enthusiastically receptive.[51] Quick calculations showed that x-ray intensities one might expect from galactic sources would be well below the detection limit for existing instruments—doubtless explaining why Naval Research Laboratory searches for x-ray sources had not found any. In 1960 NASA provided support to Giacconi and his colleagues and they prepared to fly sufficiently sensitive instruments in sounding rockets.

After one failure, the group succeeded in getting an Aerobee rocket to an altitude of 230 kilometers, on 12 June 1962. Although the planned objective of the flight was to look for solar-induced x-radiation from the moon, that objective was completely eclipsed by the excitement of detecting an object in the sky that was apparently emitting x-rays at a rate many, many orders of magnitude greater than the sun. The sheer intensity of the source gave one pause, and the experimenters spent a considerable time reviewing their results before announcing them in late summer. In two additional flights, October 1962 and June 1963, Giacconi's group was able to confirm the original findings and discover additional sources, again detecting a strong isolated source near the center of the galaxy.[52] There was also a diffuse isotropic background which one supposed could be extragalactic in origin.

Within a few months of Giacconi's original announcement, the Naval Research Laboratory experimenters had confirmed the existence of these sources by independent observations. In April 1963 the NRL group made an important contribution by pinpointing the source near the galactic center more precisely. Since it lay in the constellation Scorpio, the source was named Sco X-1. NRL also detected a somewhat weaker source, about 1/10 the strength of Sco X-1, in the general vicinity of the Crab Nebula.[53] During the rest of the decade additional sources were discovered until by 1970 several dozen were known. Efforts were made to discover just what these strange objects were. In particular, it was felt especially important to identify the sources with objects already known in the visible or radio wavelengths. The reason for wanting to tie x-ray sources in with known objects was quite simple. It seemed clear that the extremely intense x-radiation from these sources had to be connected with the basic energetics of the objects. It could be very important then to compare the x-radiation with that in other wavelengths, a comparison that might better reveal the true nature of the phenomenon.

As an illustration consider the Crab Nebula.[54] It is believed to be the remnant of a supernova explosion. The material of the nebula consists of the debris ejected by the exploding star into the surrounding space. At present this debris, which is expanding at about 1000 kilometers per second, fills a roughly ellipsoidal volume with the major diameter about six light-years—where a light-year is 9500 trillion km. Radiation from the Crab Nebula was observed to be strongly polarized across the whole spectrum, probably resulting from electrons revolving about the lines of force of a magnetic field. At the densities that appeared to exist in the cloud, the lower-energy electrons that would produce the radio wavelengths observed could last for thousands of years, but those responsible for optical wavelengths would be depleted in a few hundred years and those producing the x-rays in less than a year. Hence there had to be a continuous resupplying of energy to maintain the observed radiations.

This remained a mystery until the discovery of a pulsar in the Crab. The end product of a supernova explosion was expected to be a very dense neutron star, in which a mass comparable to that of the sun would be compressed into a ball about 10 km in diameter. There were reasons to believe that the pulsations from which a pulsar took its name were generated by the rapid spinning of the star. When it was noted that the period of the Crab pulsar was lengthening at the rate of about one part in 2000 per year, a possible source for the resupply of energy to the nebula leapt to mind. The slowdown of the neutron star's rotation corresponded to a considerable amount of energy, and calculations soon showed that the amount was adequate to provide the energy being released by the nebular gases. One speculation had it that the spinning magnetic field of the pulsar accelerated gases to relativistic speeds at which they could escape from the star's magnetosphere into the nebula, carrying their newly acquired energy with them.

This example illustrates the importance of being able to observe and visualize the object that was radiating the x-rays. As a consequence there was a continuing effort to identify the other x-ray sources with known objects. But, except for Sco X-1, the efforts were unsuccessful. As for Sco X-1, in an experiment in March 1966, Giacconi's group showed that the angular size of the source could not exceed 20 arcseconds. The new positional data were communicated to workers at the Tokyo Observatory and at Mt. Wilson and Palomar Observatories. On 17–18 June 1966 the Tokyo observers found a blue star of 13th magnitude, which further study identified with the x-ray source. This result confirmed that stars existed that emitted 1000 times as much power in x-rays as in the visible wavelengths. As Giacconi wrote: "Sco X-1 was a type of stellar object radically different from any previously known and whose existence had not and could not be foreseen on the basis of observations in the visible and radio."[55]

But that was as far as one appeared able to go with the sounding rockets. Longer-term observations with very sensitive detectors were required to advance the field further. These were supplied by *Explorer 42*. Launched from an Italian platform in the Indian Ocean off the coast of Kenya on 12 December 1970, Kenyan Independence Day, the satellite was at once named *Uhuru*, the Swahili word for independence. *Uhuru* provided the breakthrough astronomers were looking for. When issued sometime afterward, the *Uhuru* catalog listed 161 x-ray sources. It was clear that the strongest sources had to have x-ray luminosities at least 1000 times the luminosity of the sun. Even the weakest x-ray source was 20 times more intense than the meter-band radiation from the strongest radio source. Thirty-four sources had been identified with known objects. Of the sources known to lie within the galaxy, one was at the galactic center, seven were supernova remnants, and six were binary stars. Outside the galaxy were sources associated with ordinary galaxies, giant radio galaxies, Seyfert galaxies, galactic clusters, and quasars. And much of the diffuse background x-radiation had been shown to come from outside the local galaxy.[56]

One of the most exciting results from x-ray astronomy came from the realization that some at least of the sources were binary stars—two stars revolving around each other (fig. 65)—in which one of the companions was a very dense star, either (1) a white dwarf, a mass comparable to that of the sun compressed into a sphere about the size of Earth; or (2) a neutron star, of mass exceeding 1.4 solar masses, compressed into a ball about 10 km in radius; or (3) a black hole in space. A white dwarf is the end product of a star about the size of the sun, after it has used up its nuclear fuel and can no longer avoid collapsing under gravity to a planetary size. The neutron star is the end product of a massive star that, after burning up its nuclear fuel, undergoes a violent implosion caused by gravity, following which a

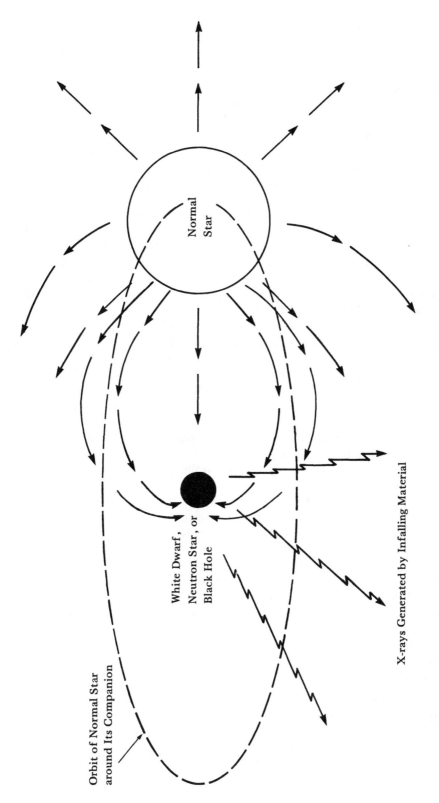

Normal
Star

White Dwarf,
Neutron Star, or
Black Hole

X-rays Generated by Infalling Material

Orbit of Normal Star
around Its Companion

Figure 65. Binary x-ray star. Material from the larger, much less dense companion is drawn toward the extremely dense smaller companion and accelerated to velocities sufficient to produce x-rays by collision with the ambient gases.

great deal of the star's material rebounds from the implosion to be blown out in a supernova explosion, leaving behind an extremely dense object consisting of neutrons. But if the residual mass after the supernova explosion is greater than a certain critical value, the gravitational contraction of the star does not stop even at the neutron star stage. Instead the star continues to contract indefinitely, pulling the matter tighter and tighter together until the object disappears into a deep gravitational well out of which neither matter nor electromagnetic radiation can escape because of the intense gravitational fields there. Hence the name "black hole."

The binary nature of some of the x-ray objects could be deduced from the doppler shifts in the light from the ordinary companion, the shift being toward the blue as the star moved toward the observer in its orbit, and toward the red as the star moved away. If the stars eclipsed each other the binary nature would show up in a periodical disappearance of the x-rays when the emitter was hidden by the other star and reappearance when the emitter emerged from eclipse.

After careful study astronomers finally concluded that the x-rays were generated by material from the ordinary companion's being pulled into the gravitational well of the degenerate star. If the gravitational attraction were sufficiently strong, then the gas would be accelerated to such velocities that the gas would emit x-rays as particles collided. It seemed that white dwarfs would not provide sufficient gravity to accomplish this, so one was left with the conclusion that the degenerate companion in binary x-ray sources was either a neutron star or a black hole in space. In most cases it appeared that the companion was a neutron star, but the source Cyg X-1, in the constellation Cygnus, could be a black hole. If so, it was the first such object to be detected in the universe.[57]

The possibility that a black hole had at last been discovered emphasized the fundamental importance to astronomy of the new field of x-ray and gamma-ray astronomy. Gradually scientists had begun to talk about their work as *high-energy astronomy*, not only because they were working at the high-energy end of the wavelength spectrum, but more significantly because their observations were showing that throughout the universe extremely violent events were rather common, involving enormous quantities of energy and tremendous rates of energy production. And among these energetic events were those occurring during the last stages of a star's evolution, stages in which neutron stars and black holes were created, with intense x-ray emissions. Speculating on the philosophical implications, Giacconi showed excitement:

> The existence of a black hole in the X-ray binary Cyg X-1 has profound implications for all of astronomy. Once one such object is shown to exist, then this immediately raises the possibility that many more may be present in all kinds of different astrophysical settings. Super-massive black holes may exist at the center of active galaxies . . . and explain the very large energy

emission from objects such as quasars. Small black holes of masses [very much smaller than the mass of the sun] may have been created at the instant of the primeval explosion. . . . In black holes matter has returned to condition similar to the primordial state from which the Universe was created. The potential scientific and intellectual returns from this research are clearly staggering.[58]

Should one then conclude that rocket and satellite astronomy had by the early 1970s generated a scientific revolution in the field of astronomy? The answer may well be yes, although many of the strange concepts that were being dealt with had been considered decades before.[59] In any event, it is probably too early to make the case. Certainly these topics, concerning the interplay of energy and matter on a cosmological scale, are fundamental; and if anywhere in the space program one might expect a scientific revolution to emerge, it would be here. But it should also be noted that if any such revolution is to arise, it would almost certainly come from a cooperation between ground-based and space astronomy.

Solar Physics

The sun was a most important target of space science investigations for at least two reasons. First, the sun's radiation supports life on Earth and controls the behavior of the atmosphere. For meteorology it was important to know the sun's spectrum in the visible, infrared, and near-ultraviolet wavelengths. To understand the various physical processes occurring in the upper atmosphere, a detailed knowledge of the solar spectrum in the ultraviolet and x-ray wavelengths was essential. The reader will recall the overriding importance that S. K. Mitra, in his 1947 assessment of major upper-atmospheric problems, gave to learning about the electromagnetic radiations from the sun (pp. 59–60). For this reason many sounding rocket experimenters devoted much of their time to photographing and analyzing the solar spectrum both within and beyond the atmosphere. Some of their work before the creation of NASA was discussed in chapter 6. Finally, with the discovery of the magnetosphere and the solar wind the importance of the particle radiations from the sun for sun-earth relationships became apparent, a topic that was treated in chapter 11. Thus, solar physics was of central importance in the exploration of the solar system.

But the sun was important also to astronomy, to the investigation of the universe. Although an average star, unspectacular in comparison with many of the strange objects that astronomers were uncovering in their probing of the cosmos, nevertheless it is a star, and it is close by. The next star, Proxima Centauri, is 4.3 light-years (400 trillion kilometers) away, while most of the stars in the galaxy are many tens of thousands of light-years distant. Stars in other galaxies are millions and even billions of light-years from earth. So the sun afforded the only opportunity for scientists to study stellar physics with a model that could be observed in great detail.

Because of its nearness and its importance, astronomers amassed a great deal of data and theory about the sun in the years before rockets and satellites.[60] What they learned came almost entirely from observations in the visible, with only occasional glimpses from mountain tops and balloons at the shorter wavelengths. But, as space observations showed, much of solar activity, particularly that associated with the sunspot cycle, solar flares, and the corona involved the short wavelengths in essential ways. Sounding rocket and satellite measurements were, accordingly, able to round out the picture in important ways.

To understand the significance of these contributions, a brief summary of the principal features of the sun may be helpful.[61] The visible disk of the sun is called the photosphere (fig. 66). It is a very thin layer of one to several hundred kilometers thickness, from which comes most of the radiation one sees on Earth. The effective temperature of the solar disk is about 5800 K. Above the photosphere lies what may be called the solar atmosphere; below it, the solar interior.

The sun's energy is generated in the interior, from the nuclear burning of hydrogen to form helium in a central core of about one-fourth the solar radius and one-half the solar mass. Here, at temperatures of 15×10^6 K, some 99 percent of the sun's radiated energy is released. This energy diffuses outward from the core, colliding repeatedly with the hydrogen and helium of the sun, being absorbed and reemitted many times before reaching the surface. In this process the individual photon energies continually decrease, changing the radiation from gamma rays to x-rays to ultraviolet light and finally to visible light as it emerges from the sun's surface.

From the core of the sun to near the surface, energy is thus transported mainly by radiation. But toward the surface, between roughly 0.8 and 0.9 of the solar radius, convection becomes the principal mode of transporting energy toward the surface. The existence of the convection zone is evidenced by the mottled appearance of the photosphere in high-resolution photographs. This mottling, or granulation as it is called, consists of cells of about 1800-km diameter which last about 10 minutes on the average and which are thought to be associated with turbulent convection just beneath the photosphere. A larger scale system of surface motion—20 times the size of the granulation cells, called supergranulation—is believed to be much more deeply rooted in the convection zone.

Analogous to Earth's upper atmosphere is the chromosphere, or upper atmosphere, of the sun. The chromosphere overlies the photosphere and is about 2500 km thick. While the density drop across the photosphere is less than an order of magnitude, density in the chromosphere decreases by four orders of magnitude from the top of the photosphere to the top of the chromosphere.

Not long ago, photospheric and chromospheric temperatures were extremely puzzling to astronomers, dropping from about 6600 K at the base

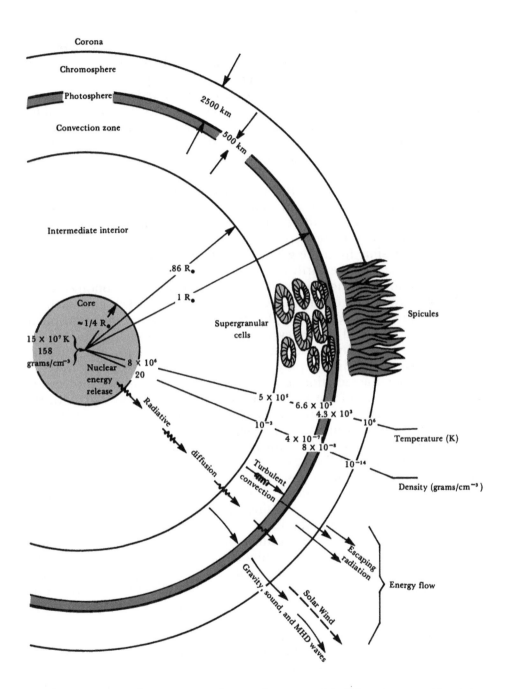

Figure 66. Idealized structure of the sun (not to scale). There is a complex inter-play among the different regions of the sun. Edward G. Gibson, The Quiet Sun, *NASA SP-303 (1973), p. 11, fig. 2-3.*

of the photosphere to around 4300 at the base of the chromosphere, and then rising through the chromosphere, at first slowly but then very steeply to between 500 000 and 1 000 000 K at the top. This temperature curve posed a problem, for it was assumed that the corona derived its heat from the chromosphere, yet that would imply that heat was flowing from a colder region to a hotter one, contrary to the laws of thermodynamics. As late as 1972 Leo Goldberg, director of Kitt Peak National Observatory, pointed to this phenomenon as "the most important unsolved mystery surrounding the quiet sun."[62]

Above the chromosphere lies the corona, the sun's exosphere. Here 1 000 000 K temperatures prevail, and an important problem facing the solar physicist was to explain how the corona gets its energy. Although the corona is extremely hot and very active, its density is so very low that it is not normally visible from the ground, where it is completely obscured by scattered sunlight in the earth's atmosphere. Only during solar eclipses, with the moon blocking out the sun's disk, could the astronomer get a good look at the entire corona. One of the benefits of rockets and satellites was to permit carrying coronagraphs above the light-scattering atmosphere where the corona could be seen even in the absence of a solar eclipse.

Much of solar physics concerns the interplay among the different regions of the sun. This interplay, however, can be followed only in terms of its effect upon the radiations emitted from those regions. For this reason, one of the first tasks of the astronomer was to obtain good spectra of the sun and their variation with time. Regions from which radiations of highly ionized atoms came would be hot regions, and temperatures could be estimated. The magnetic field intensities, for example in sunspots, could be estimated from the splitting of lines emitted within the field. If a cooler gas overlay a hotter, similar gas, the cooler gas would absorb some of the light emitted by the hotter one. This would produce reversals in the emission lines of the hotter gas, generating the famous Fraunhofer lines of the solar spectrum discovered in the 19th century. By piecing together information of this kind, the locations of different gases relative to each other and their temperatures could be determined. Changes in magnetic field that occurred in association with solar activity, such as the appearance of solar flares, could be followed. Changes were important, since there were strong indications that magnetic fields were the source of much of the energy in solar flares.

These techniques were, of course, applicable in the visible wavelengths and were employed to the fullest by the ground-based astronomer. The space astronomer simply provided an additional handle on things by furnishing spectral data in the ultraviolet and x-ray wavelengths. And these data began to accumulate from the very earliest sounding rocket flights. Year by year, flight by flight, they were added to until by the end of the decade the solar spectrum was known in great detail from visible through the ultraviolet wavelengths and into the x-rays.[63]

A powerful technique for study of the sun is that of imaging the sun in a single line; for example, the red line emitted by hydrogen known as hydrogen alpha. In such spectroheliograms, as they are called, one can see the structure and activity of the sun associated with that line. Spectroheliograms taken in hydrogen, calcium, and other lines in the visible have long been an effective tool for the study of solar activity. Members of the Naval Research Laboratory group pioneered the use of this technique in space astronomy, where it was possible to get spectroheliograms in both the ultraviolet and x-rays.[64] These, taken with photographs in the visible, gave a powerful means of discerning and analyzing active regions on the sun. Sequences of such images taken over many days, or at intervals of 27 days, the solar rotation period, permitted one to follow the evolution of flares and other features on the sun. It was in this sort of imaging that Skylab was particularly fruitful.

During the decade and a half that was climaxed by the Skylab solar observations, solar physics progressed rapidly, advanced by a combination of ground-based and space astronomy. In the shorter wavelengths the sun was found to be extremely patchy (fig. 67), a patchiness that extended into the visible wavelengths as well.[65] The sequence of events in a solar flare could be followed in wavelengths all the way from x-rays through the ultraviolet and visible into the radio-wave region, and related to motions of electrons and protons associated with the flare.[66] Contrary to previous expectations fostered by ground-based pictures during solar eclipses, the corona turned out to be not even nearly homogeneous. X-ray images of the corona especially showed a great deal of structure (fig. 67). Quite surprising were large-scale dark regions of the corona—which came to be called dark holes—and hundreds of coronal bright spots. The holes appeared to be devoid of hot matter and to be associated with diverging magnetic field lines of a single polarity. If the magnetic field lines were open, these holes could be a source of particles in the solar wind.[67]

The bright spots were observed to be uniformly distributed over the solar disk. They were typically about 20000 km in diameter, and of a temperature about 1.6×10^6 K. They appeared to be magnetically confined, and one speculated that they might be an important link in the explanation of the sun's magnetic field.[68]

The interlocking features of the lower solar atmosphere and the corona visible in satellite images of the sun provided hints as to how the sun might heat the corona to the extreme temperatures that were observed. Gravity waves and acoustical waves might carry energy upward from the convective regions below the photosphere into the corona. This explanation would remove the mystery of the steep temperature curve in the

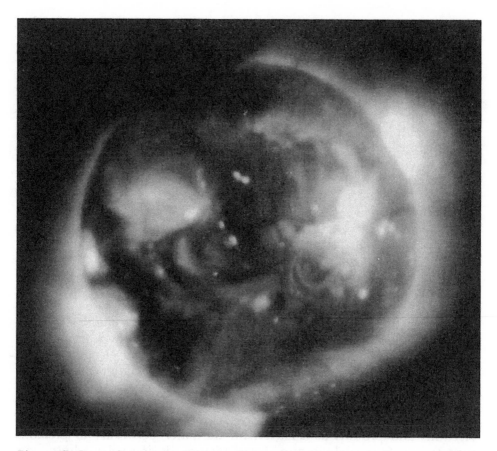

Figure 67. Coronal structure of the sun. X-ray pictures of the sun show a great deal of structure in the solar corona, including dark coronal holes and hundreds of intense bright spots. The x-ray photo above was taken by the Skylab Apollo telescope mount 28 May 1973 in an American Science and Engineering, Inc., experiment. NASA photo. See also Giuseppe Vaiana and Wallace Tucker in X-ray Astronomy, ed. R. Giacconi and H. Gursky (Dordrecht-Holland: D. Reidel Publishing Co., 1974), pp. 170–71, fig. 5.1a and 5.1b.

chromosphere. It would not be the chromosphere that was heating the corona in violation of the laws of thermodynamics. On the contrary, the corona, heated by energy from within the sun, would itself be heating the top of the chromosphere.

The importance of rocket and satellite solar astronomy lay in the integrated attack that the researcher could now make in seeking to understand the nearest star, an integrated attack made possible by opening up the window that the earth's atmosphere had so long kept shut. It was an importance attested to by the large numbers of solar physicists who bent to the task of assimilating the new wealth of data.

By the end of the 1960s the early years of space science were well behind. More than a dozen disciplines and subdisciplines had found sounding rockets and spacecraft to be powerful tools for scientific research. Thousands of investigators turned to these tools to help solve important problems. Moreover, while the disciplines to which the new tools could contribute were many and varied, there was a clearly discernible melding of groups of disciplines into two major fields: the exploration of the solar system and the investigation of the universe. The pursuit of these two main objectives would grow in intensity as space science moved into the 1970s—in spite of fears prevalent in the late 1960s that support for space science was waning. The new decade would witness the scientific missions of Apollo to the moon, the remarkable solar astronomy from Skylab, breakthroughs in x-ray astronomy, and the serious start of a survey of all the important bodies of the solar system. It was eminently clear that space scientists would be important clients of the Space Shuttle, which was intended to introduce a new era in space activities. Because of their accomplishments, the scientists could legitimately ask that the Shuttle be tailored as much to their requirements as to other space needs.

Part VI

Future Course

Nature that framed us of four elements,
Warring within our breasts for regiment,
Doth teach us all to have aspiring minds:
Our souls, whose faculties can comprehend
The wondrous Architecture of the world:
And measure every wandering planet's course,
Still climbing after knowledge infinite. . . .

Christopher Marlowe, Conquests of Tamburlaine

So many worlds, so much to do,
So little done, such things to be.

Alfred, Lord Tennyson, In Memoriam

21

Objectives, Plans, and Budgets

As the peak of labor on Apollo passed, following the middle of the decade, and as NASA completed more and more of the projects undertaken in the first years of the agency, the question of NASA's future course assumed a growing importance in the minds of NASA managers. For a decade the quest for world leadership in space had helped to sustain NASA's program. But success brought a particular erosion of that support. For, when it became clear that the United States was at the very least fully competitive with the USSR, and more likely was well ahead of them in space research and engineering, the initial motivation dwindled. There began a reassessment of NASA's mission, particularly by the executive side of government. Coming at a time of national reassessment of priorities, of concern about civil rights and student unrest, of disenchantment with the Vietnam War, and a shaky economic situation, the reassessment was bound to affect the agency.

To many who had a say about NASA's budget—especially in the Nixon administration, but actually well before then—the precise nature and importance of NASA's mission were not at all clear. The strongest challenges to NASA's role came in space applications, where by law other agencies had the prime responsibilities. Meteorological and oceanographic applications now came under the National Oceanic and Atmospheric Administration, which had absorbed the former Weather Bureau and much of the Navy's oceanographic activities along with a number of other responsibilities.[1] In communications other agencies called the shots: for national policy, the Office of Telecommunications in the White House; for commercial uses, the Federal Communications Commission, the Communications Satellite Corporation, and private industry; for regulation, the Federal Communications Commission; for applications to education and health, the Department of Health, Education, and Welfare; and for use in commercial shipping, the Maritime Commission.[2] The primary mission in agriculture and forestry—which satellite observations promised to aid substantially—belonged to the Department of Agriculture. Any use of satellites for the exploration and survey of mineral resources fell squarely into the legally assigned mission of the U.S. Geological Survey of the Department of the Interior. As a consequence any payoff from space investments that NASA might seek to realize in space applications would have to be sold in the form of a service to another agency within whose purview the specific application fell.

Even aeronautics—the primary activity of the former National Advisory Committee for Aeronautics, which formed the nucleus for NASA, and an activity that remained a strong component of the NASA program—belonged primarily to others; namely, the Department of Defense, the Federal Aviation Administration, and industry.

Only three areas could NASA claim as its own: the development of space techniques and hardware; space exploration; and space science. But in the difficult climate existing 10–15 years after the start of the space program, even these had a hard row to hoe. For example, to assign all development of space techniques and hardware to NASA (excepting, of course, the substantial amount the military did) ran counter to the widely held view that a user should develop his own hardware to meet a specifically perceived need. There are many virtues to this point of view. Certainly a prospective user would be motivated to tailor his research and development to the actual need and to be properly attentive to keeping costs down. Also, the actual user could be assumed to know best exactly what was required for his application. But the large-scale, highly specialized, very expensive test and launching equipment and the large teams that were required for space development and operations argued for assigning the research, development, and operations to a single agency. For each user to duplicate the personnel and facilities would be extremely wasteful. There were accordingly strong pressures on NASA to assume a largely service role in support of the many users interested in applying space methods to their missions. The forces in this direction outweighed the natural desires of the different agencies to provide their own services, and in the balance between the two conflicting pressures NASA maintained an uncertain hold on a role in the field of space applications.

In a period of retrenchment NASA found that role particularly difficult. NASA was expected to perform the necessary advanced research for prospective applications. But in the late 1960s it was difficult to get administration approval for such advanced research in spite of vigorous urging from many congressional quarters for NASA to do more applications work. Before starting any new applications projects, the Office of Management and Budget wanted from potential users not merely pious words in support, but assurances that there were genuine plans to use the new methods, not merely as a *supplement* to old methods but actually as a more efficient *replacement* for some of them. Potential users might underwrite specific and clearly realizable applications, but were usually very reluctant to support the advanced engineering and development needed to establish the feasibility of potential applications. Under the circumstances the administration was even less ready to approve the advanced work. This was particularly true when the development, as with earth-resources surveys from space, was likely to introduce large new expenditures into the national budget.

The second NASA mission, space exploration—by which was meant exploration of the moon and planets by men—also was very difficult to support during those later years. Having proved our mettle by being the first to explore the moon, it was not perceived as necessary to prove ourselves further, at least for the time being, by going on to the planets. Nor was the case for permanent earth-orbiting space stations regarded at the time as persuasive. Manned flight in the Space Shuttle and Spacelab—which in the early 1970s gained somewhat grudging support (see pp. 389–91)—was seen as enough, and to some more than enough, for the time being.

There remained, then, the third NASA mission, space science. Even here the situation was not clear, since one could apply to science the same argument that was being applied to the applications areas. The primary mission in science had long since belonged to another agency, the National Science Foundation.* But few seemed to wish to press this argument, since the existence of a space science program in NASA served to funnel large amounts of money into science without those dollars having to compete with the funds available through the Science Foundation. The highly specialized character of the tools of space research, plus the mental anguish that would arise if space science budgets had to compete with other science budgets, together with NASA's practice of providing substantial support to science in the universities, appear to have led the nation's science community to agree that space science was properly NASA's. Both the administration and Congress went along.

The searching scrutiny of NASA's role that took place in the late 1960s and early 1970s, painful at times to those in the agency, in the end proved salutary. Out of the probing emerged an acceptance of a continuing role for the agency in which science, applications, and exploration would all play a part. Freed at last from an uneasy dependence on a passing sense of urgency over the nation's technological strength relative to that of the USSR, NASA's position in the 1970s could be intrinsically stronger. During the 1960s the fundamental contribution that space could make to a long list of important practical applications had become plain, and there could be no question but that these applications would be developed in the course of time. Not spurred on by the need to compete with the Soviet Union, the pace might be slower, but it would be more assured. And, like

---

*And in life sciences, to the National Institutes of Health.

its predecessor National Advisory Committee for Aeronautics, NASA would have a sizable role in providing important services to other agencies.

Likewise the breadth of space science, already apparent in the year following Sputnik, was abundantly clear, and its importance to the continuing development of the country's technological strength recognized. Again, the pace would certainly be measured, the smaller projects favored, the larger projects thoroughly scrutinized before being accepted, and extremely large and costly projects avoided. But within those limits it would be possible to put high-precision astronomical telescopes in orbit and to explore the farthest reaches of the solar system.

Space exploration, too, could be expected to continue, but at a very much reduced pace. While the planets might continue to beckon, astronauts would have to await the orderly development of the means and a still-to-be-awakened national desire to explore beyond the moon. Meantime, NASA's major attention in the field of manned spaceflight would be to create the Space Shuttle and its accompanying equipment and facilities. The Space Shuttle would make flight into space easier, more routine, and more economical. Its versatility and affordability would make the Shuttle the key to the future of America in space. Because the Shuttle would replace a great many of the previously used, expendable launch vehicles, and because the Shuttle would fundamentally change the complexion of space operations, the 1970s became a decade of transition for NASA and those engaged in space research and development.

PLANS

In the first years after Sputnik, when the space program objectives were clear—or at least thought to be clear—planning was relatively straightforward. Administrator Glennan had a small planning group in headquarters under Homer J. Stewart, who earlier had chaired the Defense Department committee that had chosen the Navy's Vanguard for the International Geophysical Year satellite program. In December 1959 Glennan's planning group turned out a secret document entitled "NASA Long Range Plan," a confidential version of which was called the "NASA Ten Year Plan."[3] These early documents were directed more at estimating what advancing technology might permit in the distant future—a decade or more away—than at establishing a true plan. To create a valid plan in the usual sense, a firmer tie would have to be made with the current program and its prospective evolution into the immediate future.

Like most such products of centralized planning, these documents aroused the criticism of the program divisions and the centers, which felt that the projections did not do justice to their own recommendations. But the papers served to focus the attention of the agency on what was being thought about at the top, and central planning continued, later with the aid of John Hagen, radio astronomer who had directed Vanguard at the Naval Research Laboratory, and Abraham Hyatt, pioneering propulsion engineer.

Space science documents described in chapter 8 provided one source of material for the central plan.[4] But scientists were by nature opposed to long-range plans. Many felt that NASA's planning did violence to the way in which scientists worked. It was repeatedly pointed out that what might now be considered a very important project for 10 years later could lose its importance in the light of discoveries made in the interim. NASA managers found it difficult to get scientists to think seriously about scientific plans on the scale of 10 or more years ahead.

Instead, the scientists preferred to indicate broad areas of research that were likely to be important in distant years and to identify specific projects only for the immediate future. To accommodate the need of the agency for long-range planning while not pressing the scientists to be more specific in the long term than they felt they could legitimately be, space science planners evolved a style that differed from that in other areas in the agency. Instead of labeling their documents as specific long-range plans, they began to use such phrases as *long-range planning* or *long-range thinking* in titling the papers.[5] By the fall of 1960 the space science office had settled into a routine of periodically issuing documents with titles suggesting a planning process rather than a firm plan.[6] In September 1962, the author addressed a memorandum to the Space Sciences Steering Committee, its subcommittees, and space science division directors describing the long-range and short-range planning process to be followed for space science and specifying procedures for keeping the planning timely.[7] An important tool of this planning process was the *"Space Science Prospectus,"* which was updated at least once a year.[8]

Described as a source document for space science planners, the prospectus contained a large number of possible projects or scientific investigations that appeared sufficiently important to consider including in the program. The prospectus, however, contained many more projects than could be undertaken with the expected budgets. Nevertheless, to make the prospectus much more than just a list of potentially desirable projects—a mere "wish list" as some put it—the projects set forth in the prospectus were studied and analyzed to determine costs, manpower, schedules, launch vehicles, facilities, and supporting services that would be needed to carry them out.

Each year when the budget was prepared the prospectus was drawn upon for projects to put into the budget request. In the process, projects most likely to be candidates for the next year's budget were also identified. In this way the prospectus, while not a plan, became an important element in the space science planning process. Moreover, it furnished a mechanism for the scientists to engage in the long-range thinking of the agency without doing too much violence to their natural reluctance to specify too far ahead.

NASA Administrator James Webb did not object to the Office of Space Science and Applications' use of the prospectus. But he was not in favor of publishing long-range plans, in spite of constant congressional pressure to get them. Webb preferred to reveal the agency's course year by year in the annual budget proposals. As he stated it, in putting out the current year's proposals one gears up to do battle for them. In the defense of the budget one has the immediate assistance of those ready to support the program. But publishing a plan that goes much beyond the current year invites adversaries to shoot the agency down at their leisure. Friends and supporters aren't prepared to come forward to defend the agency in what must for the moment seem largely an academic exercise. Meantime, enemies will seize upon different aspects of the plan—often out of context—to challenge and embarrass the agency.

Not being a true plan, the space science prospectus did not afford detractors the kind of leverage that a fixed plan would have. As a consequence Webb permitted the document to be updated and issued each year, although he periodically called attention to the dangers of being too specific too early.

When the author became associate administrator, the various program offices had become accustomed to developing their own plans without too much consideration of the planning of the other offices. Webb asked that an agency-wide planning activity be developed. Working with the Planning Steering Group created for the purpose, the author intended to create a NASA prospectus much along the model of that used in space sciences. It was estimated that perhaps as much as five years would be required to do this; but before getting beyond the initial stages administrations changed, and the Republicans called for a specific space plan. Moreover, Thomas Paine, who took over from Webb, favored publishing specific plans and was willing to stand up and fight to defend them. Paine's view was that leaving options open for the future was simply an indication of not having thought through those options. Under Paine, and later under the fourth administrator, James C. Fletcher, NASA began again to develop and publish long-range plans for the agency and to use them in preparing short-range plans and budget proposals. Paine's plans turned out to be too sweeping and expensive to receive administration support, while Fletcher's more modest, almost constant-dollar-level plans did gain Nixon's backing and helped to launch NASA on the Space Shuttle development program.

BUDGETS

To the scientists in space program management, nothing could be duller than an endless round of budgets, appropriations, obligations, cost accruals, and expenditures. Hence it came as something of a surprise when one of the author's colleagues on the financial side of things began to commiserate that the fates had tricked so many people into the boring, impersonal, uninspiring field of science, when by a more happy stroke of fortune they might have been led into the vital, intensely human, and real world of budget and finance!

The phrase "real world of budget and finance" struck a responsive chord, for clearly the resources made available to NASA determined what the agency could do. Moreover, those who did deal full time with the budgets and expenses of the agency were undoubtedly the most completely informed as to what NASA was doing, at least in a general sense, and probably had as good an idea as anyone as to how all the parts fitted together into a total program. Through a detailed analysis of the agency's budget requests, appropriations, and expenditures the historian can get a comprehensive picture of what NASA was up to, and at the very least can put together a complete framework around which to weave the very human story of the nation's venture into space. Although such an analysis is beyond the author's interests and powers of endurance, a few general observations are in order.

As it was, the fates that had led scientists into NASA Headquarters had provided amply for their participation in the "real world of budget and finance." Work with budgets never ceased. It was an essential part of converting plans into reality. In the spring, even as the defense before Congress of the current budget request was getting under way, the agency would begin exchanges with the Bureau of the Budget—which became the Office of Management and Budget under Nixon—as to the likely acceptable level of the next budget request, for the period beginning some 16 months later. Throughout the spring and summer the detail of this exchange would grow until by fall the bureau would have in its hands a complete budget proposal. The proposal, of course, had grown out of the program planning that went on continually in the agency. During the fall and winter, the final budget proposal would be developed in sometimes heated discussion, often with many compromises, between NASA and the administration. In late January, as part of the now enormous national budget, the space program request would be sent to the Congress for review, authorization, and appropriation.[9]

In the midst of this process, hearings on the previous budget request had been going on, and in the summer or autumn Congress had authorized new obligations for the program and had appropriated funds. Thus, throughout the year NASA managers worked intimately with three separate budgets: (1) conducting the current year's program corresponding to the recently authorized budget, (2) defending a budget request for the next fiscal year, and (3) preparing still another budget request for the fiscal year after that. Such activities kept the scientists-turned-managers away from the science they would have preferred to do; but, along with planning and formulating a program, budgeting was an essential element of the head-quarters management job.

The business of gaining the necessary budgets involved a great deal of salesmanship and political savvy. An important part of the work of NASA's top management was to develop and preserve a climate in which the lower echelons could sell their wares. Much of this responsibility fell directly on the administrator, whose relations with the president and other administration officials, and with leaders in the Congress, had a determining influence on how successful the agency would be.

James Webb used to emphasize that the way to sell a program was to get the support of the president. Having that, all the rest moved along in more or less orderly fashion—unless, of course, the agency found itself in the middle in one of the classic confrontations between the legislative and executive forces such as did occur during the 1960s when the Congress tried to recapture some of the initiative that seemed to have passed to the White House. In the course of NASA's history the complexion of presidential backing varied widely. President Eisenhower, a lukewarm supporter of the space program, wished to keep it at a relatively modest level. His choice for first administrator, T. Keith Glennan, kept a tight rein on the newly evolving program. President Kennedy gave strong support, particularly after he had personally proposed the Apollo program to Congress. Under Kennedy, NASA's program and expenditures grew rapidly toward the peak of $6 billion a year during Lyndon Johnson's administration.[10] As one of the architects of the National Aeronautics and Space Act of 1958, Johnson brought with him to the vice presidency and later to the presidency a built-in commitment to a vigorous national space effort. This commitment lasted throughout his incumbency in the Oval Office, although in the last year the turmoil and emotional toll of the times had begun to weaken his original enthusiasm, and to Webb the president's backing of the space program seemed at times to become indifferent. President Nixon had no binding intellectual commitment to space, or at any rate none not easily overridden by political expediency—even though he had been one of the first to endorse the early efforts to establish a national space program. Nixon enjoyed and took political advantage of each Apollo mission and other space successes, and shrewdly considered the political impact of any major space program presented for his approval and backing.

The different flavors of presidential attitude toward the space program were felt in the discussions that NASA went through each year with the Bureau of the Budget or the Office of Management and Budget. In the early years a reasonably well-planned budget secured fairly ready acceptance in the administration. Major issues sparked by the country's sudden precipitation into this new arena included the question of how soon the United

States could close the launch vehicle gap with the Soviet Union and the relative roles of the NASA and military space programs.

Congress, of course, had the final say as to how much money would be authorized and appropriated for NASA. In those first years NASA had little trouble in getting its budgets passed. In fact, throughout the years Congress consistently gave strong backing to the space program even when, in later years, paring down parts of the budget.

In the period immediately following launch of the first Sputnik, committee members would listen with rapt attention and undisguised enthusiasm to description of plans and accomplishments—and then give NASA pretty much what it asked for. It was a learning period for the legislators as much as it was for NASA. As experience and understanding grew, lawmakers' questions became more pointed and penetrating, and no longer was there an inclination to accept a budget simply on the grounds that NASA said it was required. But while increasingly critical, the Congress remained basically supportive throughout the years.

NASA budgets for 1958 (fiscal 1959) through 1976 are given in table 8 and figure 68. A comparison of the total space budget with the part assigned to research and development for space science appears in table 9. For the sake of comparison, budgets for space activities in the Department of Defense and other agencies are included in table 8.[11]

The simplified numbers and graphs cannot give a true picture of the agency's funding structure. For example, a good amount of space science was supported with funds in the manned spaceflight budget, since the exploration of the moon necessarily included a great deal of scientific investigation. Likewise, much advanced research and technology was important to space science and could properly be charged to that activity if one chose to do so. But, with these limitations in mind, it is still possible to derive some valid impressions about the support that space science received through the years.

First, while never a major part of NASA's total budget, space science funding was nevertheless an appreciable part of the total, at times accounting for as much as 20%. Actually, as Webb continually pointed out, invidious comparisons of absolute or relative numbers did not make sense, because, while manned spaceflight did indeed enjoy much greater funding than did space science—as the scientific community repeatedly complained—that did not imply a lack of suitable support for science. The Gemini and Apollo projects simply cost more money, and if the nation was going to have a manned spaceflight program, it had to pay the necessary costs. The proper question to ask was not whether manned spaceflight was getting more money than space science, but whether space science was getting the funding it needed. Webb would add that even if the scientists should manage to get the manned spaceflight program canceled—as Abelson and others

Table 8
United States Space Budget

(18-year budget summary—budget authority in millions of dollars)

Fiscal Year	NASA Total	NASA Space*	Department of Defense	ERDA	Com- merce	Interior	Agri- culture	NSF	Total Space
1959	305.4	235.4	489.5	34.3					759.2
1960	523.6	461.5	590.6	43.3				0.1	1065.8
1961	964.0	926.0	813.9	67.7				.6	1808.2
1962	1824.9	1796.8	1298.2	147.8	50.7			1.3	3294.8
1963	3673.0	3626.0	1549.9	213.9	43.2			1.5	5434.5
1964	5099.7	5046.3	1599.3	210.0	2.8			3.0	6861.4
1965	5249.7	5167.6	1573.9	228.6	12.2			3.2	6985.5
1966	5174.9	5094.5	1688.8	186.8	26.5			3.2	6999.8
1967	4967.6	4862.2	1663.6	183.6	29.3			2.8	6741.5
1968	4588.8	4452.5	1921.8	145.1	28.1	0.2	0.5	3.2	6551.4
1969	3990.9	3822.0	2013.0	118.0	20.0	0.2	0.7	1.9	5975.8
1970	3745.8	3547.0	1678.4	102.8	8.0	1.1	0.8	2.4	5340.5
1971	3311.2	3101.3	1512.3	94.8	27.4	1.9	0.8	2.4	4740.9
1972	3306.6	3071.0	1407.0	55.2	31.3	5.8	1.6	2.8	4574.7
1973	3406.2	3093.2	1623.0	54.2	39.7	10.3	1.9	2.6	4824.8
1974	3036.9	2758.5	1766.0	41.7	60.2	9.0	3.1	1.8	4640.3
1975	3229.1	2915.3	1892.4	29.6	64.4	8.3	2.3	2.0	4914.3
1976	3550.3	3226.9	1983.3	23.3	71.5	10.4	3.6	2.4	5321.4

SOURCE: *Aeronautics and Space Report of the President, 1976 Activities* (Washington: NASA, 1977), p. 107.

*Excludes amounts for air transportation.

[May not add, because of rounding]

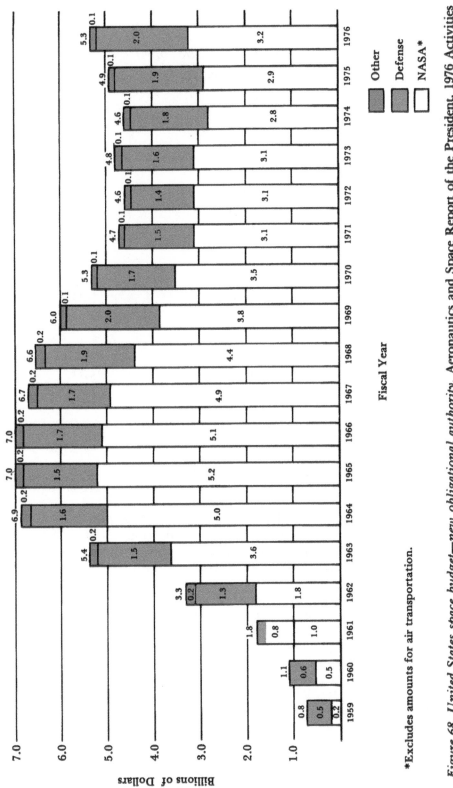

Billions of Dollars

Fiscal Year

■ Other
■ Defense
□ NASA*

*Excludes amounts for air transportation.

Figure 68. United States space budget—new obligational authority. Aeronautics and Space Report of the President, 1976 Activities (1977), p. 107; ibid., 1966 (1967), p. 166.

Table 9
NASA Budget—Space Sciences Research and Development
(millions of dollars)

Fiscal Year	Total Space Budget Authority	Space Sciences Research and Development Portion*
1963 and before	7045.7	1349.2
1964	5046.3	617.5
1965	5167.6	621.6
1966	5094.5	664.9
1967	4862.2	511.9
1968	4452.5	452.6
1969	3822.0	356.5
1970	3547.0	396.7
1971	3101.3	398.7
1972	3071.0	552.5
1973	3093.2	678.2
1974	2758.5	602.0

*Space science was also assigned additional funding for construction of facilities, research and program management, and administrative operations.

would have liked[12]—the monies would not be reassigned to the science program, which would continue to have to justify its budget on its own merits.

Certainly the funding available to space science was enough to pay for a great deal of scientific research. The tens and hundreds of millions of dollars per year available in NASA's appropriations for science were a far cry from the one or two millions per year with which the Rocket and Satellite Panel had to make do. In fact, the amount of money going into space science was so large in comparison with other science budgets—for example, NASA's funding of space astronomy equaled or sometimes exceeded the National Science Foundation's entire budget for ground-based astronomy—that many scientists were greatly concerned. But space scientists, spurred on by the growing number of exciting problems that the field had to offer, did not hesitate to complain about not getting their fair share of the space budget.

When the Apollo program was introduced, the upward slope of the space science budget lessened appreciably. This point was not missed by the scientists who felt that the rapid rate of increase in the space science budget would have continued had not the Apollo program imposed its great demands. The point could not be proved—and in fact there were those who thought that the NASA budget, including that for space science, would soon have leveled off had it not been for the sustaining influence of the manned spaceflight program. Both Webb and the Apollo people were convinced that the Apollo budget helped to keep the other budgets up. With this in mind, as an aid to justifying budget requests, the practice developed of dividing NASA's total request into three parts: Apollo and related manned spaceflight work; other programs that supported Apollo, such as unmanned lunar exploration, studies of the space environment, and solar physics; and the remaining NASA program.

For the scientists, especially those who were opposed to the Apollo program, this practice of justifying a substantial part of the space science program on the basis of what it could do for Apollo was anathema. In their view space science, like space applications, was one of the intrinsically valuable components of the space program, justifiable on its own merits. Moreover, during the long period of preparation for the manned lunar missions, most of the substantive achievements of NASA came from the applications and space science programs, not from manned spaceflight.[13] But for NASA management it was a matter of practical politics, of recognizing the realities of life: better to assign science to a service role and get the money to carry it out than to risk a loss in total funding just to keep the science pure.

The scientists had a stronger reason for complaint when programs were actually cut short for lack of sufficient funding. The lunar projects Ranger, Surveyor, and Lunar Orbiter, all terminated just as they were getting into full swing, were cases in point.[14] So was the cancellation of the Advanced Orbiting Solar Observatory, which caused almost a hundred solar physicists to petition NASA for better support.[15] But as Administrator James Webb and Associate Administrator Robert Seamans could point out, such actions were not arbitrary or whimsical, nor were they antiscience in nature. In fact, Webb was one of the strongest supporters of a balanced space program. When, in 1962, the Apollo program needed an additional $400 million, President Kennedy seemed ready to accept a suggestion that the funds be taken from other parts of the NASA budget. To do so, however, would have crippled the space science and applications programs, and Webb refused to go along. Mr. Webb told the author that he had indicated to the president an unwillingness to continue as administrator of a program that did not have a proper balance among space science, applications, technology, and manned spaceflight. In a letter to the president, Webb offered to wait until the next budget to request the additional funds for Apollo, a compromise that was accepted.[16] Throughout his tenure Webb continued to give strong backing to space science, but he also refused to accept as a valid complaint the grumbling of scientists that manned spaceflight was getting most of the NASA dollars.

Given the need to keep within imposed budget limitations, there was a logic to the cuts made in the space science program during the mid-1960s. For the lunar missions, it was pointed out that while the investigations would cease for a while, nevertheless when Apollo flights began the lunar studies could be picked up again with the added power provided by the personal presence of astronauts on the moon. The reasoning was legitimate, but not acceptable to many scientists who felt that unmanned investigation of the moon was more economical and more versatile, hence more sensible. Nevertheless, NASA managers had to insist that Apollo was a national commitment, entered into for many reasons and not primarily for science, and that the most desirable total program would be one that made effective use of Apollo for science as well as for other purposes. The case of the Advanced Orbiting Solar Observatory was different, in that Skylab's Apollo Telescope Mount, which replaced the observatory, would not duplicate what AOSO could have done. One could show that the unmanned spacecraft was needed for the more advanced investigations of the sun requiring long-duration monitoring of solar activity for a substantial fraction of a sunspot cycle, and high spatial, temporal, and spectral resolutions not afforded by either the first solar observatory satellites or the Skylab telescopes to come. Although Skylab in the manned program did provide a means for some excellent solar research, the need for the advanced, long-duration observatory persisted, and in the course of time much of what the Advanced Orbiting Solar Observatory would have done was accomplished in a continuing series of improved solar satellites: OSOs G-K.[17]

But, whatever complaint there might have been about either the absolute or relative level of the space science budget within the agency's total, there can be little doubt that it represented a substantial program. Through the 1960s and into the 1970s support for science in the space program remained, perhaps steadier in the Congress than on the executive side. But the backing was by no means unquestioning. Congressman Joseph Karth, who for most of the 1960s was chairman of the Subcommittee on Space Science and Applications of the House Committee on Science and Astronautics, was a powerful advocate of the science and applications programs. He was an equally formidable inquisitor of NASA representatives who appeared before his subcommittee. No detail of the budget seemed too small for his eye or his interest, and space science managers each year supplied the subcommittee with reams of testimony and written reports in justification of the space science budgets.[18] For the historian interested in the evolution and progress of the program, the printed records of both the House and Senate authorization committees make informative, if somewhat dreary, reading.

When Apollo passed its peak funding in fiscal 1966 and began to decline, drawing the total NASA budget down at the same time, much of NASA's planning was directed toward finding new programs and projects to follow the lunar missions. The initial failure of that effort has already been described, and for a while the worrisome decline raised doubts as to just what the future of the space program might be. It was during this period that a number of scientists—among them James Van Allen and Thomas Gold, physicist at Cornell University—suggested that manned spaceflight could be greatly reduced or dispensed with.[19] For $2 billion a year a substantial program of primarily science and applications could be carried out. The overriding theme in the debates was economy; and until NASA satisfactorily addressed this issue the budget continued to decline.

Two factors turned the tide. First, even though the manned spaceflight program, mostly because of the tremendous expense of Apollo, was the principal target of attack, there was an underlying reluctance in Congress, in the administration, and even among many of the scientists, to forego something that had brought so much prestige and acclaim to the United States. Second, once NASA had become willing to let go of the Saturn-Apollo line and to stop pushing for an early program to build permanent space stations in orbit or on the moon, the Space Shuttle could be presented as a means of greatly reducing the costs of space operations. In the role of a service to the rest of the space program, manned spaceflight once again became salable. The decline in NASA's budget stopped, and after falling to around $3.25 billion began slowly to climb again as the Shuttle program got under way.[20]

<h2>THE SPACE SHUTTLE</h2>

As the 1970s began, the Space Shuttle, scheduled for the 1980s, looked very much like the keystone of the future for the space program. Nevertheless, even experts were hard put to follow all the ins and outs of the seemingly infinite variety of tradeoffs between technical and economic factors that had to be considered in arriving at the final design of the new space vehicle. Bypassing all of that, plans boiled down to the following. The Shuttle would be able to:

- Launch to near-earth orbit the kinds of payload that the previous expendable space launch vehicles could.

- Place in near-earth orbit payloads weighing 10 tons or more. This would make possible the launching of a large space telescope, which was of considerable interest to scientists. Heavy payloads for high-energy astronomy would also be possible.

387

- Recover such heavy payloads from orbit and return them to the ground. Refurbishing and updating of expensive spacecraft and equipment for reuse would then be possible.

- Carry experimenters with a minimum of spaceflight training into orbit and back. Only the pilot and copilot would have to be fully qualified astronauts.

- Remain on orbit for several days or even several weeks, operating in effect as a temporary space station.

- Carry into orbit and return to earth an outfitted laboratory for the performance of experiments in the space environment. Investigators would go aloft to conduct the experiments.

The development cost, spread over the decade of the 1970s, was estimated at some $5 billion (1971 dollars). The operating cost per flight, including refurbishment, was expected to be on the order of $10 million (again in 1971 dollars).[21]

The idea of a shuttle to space was not new. Various schemes for using lifting bodies to return passengers to earth after flight in space were floating around in the 1950s, but the United States was not then prepared to make much of them. With imagination one could visualize the X-15 as an early step toward a manned space launcher.[22] The Air Force's Dyna-Soar, which was never completed, would have been still another step.[23] But none of these posed the challenges that an operational space plane would. The problems of aerodynamics, structure, thermal protection, and guidance and control in creating a vehicle that would go into orbit like a space launcher and thereafter return to earth and land like an airplane were intimidating. A great deal of work had to be done before one could seriously contemplate proceeding with the project. But after additional years of experience in high-speed flight with the X-15 and Apollo programs, in 1968 and 1969 a number of NASA members including George Mueller, head of the Office of Manned Space Flight, were ready to promote a space shuttle. Even so, many industrial representatives were frank to say that they were not sure the project could yet be pulled off, considering that not only did the technical problems have to be solved, but it all had to be done cheaply by aerospace standards.

And that factor of economy lay at the heart of the shuttle's salability. After the very expensive Apollo, in the midst of a period of economic recession, inflation, and dwindling balance of trade, the country was not about to support another costly space project unless it had some clearly foreseeable practical benefits. National concern with issues other than space had permitted the NASA budget to fall from its peak of a little more than $5 billion in the mid-1960s to about three-fifths the peak value at the end of the decade. The agency would be fortunate if it could keep the budget from going even lower. As has been seen, sentiment in part of the space community to continue with an extensive use of Apollo hardware led to the abortive planning of an Apollo Applications program. Administrator Thomas Paine and Vice President Spiro Agnew would have liked the country to send astronauts to the planets, but that simply wasn't in the cards. Paine and others would also have favored establishing a very large, permanent space station in orbit. As both Paine and Abe Silverstein described the proposition, that would be "the next logical step" in the development of space for man's use. Much of the necessary experience and know-how had already been acquired in the Gemini and Apollo programs, and it was simply a matter of deciding to make a space station and then doing it.

But there was a flaw in this reasoning. A space station would require frequent logistic flights. With Saturn and Apollo hardware, these would entail enormous expense, hardly the kind of economy that was being demanded.

It took a while, but gradually the message came through. Reluctantly space program managers let go of their more exotic dreams and turned attention to discerning what the country might be willing to support. It became clear that the space program for the foreseeable future would have to emphasize specific returns for the large investments that had been made. As they had repeatedly emphasized throughout the 1960s, members of Congress would favor a strong effort on applications. Also, there appeared to be a continuing support for a substantial space science program. Technology that would clearly be helpful in tackling problems on earth was also a salable item. But whatever was undertaken would have to be done at a much lower cost than hitherto. It appeared that only by becoming much more efficient in the use of dollars could the space program continue in any shape comparable to that of the 1960s.

That was perhaps the major issue in the years of dicussion that preceded the decision finally to build the Space Shuttle. The operational capabilities proposed for the new craft were very attractive, to Europeans as well as to Americans, and captured the interest of many scientists. Unlike Apollo, which most of the scientific community appeared to oppose at the beginning, the Space Shuttle had the interest and at least the tentative support of some leading scientists. Even as James A. Van Allen and Thomas Gold spoke out against a shuttle program, many of their colleagues gave it their conditional endorsement.[24]

During the summer of 1970 the National Academy of Sciences made a study of priorities in the nation's space science program.[25] Inevitably the Space Shuttle came in for much discussion. In a lecture to the study participants, Hermann Bondi, head of the European Space Research Organization, expressed his support for the Shuttle, a view that may have reflected the growing interest of the Europeans in cooperating with the United States on some aspect of a shuttle program.[26] One could detect among many of the American scientists a decided interest. But their support was contingent upon a number of conditions.

The study participants made much of the fact that they did not want to get again into a large-scale, manned spaceflight program. They made it plain that they had found much of their experience with the Apollo program distasteful. Hence, if the Shuttle program were to be merely a means to continue a manned spaceflight activity, it would forfeit their interest. In the scientists' view the Shuttle should be developed and operated as a tool to support the country's principal objectives in space, one of which was space science. Astronauts would, of course, fly the Shuttle, and on some missions other passengers might go along; but the controlling elements on each flight should be the technology, applications, or science objectives of the mission.

If a proper perspective were maintained on the agency's objectives, that would mitigate the effect on other programs of the large budgets that would be needed to develop the Shuttle. With the proper perspective, the agency would devote the necessary funds to continuing a strong space science program and to preparing in advance for the use of the Shuttle when it became operational. But the scientists were worried that NASA might not cherish the proper perspective. They had seen the large budgets for Apollo force the curtailment of the Ranger and Surveyor projects and the cancellation of the Advanced Orbiting Solar Observatory. As if to emphasize the point, at the time of the summer study the scientists were wrestling with the impact the expensive Viking—itself a space science project strongly endorsed some years before by a different Academy of Sciences study—was having on other projects favored by space scientists, such as Pioneer missions to Venus.[27]

These two themes—that the Shuttle should be considered a tool and used as a tool to support space science and that its development and deployment should not be allowed to cripple NASA's other programs—scientists kept reiterating in the Space Science Board, in NASA's Space Program Advisory Council and its committees, and in numerous NASA working groups. There were many aspects to the related issues. If the Shuttle were to be a useful tool, it had to be easy to use. In the view of the scientists, that would call for sharply less documentation and testing of equipment than had been required in the Apollo program; schedules also had to be streamlined. As Deputy Administrator Low told the author, it had cost 10 times as much to prepare a magnetometer for the Apollo project as it had to prepare a similar one for an unmanned project. In fact, the main point was that the tool should be made to fit the hand, not the hand distorted to fit the tool.

Because of these concerns, in October 1972 NASA put together a Shuttle users group to discuss periodically with the administrator and various program managers how to use the Shuttle when it came into being.[28] Many groups were already wrestling with how to build the Shuttle and what to use it for, but no group was adequately addressing itself to the question of how it could be run to make the most of its potential. The single most important recommendation to come out of the meetings of this panel was to operate the Shuttle in such a way that the tool did not overshadow the application.

Pursuing both questions, what to do with the Shuttle and how to operate it so as best to serve its users, NASA sponsored still another in the long chain of summer studies on major issues facing the agency. This study was conducted at Wood's Hole, Massachusetts, in July 1973.[29] Again the National Academy of Sciences conducted the study, most of which was devoted to what the Shuttle, particularly the Spacelab it would carry on board, could do for the space program. By that time the scientists had developed a restrained, somewhat worried interest in the vehicle. There was more willingness than hitherto to assume that perhaps the craft could be developed and flown in such a way as to bring down the costs of space missions. There remained still the question of whether it really would be operated as a tool rather than as an end in itself.

The scientists' fears in this matter were revived by NASA's insistence that a great deal of attention be paid to how the Spacelab, which the Europeans were developing for the program at a projected cost of several hundred million dollars, would be used. At the time most of the scientists could see little use for Spacelab and wondered if they were going to be pressured into using it simply to keep man-in-space in the picture. Although the life scientists and atmospheric physicists expressed interest in Spacelab, most of the study participants insisted that they would like to use the Shuttle as a truck to carry payloads into space, including the very heavy ones like space telescopes and high-energy astronomy payloads.

The discussions brought into stark relief another very serious problem. The Shuttle itself would be capable of placing payloads in near-earth orbits; but that would take care of only part of the missions the scientists wanted flown. At one end there were the very small payloads of the kinds that had gone into sounding rockets. Study participants just did not believe that the sounding-rocket class of payload could be accommodated economically within the Shuttle cost structure. Nor, for that matter, did the Shuttle appear to be appropriate for small satellites of the kind that Scouts had been launching, especially payloads that had to go into unusual orbits or trajectories. Would provision be made to keep sounding rockets and a small expendable vehicle like Scout for these requirements?

Also, what about payloads that were headed for synchronous or other high-altitude orbits, or for escape trajectories to the moon and planets?

How would these be launched? If the Shuttle were to be used for the initial boost from the earth's surface, suitable upper stages would still be required to carry the payloads beyond the low-altitude orbit. Was NASA going to ensure that suitable upper staging would be ready for use with the Shuttle, or would there be an undesirable hiatus in such missions when the Shuttle came into operation?

These questions NASA would have to address itself to as the space program moved through the transition period of the 1970s to the 1980s when the Space Shuttle would become the country's principal space booster. If the various collateral requirements were met, the Shuttle had a rosy future in prospect. If they were not met, NASA could expect trouble with its clients.

The early years of American space science may be taken to be the 1950s and 1960s in which first sounding rockets and then satellites and space probes were used to extend scientific research into outer space. Space vehicles were expendable, new ones being required for each new mission. The decision in 1970 to proceed with the development of a reusable Space Shuttle signaled the end of the era in which only expendable boosters were used. It did not, however, signal the end of expendable rockets, since the Shuttle would probably not meet all near-earth launcher requirements and would certainly have to use additional stages to send spacecraft beyond low-altitude earth orbits.

Nevertheless, the decision inaugurated a period of transition for the space program from conventional methods to the use of the Shuttle. During the period of transition space science and applications programs would continue much as in the past, but in parallel much work would be under way to prepare for the use of the Shuttle. If the Shuttle did perform as promised and did prove to be economical, it could be highly useful for space science. Its usefulness would depend on whether the program were operated so as to support the scientific objectives properly.

Once the Shuttle program was under way, it remained to see how well the engineers could do in creating the vehicle and how wise NASA managers would be in using it.

22

Review and Assessment

Iamque opus exegi, quod nec Iovis ira, nec ignis,
Nec poterit ferrum, nec edax abolere vetustas.

Ovid, Metamorphoses

Viewing events in retrospect one cannot but be impressed with the seeming inexorability of human progress toward spaceflight, particularly in the 20th century. There is a temptation to claim that once Tsiolkovsky, Goddard, Oberth, von Braun, and their followers took aim at outer space, the large rocket and spaceflight were inevitable. Certainly by the time *Sputnik 1* went into orbit, a substantial groundwork had been laid by a large number of pioneers working assiduously through many decades.

But the character of the space program that emerged in the late 1950s and 1960s was not so predictable. Many, if not most, of the early workers were primarily interested in interplanetary travel and high-altitude research, but for the most part had to rely on the military for support. In providing support the services naturally were considering the potential military uses of space, and indeed the first major rocket to go into operation was a weapon, the V-2. Because of the importance of atmospheric and ionospheric data for applications of radio and radar, and in the design, construction, and operation of various military systems, the services supported a considerable amount of high-altitude rocket research during the 1940s and 1950s. In the normal course of events one could thus visualize a U.S. space program, including space science, as evolving over the years, emerging quietly as a part of military research and development. Under such circumstances the ability of space scientists to devote their research primarily to the most important scientific problems would have been hampered by the requirement to contribute in a demonstrable way to more immediate military needs. In addition, as the experiences of the Upper Atmosphere Rocket Research Panel during the 1940s and 1950s showed, there would have been the constant threat of being pulled under the cloak of military secrecy—a restriction fundamentally incompatible with the scientific process.

Such limitations on the U.S. space program were avoided when the administration and Congress, reacting to the Sputnik challenge, decided that in the best interests of the country most of the space program should be conducted openly under civilian auspices. Moreover the vagueness and grand sweep of the National Aeronautics and Space Act of 1958 gave the NASA administrator a great deal of flexibility in specifying the content of the NASA program. As one consequence, under NASA management the space science program became very much a creature of the nation's interested scientists.

When the Soviet Union surprised the world by launching the first artificial satellite into orbit, the shocked reaction of the United States tended to distort the country's perception of what was happening. The weight of *Sputnik 2* and *3* showed how advanced the USSR was in rocket payload capability, and it was easy to focus on this factor while underestimating the importance of the work that the United States had already done in the field. Looking back, it is now clear that America, while lagging in rocket propulsion, was more than competitive in communications, tracking, and telemetry, in guidance and control, and in sounding rocket research. Taking all factors into consideration the imbalance was not so great as had been imagined. Proceeding from its substantial state of readiness the United States built an enviable record of success in space over the next dozen years, culminating with the Apollo missions to the moon.

Space science contributed its share to the overall success. Indeed, for most of the 1960s applications and science missions provided most of the return on the nation's investment in space, and it was not until the Apollo lunar flights that the manned spaceflight program began to generate the prodigious quantities of data that continued to flow from it during the first half of the 1970s.

One can use several criteria in assessing the success or failure of the space science program. The simplest is whether the program achieved what its planners set out to do. By this criterion the space science program must be adjudged successful. In every area—earth and planetary sciences, solar physics, stellar astronomy and cosmology, and to a smaller extent biology—substantial progress was made, bringing a number of important discoveries. Successful unmanned scientific spacecraft missions were legion, including thousands of sounding rockets; dozens of Explorer satellites; solar, geophysical, and astronomical observatories; Pioneer space probes; Ranger, Lunar Orbiter, and Surveyor spacecraft to the moon; and Mariners to Mars and Venus. Sharing in some of these successes were many other countries taking part in a quite extensive international cooperative program.

A more substantive criterion of success is whether what was achieved was worthwhile. This is more difficult to judge, but that hundreds of first-rate scientists chose to devote their personal careers, or a substantial part of them, to space science is evidence of the program's success. The numbers of scientists working in the field and the voices of scientists raised in strong support of important projects and equally strong protest against proposed cuts had to be important considerations to the administration and Congress in deciding the extent of support to accord to space science.

Success in the space science program was not bought without some failures. Indeed, for the first two years failures seemed at times to eclipse successes, although before the end of the 1960s the success rate had risen well into the 90 percent range. Both failures and successes had their lessons to convey, and there was much to be learned by participants in the space science program, not only of a scientific nature but also concerning organization and management, and the perplexities of human relations.

ORGANIZATION AND MANAGEMENT

A great many of NASA's working hours were taken up in problems of management. Patient attention to detail was required to make the agency's complex projects succeed. The space team did a good job, evoking worldwide praise for NASA. But it must be remembered that the team was more than a single agency, consisting as it did of thousands of engineers, technicians, laborers, scientists, and administrators from government, industry, universities, the military, and even other countries. Furthermore, accomplishments were much more labored than one might suppose from a distance. The picture of a well-oiled machine purring along without a clank or a clatter is inappropriate. The space program endured the same kinds of personnel problems, development snags, labor disputes, schedules missed, cost overruns, failures and temporary setbacks, and management mistakes that were the experience of the military and industry in the large weapon projects that might be pointed to as the closest analog to what NASA was trying to accomplish.

That NASA had to struggle through the same difficulties that beset other large-scale programs in no way diminished the luster of space achievements. On the contrary, to meet and overcome such difficulties was the nature of the task. NASA was eclectic in its approach, borrowing management ideas from various sources, especially the military. The agency was willing to experiment, to pioneer in the use of new management techniques in government-industry relations, incentive contracting, project planning, technical and cost reporting, management reviews, and quality assessment and control. By remaining flexible, reorganizing several times in the course of a decade, it was possible to accommodate changing needs of the program.

The management style of the agency reflected those of the several administrators who stood at the NASA helm through the 1960s. The first administrator, T. Keith Glennan, came to NASA with a controlled enthusiasm for space that served to prevent any explosive growth through over-reaction to Sputnik. Glennan's measured pace elicited a steady pressure from numerous quarters to move faster, particularly to get on to the planets, which in the minds of many scientists were taking on new importance with the possibility of investigating them at close range. In retrospect the situation seems to have been ideal, with a positive leadership setting forth on a substantive program, and a strong followership ready to go along and even to move faster and farther given the opportunity to do so. In this climate Glennan was able to set the agency upon the course that it followed for many years afterward.

In February 1961, James E. Webb became the second administrator of NASA. The approval by President Kennedy and the Congress of the Apollo project gave Webb the opportunity to step up the pace of the space program. All aspects of space science were expanded. A primary concern of Webb, which characterized his style of management, was to maintain the independence of action of the agency. While working to build up the program, he was also careful to avoid becoming the captive of any group in industry, the administration, or the Congress.

Under Webb's vigorous leadership the agency's followership grew steadily and, by keeping a balanced program even under high-level pressure, to concentrate more on the Apollo mission at the expense of other parts of the program, the administrator maintained a broad base of support. Then tragedy struck, a fire in the Apollo capsule killing three astronauts—three of the nation's heroes. Had it not been for the race with the Soviet Union and the severe blow to U.S. prestige in the world that a failure to follow through on the Apollo commitment would have entailed, the lunar venture might well have ended at that point. As it was it took many agonizing months and Webb's considerable administrative and political skill to redress the situation, to pick up the pieces and move on again toward the lunar landing still years away. But from that point on support for the agency was permanently weakened, more tentative, more questioning. So, when the muddy planning for an Apollo Applications program to follow the manned lunar missions looked to outsiders more like an attempt on NASA's part merely to keep the Saturn and Apollo teams in business rather than to serve any genuine need, the necessary support could not be developed. While resistance was general, it was especially strong among the scientists, who protested that as far as science was concerned, the prodigious sums being asked for Apollo Applications could better be spent on any of a large number of important, unmanned scientific investigations.

In this climate NASA leadership faltered. Finding in his contacts with the administration, the legislators, and industry no strong support for large new initiatives in space, Webb shied away from making any specific proposals. He chose rather to encourage the nation to debate what the country's future in space should be, hoping that the agency could get some guidance from such a debate. But the country did not move to fill the leadership vacuum left by NASA, and no great debate took place. It was left squarely up to NASA to recapture the leadership it had temporarily relinquished.

In contrast to his predecessor the third administrator, Thomas O. Paine, was eager to strike out on bold new paths, optimistic that he could generate the necessary support. Paine made the courageous decision to proceed with the *Apollo 8* flight in December 1968, at a time when there were growing concerns and doubts about the ability of Apollo to accomplish its objectives and much fear that a serious failure in an early lunar mission might lead to a strong reaction against continuing the project. The outstanding success of *Apollo 8* completely altered the mental climate for a while and set Apollo firmly on its final course to success. But, later, when Paine campaigned unrelentingly in the Nixon administration for a large-scale space program costing $8 billion or more a year, including shuttles, space stations, and manned spaceflight to the planets, he found himself completely out of tune with the conservative, budget-conscious mood of the time. In the face of distressing societal problems that impinged on the daily life and the pocketbook of the average citizen, the country was not in a mood to "swashbuckle," as Paine had put it. Much of NASA's followership again shied away.

The fourth administrator, James C. Fletcher, who took over on 27 April 1971, recaptured the NASA followership with a policy of moderation and cost consciousness. An effort was made to project an image of applying space knowledge and capabilities to problems of concern to the man on the ground, and to do it economically. The Space Shuttle was sold largely on the basis that it would make it possible to use space more effectively and at far less cost than with conventional launch vehicles and space hardware. Fletcher's style was more like that of Glennan; his willingness to proceed at a measured pace, as Glennan had sought to do, made his approach acceptable. The image of conservatism and public responsibility that he projected made it possible for Fletcher to discuss publicly future exciting adventures that had appealed to Paine, like sending men to the planets or building space outposts in orbit or on the moon.

Under each of its administrators NASA had, of course, to engage in the usual activities of management. These included—in the jargon of the government manager—planning, programming, budgeting, and execution. Space science managers could no more escape these necessities than could any others, but differences of approach were worthy of note.

It is customary for a large-scale operation to maintain a series of plans for the activity—short term, intermediate, and long range. In theory the short-term plans are those largely in effect or being carried out, the intermediate plans those that are to be used in formulating the next budget proposals, while the long-range plans serve as a guide into the more distant future. Properly worked out plans should include not only the objectives to achieve, but also suitable estimates of specific projects, their feasibility and promising approaches, funding, manpower, and facility requirements, schedules, an appraisal of the availability of suitable contractors, and some thought about organizational and management setups. Shorter-term plans would, of course, furnish such detail in greater depth than would long-range plans, which for the quite distant future might become rather general in treatment.

When NASA began operations, Administrator Glennan required the agency to maintain both short-term and long-range plans. As did the other offices, the space science division contributed to those plans. The second administrator, James E. Webb, however, while requiring adequate planning on the part of the agency, did not favor publishing specific plans. His concern was that the issuance of specific plans for the more distant future would call forth attacks from NASA's opponents when neither the agency nor its supporters were prepared to engage in a suitable defense of the plans. Webb preferred to publish specific plans as he requested the next year's budget, at which point the agency was prepared to put forth a strong defense of its proposals. Webb's approach placed upon the different offices in the agency the responsibility to maintain an adequate planning activity while refraining from publishing specific long-range plans.

While there was something to gain in not revealing NASA's intentions too early, there were also disadvantages. Potential participants in the program needed to know what was in prospect, so that they might plan and make proposals to NASA. In space science, especially, managers felt the need to inform individual scientists of the opportunities that lay ahead so that they might plan and work on experiments that often took years of advance preparation. Similarly there was a need to keep industry informed of the kinds of spacecraft and instrumentation contractors might be called on to provide. To meet the need for advance information on likely future space science projects while at the same time not committing themselves to specific future plans, space science managers devised what they called a space science *prospectus.*

The prospectus differed from an actual plan in that for each area or discipline the prospectus listed a variety of possible choices for future programs and projects. The choices were studied and analyzed in sufficient depth to ensure that they were feasible and to afford a suitable estimate of funding, manpower, and other requirements. In theory, the prospectus provided NASA people, industry, and outside scientists useful information about what NASA had in mind for the future without drawing the fire of critics that a firm plan might occasion. The prospectus did prove to be a useful planning device, and in the last two years of Webb's administration the author and some of his colleagues worked on such a prospectus for the whole agency. For a variety of reasons this effort did not succeed, the most important of which probably was that Thomas Paine, who became admin-istrator after Webb left, strongly favored specific plans and was willing to battle for a bold, long-range program for the agency.

In the jargon of government workers, programming is the process of putting together individual elements of a plan into a properly integrated program for an office or the agency to undertake. Then budgeting is figur-ing out the funds and other resources according to time required to carry out the proposed program. For space science managers, one aspect of planning and programming differed from the approach of other offices in NASA. That was the conscious effort to make the space program the crea-ture of the nation's scientific community.

To achieve this end it was necessary to bring large numbers of outside scientists into the planning in some way that made their input effective, while NASA still made the required decisions. There was a narrow path to tread here, for the scientists would gladly have wielded the authority while leaving to NASA the responsibility for the actions taken. NASA managers took the approach of including the thinking of a series of advisory com-mittees in their planning and programming. It was not an easy process to sustain, since advisers could never hope to be as fully informed of all the issues as NASA employees working full-time on the job. Moreover, at times other than scientific issues forced decisions that were unpalatable. In mak-ing such decisions NASA managers could not always get the help they needed from the scientific community, since scientists were reluctant to set priorities between different disciplines. Hence, when budget restrictions required a choice between projects in differing disciplines the onus landed on NASA people. Not until the end of the 1960s did outside scientists finally face up squarely to the problem of giving NASA specific advice on setting priorities among various disciplines, as well as within a specific one.

Nevertheless, except for this one lack, the scientific community supplied NASA with much advice on space science programs and projects, to the extent that the NASA space science program could genuinely be described as a program of the scientists. Supporting this program NASA was able to obtain sizable budgets, particularly during the first half of the 1960s. At the peak of support for NASA in the middle of the decade, space science was enjoying the lion's share of a science and applications budget that ap-proached $1 billion a year and, although funding declined sharply toward the end of the decade, space science continued to command resources in the neighborhood of $500 million annually.

As for execution, the space science program relied on NASA centers, industry, and the universities. For most of the 1960s the Office of Space Science and Applications was assigned the responsibility for the Goddard Space Flight Center, Jet Propulsion Laboratory, and Wallops Station. Most of the internal support for space science was obtained from these centers, but every other NASA center also provided support to the science program. In general, relations between NASA Headquarters and the centers were effective, but at times, particularly in the early years, there were severe strains. Illustrating these were difficulties with the Goddard Space Flight Center and with the Jet Propulsion Laboratory. The problems were similar, arising from conflict between the center's desire for autonomy and headquarters' responsibility to represent the agency to the administration and the Congress. But the circumstances were different in that Goddard was a Civil Service center while JPL was a contractor to NASA. In both cases accommodation on both sides was required to overcome the difficulties. With Goddard, headquarters had to take care to keep to its own job of *program* management, leaving the center free to handle the management of *projects* assigned to it. As for JPL, the laboratory had to recognize its responsibility as a NASA contractor to follow NASA direction, while NASA had to leave JPL sufficient leeway to exercise its own judgment with regard to basic research.

While the space science and manned spaceflight programs supported each other—the former furnishing advance information on the moon for the design of hardware and planning of mission operations, the latter eventually providing the most powerful method of investigating the moon—nevertheless there were serious strains for a variety of reasons. Many scientists were unconvinced of the worth of the manned spaceflight program in general or of the lunar landing project in particular. To these persons it seemed clear that a much greater return in scientific data could be had, sooner, in an unmanned program of far smaller cost. Leading members of the scientific establishment stated unequivocally that the real substance of the space program lay in science and applications. Accordingly it rankled that top priority and huge funds were accorded the Apollo project, whose principal missions were the better part of a decade away, while valuable scientific projects that one knew how to do and that would yield important data quickly had to wait for later funding. The distress increased when Apollo needs threatened ongoing projects, as happened from time to time.

A subtle complication arose when the agency urged scientists to put experiments in Gemini and Apollo flights, but then did not accord the experiments the kind of support or level of priority the investigators felt they deserved. One can appreciate the views of the Apollo managers, since they were attempting to achieve something never done before, something very difficult, very hazardous, and also important to the country's image in the world. Nevertheless there were many who felt that the Apollo engineers indulged in overkill, thereby precluding a great deal of valuable science that might otherwise have been done. Eugene Shoemaker, a geologist from the U.S. Geological Survey, provided an extreme example in this respect. For many years Shoemaker worked intimately on the Apollo project, helping to train astronauts and to prepare for scientific investigations during the Apollo landings on the moon. Yet after the first successful landing he left in disgust to spend more than a year excoriating NASA for shortsightedness with regard to Apollo science.

The strains between space scientists and the Apollo people were exacerbated by the fact that Apollo was principally an engineering project. Engineers and scientists differ fundamentally in their outlook and approach to their jobs. To engineers, trained in highly disciplined teamwork, the independence and individualism of the successful scientist looks like anarchy. To overcome these basic differences requires conscious and continuing attention from management. In the Office of Space Science and Applications an organizational device was used for many years to try to alleviate this problem. Instead of gathering the scientists into a single research group and the engineers into a separate service group, which is a traditional arrangement, engineers and scientists were intimately mixed in a number of smaller units. As head of the office, the author, himself a scientist, chose an engineer as his deputy. In the division for geophysics and astronomy, initially a scientist was in charge, with an engineer as deputy. Later when the scientist was promoted, the engineer became the head and chose a scientist as deputy. Scientists and engineers were paired at all levels throughout the organization. The arrangement sometimes evoked the criticism that it generated a collection of little "baronies" in the office, yet the organization appeared to promote its intended objective. Engineers and scientists came to appreciate each other's problems and to share enthusiasm for each other's triumphs. Several times in the course of the decade space science management considered the possibility of returning to the more traditional arrangement, only to reaffirm the original choice.

The effort to solve the problems that the Office of Space Science and Applications and the Office of Manned Space Flight had in working together by setting up a special Manned Space Science Division was less successful. For one thing, the problems were more severe. Manned Space Flight had the priority, and even was assigned the funds for the manned space science for which the Office of Space Science and Applications was given the responsibility. Thus two fundamental management errors stood in the way. The Manned Space Science Division had two bosses to try to satisfy, which is universally recognized as unsatisfactory. Second, the Office of Manned Space Flight had the money for (hence in practice control of) manned space science. As a consequence the Office of Space Science and Applications long felt frustrated in putting together the kind of manned space science program the scientific community desired. Not until the initial lunar landing had taken place and the primary remaining motive for any further Apollo missions was science—to explore and investigate the moon—did these problems begin to resolve themselves. At that point lunar scientists, in a tremendous surge of interest, working for the most part directly with the Johnson Space Center, generated the kind of science program they had long sought.

In the course of the space science program NASA managers relearned a number of management lessons others had learned before, such as not assigning two bosses to the same group and not assigning the money for one program to the control of another office. It ought almost to be axiomatic that objectives should be clear and reasonable, yet with the Centaur program NASA put itself through a period of considerable strain by trying to make the as-yet-undeveloped rocket stage satisfy at least four different sets of requirements. Only when the development was directed toward a single set of requirements, those of the Surveyor lunar spacecraft, did Centaur move smoothly toward its first successful flights. Once developed, Centaur was uprated to satisfy additional requirements.

Again, it should be clear that attempting too big a step in a new development is unwise. While the intended objectives may ultimately be achieved, the cost of overreaching can be too great. This point was illustrated by the Orbiting Astronomical Observatory in which snags were encountered in developing the guidance and control system that took inordinate amounts of time and money to solve. Moreover, the seven-year-long development time for the observatory adversely affected experimenters who had to mark time with their experimental programs while the spacecraft was being developed. In this connection it should be noted that a number of scientists had advised NASA to fly a less complicated observatory first.

To avoid such harmful overextension and costly overruns, NASA management introduced the device of phased project planning. While its use was rather fuzzy in NASA, nevertheless the policy of requiring a careful review and assessment of the size and appropriateness of steps to be taken in NASA projects was beneficial.

INDIVIDUAL AND INSTITUTIONAL RELATIONS

Thousands of individuals and institutions were required to carry out the space program. NASA's relations with these took on many forms, some of them simple and uncomplicated, some very complex. All were important management concerns.

Basic to maintaining the necessary political support for the program were the often delicate and subtle relations with the administration and the Congress, neither of which has been dealt with in any depth in this book. As for space science, on the legislative side the program had both the strong support and continuing criticism of the Subcommittee on Space Science and Applications of the House Committee on Science and Astronautics, with more general support and somewhat less penetrating criticism from the Senate Committee on Aeronautical and Space Sciences. Within the

administration, primary attention came from the president's science adviser
and the President's Science Advisory Committee, especially from the Space
Science and Technology Panel of PSAC. Although a number of the PSAC
members had devoted considerable time and effort helping to establish
NASA with a strong scientific flavor, they did not choose to devote their
own personal careers to space research. The science panel did, however,
keep a watchful eye on the agency, especially for the first few years.
Although the Science Advisory Committee and its panel had no direct
authority over NASA, their position in the White House gave considerable
weight to their views, and at times they served as effective safety valves for
the scientific community when space scientists felt that their needs were
not receiving the proper attention. The letter to Kistiakowsky from Lloyd
Berkner, chairman of the Space Science Board, in November 1959 expressing concern that NASA might neglect ground-based scientific research
related to space in favor of only flight experiments, and also might not
publish scientific results in the open literature, illustrates the point. As a
second example, astronomers sought similar help from the space science
panel when they were dissatisfied with how work on an orbiting astronomical satellite was progressing. In both cases the science adviser and the
panel used their good offices with NASA to help clear the air. More substantive was one science adviser's assistance in breaking the deadlock over
classification that threatened to damage a long-planned program of international cooperation in geodesy.

After the first few years the science panel was more or less quiescent
until widespread dissatisfaction over NASA's planning of the Apollo Applications program stirred the panel to renewed activity. Its concern and
recommendations did much to help steer the thinking on Apollo Applications out of its preoccupation with merely keeping Saturn and Apollo alive
toward the more acceptable Skylab program.

Also not treated in depth in this narrative were NASA's vital relationships with other government agencies. A particularly intimate partnership
with the Department of Defense and the military services was essential.
Indeed, mutual assistance between the two agencies was required in the
NASA Act. But in other areas not specifically addressed in the NASA legislation, NASA also needed to work closely with sister government agencies.
The use of satellites for monitoring the weather, of major importance to
the Weather Bureau of the Department of Commerce, was one example.
Others were the use of satellites for making a worldwide geological survey
and for monitoring changing land use patterns, both of concern to the
Department of the Interior. It soon became apparent that satellites could be
of assistance in surveying and monitoring agricultural crops, forests, and
grazing lands, bringing NASA and the Department of Agriculture together.
In similar fashion the potential contributions of satellites to communications, air and marine navigation, and air traffic control invited still other

associations between NASA and the rest of the government establishment. Associations with private activities were legion, as industry designed, built, and operated most of the hardware that made space science, space applications, and space exploration possible.

NASA's relationship with the National Academy of Sciences, through the Space Science Board especially, was inherited from the International Geophysical Year along with the IGY sounding rocket and satellite programs that gave the agency its headstart in science. As the nation's most prestigious scientific body, the academy's advice carried extra weight and its support to the space program special significance. While there were rough spots in the association, on the whole the relationship was intimate and productive, and through the years Space Science Board recommendations, including those from a long list of special summer studies, weighed heavily in NASA's planning and programming.

But NASA soon learned that the scientific community was not monolithic, and that often important groups of researchers took exception to specific recommendations of the academy. Thus, while still placing high value on advice from the Space Science Board, NASA managers came to feel the need for a closer association with a broad segment of the scientific community. To this end the agency made use of a series of advisory groups, which throughout the 1960s proved to be a powerful means of involving outside scientists intimately in the planning and conduct of the space science program. At times more than 200 of the leading workers in space science were on NASA committees and working groups. Also, by periodically replacing a portion of these advisers with new recruits, NASA was able to keep infusing new thinking into the system.

The first advisory groups were subcommittees of the Space Sciences (later the Space Science and Applications) Steering Committee, which consisted of key managers of the NASA program. These subcommittees were highly specialized, furnishing advice in essentially a single discipline, such as particles and fields. They advised on program content, on the selection of experiments and experimenters for flight missions, and on what laboratory work to support to ensure a proper groundwork for future space missions. The disciplinary subcommittees were very effective, but in time scientists began to complain that there was a need for less specialized advisory bodies and a broader participation of the community. The Astronomy Missions Board and the Lunar and Planetary Missions Board were established to meet this criticism, while retaining the subcommittees of the Steering Committee. When the missions boards came to feel that even they did not always have enough perspective on the NASA planning and programming, the Space Program Advisory Council, with subsidiary committees to the council, and subsidiary panels to the committees, was brought into being.

Since plans and programs began to take shape in the division and lower levels of the organization, that suggested advisory groups would be most effective if they worked directly with the divisions—and this was done. At the same time advisory groups traditionally feel that they ought to make their views known directly to top management. Accordingly, while advisory groups were established to work with the divisions, they were also asked to report findings and recommendations directly to associate administrator and administrator levels.

This multipronged connection with NASA management worked well with the missions boards, but not so well with the Space Program Advisory Council. In the advisory council organization there was too much layering, and the divisions lost touch with the council, feeling little kinship with the council's committees and panels. To the divisions the council appeared too much as a special group for the administrator, so that the divisions began to set up advisory groups of their own with which they could work with the necessary intimacy. Even for those who had close contact with the council, the ponderous, multitiered structure was burdensome, requiring more labor to make the mechanism work than ought to be true with a properly functioning advisory arrangement.

After several years of operation it appeared, to the author at least, that the advisory structure should be streamlined, particularly to reestablish its usefulness to the division levels as well as to the administrator.

To work effectively with the scientific community NASA management considered it essential to have scientists both in headquarters and in the centers, center scientists also being experimenters in the program. Under the circumstances outside scientists found themselves both allies to NASA, helping to plan the space science program and writing and speaking in its defense, and competitors to the agency as they vied with scientists in the centers for space science dollars and for rides for their experiments on the satellites and space probes. To avoid such a situation, representatives of the Space Science Board had originally urged NASA to stick to engineering and operations, leaving the science entirely to the universities and other outside research groups. To a number of persons in NASA that proposition appeared too simplistic, and it did not seem that engineers bent primarily on producing and operating space hardware could always be counted on to work effectively with the scientists without some internal scientific guidance. As a consequence the agency proceeded to build up a small collection of scientists in the centers and in headquarters. To avoid severe conflict of interest for center scientists, headquarters was given the task of selecting experiments and experimenters to be supported in the space science program. In theory, at least, the center experimenters would have to compete on equal terms with the outside scientists for support.

Experience suggests that the NASA decision in this matter was the right one. The difficulty between scientists and engineers in the manned spaceflight program during the period of hardware development and test flights showed the importance of having a scientist's ear within the organization, to which outside scientists could turn. Struggles with Jet Propulsion Laboratory engineers made the same point. On a more positive note, scientists within NASA working full-time on the task of furthering space science began to conceive and bring into being highly useful spacecraft—like the Interplanetary Monitoring Platform and the Radio Astronomy Explorer. These were also a boon to the outside scientists, who put their experiments aboard such spacecraft but could not have afforded the time to conceive or create them.

In making the outside scientific community such an important part of the space science program, NASA managers had to recognize certain fundamental facts. For one thing, continuity of support to a researcher was essential. Most investigations in the science program were long term. Most experimenters had in mind an important problem or group of problems to solve—concerning the upper atmosphere, or interplanetary space, or the sun, for example—and this would usually require many years and many sets of measurements. It became incumbent on NASA to provide continuing support to these investigators and their groups in order that they might carry their work to a proper conclusion. Since many of the experimenters were in universities, it was necessary to accommodate funding to the university's special situation. A sudden withdrawal of NASA dollars from a university research group funded only by NASA could be catastrophic, particularly for students working toward a degree. The step-funding approach devised by NASA's university office was a highly acceptable way of funding research in the universities, allowing as it did at least two full years to phase down a research program that NASA could no longer support.

Continuity of support to investigators also called for NASA to follow through on productive projects. When the agency had created an especially effective tool—like Ranger or Surveyor—scientists assumed that NASA would make that tool available long enough for a reasonably complete series of investigations. This is, in fact, where NASA had some of its most serious confrontations with scientists. In the scientists' view Ranger, Lunar Orbiter, and Surveyor were all terminated much too soon, when the investigators still had in mind a long list of important problems on which to use them.

The amortization of an expensive development over a long period of continued use makes good economic sense. The follow-through on the scientific investigations for which the equipment was produced in the first place makes good scientific sense. That was the very reasoning that NASA later applied to the justification of the Space Shuttle.

Related to continuity of support was the scientists' preference for smaller spacecraft and projects, a preference that continued in evidence throughout the 1960s. With small spacecraft of relatively short lead-times, experimenters could more easily follow up on new discoveries than they could with large, complicated spacecraft which took many years of preparation and which to a large extent froze an investigation into a specific line for a considerable time. Moreover, large projects—like Apollo and Viking—were very expensive and in times of tight budgets threatened, sometimes actually precluded, smaller projects. Yet some investigations required the more complex, more expensive spacecraft—like planetary landers and astronomical observatories. When budgets permitted both, the scientific community usually was glad to have the more versatile spacecraft, as long as support was also provided for the smaller projects. When both could not be supported, it is safe to say that most space scientists would opt for a varied program of smaller, cheaper projects.

On the whole NASA dealt most effectively with outside scientists in their own universities. There the researchers continued their teaching, producing new talent and drawing many of their students into the space science program. From time to time the agency considered setting up special institutes to draw investigators more closely into the program. But there were difficulties with institutes. An institute was an additional source of overhead and could easily tie up one-half to several million dollars a year. Moreover institutes could create undesirable competition with the universities for top-notch scientists, who might better be left in the educational system to teach the next generation. But there were also advantages, and NASA did set up two institutes. The first, the Goddard Space Flight Center's Institute for Space Studies in New York City, was a genuine success primarily because of its director, Robert Jastrow. By quickly establishing close working arrangements with nearby universities like Columbia and Princeton, Jastrow drew outstanding doctoral candidates into the institute's program. In this move he simultaneously removed the element of competition with the universities, replacing it with a mutually profitable partnership. The Lunar Science Institute in Houston was a more difficult proposition. Although regarded as a boon by foreign scientists working with the Apollo program and although useful as an interface between the scientific community and the Johnson Space·Center, it is likely that its benefits might have been achieved more cheaply some other way.

Finally, NASA's ties did not stop at the nation's borders. The extensive program of international cooperation in space science brought with it numerous relationships with foreign academies of science, research institutions, and individual scientists. The effectiveness of the international science program may be attributed to a few guiding principles established at the start. These were: to engage only in programs of genuine substance and of mutual interest, to share (not necessarily equally) in the conduct of the program without an exchange of funds, and to publish the results in the open literature. In a program in which a nation was paying its own way, the cooperating country would take a deeper interest and could take greater pride than in one for which the United States paid all the costs. Cooperation with the Soviet Union was always difficult. Only in the 1970s when the USSR felt it could deal with the United States on more or less equal terms and that it had something substantive to gain—as in the Apollo-Soyuz mission—did the difficulties abate somewhat. In the mid-1970s it remained to see how much more cooperation with the Soviet Union would be possible in the approaching era of space shuttles and orbiting space stations.

THE SCIENTIFIC PROCESS AND SPACE SCIENCE

The evolution of the space science program furnished a good example of the scientific process in operation. Out of advancing technology came rockets and spacecraft which, even before they were developed, were envisioned as powerful scientific tools. As soon as large enough rockets were available, they were put to work in high-altitude research. When the space program was formally established, researchers working on problems of the atmosphere and space naturally gravitated to the new tools. The phrase *space science* came to mean scientific research made possible or significantly aided by rockets and spacecraft.

The rapid growth of research stemmed from the remarkable range of scientific disciplines to which rockets, satellites, and space probes could contribute. Although many space science results would have practical importance in such areas as meteorology, geodesy, aircraft and spacecraft design, communications, navigation, and earth-resource surveys, still the field was largely pure science, pursued primarily to advance man's knowledge and understanding of his universe. It is pertinent, then, to ask how space science affected science, particularly the disciplines to which it could best contribute.

Particularly noteworthy was the progress made in the earth and planetary sciences. Here the impact of space science was profound, generating a fruitful partnership among astronomers, physicists, and earth scientists. No longer was the geophysicist confined to a study of only one body of the solar system. No longer was the study of the planets solely a venture of the astronomers. The dearth of new data that had led planetary studies into the doldrums and even disrepute among astronomers, gave way to a sudden flood of new information that reawakened the astronomer's interest. Geophysical, geochemical, and geologic data on the moon and planets that poured in from astronauts and instrumented spacecraft—Explorers, Mariners, Pioneers, Rangers, Surveyors, and Lunar Orbiters—afforded earth scientists the opportunity to begin the serious development of a science of comparative planetology.

Equally exciting was progress in space astronomy, where rockets and satellites made possible the observation of the sun and the cosmos in wavelengths not observable at the ground. Inasmuch as current theories of the origin, evolution, and demise of celestial objects indicated that most of the information on these objects would be manifested in the hitherto hidden wavelengths, rockets and satellites were in a position to make a tremendous contribution to astronomy, particularly in a period when there were many fundamental questions to answer in connection with phenomena such as radio and Seyfert galaxies, galactic nuclei, quasars, pulsars, neutron stars, and black holes in space. The expectations were borne out in the ultraviolet and x-ray measurements of the sun, and in the discovery and investigation of hundreds of x-ray sources in the sky. The solar observations produced a number of surprises, particularly the x-ray pictures showing considerable structure in the solar corona. As for celestial x-ray sources, they introduced a new field of high-energy astronomy which no one doubted would be intimately involved in answering important questions about fundamental processes in the universe.

As for the life sciences, the most significant results came from the manned spaceflight program, from biomedical studies that have not been dealt with in this book. With increasing productivity, Gemini, Apollo, and Skylab all contributed to an understanding of the effects of prolonged exposure to the space environment, particularly weightlessness, on human physiology and performance. In contrast, during the 1960s exobiology remained earthbound. No indigenous life was found on the moon, nor was any chemical evolution toward the formation of life found. Even in the mid-1970s, after two Viking landers failed to detect any evidence of life on Mars, the question of life on the Red Planet remained open.

Surveying what was accomplished in space science in its first decade and a half, it is clear that the rocket and spacecraft were revolutionary tools, making possible researches that could not have been carried out without them. Great quantities of valuable data flowed from space-borne instrumentation, and innumerable discoveries were made, greatly extending and enriching scientific paradigms in earth and planetary sciences and in astronomy. In the broadest terms, however, the paradigms of space science in the mid-1970s were compatible with those of the 1950s, in that no change in fundamental physical concepts and laws had been forced by the discoveries. From this point of view, then, the first decade and a half of space science was normal science.

But a more restrictive view is perhaps more appropriate. Within individual disciplines many scientists regarded space science as revolutionary. A case in point was the abandonment of what had previously been accepted as the basic hypothesis of geodesy, and the rise in importance of spherical harmonic techniques in the study of the earth's gravitational field. There were other minirevolutions. One was from the discovery of the earth's magnetosphere, not suspected beforehand, and the emergence of a new discipline of magnetospheric physics. Knowledge of the extensive evolution of the lunar surface after its formation produced a revolutionary change in the lunar paradigm. Perhaps, too, the discovery and characterization of celestial x-ray sources, which had been missed in previous astrophysical theory, presaged a revolutionary change in astrophysical paradigms.

FUTURE COURSE

Vannevar Bush characterized science as the "endless frontier." Science showed space to be another endless frontier. The allure of these two in combination imparted a natural impetus to space science in its early years. Benefiting from the powerful political forces of the Cold War and the concern generated in the United States by the Soviet launching of Sputnik in October 1957, scientists in the United States were given resources by the nation sufficient to tackle an impressive array of problems not previously tractable. By the time of the Apollo missions the number of space scientists around the world had risen into the thousands.

But so vast a subject as science in space and the science of space could hardly be more than touched upon in one or two decades. While magnetospheric physicists might speak of their results in the investigation of Earth's magnetosphere as comprehensive, not one would think of the subject as closed. There still remained in the early 1970s the problem of understanding the processes and complex interrelationships. Also there were the magnetospheres of the sun and Jupiter, and perhaps of other planets, to investigate, the study of which would inevitably turn attention to the magnetospheres of stars and planets beyond the solar system.

While comparative planetology had quickly revolutionized the earth sciences, expanding their scope from Earth to the solar system, here again the new discipline had hardly reached its adolescence as unmanned spacecraft of the 1970s began their probing of the major planets and their satellites. Other decades would have to pass before comparative planetology could be said to have matured.

Although the failure fo find life on Mars in the first Viking missions was disappointing, it seemed clear that interest in exobiology would continue. For one thing, many scientists believed that the processes leading to the formation of life are inexorable, and that there must be innumerable examples of extraterrestrial life to be discovered if only one knew how to find them. This belief would lead to various schemes to communicate by electromagnetic means with living beings beyond the solar system. Within the solar system, even if only Earth had living beings, still the chemical evolution of the other planets and satellites would be important in studying evolutionary steps toward life.

With the discovery of x-ray sources, space science made a unique contribution to the newly emerging field of high-energy astronomy. While some might label the evolution of x-ray astronomy in the 1960s as revolutionary, others would feel that the most significant contributions of space astronomy still lay ahead.

Thus, space science in the 1970s retained a considerable momentum, with the prospect of challenging and important problems to work on for the foreseeable future. For a few years the diminishing urgency of the space program appeared to pose a threat to space science. But, with the decision to proceed with the development of the Space Shuttle, a renewed commitment to space science seemed ensured. It would not be an easy road, and all the signs indicated that in the future the need for specific space projects would be carefully weighed by both administration and Congress. But few doubted that the program, including space science, would continue at some pace. Indeed, there seemed little doubt that at some time men would land on the planets, as they had once landed on the moon.

Appendixes
Bibliographic Essay
Source Notes
Index

Appendix A
Membership of Rocket and Satellite Research Panel

V-2 Upper Atmosphere Panel and V-2 Upper Atmosphere Research Panel until March 1948. From March 1948 until 29 April 1957, Upper Atmosphere Rocket Research Panel. After 29 April 1957, Rocket and Satellite Research Panel. The panel never disbanded; in 1960 it simply ceased operating.

Name and Original Institution	1946	47	48	49	50	51	52	53	54	55	56	57	58	59	60	61	Institution Early Years of NASA (1958–1960)*
Ernst Krause NRL	C	CR															Industry
C. F. Green G.E.		M	M	M	M	M	M	M	M	M	M	M	M	M	M	M	Industry
K. H. Kingdon G.E.	MR																Industry
G. K. Megerian G.E.	S		S	S	S	S	S	S	S	S	S	S	S	S	S	S	Industry/univ.
M. H. Nichols Princeton, U. Michigan		MR								M	M	M	M	M	M	M	University
F. L. Whipple Harvard		M	M	M	M	M	M	M	M	M	M	M	M	M	M	M	University
W. G. Dow U. Michigan		M	M	M	M	M	M	M	M	M	M	M	M	M	M	M	University
M. J. E. Golay SCEL	M	M	MR														Government
J. A. Van Allen APL	M	MC	C	C	C	C	C	C	C	C	C	C	M	M	M	M	University
M. D. O'Day AFCRL		M	M	M	M	M	M	M	M	M	M	M	M	M	M	M	Government
Newburn Smith NBS	M	M	MR														
H. E. Newell NRL		M	M	M	M	M	M	M	M	M	M	M	M	C	C	CM	NASA

Name / Organization	1	2	3	4	5	6	7	8	9	10	11	Status	Affiliation
W. H. Pickering — JPL	M	M	M	M	M	M	M	M	M	M	M	M	University
L. A. Delsasso — BRL	M	M	M	M	M	M	M	M	M	M		M	Government
M. J. Ference, Jr. — SCEL	M	M	M	M	M	MR	MR	M	M	M		M	Industry†
P. H. Wyckoff — AFCRL	M	M	M	M	M	M	M	M	M			M	Government
R. A. Weiss — SCEL				M	MR	M							Government
W. G. Stroud — SCEL	M	M	M	M	M	M						MC	NASA
L. M. Jones — U. Michigan	M	M	M	M	M	M	M					M	University
W. Berning — BRL	M	M	M	M	M	M	M					M	Government
J. W. Townsend, Jr. — NRL	M	M	M	M	M	M	M					M	NASA
K. Ehricke — Industry	M	M	M	M	M	M						M	Industry
M. Greenberg — AFCRL	M	M	M	M	M	M						M	Industry
W. W. Kellogg — UCLA	M	M	M	M	M	M						M	University
H. Strughold — Randolph AFB	M	M	M	M	M	M						M	Government
E. Stuhlinger — ABMA	M	M	M	M	M	M						M	NASA
W. von Braun — ABMA	M	M	M	M	M	M						M	NASA
N. Spencer — U. Michigan	M	M	M	M	M	M						M	NASA
J. Kaplan — UCLA	M	M	M	M	M	M						M	University
K. Stehling — NRL	M	M	M	M	M	M						M	NASA
H. J. Stewart — JPL	M	M	M	M	M	M	M					M	University
M. Zelikoff — AFCRC	M	M	M	M	M	M						M	Industry

415

Rocket and Satellite Research Panel—Continued

Name and Original Institution	1946	47	48	49	50	51	52	53	54	55	56	57	58	59	60	61	Institution Early Years of NASA
M. Rosen NRL													M	M	M	M	NASA
R. Tousey NRL													M	M	M	M	Government
H. Friedman NRL													M	M	M	M	Government
E. R. Manring AFCRC													M	M	M	M	Government
R. Slavin AFCRC													M	M	M	M	Government
V. Suomi U. Wisconsin													M	M	M	M	University
R. Porter G.E.													M	M	M	M	Industry
F. L. Bartman U. Michigan														M	M	M	University
S. Chapman G.I., U. Alaska														M	M	M	University
T. Chubb NRL														M	M	M	Government
J. Clark NASA														M	M	M	NASA
R. P. Haviland G.E.														M	M	M	Industry
R. Jastrow NASA														M	M	M	NASA
H. K. Kallman RAND														M	M	M	Industry

Name	Affiliation			Program
J. E. Kupperian	NASA	M	M	NASA
H. E. LaGow	NASA	M	M	NASA
L. H. Meridith	NASA	M	M	NASA
J. A. O'Keefe	NASA	M	M	NASA
G. Schilling	NASA	M	M	NASA
J. C. Seddon	NASA	M	M	NASA
M. Stoller	NASA	M	M	NASA
E. H. Vestine	RAND	M	M	Industry
D. D. Woodbridge	ABMA	M	M	NASA
W. Nordberg	NASA	M	M	NASA
M. Dubin	NASA	M	M	NASA
H. auf'm Kampe	SRDL	M	M	M

M = Member. C = Chairman. S = Secretary. R = Resigned.

ABMA = Army Ballistic Missile Agency.
AFB = Air Force Base.
AFCRC = Air Force Cambridge Research Center.
AFCRL = Air Force Cambridge Research Laboratories.
APL = Applied Physics Laboratory, Johns Hopkins University.
BRL = Ballistic Research Laboratory, Aberdeen Proving Ground.
G.E. = General Electric Company.
G.I. = Geophysical Institute, University of Alaska.

* In the space program unless otherwise indicated.

† Not in the space program.

JPL = Jet Propulsion Laboratory.
NBS = National Bureau of Standards.
NRL = Naval Research Laboratory.
SCEL = Signal Corps Engineering Laboratory.
SRDL = Signal Research and Development Laboratories.
UCLA = University of California at Los Angeles.

Appendix B
Typical Distribution of RSRP Reports

ROCKET AND SATELLITE RESEARCH PANEL MINUTES
(For 27 Feb. 1946 Meeting)

Distribution

(3*) Dr. W. G. Dow
Dept. of Electrical Engr.
University of Michigan
Ann Arbor, Michigan

(3) Director
Evans Signal Laboratory
Belmar, N.J.
Attn: Dr. M. J. E. Golay

(3) Dr. Charles F. Green
General Electric Co.
Aero. and Marine Engr. Div.
Schenectady, N.Y.

(3) Dr. K. H. Kingdon
Research Laboratory
General Electric Co.
Schenectady, N.Y.

(3) Major E. Kotcher
Pilotless Aircraft Br. TSESA-7
Air Technical Service Command
Wright Field
Dayton, Ohio

(3) Dr. E. H. Krause
Naval Research Laboratory
Washington, D.C.

Mr. G. K. Megerian
General Electric Co.
Aero. and Marine Engr. Div.
Schenectady, N.Y.

(3) Dr. M. H. Nichols
Palmer Physical Laboratory
Princeton University
Princeton, N.J.

(3) Dr. Calvin N. Warfield
NACA
Langley Field, Virginia

(3) Dr. J. A. Van Allen
Applied Physics Laboratory
Johns Hopkins University
Silver Spring, Maryland

(3) Dr. Fred L. Whipple
Harvard College Observatory
60 Garden Street
Cambridge, Mass.

(1) Commanding Officer
Signal Corps Engineering Labs.
Bradley Beach, N.J.
Attn: Lt. Col. H. A. Zahl

(2) Dr. R. W. Porter
General Electric Co.
Aero. and Marine Eng. Div.
Schenectady, N.Y.

(1) Dr. S. Ramo
General Electric Co. Rep.
Guggenheim Aeronautical Lab.
Cal. Inst. of Tech.
Pasadena, Cal.

(1) Mr. J. M. McAllister
c/o C. J. Thompson
White Sands Proving Grounds, N.M.

(6) Office of Research and Invention
Navy Department
Washington 25, D.C.

(3) Chief of Naval Operations
Navy Dept.,
Washington 25, D.C.
Attn: Code Op-06

---

*Number of copies.

418

(1) The Chief of the Bureau of Ships
     Navy Department
     Washington 25, D.C.
     Attn: Code 330

(5) The Chief of Bureau of Aeronautics
     Navy Department
     Washington 25, D.C.

     Also codes Aer-E-18
                 Aer-EL
                 Aer-EL-35
                 Aer-SI-50DAS

(5) The Chief of the Bureau of Ordnance
     Navy Department
     Washington 25, D.C.

     Also codes Re
                 Re-9

(3) Guided Missiles Committee of the
        Joint Committee for New Weapons
     Joint Chiefs of Staff
     1901 Constitution Avenue
     Washington 25, D.C.
     Attn: Dr. David Langmuir, Secy.

(6) Commanding General
     Army Service Forces
     Ordnance Department
     Rocket Development Division
     Pentagon Bldg.
     Washington 25, D.C.
     Attn: Col. J. G. Bain

(1) The Chief of Naval Operations
     Navy Department
     Washington 25, D.C.
     Attn: Code Op-413

(2) The Director
     Ballistics Laboratory
     California Institute of Technology
     Pasadena, Calif.
     Attn: Dr. Malina

(3) Commanding General, Army Air
        Forces
     Washington 25, D.C.
     Attn: Miss L. Diamond
     Office of Air Communications
        Officers

(3) Office of the Chief Signal Officer
     Army Service Forces
     Washington 25, D.C.

(2) Commanding General
     ATSC
     Wright Field, Dayton, Ohio
     Attn: Major N. R. Scott, TSELP

(1) Commanding Officer
     Research and Development Service
     ASF Sub Office (Rockets)
     California Institute of Technology
     Pasadena, Calif.

(2) National Bureau of Standards
     Connecticut Ave. and Van Ness St.
     Washington 8, D.C.

(1) The Weather Bureau
     M St., between 24th and 25th Sts.,
     Washington 7, D.C.

(1) Mass. Institute of Technology
     Cambridge, Mass.
     Attn: Dr. Zdenek Kopal

Appendix B

ROCKET AND SATELLITE RESEARCH PANEL MINUTES
(For 31 May 1956 Meeting)

(A)

Distribution

(1) Mr. W. W. Berning
Ballistics Research Lab.
Aberdeen Proving Ground
Maryland

(1) Dr. L. A. Delsasso
Ballistics Research Lab.
Aberdeen Proving Ground
Maryland

(2) Dr. W. G. Dow
University of Michigan
Dept. of Elec. Engineering
Ann Arbor, Michigan

(3) Dr. Charles F. Green
207 Highgate Rd.
Ithaca, N.Y.

(1) Mr. L. M. Jones
University of Michigan
Engineering Research Institute
Ann Arbor, Michigan

(7) Dr. H. E. Newell, Jr.
Naval Research Laboratory
Rocket Sonde Research Section
Code 7150
Washington, D.C.

(1) Commander
AF Cambridge Research Center
224 Albany St.
Cambridge, Mass.
Attn: CRHU: Dr. M. D. O'Day

(1) Dr. W. H. Pickering, Director
Jet Propulsion Laboratory
4800 Oak Grove Drive
Pasadena 3, California

(1) Mr. W. G. Stroud
Signal Corps Engr. Laboratory
Belmar, New Jersey

(1) Mr. J. W. Townsend
Naval Research Laboratory
Rocket Sonde Research Section
Code 7150
Washington, D.C.

(3) Dr. J. A. Van Allen
State University of Iowa
Dept. of Physics
Iowa City, Iowa

(2) Dr. Fred L. Whipple
Harvard College Observatory
60 Garden St.
Cambridge, Mass.

(1) Office, Chief of Ordnance
Research & Development Div.
Rocket Branch
Washington 25, D.C.
Attn: Col. C.W. Eifler, ORDTU

(2) Commander
AF Cambridge Research Center
224 Albany St.
Cambridge 39, Mass.
Attn: Mr. P. H. Wyckoff

(6) Office of Naval Research
Navy Department
Washington 25, D.C.

(3) Chief of Naval Operations
Navy Department
Washington 25, D.C.

(1) Chief of the Bureau of Ships
Navy Department
Washington 25, D.C.
Attn: Code 320

(1) Chief BuAer
Navy Department
Washington 25, D.C.
Attn: Aer-SI-5

(3) The Chief of the Bu. of Ord.
Navy Department
Washington 25, D.C.

Also Codes RE-9
RE

(B)

(1) Coordinating Committee on
   Guided Missiles
   Office, Assistant Secretary
   of Defense (R&D)
   Room 3E130 The Pentagon
   Washington 25, D.C.

(1) Coordinating Committee on
   General Sciences
   Office, Assistant Secretary
   of Defense (R&D)
   Room 3D137 The Pentagon
   Washington 25, D.C.

(6) Headquarters USAF
   Director of Research & Development
   Attn: Research Div. (AFMRS-4)
   Washington 25, D.C.

(3) Commander
   AF Cambridge Research Center
   224 Albany St.
   Cambridge 39, Mass.
   Attn: Dr. H. D. Edwards
         Mr. R. A. Minzner, CRHMS
         Mr. R. M. Slavin

(1) Commander
   AF Cambridge Research Center
   224 Albany St.
   Cambridge 39, Mass.
   Attn: CRHLP

(4) Commander
   Wright Air Development Center
   Attn: WCLDE
   Wright-Patterson AFB, Ohio

(1) Commander
   Air Research & Development
   Command
   Attn: RDTRG
   P.O. Box 1395
   Baltimore 3, Maryland

(1) Commanding Officer
   U.S. Naval Ordnance Missile
   Test Facility
   White Sands Proving Ground
   Las Cruces, New Mexico

(1) Prof. B. Gutenberg
   California Inst. of Tech.
   Pasadena, California

(1) NRL Project Officer
   U.S. Naval Ordnance Missile
   Test Facility
   White Sands Proving Ground
   Las Cruces, New Mexico

(1) Commanding General
   White Sands Proving Ground
   Las Cruces, New Mexico
   Attn: Technical Staff,
         Dr. W. H. Clohessy

(1) Chief of Naval Research
   Department of the Navy
   Washington 25, D.C.
   Attn: Code 463

(1) Office of Chief Signal Officer
   Engr. & Technical Division
   National Defense Bldg.
   Washington 25, D.C.
   Attn: SIGGG-M

(1) Mr. A. G. McNish
   National Bureau of Standards
   Room 302, The Manse
   Washington 25, D.C.

(1) Commander
   Air Weather Service
   MATS, USAF
   Attn: Directorate of
          Scientific Services
   Washington 25, D.C.

(1) Scientific Services Division
   U.S. Weather Bureau
   Washington 25, D.C.

(1) Commander
   6580th Test Squadron (Special)
   Holloman Air Dev. Center
   Holloman Air Force Base
   New Mexico

(1) Commander
   Wright Air Development Center
   Attn: WESM (AB Funderburg)
   Wright-Patterson AFB, Ohio

(1) Chief
   Army Field Forces
   Ft. Monroe, Virginia

(C)

(1) Office of Ordnance Research
    U.S. Army
    Box CM, Duke Station
    Durham, N.C.
    Attn: Dr. S. Githens
        Physical Sciences Div.

(1) Professor J. Kaplan
    University of California
    Dept. of Physics
    Los Angeles 24, California

(1) University of California
    Institute of Engr. Research
    205-T3
    Berkeley 4, California
    Attn: Mr. G. J. Maslach

(1) Mr. T. B. Walker
    Liquid Engine Division
    Aerojet General Corporation
    Azusa, California

(3) Dr. Smith J. DeFrance
    Director
    NACA, Ames Aeronautical Laboratory
    Moffett Field, California

(1) Dr. W. W. Kellogg
    The Rand Corporation
    1700 Main St.
    Santa Monica, California

(1) Radio Propagation Physics Div.
    National Bureau of Standards
    Boulder, Colorado
    Attn: Upper Atmosphere
        Research Section

(1) Mr. N. W. Spencer
    University of Michigan
    Dept. of Electrical Engr.
    Ann Arbor, Michigan

(1) Dr. H. A. Zahl
    Director of Research
    Signal Corps Engineering Lab.
    Fort Monmouth, New Jersey

(1) Dr. H. W. Batchelor
    Camp Detrick
    Frederick, Maryland

(4) Colonel H. C. Beaman
    AF Dev. Field Representative
    Applied Physics Laboratory
    Johns Hopkins University
    Silver Spring, Maryland

(1) Mr. H. E. Norton
    Building 1, Pr. Grd. Div.
    Army Medical Center
    Maryland

(1) The Commanding Officer
    U.S. Naval Proving Ground
    Dahlgren, Virginia

(2) Mr. W. J. O'Sullivan
    Pilotless Aircraft Res. Div.
    Langley Aeronautical Lab
    Langley Field, Virginia

(1) Director of Laboratory
    Watertown Arsenal
    Watertown, Mass.
    Attn: Col. P. Gillon

(1) Prof. S. Fred Singer
    Dept. of Physics
    University of Maryland
    College Park, Maryland

(3) British Washington Guided
        Missile Committee
    British Joint Services Mission
    1800 K St., N.W.
    Washington, D.C.
    Attn: Mr. G. L. Hutchinson, Secr.

(1) U.S. Marine Corps
    Guided Missile Liaison Officer
    Naval Ord. Missile Test Facility
    White Sands Proving Ground
    Las Cruces, New Mexico

(1) Mr. I. M. Levitt, Director
    The Fels Planetarium
    Franklin Institute
    Philadelphia, Pa.

(1) Dr. Wernher Von Braun
    Redstone Arsenal
    Huntsville, Alabama

(D)

(1) Mr. R. P. Haviland
General Electric Co.
Missile & Ordnance Systems Dept.
3198 Chestnut St.
Philadelphia, Pa.

(1) Mr. G. K. Megerian
General Electric Co.
Aircraft Accessory Turbine Dept.
950 Western Ave.
Lynn, Mass.

(1) Dr. S. Chapman
Geophysical Institute
College, Alaska

(1) Professor H. S. W. Massey
University College London
Gower Street WC1
London, England

(1) Dr. W. J. G. Beynon
Dept. of Physics
University College, Swansea
Great Britain

(1) Prof. F. A. Paneth
Max-Planck-Institut für Chemie
Mainz/Rhein
Saarstrasse 23
Western Germany

(1) Dr. M. Nicolet
Geophysical Year Secretariat
3 Avenue Circulaire
UCCLE, Belgium

(1) Professor E. Vassey
Physique de l'Atmosphère
1 Quai Branly
Paris, France

(1) Mr. J. B. Kendrick
Aerophysics Development Corp.
P. O. Box 257
Pacific Palisades, California

(1) Chief of Naval Operations
(OP-51)
Department of the Navy
Washington 25, D.C.

(2) Capt. J. M. Armstrong, R.A.N.
Defence Production and Supply
    Representative
Australian Embassy
Washington, D.C.

(2) Dr. H. Itokawa
Institute of Industrial Science
University of Tokyo
Chiba City, Japan

(2) Dr. D. C. Rose
National Research Council
100 Sussex Drive
Ottawa, Canada

(1) Dr. H. D. Edwards
Lockheed Aircraft Corp.
Engineering Division
Marietta, Georgia

(3) Col. Richard C. Gibson
DCS/Operations
Holloman Air Development Center
Holloman Air Force Base
New Mexico

Appendix C
Meetings of Rocket and Satellite Research Panel

Meeting	Date	Place	Remarks
0	16 Jan. 46	NRL	Preliminary, exploratory discussion.
1	27 Feb. 46	PU	Organizing meeting.
2	27 Mar. 46	NRL	
3	24 Apr. 46	NRL	Now called V-2 Upper Atmosphere Panel. Panel begins practice of hearing reports on firings and research results.
4	3 June 46	APL	
5	9 July 46	GE	
6	5 Sept. 46	WL	Now called V-2 Upper Atmosphere Research Panel.
7	4 Nov. 46	ESL	
8	28 Jan. 47	NRL	
9	25 Mar. 47	PF	JRDB requests long-range plans from panel.
10	7 May 47	APL	
11	3 July 47	WSPG	
12	1 Oct. 47	GE	Aerobee test firings have started. Panel promotes symposium on high-altitude physical research by rockets, to be held at American Physical Society meetings in Chicago, 29–31 Dec. 1947.
13	29 Dec. 47	Chi.	Krause resigns; Van Allen elected chairman. Office of Chief of Ordnance proposes panel consider broadening its scope.
14	28 Jan. 48	NRL	
15	18 Mar. 48	ERL	Name changed to Upper Atmosphere Rocket Research Panel.
16	28 Apr. 48	ESL	
17	16 June 48	APL	
18	29 Sept. 48	WSPG	
19	5 Jan. 49	CIT	
20	21 Apr. 49	UM	Van Allen has been using Aerobees fired from shipboard to extend geographic coverage of his research.
21	3 Aug. 49	HCO	Viking development is under way, and a first firing has been made.
22	26 Oct. 49	NRL	
23	14 Feb. 50	APL	

No.	Date	Loc.	Description
24	20 Apr. 50	NRL	Panel plans a coordinated set of high-altitude temperature experiments.
25	14 June 50	UC	Sydney Chapman, British scientist, attends.
26	8 Sept. 50	GE	Future research requirements and need for higher altitude vehicles considered.
27	31 Jan. 51	NRL	Panel begins discussions that lead to publication of panel paper on properties of upper atmosphere in *Physical Review*.
28	25 Apr. 51	NRL	
29	14, 15 Aug. 51	URI	Panel conducts seminar on properties of atmosphere at high altitudes. Panel has been giving extensive consideration to sounding rocket firings at other locations than White Sands; Fort Churchill, Canada, is one of the possibilities being considered.
30	24 Oct. 51	UCh	Panel plans to accept invitation from Gassiot Committee to join symposium on upper atmosphere at Oxford in August 1953.
31	8 Jan. 52	SUI	
32	30 Apr. 52	MN	Van Allen reports on use of balloon-launched rockets, called Rockoons. Panel is planning a symposium on rocket ionospheric studies.
33	7 Oct. 52	TN	
34	29, 30 Jan. 53	PAFB	Panel is going in depth into plans for coordinated northern latitude firings.
35	29 Apr. 53	TN	
36	7 Oct. 53	AFCRC	Panel reviews results of international symposium on upper atmospheric research held at Oxford the preceding August. Panel discusses participation in IGY.
37	4 Feb. 54	MN	Special Committee for the IGY (SCIGY) to work on Arctic firings is appointed.
38	29 Apr. 54	MN	Plans progressing for IGY rocket program.
39	8, 9 Sept. 54	NRL	Panel develops budget for IGY program.
40	3 Feb. 55	JPL	Panel votes to offer SCIGY to Technical Panel on Rocketry of National Academy of Sciences.
41	2 June 55	TN	Panel data on upper atmosphere has been used in preparing proposed extension to ICAO Standard Atmosphere used in aeronautical design work.
42	27 Oct. 55	BRL	Van Allen reports on Rockoon firings in auroral zone. Panel is planning symposium on scientific uses of earth satellites.
43	26, 27 Jan. 56	UM	Symposium on scientific uses of earth satellites.
44	31 May 56	P	Panel hears reports on Japanese, Australian, British, and French rocket programs. IGY satellite plans are discussed.
45	17 Dec. 56	NRC	Rocket firings are under way at Fort Churchill.
46	29 Apr. 57	NRL	Panel changes its name to Rocket and Satellite Research Panel.
47	19 Sept. 57	AFCRC	Committee on the Occupation of Outer Space formed.
48	13, 14 Nov. 57	UM	Meeting devoted to report of COS, and to discussion of future of RSRP.
49	6 Dec. 57	NAS	Meeting devoted to planning RSRP's promotion of a National Space Establishment. Panel has been enlarged—about double.
50	19 Dec. 57	UCh	Meeting devoted to planning RSRP's activity in support of National Space Establishment.
51	8 Jan. 58	NAS	Meeting devoted to the promotion of National Space Establishment.
52	14 Feb. 58	MN	Meeting hears reports of progress on promoting National Space Establishment.

Meetings of Rocket and Satellite Research Panel—Continued

Meeting	Date	Place	Remarks
53	2 Apr. 58	NAS	
54 (1959—1)	29 Jan. 59	NASA	Panel discusses its future role; decides on series of colloquia.
55 (1959—2)	10 Apr. 59	NASA	Colloquium on Van Allen Radiation Belt.
56 (1959—3)	15 June 59	NAS	Symposium on IGY rocket and Satellite results.
57 (1959—4)	6 Nov. 59	NAS	Colloquium on ionosphere.
58 (1960—1)	17 Feb. 60	UM	Colloquium on magnetic storms and their relation to rocket and satellite research. Panel adopts formal constitution.
59 (1960—2)	18, 19 May 60	SH	Review of panel firings and results.

PANEL SUSPENDS OPERATIONS

Meeting	Date	Place	Remarks
60 (1968—1)	2 Feb. 68	JPL	Primarily to renew acquaintances. Secretary proposes to turn over panel files, when he finishes with them, to National Air and Space Museum, Smithsonian Institution, for archiving.

AFCRC = Air Force Cambridge Research Center.
APL = Applied Physics Laboratory, Johns Hopkins University, Silver Spring, Maryland.
BRL = Ballistic Research Laboratories, Aberdeen Proving Ground, Aberdeen, Maryland.
Chi. = Chicago.
CIT = California Institute of Technology.
ERL = Electronics Research Laboratories, Watertown, Massachusetts.
ESL = Evans Signal Laboratory, New Jersey.
GE = General Electric Company, Schenectady, New York.
HCO = Harvard College Observatory, Cambridge, Massachusetts.
ICAO = International Civil Aviation Organization.
JPL = Jet Propulsion Laboratory, Pasadena, California.
MN = Main Navy Building, Washington, D.C.
NRL = Naval Research Laboratory, Washington, D.C.

PAFB = Patrick Air Force Base, Florida.
PF = Paterson Field, Dayton, Ohio.
PU = Princeton University.
SH = Statler-Hilton Hotel, Boston, Massachusetts.
SUI = State University of Iowa, Iowa City, Iowa.
TN = T-3 Navy Building, Washington, D.C.
UCh = University of Chicago.
UC = University of Colorado.
UM = University of Michigan.
URI = University of Rhode Island.
WL = Watson Laboratory, Cambridge, Massachusetts.
WSPG = White Sands Proving Ground, Las Cruces, New Mexico.

Appendix D
RSRP Proposals for a National Space Program

A NATIONAL MISSION TO EXPLORE OUTER SPACE
A Proposal by the Rocket and Satellite Research Panel

Mission

In the interest of human progress and our national welfare, it is proposed that a national project be established with the mission of carrying out the scientific exploration and eventual habitation of outer space. It is imperative that the nation do so to increase its scientific and technological strength.

The present state of rocket technology makes it possible to initiate such a project. First steps in this direction are already being taken in the form of rocket soundings, the development of long range missiles, and the launching of artificial earth satellites. These first steps will inevitably be followed by attempts to place man in space. It is only proper for this country to establish its leadership in this direction.

To carry out the objectives of the stated mission it is recommended that a National Space Establishment be created. This Establishment in carrying out its mission shall have the authority, responsibility, and accountability to conduct the theoretical, experimental, developmental and operational work necessary, making best use of the academic, industrial and military resources of the nation.

Significance and Nature of the Mission

The proposed mission is of the broadest significance to our people. Both the performance and the accomplishments will produce a beneficial impact on the life of the nation. The mission encompasses the most advanced aspects of research and engineering on the one hand and a great promise of practical benefit on the other. Some of the fields of research and engineering involved are: materials, propulsion, electronics and communications, meteorology, the life sciences, psychology, physics, chemistry, geo-physics, astronomy, astrophysics, astronautics, and cosmology. In fact, virtually all aspects of our technology and science will contribute to and profit from the effort required to carry out the mission.

The project will lead to both foreseeable and unpredictable applications. For example, weather patrol satellites providing a basis for vastly improved long and short range weather forecasting will be important to agriculture, commerce, industry, military operations, and the saving of life and property. Satellite radio relay stations will make possible continuous dependable global communications networks. It is to be expected that new and improved energy sources will be developed. One of the most important practical contributions will be the stimulation of science education in our country and throughout the free world.

Organization of the Mission

It is essential that the National Space Establishment be scientific in nature and in concept and be under civilian leadership and direction. It should be organized within the Executive branch of the Government taking full account of the requirements of the Department of Defense in the field of space research and engineering to insure that the National Space Establishment contributes its maximum to the national security. The Estab-

427

lishment should be staffed and operated on the basis of a salary and wage scale suitable to its needs.

The National Space Establishment must have within its own structure a strong, experienced staff with the necessary facilities for the research, development, and operations necessary to maintain competency in the full range of tasks to accomplish its mission. At the same time, the National Space Establishment must have substantial contractual assistance, drawing extensively upon universities, research organizations, industry, and upon the military for basic science, technological know how, production, logistic support, and facilities.

The cost of the enterprise will be comparable to the governmental expenditure in the field of atomic energy and should be funded on a long-term basis, not dependent upon direct military appropriations nor upon any one of the armed services.

Tasks under the Mission

The proposed mission for the National Space Establishment has two different but complementary phases. On the one hand there is space research, per se, which is concerned with such questions as the properties of the upper atmosphere, the nature and intensity of electromagnetic and corpuscular radiations from the sun, the character and distribution of matter in space, and the electric, magnetic, and gravitational fields within the solar system. On the other hand there is the exploration of space by manned expeditions. In this phase of the mission the Establishment will be concerned with the problems of placing man in space with adequate provisions for his survival and safe return. This will involve investigations into such areas as the creation and maintenance of viable atmospheres, protection of passengers against the stresses of high accelerations and radiations in space, psychological studies, biophysics research, and astronautics. Although the two phases have their different aspects, the successful and effective accomplishment of the total mission requires that they be inseparable and conducted in close contact with complete unity of purpose.

There are specific steps that the National Space Establishment should undertake immediately. There should be a strong re-enforcement of the upper-air rocket sounding program. Immediate attention should be given to a continuing program involving unmanned earth satellites. Planning and preparation should be initiated with regard to vehicles and other problems of manned rocket flights as a first step towards manned satellite stations. The development of instrumentation for physical and life-sciences experiments in rockets and satellites should be extended, making full use of what has already been accomplished. Studies of man himself relative to the projected expeditions into space should be strengthened. The studies of problems relative to lunar flight should be taken up by the National Space Establishment.

Concluding Remarks

The Rocket and Satellite Research Panel is absolutely convinced that there are compelling reasons for our nation to undertake the scientific exploration and habitation of outer space. Our past efforts have brought us to this new and challenging scientific frontier. A wholehearted and imaginative acceptance of this challenge will strengthen our national science, prestige, and defense. It will inevitably lead to a wide variety of practical benefits in commerce and industry. The National Space Establishment will unify the efforts and contributions of science, industry, and the military to space research, and will help to draw the youth of our country into science.

The magnitude of the venture will require a strong dedication of purpose on the part of our people. The country must provide the necessary resources and money to accomplish the mission. This means, among other things, an expenditure of some 10 billion dollars over the next decade.

The Rocket and Satellite Research Panel has devoted itself for the last ten years to pioneering the nation's effort in the research exploration of the threshold of space. The Panel is dedicated to continuing these activities and looks forward to participating in the actual accomplishment of the stated mission.

The Rocket and Satellite Research Panel

Berning, W. W.	Army Ballistics Research Lab.
Delsasso, L. A.	Army Ballistics Research Lab.
Dow, W. G.	University of Michigan
Ehricke, K.	Convair Corp.
Ference, M.	Ford Research Laboratory
Green, C. F.	General Electric Co.
Greenberg, M.	AF Cambridge Research Center
Jones, L. M.	University of Michigan
Kaplan, J.	University of California
Kellogg, W. W.	Rand Corp.
Newell, H. E.	Naval Research Laboratory
Nichols, M. H.	University of Michigan
O'Day, M. D.	AF Cambridge Research Center
Pickering, W. H.	Jet Propulsion Laboratory
Spencer, N. W.	University of Michigan
Stehling, K.	Naval Research Laboratory
Stewart, H. J.	Jet Propulsion Laboratory
Stroud, W. G.	Army Signal Engineering Lab.
Strughold, H.	Randolph AFB
Stuhlinger, E.	Army Ballistic Missile Agency
Townsend, J. W.	Naval Research Laboratory
Van Allen, J. A. Chairman	University of Iowa
Von Braun, W.	Army Ballistic Missile Agency
Whipple, F. L.	Smithsonian Astrophysical Obs.
Wyckoff, P. H.	AF Cambridge Research Center
Zelikoff, M.	AF Cambridge Research Center
Megerian, G. K. Secretary	General Electric Co.

November 21, 1957

27 December 1957

NATIONAL SPACE ESTABLISHMENT
A Proposal of the Rocket and Satellite Research Panel

Summary of Proposal

It is proposed that there be created a unified *National Space Establishment* for the purpose of carrying out the scientific exploration and eventual habitation of outer space.

It is imperative that the United States establish and maintain scientific and technological leadership in outer space research in the interests of long-term human progress and national survival.

1. Role

The role of the National Space Establishment shall be to unify and to greatly expand the national effort in outer space research, specifically excluding areas of immediate military urgency (e.g., the development, production and fielding of intercontinental and intermediate-range ballistic missiles).

2. Mission

The broad mission of the National Space Establishment shall be to establish United States leadership in space research by 1960 and to maintain it thereafter.

Accomplishment of this mission requires the following specific achievements:

(a) An intensified program of scientific soundings with high altitude rockets, immediately.
(b) An intensified program of scientific and technical developments with small instrumented satellites of the earth, immediately.
(c) Impact on the moon with non-survival of apparatus, by 1959.
(d) Placing an instrumented satellite in an orbit about the moon, by 1960.
(e) Impact on the moon with survival of scientific instruments, by 1960.
(f) Returnable, manned satellites in flight around the earth, by 1962.
(g) Manned circumnavigation of the moon with return to the earth, by 1965.
(h) Manned permanent satellite, by 1965.
(i) Manned expedition to the moon by one or two men, by 1968.
(j) Manned expedition to the moon by a sizeable party of men, by 1971.

A thorough analysis of existing capabilities shows that all of these objectives are within reach of a unified, vigorous national effort.

3. Funds Required

A detailed analysis shows that the accomplishment of the basic mission will require a national expenditure of ten billion dollars over the next decade.

4. Administrative Status of National Space Establishment

(a) It is strongly desirable that the N.S.E. be given statutory status as an independent agency in order that its work can be freely directed toward broad cultural, scientific and commercial objectives. Such objectives far transcend the short term, though vitally important, military rocket missions of the Department of Defense.
(b) If the proper creation of an independent agency is judged to require an intolerable delay, then it is believed that statutory existence under the Secretary of Defense (but

not within the jurisdiction of any one of the military services) will be a workable arrangement for the immediate future. But in this event, it is urged tht the "charter" of the agency explicitly provide for its independence as soon as its stature and achievements make this advisable.

(c) It is explicitly advised that the National Space Establishment not be placed within the jurisdiction of any one of the three military services. There are many reasons, growing out of extensive professional experience, for this view. The military services are basically operating agencies, not research ones. The research talent of any branch of the military services is almost inevitably turned toward helping meet short term, limited objectives. Such a point-of-view would assure the failure of a National Space Establishment in its broad mission—which is truly a national one, far beyond the mission of any one of the services or of the Department of Defense taken as a whole. During the early phases of space research, it is evident that existing facilities and missile technology of the Department of Defense can make enormous contributions. The National Space Establishment must be set up in such a way that it enjoys the unqualified support of all three services, and not merely one of them. Such a situation is believed to be possible only if the N.S.E. is an independent agency from the outset or if it is directly responsible only to the Secretary of Defense during its early years—with the clear prospect of independence at the earliest possible date.

(d) There must be clear channels for mutual cooperation between the proposed N.S.E. and all levels of the Department of Defense, in order to assure no jeopardy of short term, vital military need on the one hand and in order to assure maximum rate of advance of space research on the other.

5. Remarks on the Long Range Importance of Space Research

It is already clear that international leadership hinges, to a very great extent, on pre-eminence in scientific and technological matters.

Space research will contribute enormously to the educational, cultural, and intellectual character of the people of the United States and of the world. Indeed, the exploration and eventual habitation of outer space are the finest examples of the "Endless Frontier". It is for such bold endeavors that the highest motives of men should be invoked.

There will be a rich and continuing harvest of important practical applications as the work proceeds. Some of these can already be foreseen—reliable short term and long term meteorological forecasts, with all the agricultural and commercial advantages that these imply; rapid, long range radio communications of great capacity and reliability; aids to navigation and to long range surveying; television relays; new medical and biological knowledge, etc. And these will be only the beginning. Many of these applications will be of military value; but their greater value will be to the civilian community at large. (To use a homely example, the telephone is certainly a valuable military device, but its importance to the civilian population is vastly greater.)

6. Availability of the Rocket and Satellite Research Panel for Consultation and Participation

The Rocket and Satellite Research Panel comprises a broad membership of persons of extensive experience in all aspects of the proposed program of outer space research. Its members are professionally dedicated to national leadership in this field. They offer their services, individually and collectively, in the conduct of the broad mission of the National Space Establishment.

The Rocket and Satellite Research Panel

Berning, W. W.	Army Ballistics Research Laboratory
Delsasso, L. A.	Army Ballistics Research Laboratory

Dow, W. G.	University of Michigan
Ehricke, K.	Convair Corporation
Ference, M.	Ford Research Laboratory
Green, C. F.	General Electric Company
Greenberg, M.	Air Force Cambridge Research Center
Jones, L. M.	University of Michigan
Kaplan, J.	University of California
Kellogg, W. W.	Rand Corporation
Newell, H. E.	Naval Research Laboratory
Nichols, M. H.	University of Michigan
O'Day, M. D.	Air Force Cambridge Research Center
Pickering, W. H.	Jet Propulsion Laboratory
Spencer, N. W.	University of Michigan
Stehling, K.	Naval Research Laboratory
Stewart, H. J.	Jet Propulsion Laboratory
Stroud, W. G.	Army Signal Engineering Laboratory
Strughold, H.	Randolph Air Force Base
Stuhlinger, E.	Army Ballistic Missile Agency
Townsend, J. W.	Naval Research Laboratory
Van Allen, J. A. Chairman	University of Iowa
Von Braun, W.	Army Ballistic Missile Agency
Whipple, F. L.	Smithsonian Astrophysical Obs.
Wyckoff, P. H.	Air Force Cambridge Research Center
Zelikoff, M.	Air Force Cambridge Research Center
Megerian, G. K. Secretary	General Electric Company

Appendix E
Original Membership of United States National Committee for the International Geophysical Year

Joseph Kaplan	University of California, Chairman
A. H. Shapley	National Bureau of Standards, Vice Chairman
N. C. Gerson	Air Force Cambridge Research Center, Recording Secretary
L. H. Adams	Carnegie Institution of Washington
H. G. Booker	Cornell University
Lyman J. Briggs	National Geographic Society
G. M. Clemence	U.S. Naval Observatory
John A. Fleming	American Geophysical Union
Lawrence M. Gould	Carleton College
F. W. Reichelderfer	U.S. Weather Bureau
E. B. Roberts	U.S. Coast and Geodetic Survey
Paul A. Siple	Department of the Army
A. F. Spilhaus	University of Minnesota
Merle A. Tuve	Carnegie Institution of Washington
A. L. Washburn	Dartmouth College

Ex Officio

Wallace W. Atwood, Jr.	National Academy of Sciences
Lloyd V. Berkner	Associated Universities, Inc.
William G. Rubey	National Research Council
Walter Rudolph	Science Adviser's Office, Dept. of State
K. Stephenson	National Science Foundation, liaison
Hugh Odishaw	Administrative Secretary

Appendix F
Membership of the Space Science Board, 1958–1972

Member	1950s		1960s										1970s		
	8	9	0	1	2	3	4	5	6	7	8	9	0	1	2
Lloyd V. Berkner Associated Universities	C	C	C	C	M	M	M								
Harrison S. Brown Cal. Inst. Tech.	M	M	M	M	M	M									
Leo Goldberg U. Michigan	M	M	M	M	M	M									
H. Keffer Hartline Rockefeller Inst.	M	M	M												
Donald F. Hornig Princeton U.	M	M	M												
W. A. Noyes U. Rochester	M														
Richard W. Porter Gen. Electric	M	M	M	M	M	M	M	M	M	M	M	M			
Bruno B. Rossi Mass. Inst. Tech.	M	M	M	M	M	M									
Alan H. Shapley Nat. Bur. Standards	M	M	M												
John A. Simpson U. Chicago	M	M	M	M	M	M	M	M							
S. S. Stevens Harvard U.	M	M													
Harold C. Urey U. Cal., La Jolla	M	M	M												
James A. Van Allen State U. Iowa	M	M	M	M	M	M	M	M	M	M	M	M			
O. G. Villard, Jr. Stanford U.	M	M	M												
Harry Wexler U.S. Weather Bur.	M	M	M												

Membership of the Space Science Board, 1958–1972 (Continued)

Member	1950s		1960s										1970s		
	8	9	0	1	2	3	4	5	6	7	8	9	0	1	2
George P. Wollard* U. Wisconsin	M	M	M	M	M	M	M	M	M	M	M				
Joshua Lederberg Stanford U.		M	M	M											
William W. Kellogg U. Cal., Los Angeles			M	M	M	M	M								
Christian J. Lambertson			M	M											
Colin S. Pittendrigh Princeton U.			M	M		M		M	M						
John Findlay Nat. Radio Astronomy Observatory				M	M	M	M	M	M	M	M	M	M		
Herbert Friedman Nav. Res. Lab.				M	M	M	M	M	M	M	M	M	M	M	M
Gordon J. F. MacDonald U. Cal., Los Angeles				M	M	M	M	M	M	M	M	M			
H. H. Hess Princeton U.						C	C	C	C	C	C	C			
Allan H. Brown U. Pennsylvania					M	M	M	M	M	M	M	M			
Martin A. Pomerantz Bartol Research Foundation, Franklin Inst.					M_c	M_c	M_c	M_c	M_c	M_c	M_c				
Nicholas U. Mayall Kitt Peak Nat. Observatory							M	M	M	M	M	M			
Luis W. Alvarez U. Cal., Berkeley								M	M	M	M	M			
Loren D. Carlson U. Cal.								M	M	M	M	M			
Courtland D. Perkins Princeton U.								M	M	M	M	M			
Leonard I. Schiff Stanford U.								M	M	M	M	M			
John S. Hall Lowell Observatory										M	M	M	M		
Francis S. Johnson U. Texas at Dallas										M	M	M	M	M	M

Membership of the Space Science Board, 1958—1972 (Continued)

Member	1950s		1960s										1970s		
	8	9	0	1	2	3	4	5	6	7	8	9	0	1	2
William M. Kaula U. Cal., Los Angeles										M	M	M	M		
Donald U. Wise Franklin & Marshall Coll.								M	M						
Donald B. Lindsley U. Cal.										M	M				
William W. Rubey U. Cal., Berkeley												M	M	M	
Roman Smoluchowski Princeton U.												M	M	M	M
Wolf Vishniac U. Rochester												M	M	M	
Charles H. Townes Mass. Inst. Tech., then U. Cal., Berkeley													C	C	C
James R. Arnold U. Cal.													M	M	M
Harry Eagle Albert Einstein College of Medicine													M	M	M
William A. Fowler Cal. Inst. Tech.													M	M	M
Richard M. Goody Harvard U.													M	M	M
Brian O'Brien													M	M	M
Raymond L. Bisplinghoff Mass. Inst. Tech.													M_c		
E. Margaret Burbidge U. Cal.														M	M
Norman H. Horowitz Cal. Inst. Tech.														M	M
Edward H. Kass Harvard U.														M	M
Philip Morrison Mass. Inst. Tech.														M	M
Willis M. Hawkins Lockheed Aircraft														M_c	M_c

Membership of the Space Science Board, 1958—1972 (Continued)

Member	1950s		1960s										1970s		
	8	9	0	1	2	3	4	5	6	7	8	9	0	1	2
Robert E. Danielson Princeton U.															M
Joshua Menkes U. Colorado															M
Robert A. Phinney Princeton U.															M

C   = Chairman
M   = Member
M_e = Member, ex officio
      *Added after first meeting of board.

Appendix G
Reports from Space Science Board Summer Studies

Representative Reports, 1962–1976

1962   *A Review of Space Research*

1965   *Space Research: Directions for the Future*

1968   *Planetary Exploration: 1968–1975*

1969   *The Outer Solar System: A Program for Exploration*

1969   *Lunar Exploration: A Strategy for Research 1969–1975*

1970   *Venus Strategy for Exploration*

1970   *Priorities for Space Research 1971–1980*

1971   *Outer Planets Exploration 1972–1985*

1973   *Scientific Uses of the Space Shuttle*

1974   "Future Exploration of Mars" in *Opportunities and Choices in Space Science, 1974*

1975   "Infrared and Submillimeter Astronomy" in *Report on Space Science, 1975*

1975   "Solar Physics" in *Report on Space Science, 1975*

1976   *Institutional Arrangements for the Space Telescope*

Appendix H
Advisers Attending NASA's
First University Program Conference
14 August 1961

(Nongovernmental Participants)

Preston Bassett, president (retired), Sperry Corporation
Francis X. Bradley, Jr., research administrator, Notre Dame
George E. Corcoran, Dept. of Electrical Engineering, University of Maryland
Henry Eyring, dean of Graduate School, University of Utah
L. E. Grinter, dean of Graduate School, University of Florida
James Harlow, dean of College of Education, University of Oklahoma
John C. Honey, executive associate, Carnegie Corporation of New York
Ralph Knutti, associate director for extramural programs, National Institute of Arthritis
    and Metabolic Diseases, National Institutes of Health
Charles B. Officer, president, Marine Geophysics Services Corporation, Houston, Texas
Richard W. Poole, Dept. of Economics, Oklahoma State University
Alex S. Pow, director for Contract and Grant Development, University of Alabama
Guenter Schwarz, Dept. of Physics, Florida State University
Martin Schwarzschild, professor of Astronomy, Princeton University
John Simpson, professor of Physics, Enrico Fermi Institute for Nuclear Studies, University
    of Chicago
Stafford L. Warren, dean of Medical School, University of California, L.A.
Marsh W. White, Dept. of Physics, Pennsylvania State University
Emery Wine, Municipal Manpower Commission

Appendix I
Meeting of Physicists at Airlie House, Warrenton, Virginia, 20, 21 June 1963

(As a result of this meeting the attendees agreed to serve as NASA's first Physics Committee.)

University Participants

Dr. Luis W. Alvarez, University of California, Berkeley
*Dr. Robert H. Dicke, Princeton University
Professor Freeman J. Dyson, Princeton University
Dr. William M. Fairbank, Stanford University
*Dr. William A. Fowler, California Institute of Technology
Dr. Murray Gell-Mann, California Institute of Technology
Dr. Robert Hofstadter, Stanford University
Dr. Polykarp Kusch, Columbia University
Dr. Willis E. Lamb, Jr., Yale University
Dr. Robert B. Leighton, California Institute of Technology
Dr. Willard F. Libby, University of California, Los Angeles
*Dr. Gordon J. F. MacDonald, University of California, Los Angeles
Dr. Eugene Parker, University of Chicago
Dr. Norman F. Ramsey, Jr., Harvard University
*Dr. Bruno B. Rossi, Massachusetts Institute of Technology
Dr. Emelio Gino Segre, University of California, Berkeley
*Dr. Lyman Spitzer, Jr., Princeton University
Dr. James A. Van Allen, State University of Iowa

---

*Speakers

NASA Participants

Mr. James Webb, Headquarters
Dr. Hugh L. Dryden, Headquarters
Dr. Homer Newell, Headquarters
Dr. John Clark, Headquarters
Dr. Robert Jastrow, Goddard Institute of Space Studies, N.Y.
Dr. Urner Liddel, Headquarters
Dr. Harry Goett, Goddard Space Flight Center

Appendix J
Statement by President Nixon on the Space Program
7 March 1970

Released from the Office of the White House
Press Secretary, Key Biscayne, Florida

Over the last decade, the principal goal of our nation's space program has been the Moon. By the end of that decade men from our planet had traveled to the Moon on four occasions and twice they had walked on its surface. With these unforgettable experiences, we have gained a new perspective of ourselves and our world.

I believe these accomplishments should help us gain a new perspective of our space program as well. Having completed that long stride into the future which has been our objective for the past decade, we must now define new goals which make sense for the Seventies. We must build on the successes of the past, always reaching out for new achievements. But we must also recognize that many critical problems here on this planet make high priority demands on our attention and our resources. By no means should we allow our space program to stagnate. But—with the entire future and the entire universe before us—we should not try to do everything at once. Our approach to space must continue to be bold—but it must also be balanced.

When this Administration came into office, there were no clear, comprehensive plans for our space program after the first Apollo landing. To help remedy this situation, I established in February of 1969 a Space Task Group, headed by the Vice President, to study possibilities for the future of that program. Their report was presented to me in September. After reviewing that report and considering our national priorities, I have reached a number of conclusions concerning the future pace and direction of the nation's space efforts. The budget recommendations which I have sent to the Congress for Fiscal Year 1971 are based on these conclusions.

Three General Purposes

In my judgment, three general purposes should guide our space program.

One purpose is exploration. From time immemorial, man has insisted on venturing into the unknown despite his inability to predict precisely the value of any given exploration. He has been willing to take risks, willing to be surprised, willing to adapt to new experiences. Man has come to feel that such quests are worthwhile in and of themselves—for they represent one way in which he expands his vision and expresses the human spirit. A great nation must always be an exploring nation if it wishes to remain great.

A second purpose of our space program is scientific knowledge—a greater systematic understanding about ourselves and our universe. With each of our space ventures, man's total information about nature has been dramatically expanded; the human race was able to learn more about the Moon and Mars in a few hours last summer than had been learned in all the centuries that had gone before. The people who perform this important work are not only those who walk in spacesuits while millions watch or those who launch powerful rockets in a burst of flame. Much of our scientific progress comes in laboratories and offices, where dedicated, inquiring men and women decipher new facts and add them to old ones in ways which reveal new truths. The abilities of these scientists constitute one of our

441

most valuable national resources. I believe that our space program should help these people in their work and should be attentive to their suggestions.

A third purpose of the United States space effort is that of practical application—turning the lessons we learn in space to the early benefit of life on Earth. Examples of such lessons are manifold; they range from new medical insights to new methods of communication, from better weather forecasts to new management techniques and new ways of providing energy. But these lessons will not apply themselves; we must make a concerted effort to see that the results of our space research are used to the maximum advantage of the human community.

A Continuing Process

We must see our space effort, then, not only as an adventure of today but also as an investment in tomorrow. We did not go to the Moon merely for the sport of it. To be sure, those undertakings have provided an exciting adventure for all mankind and we are proud that it was our nation that met this challenge. But the most important thing about man's first footsteps on the Moon is what they promise for the future.

We must realize that space activities will be a part of our lives for the rest of time. We must think of them as part of a continuing process—one which will go on day in and day out, year in and year out—and not as a series of separate leaps, each requiring a massive concentration of energy and will and accomplished on a crash timetable. Our space program should not be planned in a rigid manner, decade by decade, but on a continuing flexible basis, one which takes into account our changing needs and our expanding knowledge.

We must also realize that space expenditures must take their proper place within a rigorous system of national priorities. What we do in space from here on in must become a normal and regular part of our national life and must therefore be planned in conjunction with all of the other undertakings which are also important to us. The space budget which I have sent to Congress for Fiscal Year 1971 is lower than the budget for Fiscal Year 1970, a condition which reflects the fiscal constraints under which we presently operate and the competing demands of other programs. I am confident, however, that the funding I have proposed will allow our space program to make steady and impressive progress.

Six Specific Objectives

With these general considerations in mind, I have concluded that our space program should work toward the following specific objectives:

1. We should continue to *explore the Moon.* Future Apollo manned lunar landings will be spaced so as to maximize our scientific return from each mission, always providing, of course, for the safety of those who undertake these ventures. Our decisions about manned and unmanned lunar voyages beyond the Apollo program will be based on the results of these missions.

2. We should move ahead with bold exploration of the planets and the universe. In the next few years, scientific satellites of many types will be launched into Earth orbit to bring us new information about the universe, the solar system, and even our own planet. During the next decade, we will also launch unmanned spacecraft to all the planets of our solar system, including an unmanned vehicle which will be sent to land on Mars and to investigate its surface. In the late 1970s, the "Grand Tour" missions will study the mysterious outer planets of the solar system—Jupiter, Saturn, Uranus, Neptune, and Pluto. The positions of the planets at that time will give us a unique opportunity to launch missions which can visit several of them on a single flight of over three billion miles. Preparations for this program will begin in 1972.

There is one major but longer range goal we should keep in mind as we proceed with our exploration of the planets. As a part of this program we will eventually send men to explore the planet Mars.

3. We should work to *reduce substantially the cost of space operations*. Our present rocket technology will provide a reliable launch capability for some time. But as we build for the longer-range future, we must devise less costly and less complicated ways of transporting payloads into space. Such a capability—designed so that it will be suitable for a wide range of scientific, defense and commercial uses—can help us realize important economies in all aspects of our space program. We are currently examining in greater detail the feasibility of re-usable space shuttles as one way of achieving this objective.

4. We should seek to *extend man's capability to live and work in space*. The Experimental Space Station (XSS)—a large orbiting workship—will be an important part of this effort. We are now building such a station—using systems originally developed for the Apollo program—and plan to begin using it for operational missions in the next few years. We expect that men will be working in space for months at a time during the coming decade.

We have much to learn about what man can and cannot do in space. On the basis of our experience with the XSS, we will decide when and how to develop longer-lived space stations. Flexible, long-lived space station modules could provide a multi-purpose space platform for the longer-range future and ultimately become a building block for manned interplanetary travel.

5. We should *hasten and expand the practical applications of space technology*. The development of earth resources satellites—platforms which can help in such varied tasks as surveying crops, locating mineral deposits and measuring water resources—will enable us to assess our environment and use our resources more effectively. We should continue to pursue other applications of space-related technology in a wide variety of fields, including meteorology, communications, navigation, air traffic control, education and national defense. The very act of reaching into space can help man improve the quality of life on Earth.

6. We should *encourage greater international cooperation in space*. In my address to the United Nations last September, I indicated that the United States will take positive, concrete steps "toward internationalizing man's epic venture into space—an adventure that belongs not to one nation but to all mankind." I believe that both the adventures and the applications of space missions should be shared by all peoples. Our progress will be faster and our accomplishments will be greater if nations will join together in this effort, both in contributing the resources and in enjoying the benefits. Unmanned scientific payloads from other nations already make use of our space launch capability on a cost-shared basis; we look forward to the day when these arrangements can be extended to larger applications satellites and astronaut crews. The Administrator of NASA recently met with the space authorities of Western Europe, Canada, Japan and Australia in an effort to find ways in which we can cooperate more effectively in space.

* * *

It is important, I believe, that the space program of the United States meet these six objectives. A program which achieves these goals will be a balanced space program, one which will extend our capabilities and knowledge and one which will put our new learning to work for the immediate benefit of all people.

As we enter a new decade, we are conscious of the fact that man is also entering a new historic era. For the first time, he has reached beyond his planet; for the rest of time, we will think of ourselves as men *from* the planet Earth. It is my hope that as we go forward with our space program, we can plan and work in a way which makes us proud *both* of the planet from which we come *and* of our ability to travel beyond it.

Appendix K
Glossary

This glossary contains terms used in the narrative that may not be familiar to some readers. For terms not in the glossary, the reader may find the following helpful: William H. Allen, ed., *Dictionary of Technical Terms for Aerospace Use,* NASA SP-7 (Washington, 1965).

absorption coefficient. A numerical quantity that indicates the relative effectiveness of a material substance such as a gas in absorbing electromagnetic radiation.

airglow. The quasi-steady radiant emission from the upper atmosphere as distinguished from the sporadic emission of the auroras. See discussion p. 66–67.

alpha particle. The nucleus of the helium atom. The alpha particle has a double positive charge.

anisotropic. Varying with direction. Radiation that varies in intensity or nature with direction is said to be *anisotropic.*

anomalous propagation of sound. Propagation of sound waves that takes appreciably longer than the expected time to reach an observer, presumably because the sound did not follow a direct path from the origin to the observer.

applied science. Scientific research that is intended to furnish information and data that will aid in achieving specific practical applications or in developing a desired technology.

apogee. The point in the orbit of a satellite at which the satellite is at its greatest distance from the center of the earth.

artificial satellite. A man-made satellite of the earth, moon, or planet, as contrasted with a natural satellite like the earth's moon.

atmosphere of pressure. Roughly the pressure of the earth's atmosphere at sea level. By definition, exactly 1.01325×10^5 newtons per square meter.

aurora. The northern and southern lights—aurora borealis and aurora australis—which are faint radiations that at times are seen at high latitudes illuminating the night sky. The aurora becomes especially pronounced at times of high solar and magnetic activity.

bands in a spectrum. Emission bands of molecules, combinations of two or more atoms radiating in characteristic groups of lines. In contrast, radiating atoms emit discrete wavelengths, called emission lines of the atom.

basic science. The effort to define basic science almost invariably comes to grief. The first question to decide is *basic to what?* If basic to some ultimate applications, then pure, applied, and mission-oriented science can all come under the heading of basic. But, if basic to science itself is the intended meaning, then it becomes a matter of what the scientist himself perceives to be most fundamental, and indeed, the phrase *fundamental science* is often used to convey this flavor.

black body. A body that absorbs all wavelengths of electromagnetic radiation. Conversely, when heated a black body emits in all wavelengths.

chromosphere of the sun. A thin layer of relatively transparent gases above the photosphere of the sun. See p. 364.

collision frequency. The number of times per second a gas particle collides with other particles of gas.

control (of a vehicle). A means of orienting, steering, or otherwise modifying the movement of a vehicle.

core of the earth. The central region of the earth, extending to about 3400 kilometers from the center. The core is very dense, consisting primarily of iron and iron sulfide. Its outer portions are molten.

corona of the sun. The outer visible envelope of the sun, lying above the chromosphere. See p. 364–66.

cosmic rays. Streams of high-energy subatomic particles that travel the solar system and bombard the earth from all directions. See pp. 67–69.

countdown. A step-by-step process leading to a climactic event such as the firing of a space launch vehicle.

crust of the earth. The outermost layer of the earth, consisting of the continents and the floor of the ocean basins. Beneath the crust is a mantle surrounding a dense core. While the central part of the core appears to be solid, its outermost portion underlying the mantle is molten.

diffusive separation. Separation into individual component parts as in an isothermal mixture of gases in a gravitational field, where the lighter parts come to predominate at the top and the heavier at the bottom—as cream separates from milk.

dipole. A system composed of two separated, equal, electric or magnetic charges of opposite sign. **Dipole field:** see chap. 6, pp. 65–66 and fig. 3.

discipline. A scientific discipline is an area of investigation in which the investigators share a common paradigm or group of paradigms, embracing a common body of theory, and techniques and often instrumentation. See discussion p. 12.

dissociation of a molecule. Separation of a molecule into component parts. If a molecule absorbs enough energy—from heating, irradiation, or an electrical discharge, for example—it may split into component parts. The reverse process, in which the individual parts join to reproduce the original molecule, is called *recombination.*

doppler shift. Changes in wavelength caused by the motion of a radiating source or of the receiver are called *doppler shifts* in the radiated wavelengths. When a radiating source, emitting either sound or electromagnetic waves, moves either toward or away from an observer, the motion affects the wavelengths as seen by the observer. If the motion of the source is toward the observer, the wavelengths are shortened so that sound is increased in pitch and light shifted toward the blue end of the spectrum. Motion away from the observer decreases the pitch of sound and shifts electromagnetic radiations toward the red.

dynamical geodesy. The study of the gravitational field of the earth and its relationship to the solid structure of the planet. Also called *physical geodesy.* See p. 186.

electron. A fundamental particle of matter carrying a single negative electric charge and having a mass 1/1840 that of a proton.

electron volt. The energy equivalent to that acquired by an electron in falling through an electric potential of 1 volt. It is equal to 1.60210×10^{-19} joule.

excitation. An atom or molecule is said to be excited when it has absorbed sufficient energy to raise it above the normal or ground-level state, but not enough to ionize it.

exobiology. The search for and study of extraterrestrial life. See pp. 274, 352–53.

exponential atmosphere. An atmosphere in which an increase in altitude by a fixed height H always decreases the pressure and density by a constant factor. See p. 60.

forbidden line. An emission line corresponding to a less probable wavelength than any given by the usual selection rules. In emitting light a radiating atom obeys the laws of quantum theory, in which certain rules, called *selection rules*, give the wavelengths an atom is most likely to emit. Under certain conditions, however, an atom may emit a less likely wavelength.

galaxy. The self-contained aggregate of stars, nebulas, gases, and dust of which the sun and its planets are members. The galaxy is one of billions of such systems, also called galaxies, which collectively compose the metagalaxy.

gamma. Magnetic flux density of 10^{-5} gauss or 10^{-9} tesla.

geodesy. The science that deals mathematically with the size and shape of the earth, the earth's external gravity field, and surveys so precise that size and shape of the earth must be taken into consideration. See pp. 186–96.

geoid. The figure of the earth as defined by the level surface that over the oceans coincides with mean sea level. See pp. 189–96.

geomagnetic equator. The great circle of the earth lying midway between the north and south poles of the earth's dipole magnetic field.

geomagnetic latitude and longitude. Analogous to geographic latitude and longitude, but referred to the dipole magnetic poles and geomagnetic equator instead of to the geographic poles and equator.

geometrical geodesy. Study, by geometrical and astronomical measurements, of the precise size and shape of the earth and accurate location of points on the surface of the earth. See p. 186.

geopotential function. That function which at each point is equal to an arbitrary constant minus the energy that would be acquired by a unit mass in falling from rest at infinity to the point in question. The gradient of the geopotential function, which can be obtained by vector calculus, yields the earth's gravitational field.

geosyncline. Layers of rocks that have been tilted from their original horizontal stratification to form a huge basin which then fills with sediments.

gradient. The rate of change of a quantity with distance in a specified direction is the *gradient* of the quantity in that direction. When the term gradient is used without specifying the direction, it is taken to mean the rate of change of the quantity in question in the direction of greatest rate of change.

guidance. The process of directing the movements of an aircraft, spacecraft, missile, or other vehicle. In general such vehicles are equipped so that they can be *controlled* to follow the guidance supplied to them.

hydrated compound. A compound in which the molecules contain water (H_2O) or hydroxyl radicals and protons (OH and H).

integration. A mathematical process used in the calculus for deriving a function from its slope. The reverse process, which yields the slope of a given function, is differentiation.

International Geophysical Year (IGY). An internationally agreed on period, July 1957 through December 1958, during which observation of worldwide geophysical phenomena was greatly increased by cooperative effort of participating nations. Activities were continued through December 1959 as International Geophysical Cooperation. See pp. 50–51.

ion. A charged atom or molecularly bound group of atoms; sometimes also a free electron or other charged subatomic particle. In the normal state atoms and molecules are electrically neutral. An atom or molecule that acquires one or more electrons becomes negatively charged and is called a *negative ion*. If the particle loses one or more electrons it becomes positively charged and is called a *positive ion*.

ionization gauge. A gauge in which the effect of an ambient gas on the electric current flow from a hot filament is used to measure a property of the gas, such as charge density.

ionosphere. Upper levels of the earth's atmosphere, extending outward from about 70 kilometers altitude and containing free electronically charged particles that reflect radio waves. See pp. 64–65.

isothermal. Of constant temperature.

isotropic radiation. Radiation that is of equal intensity in all directions.

law of sines. In trigonometry, the law which states that in a triangle the lengths of the sides are in proportion to the sines of the opposite angles. Thus, if a, b, c are the three sides of a triangle, and α, β, γ are the angles opposite to a, b, and c, respectively, then

$$a/\sin \alpha = b/\sin \beta = c/\sin \gamma .$$

level surface. A surface to which the force of gravity is everywhere perpendicular. See pp. 190–92.

liftoff. The rising of a space launch vehicle from its launching stand immediately after firing.

light-year. The distance light travels in one year, equal to 9.46055×10^{15} meters.

line in a spectrum. Light emitted or absorbed at a discrete wavelength by a radiating or absorbing atom.

Lyman alpha line. The hydrogen atom is capable of emitting (or absorbing) electromagnetic radiation in several different series of lines. One of these is known as the Lyman series. The longest wavelength of the Lyman series is at 1216 Å in the ultraviolet region, and is known as *Lyman alpha*. The next longest wavelength of the series is Lyman beta, then Lyman gamma, etc.

magnetic storm. A disturbance in the earth's magnetic field assumed to be caused by streams of particles and magnetic fields from the sun.

magnetosphere. The region of space surrounding the earth where the magnetic field plays a prominent, often controlling, role relative to particle radiations found there. See pp. 173–84.

main stage. That stage of a launch vehicle—usually the largest stage—that is used to lift the launch vehicle and its payload to a high enough altitude that the remaining stages of the vehicle can project the payload into orbit or into a space trajectory.

man-machine relations. Relations between man and a machine—such as an aircraft or spacecraft. The matching of human characteristics to a machine so as to obtain maximum efficiency or optimum conditions for operation is called *man-machine integration*. The integrated combination of man and machine in an operating unit is called a *man-machine system*.

mantle of the earth. The interior of the earth between about 3400 kilometers—the outer boundary of the core—and the crust.

mare (pl. maria). The maria are the dark regions of the moon, once thought to be seas, hence the term *mare* from the Latin for sea. The term is also applied to dark regions on Mars.

mean free path. The average distance a gas particle travels between successive collisions with other particles.

meson. An elementary particle of mass intermediate between the masses of the electron and proton.

mesosphere. Sometimes used to denote the middle atmosphere, between the stratosphere and the ionosphere.

metagalaxy. The system of all galaxies. The physical universe.

meteor. A fast moving mass from space traversing the atmosphere.

meteorite. A meteor that survived passage through the atmosphere to strike the ground.

micrometeor; micrometeorite. A meteor or meteorite that is a few hundreds of micrometers or less in diameter.

mission-oriented science. Applied science carried out in support of a specifically stated mission, such as providing the necessary weaponry for naval warfare, maintaining adequate air power, improving agriculture, developing natural resources, or predicting the weather.

monochromatic beam. A beam of radiation of a single wavelength.

nebula. A huge cloud of gas and dust in space.

neutron. An electrically neutral particle of mass essentially equal to that of the proton. A free neutron is unstable, decaying (with a half life of about 12 minutes) to an electron and a proton. (To say that the half life of a neutron is 12 minutes means that of a group of N neutrons, N/2 of them will decay before 12 minutes, and N/2 of them after 12 minutes.)

oblate ellipsoid of revolution. The surface (or solid) obtained by revolving an ellipse about its minor axis. Revolving the ellipse about its major axis produces a *prolate* ellipsoid of revolution.

447

orbit. The closed path of a moving body or particle around another object. See also *trajectory.*

ozone. The gas whose molecules consist of three oxygen atoms combined: O_3.

ozonosphere. The region of the atmosphere in the upper stratosphere and lower mesosphere where there are appreciable quantities of ozone. See pp. 61, 63, fig. 1.

perigee. For a satellite orbit, the nearest point to the center of the earth.

photosphere of the sun. The visible disk of the sun. See pp. 364–65.

physical geodesy. The study of the gravitational field of the earth and its relationship to the solid structure of the planet. Also called *dynamical geodesy.* See p. 186.

Pirani gauge. A pressure gauge in which the rate of cooling of a heated filament by the ambient gas is used to measure the pressure of the gas.

plasma. A hot gas consisting of equal numbers of positive and negative ions.

polarized light. The wave theory of light pictures light as electromagnetic vibrations in space, with electric and magnetic vectors vibrating perpendicularly to each other. Normally in a beam of light the orientations of the electric and magnetic vectors are random. But if the electric (or magnetic) vectors all vibrate parallel to a common plane, the light in the beam is said to be *plane polarized.*

programmatic science. Science carried out as part of a specific program—say, the atomic energy program, or the NASA space science program as illustrations—and constrained (loosely perhaps, but constrained nevertheless) to fit within the general confines of the program.

proton. The nucleus of the hydrogen atom. The proton carries a single positive electric charge, and has a mass of 1.673×10^{-27} kilograms.

pulsar. A star that emits radiation in equally spaced pulses.

pure science. See *uncommitted science.*

quantum theory. The theory that all electromagnetic radiation is emitted and absorbed in quanta of energy equal to $h\nu$, where h is a constant called the Planck constant after the propounder of quantum theory, and ν is the frequency of the radiation.

quasar. An exceedingly remote astronomical object which appears like a blue star but which because of the prodigious quantities of energy it radiates may turn out to be a galaxy.

radiation belt. The portion of a planetary magnetosphere that contains charged particle radiations unable to escape because of the ambient magnetic field.

radio galaxy. A galactic system that emits prominently in radio wavelengths.

recombination. The reverse of dissociation, with the separated parts of a molecule rejoining to reproduce the original molecule from which they came. For example, an electron may recombine with a positive ion to form a neutral atom or molecule; a positive ion may join with a negative ion to form a neutral molecule; or two neutral atoms might recombine to form a molecule.

recombination coefficient. A quantitative measure of the affinity of particles for recombining.

reflection coefficient. A numerical quantity that gives the proportions and manner in which a medium reflects incident radiation.

refraction. The bending of electromagnetic rays by properties of the medium traversed.

relativistic particle. A particle moving sufficiently close to the speed of light that effects of relativity, such as an increase in mass or a slowing of time, become significant.

resonance line. The longest wavelength that an atom can emit or absorb.

retrorocket. A rocket so mounted on a vehicle or spacecraft that it fires in the direction opposite to the motion of the vehicle or spacecraft. Retrorockets are used to slow down the vehicles on which they are mounted.

rocket stage. A self-propelled separable element of a rocket vehicle. In a multistage rocket, each rocket unit fires after the one behind it has used up its propellant and (normally) been discarded. See page. p. 135.

satellite. An attendant body that revolves around another body called the satellite's *primary*. By custom, the bodies revolving around the sun are called *planets*, not satellites, of the sun.

scientific discipline. An area of scientific investigation in which the investigators share a common paradigm or group of paradigms—embracing a common body of theory—and techniques and often instrumentation that stem from the underlying theoretical basis of the discipline.

sectorial harmonic. One of a series of terms representing the gravitational potential of the earth, in which the terms exhibit pronounced variation with longitude. See *spherical harmonic* and p. 192.

Seyfert galaxy. A spiral galaxy with a very compact, highly luminous nucleus.

solar wind. A continuous wind of charged particles from the sun, blowing through interplanetary space.

spacecraft. Devices, manned or unmanned, that are intended to be placed in an orbit about the earth in space, or on a trajectory to another celestial body.

space launch vehicle. A combination of rocket stages, with the necessary guidance and control equipment, used to project a spacecraft into space.

space probe. Spacecraft sent away from earth into space. See. p. 153.

space science. Scientific investigations made possible or significantly aided by rockets, satellites, and space probes. See chap. 1, especially, pp. 11–15.

spectrum. Electromagnetic radiation displayed or visualized as a function of wavelength. Thus, the rainbow is part of the spectrum of sunlight.

spherical harmonic. One of a series of terms expressing the geopotential function in sines and cosines of latitude and longitude. See pp. 192–93.

stage. See *rocket stage.*

stratosphere. The layer of the earth's atmosphere between about 16 and 50–55 kilometers altitude, lying above the troposphere and extending to the stratopause; temperature generally increases with altitude in the stratosphere. See pp. 60–62.

sunspot cycle. A cycle of variation in the total number and area of spots on the sun's surface, from a maximum to a minimum with an average period of 11 years. Since magnetic fields on the sun reverse with each such cycle, the period of a complete cycle of spottedness-plus-magnetic-condition is 22 years.

superconducting cavity. When certain materials are brought sufficiently close to absolute zero temperature, they become perfect electrical conductors capable of sustaining circulating electrical currents indefinitely without resistance losses. A cavity in which such conditions of superconductivity are maintained is a superconducting cavity. The resonance properties of such superconducting cavities permit the construction of highly accurate oscillators, or clocks.

synchronous orbit. The orbit over the earth's equator at an altitude of 36 000 km, in which the rate of revolution of the satellite around the earth equals the rate of rotation of the earth, so that the satellite always stays over the same spot on the ground.

technology. Technical know-how; the knowledge and ability to do things of a technical or engineering nature, including the field of industrial arts.

telemetering. Measuring an object or phenomenon at a distance. Radio is often used to transmit the measured data from the point of measurement to a remote observer.

trajectory. In general, the path in three dimensions (i.e., space) of a moving body. The word trajectory is often used to mean *flight path*. If the trajectory is a closed path around another object like the earth, the trajectory is called an *orbit*. When the path is not closed, the word trajectory is usually used.

troposphere. The lowest portion of the atmosphere, extending from the ground to the base of the stratosphere.

uncommitted or **pure science.** Science carried out simply to pursue what the researcher considers to be important problems, the solutions to which show promise of revealing significant information about nature and the universe. Uncommitted or pure science is not constrained by programmatic, applications, or mission objectives.

vorticity. The amount of rotational motion possessed by a fluid is called its vorticity.

zonal harmonic. One of a series of terms representing the gravitational potential of the earth. Zonal harmonics correspond to coefficients J_{nm} for which $m = 0$ and which depend only on latitude. See *spherical harmonic* and also pp. 192-93.

Bibliographic Essay

The reader has seen that the narrative of this book consists of two interwoven themes: space science itself and the activities of institutions and individuals who created and carried out the space science program. For space science the most important sources are professional papers in scientific journals, published proceedings of technical meetings, and treatises and texts dealing with space science. For the other theme a most important source is the records and files of organizations and individuals working in the program.

I naturally drew heavily on NASA records and publications. In particular I examined in detail 43 boxes of notes, letters, memoranda, reports, and formal papers that formed part of the files I had used in the Office of Space Science and Applications in managing the space science program. These records are now in the National Archives, Federal Record Center, Suitland, Maryland, accession 255-79-0649. Running sequentially through the 43 boxes is a series of numbered folders each devoted to a specific subject or time. The NASA History Office has a catalog showing the organization of these records.

In references for this book I have used notations as in the following example: NF13(193). NF is short for "Newell Files." The 13 is the box number, and the number in parentheses is that of the folder in the designated box.

Secretaries in the Office of Space Science and Applications maintained a rather complete set of files. Thus, in addition to papers originating within the office, there were also copies of key documents the originals of which would naturally be kept in other offices of the agency or in other agencies, such as the National Academy of Sciences, Office of Science and Technology, and U.S. Weather Services. Indeed, one of the great values in this set of records proved to lie in the leads it gave to many different sources of space science material.

At times, of course, papers important to the space science narrative were not to be found in the NF collection. If such a document was both of special interest and likely to be difficult to reacquire, a copy was placed in additional boxes (NF40, NF41, NF42, NF43) of the same accession number.

Several portions of the NF collection were especially pertinent to this book. In box NF28 are stored notebooks I kept during my government service, both with the Naval Research Laboratory and with NASA. Although these are likely to be more useful to me than to someone else, anyone should find them helpful in tracing the course of space science activities—particularly as seen from NASA Headquarters. Many of the notes were records of problems and issues, statements of NASA policy, decisions of NASA's top management, work assignments, and reminders of actions to be taken. As a record of such items the notes are quite comprehensive, but they are incomplete in that the follow-through on the resolution of problems and issues, the completion of work assignments, and the effectuation of requested actions are not recorded. The missing information must come from other documents, particularly the official files of NASA and other agencies.

The chronological files (boxes 11–12, 24–27) of the collection reveal the actions emanating from the office of the associate administrator for space science and applications. These reflect a large number of the difficulties and challenges faced by the science and applications programs, but the reader is cautioned that they rest upon a much greater wealth of detail to be found in the division files of those who managed the programs in lunar and planetary sciences, geophysics, astronomy, the life sciences, etc. The bias of the NF files is, of course, toward the overall office level.

The correspondence between NASA Headquarters and its centers, including the Jet Propulsion Laboratory, shows the mixture of cooperation and tension internal to the agency that characterized the space science program—indeed the whole space program. The same sort of cooperation and tension is seen in exchanges with other agencies such as the Space Science Board, the U.S. Geological Survey, the Department of Defense, and the Department of Commerce.

Much of my responsibility in NASA concerned external scientific relations of NASA—with the National Academy of Sciences, the international Committee for Space Research, the American Geophysical Union, and the Space Science and Technology Panel of the President's Science Advisory Committee, for example. As a result the NF collection is quite comprehensive with regard to these relations, and my notebooks contain a great deal on them. One exception was the file on the Rocket and Satellite Research Panel, where for some reason much of the record was lost. Because of the central role the panel played in the early history of space science, as described in chapter 4, I borrowed the official files of the panel from the executive secretary, George Megerian, who kindly granted permission to copy them for NASA's files. These copies, with copies of additional panel papers from the personal files of William Stroud and Nelson Spencer, both members of the panel, provide a comprehensive record of the panel's activities from its founding in 1946 through its last technical session in 1960. NASA's copies of the panel files are stored in boxes NF40 and NF41. Although I drew a moderately detailed overview of the panel's activities from these records, a comprehensive history of the Rocket and Satellite Research Panel is still to be written.

Many secondary references are frequently cited in the text. Quite useful for background is the NASA Special Publication (SP) series. Considerable detail on NASA's budget, manpower, organization, and facilities is given in NASA SP-4012, *NASA Historical Data Book, 1957–1968*, vol. 1, *NASA Resources*, by Jane Van Nimmen and Leonard C. Bruno with Robert L. Rosholt (Washington, 1976). Additional details on NASA's first years can be found in Robert L. Rosholt, *An Administrative History of NASA, 1958–1963*, NASA SP-4101 (Washington, 1966). Some of NASA's activities as reflected in public announcements, the news media, and other similar sources may be traced with the aid of a series of annual chronologies, *Astronautics and Aeronautics*, issued by NASA's History Office (SP-4004 through 4008, 4010, 4014–4019) starting in 1963. Similar chronological data for years before 1963 can be found in Eugene M. Emme, *Aeronautics and Astronautics: An American Chronology of Science and Technology in the Exploration of Space, 1915–1960* (Washington: NASA, 1961); *Aeronautical and Astronautical Events of 1961*, Report of the National Aeronautics and Space Administration to the House Committee on Science and Astronautics, 87th Cong., 2d sess., 7 June 1962; and *Astronautical and Aeronautical Events of 1962*, Report of the National Aeronautics and Space Administration to the House Committee on Science and Astronautics, 88th Cong., 1st sess., 12 June 1963. NASA's "Pocket Statistics," issued monthly by Headquarters, provides a variety of statistical data, including a record of NASA and Soviet launchings, the characteristics of space launch vehicles, and general budgetary information. The International Programs series, published by NASA's Office of International Affairs, gives details of NASA's international program.

An almost overwhelming wealth of detail can be found in the records of NASA's hearings before the agency's authorizing committees in the House and Senate. The hearings cover every aspect of the NASA program, both technical and administrative. Investigative hearings such as those into the Ranger failures and Centaur troubles bring out not only the kinds of difficulties NASA had to overcome in the space science program, but also the searching scrutiny under which the work had to be done.

For the space science theme, as stated, the principal sources are the technical literature. Many of these sources are cited in the chapter references, particularly those for chapters 4, 5, 6, 11, and 20. Because of its great breadth, space science finds its way into a wide variety of publications. Some, however, stand out and should be of special interest to anyone who wishes to delve into the subject. *The Annals of the International Geophysical Year*, 48 vols. (London: Pergamon Press, 1957–1970) give much of the early space science work. Especially informative is volume 12, which has the papers presented at an international symposium held in Moscow under the auspices of CSAGI, the international Committee for the International Geophysical Year.

Because so much of space science dealt with investigation of the earth, the *Journal of Geophysical Research* quickly became a favored medium for many space researchers, as did the *Journal of Atmospheric and Terrestrial Physics*. Many papers appeared in the *Physical Review*, particularly papers dealing with cosmic rays. The *Journal of the Optical Society of America* and the *Astrophysical Journal* were natural outlets for astronomical topics such as solar spectroscopy.

For quick publication of results, *Nature* was often used. NASA also worked out an arrangement with the editor of *Science* for publishing preliminary results from especially significant missions rapidly, within a week or so. At particularly productive periods, as during Apollo lunar missions, a significant proportion of *Science* was devoted to space science topics. One can trace a great deal of the space science program in its pages. Later, when *Geophysical Research Letters* was begun by the American Geophysical Union, it was also used for brief communications on early results of space science investigations.

Space Science Reviews (Dordrecht-Holland: D. Reidel Publishing Co.) and the references cited therein provide an excellent means of developing a detailed picture of virtually any space science discipline one might want to pursue. Books of the Astrophysics and Space Science Library series, also published by Reidel, give extensive treatments of specific areas of space science, such as the magnetosphere, solar physics, or x-ray astronomy. An enormous amount of information is contained in the published proceedings of the Apollo lunar science conferences sponsored annually by the Johnson Space Center, the first in January 1970. For space life sciences, one can get a good start with *Biology and the Exploration of Mars*, edited by Colin S. Pittendrigh, Wolf Vishniac, and J. P. T. Pearman, National Academy of Sciences–National Research Council publication 1296 (Washington, 1966); Elie A. Shneour and Eric A. Ottesen, compilers, *Extraterrestrial Life: An Anthology and Bibliography*, ibid., publication 1296A (1966); and a compendium prepared jointly by Soviet and U.S. scientists: Melvin Calvin and Oleg G. Gazenko, eds., *Foundations of Space Biology and Medicine*, NASA SP-374, 3 vols. (English version, Washington, 1975).

The specific sources indicated above are but a small sampling of the available literature. One does not want for detail and in-depth treatments of individual areas. But the kind of overview of the whole field of space science that asks the broader questions of how existing paradigms were affected by the research and whether any scientific revolutions were forced by space science results is another matter. As an aid in preparing chapters 6, 11, and 20, in which I have addressed myself to such questions, I sent out a questionnaire to more than a hundred leading space science investigators. I hoped to learn how the scientists themselves felt their fields of research had been affected by space methods, and whether in their view any scientific revolutions had occurred. More than 60 responded, in varying detail. Their answers provided considerable additional insight into the subject and were helpful in the writing of chapters 6, 11, and 20. But there was more in the responses than could be included in only a few chapters of this book. My treatment must be considered sketchy. The answers to the questionnaires are filed in box NF43 under the following headings: atmospheric research, ionospheric physics, particles and fields, geodesy, lunar science, planetary science, meteors and cosmic rays, solar physics, astronomy, and exobiology.

Finally, the NF collection contains most of my articles and talks from the beginning of the sounding rocket program in 1946 through the 1960s. I did not usually cite these papers, preferring to use other sources. But one can trace in them the growing knowledge produced by the space science program, and also many of the major issues encountered in the space program.

Source Notes

Chapter 1

1. Homer E. Newell, *"NASA's Space Science and Applications Programs,* NASA EP-47 (Washington, 20 Apr. 1967), especially app. 9, "What Is Science?" pp. 252-54. The publication is a reprint of statements made to the House Committee on Science and Astronautics, 28 Feb.-9 Mar. 1967, and to the Senate Committee on Aeronautical and Space Sciences, 20 Apr. 1967, in support of NASA's FY 1968 authorization requests. The material also appears in the Govt. Printing Office prints of the hearings.
2. E.g., Karl R. Popper, *The Logic of Scientific Discovery* (New York: Basic Books, 1959); James B. Conant, *Science and Common Sense* (New Haven: Yale Univ. Press, 1951); Thomas S. Kuhn, *The Structure of Scientific Revolutions,* 2d ed. (Chicago: Univ. of Chicago Press, 1970).
3. Conant, *Science and Common Sense,* p. 25.
4. Ibid., p. 45.
5. Francis Bacon, *The Advancement of Learning* (1605). Karl Popper presents one modern view of the role of induction in science; see, for example, Popper, *Logic of Scientific Discovery,* pp. 27-30, 40-42, 315. Popper points out that induction is not an element in the *logic* of science. To illustrate, from Newton's law of gravitation (and the accepted theory of mechanics) one can deduce Kepler's laws of planetary motion, in particular that in closed orbits the planets must move in elliptical orbits about the sun. But, even when a planet is observed to move in an elliptical orbit, one cannot conclude that Newton's inverse square law of gravitation holds. Another possibility would be that the planet is attracted to the sun by a force varying directly as the distance. Given such a force field it can be deduced that a planet in a closed orbit would move in an elliptical path. Additional considerations must be applied to choose between the two candidates. Thus, the inductive step from the particular to the general is not unique, forced by the logic of the situation, but requires choice among a number of likely possibilities. Of course, to be a potential candidate a proposed theory must imply, by normal deductive reasoning, the original particular the scientist is trying to "explain." This is what is meant by the assertion that whereas deduction is an essential element in the logic of science, induction is not.
6. Conant, *Science and Common Sense,* p. 71.
7. Kuhn, *Scientific Revolutions.*
8. E.g., S. K. Mitra, *The Upper Atmosphere* (Calcutta: Royal Asiatic Society of Bengal, 1947), pp. 141-327; Wilmot N. Hess and Gilbert D. Mead, eds., *Introduction to Space Science* (New York: Gordon and Breach, 1968), pp. 133-78; and Francis Delobeau, *The Environment of the Earth,* Astrophysics and Space Library, No. 28 (Dordrecht-Holland: D. Reidel Publishing Co., 1971).
9. E.g., Edward G. Gibson, *The Quiet Sun,* NASA SP-303 (Washington, 1973).
10. Herbert Butterfield, "Dante's View of the Universe," chap. 1 in Herbert Butterfield et al., *A Short History of Science* (Garden City, N.Y.: Doubleday & Co., Doubleday Anchor Books, 1959).
11. Douglas McKie, "The Birth of Modern Chemistry," chap. 9, ibid.; Conant, *Science and Common Sense,* chap. 7.
12. F. K. Richtmeyer, E. H. Kennard, and T. Lauritzen, *Introduction to Modern Physics* (New York: McGraw-Hill Book Co., 1955), chap. 2.
13. D. H. Tarling and M. P. Tarling, *Continental Drift* (London: G. Bell & Sons, 1971).
14. See, for example, Margaret Masterman, "The Nature of the Paradigm," in *Criticism and the Growth of Knowledge,* ed. I. Lakatos and A. Musgrave (Oxford: Oxford Univ. Press), pp. 59-89; Dudley Shapere, "The Structure of Scientific Revolutions," *Philosophical Review* 73 (1964): 383-94.
15. T. S. Kuhn, *Planetary Astronomy in the Development of Western Thought* (Cambridge: Harvard Univ. Press, 1957).
16. Lloyd V. Berkner and Hugh Odishaw, eds., *Science in Space* (New York: McGraw-Hill Book Co., 1961).

17. Samuel Glasstone, preface to *Sourcebook on the Space Sciences* (Princeton, N.J.: D. Van Nostrand Co., 1965), p. vii.
18. NASA Organization Chart dated 29 Jan. 1959, signed by T. Keith Glennan; 1 Nov. 1966, signed by James E. Webb; 1 Nov. 1963, signed by Webb.
19. Homer E. Newell, Jr., *High Altitude Rocket Research* (New York: Academic Press, 1953); R. L. F. Boyd and M. J. Seaton, eds., *Rocket Exploration of the Upper Atmosphere* (Oxford: Pergamon Press; New York: Interscience Publishers, 1954).
20. James A. Van Allen, ed., *Scientific Uses of Earth Satellites* (Ann Arbor: Univ. of Michigan Press, 1956).
21. Berkner and Odishaw, *Science in Space*, p. 19.
22. See, for example, *Space Science Reviews* (Dordrecht-Holland: D. Reidel Publishing Co.); Astrophysics and Space Science Library (Dordrecht-Holland: D. Reidel Publishing Co.); Wilmot Hess and Gilbert D. Mead, eds., *Introduction to Space Science* 2d ed. (New York: Gordon and Breach, 1968).

Chapter 2

1. James B. Conant, *Science and Common Sense* (New Haven: Yale Univ. Press, 1951); Alvin M. Weinberg, *Reflections on Big Science* (Cambridge, Mass.: MIT Press, 1967); Daniel S. Greenberg, *The Politics of Pure Science* (New York: New American Library, 1967; rev. ed. 1971).
2. F. Sherwood Taylor, "Scientific Developments of the Early Nineteenth Century," chap. 10 of Herbert Butterfield et al., *A Short History of Science* (Garden City, N.Y.: Doubleday & Co., Doubleday Anchor Books, 1959); H. D. Smyth, *A General Account of the Development of Methods of Using Atomic Energy for Military Purposes under the Auspices of the United States Government 1940-1945* (Washington: Govt. Printing Office, 1945); Samuel Glasstone, *Sourcebook on Atomic Energy* (New York: D. Van Nostrand Co., 1950, 2d ed. 1958).
3. Conant, *Science and Common Sense*, pp. 315-21; Vannevar Bush, *Science, the Endless Frontier: A Report to the President* (Washington: Dept. of Defense, Office of Scientific Research and Development, July 1945).
4. Illinois Institute of Technology Research Institute, *Technology in Retrospect and Critical Events in Science*, or "Traces," 2 vols. (Report prepared by IITRI for National Science Foundation, vol. 1 [summary], 15 Dec. 1968; vol. 2, 30 Jan. 1969).
5. See committee prints of annual NASA authorization hearings before House Committee on Science and Astronautics, specifically FY 1968 hearings, 90th Cong., 1st sess., on H.R. 4450, H.R. 6470, 28 Feb.-9 Mar. 1967, pp. 235-421. See also committee prints of the annual NASA authorization hearings, before Senate Committee on Aeronautical and Space Sciences, specifically the FY 1968 hearings, 90th Cong., on S. 1296, 18-20 Apr. 1967., pt. 1, pp. 365-567.
6. Bush, *Science, the Endless Frontier*, p. 83.
7. Illinois Inst. of Technology Research Inst., "Technology in Retrospect."
8. Taylor, "Scientific Developments of Nineteenth Century."

Chapter 3

1. Joseph W. Siry, "The Early History of Rocket Research," *Scientific Monthly* 71 (1950): 326-32; idem, "Rocket Research in the Twentieth Century," ibid., 408-21.
2. Samuel Glasstone, *Sourcebook on the Space Sciences* (New York: D. Van Nostrand Co., 1965), pp. 9-11; Nicolai A. Rynin, *Interplanetary Flight and Communication* (3 vols., 9 nos.), vol. 1, no. 1, *Dreams, Legends, and Early Fantasies*, trans. R. Lavoott, NASA TT F-640 (Washington, 1970), and vol. 1, no. 2 *Spacecraft in Science Fiction*, trans. Lavoott, NASA TT F-641 (1971); Eugene M. Emme, *A History of Space Flight* (New York: Holt, Rinehart and Winston, 1965), chap. 2.
3. John Lear, *Kepler's Dream* (Berkeley and Los Angeles: Univ. of California Press, 1965).
4. Emme, *Space Flight*, pp. 38-46.
5. R. M. Goody, *The Physics of the Stratosphere* (Cambridge, England: Univ. Press, 1954), pp. 1-3.
6. Arthur C. Clarke, *The Exploration of Space* (New York: Harper, 1951).

7. Siry, "Rocket Research in the Twentieth Century"; G. Edward Pendray, "Pioneer Rocket Development in the United States," *Technology and Culture 4* (1963): 384-92; Walter R. Dornberger, "The German V-2," ibid., pp. 393-409; G. A. Tokaty, "Soviet Rocket Technology," ibid., pp. 515-28; Wernher von Braun, "From Small Beginnings," in Kenneth W. Gatland, ed., *Project Satellite* (New York: British Book Centre, 1958), chap. 1; Willy Ley, *Rockets, Missiles, and Space Travel* (New York: Viking Press, 1957; 3d rev. ed., 1961); F. C. Durant III, "Rockets and Guided Missiles," in *Encyclopedia Britannica* (1967), 19: 404-24.
8. K. E. Tsiolkovskiy, "The Investigation of Universal Space by Means of Reactive Devices," in *Works on Rocket Technology by K. E. Tsiolkovskiy*, M. K. Tikhonravov et al., eds., NASA translation TT F-243 (Washington, 1964), p. 95. See also, A. A. Blagonravov, ed., *Collected works of K. E. Tsiolkovskiy*, vol. 3, *Reactive Flying Machines*, Faraday Translations, NASA TT F-237 (Washington, 1965), p. 163. The paper, published in *Vestnik vozdukhoplavania* (Herald of Aeronautics) in 1911, is described in a subtitle as "A Summary of the Works of 1903."
9. Tsiolkovskiy, "Investigation of Universal Space by Reactive Devices," p. 208, and "Cosmic Rocket Trains," pp. 249-83, in Tikhonravov et al., *Works on Rocket Technology*.
10. Tokaty, "Soviet Rocket Technology," p. 517.
11. Hermann Oberth, *Die Rakete zu den Planetenräumen* (Munich and Berlin: Oldenbourg, 1923).
12. Tokaty, "Soviet Rocket Technology."
13. Ibid., pp. 523-24.
14. Esther C. Goddard and G. Edward Pendray, eds. *The Papers of Robert H. Goddard*, 3 vols. (New York: McGraw-Hill Book Co., 1970), 1: 9, 52 n.
15. Ibid., pp. 127-52, 337-406.
16. Ibid., 2: 580-83.
17. Pendray, "Pioneer Rocket Development," p. 388.
18. H. S. Seifert, M. M. Mills, and M. Summerfield, "The Physics of Rockets," *American Journal of Physics* 15 (Jan.-June 1947): 1-140, 255-72.
19. Hsue-shen Tsien, ed., *Jet Propulsion*, a reference text prepared by the staffs of Guggenheim Aeronautical Laboratory and Jet Propulsion Laboratory, GALCIT, California Institute of Technology for [U.S. Army] Air Technical Service Command ([Pasadena: Calif. Institute of Technology], 1946).
20. James A. Van Allen, L. W. Fraser, and J. F. R. Lloyd, "Aerobee Sounding Rocket—A New Vehicle for Research in the Upper Atmosphere," *Science* 108 (31 Dec. 1948): 746.
21. Dornberger, "The V-2"; Von Braun, "From Small Beginnings"; and Erik Bergaust, *Wernher von Braun* (Washington: National Space Institute, 1976), pp. 1-97.
22. James McGovern, *Crossbow and Overcast* (New York: William Morrow & Co., 1964), pp. 141-42.
23. George K. Megerian, minutes of Rocket and Satellite Research Panel, 27 Feb. 46. (Panel minutes and reports may be found in NASA Hq. History Office.)

Chapter 4

1. George K. Megerian, Secretary, minutes of Rocket and Satellite Research Panel (mimeographed; hereafter referred to as minutes of panel), rpt. 1, 27 Feb. 1946. See also app. A, which shows how the panel membership changed with time.
2. Megerian, minutes of panel, rpt. 13, 29 Dec. 1947.
3. Homer E. Newell, Jr., *High Altitude Rocket Research* (New York: Academic Press, 1953), pp. 72-84, pp. 84-89.
4. Ibid., pp. 95-105.
5. Ibid., pp. 89-95; H. E. Newell, Jr., "Prediction and Location of Rocket Impacts at White Sands Proving Ground," Upper Atmosphere Research rpt. VIII, NRL rpt. P-3485 (Washington: Naval Research Laboratory, June 1949).
6. Newell, *High Altitude Rocket Research*, pp. 66-68.
7. Homer E. Newell, Jr., "Exploration of the Upper Atmosphere by Means of Rockets," *The Scientific Monthly* 64 (June 1947): 453-63; W. A. Baum et al., "Solar Ultraviolet Spectrum to 88 Kilometers," *Physical Review* 70 (Nov. 1946): 781-82; C. V. Strain, "Solar Spectroscopy at High Altitudes," *Sky and Telescope* 6 (Feb. 1947): 3-6; E. Durand, J. J. Oberly, and R. Tousey, "Analysis of the First Rocket Ultraviolet Solar Spectra," *Astrophysical Journal* 109 (Jan. 1949): 1-16; J. J. Hopfield and H. E. Clearman, Jr., "The Ultraviolet Spectrum of the Sun from V-2 Rockets," *Physical Review* 73 (Apr. 1948): 877-84; T. R. Burnight, "Soft Radiation in the Upper Atmosphere," *Physical Re-*

view 76 (July 1949): 165 (abstract); S. E. Golian, E. H. Krause, and G. J. Perlow, "Cosmic Radiation above 40 Miles," *Physical Review* 70 (Aug. 1946): 223-24; idem, "Additional Cosmic-Ray Measurements with the V-2 Rocket," *Physical Review* 70 (Nov. 1946): 776-77; G. J. Perlow and J. D. Shipman, Jr., "Non-Primary Cosmic-Ray Electrons above the Earth's Atmosphere," *Physical Review* 71 (Mar. 1947): 325-26; S. E. Golian and E. H. Krause, "Further Cosmic-Ray Experiments above the Atmosphere," *Physical Review* 71 (June 1947): 918-19; J. A. Van Allen and H. E. Tatel, "The Cosmic-Ray Counting Rate of a Single Geiger Counter from Ground Level to 161 Kilometers Altitude," *Physical Review* 73 (Feb. 1948): 245-51; J. A. Van Allen, "Exploratory Cosmic Ray Observations at High Altitudes by Means of Rockets," *Sky and Telescope* 7 (May 1948): 171-75; A. V. Gangnes, J. F. Jenkins, Jr., and J. A. Van Allen, "The Cosmic-Ray Intensity above the Atmosphere," *Physical Review* 75 (Jan. 1949): 57-69; J. A. Van Allen and A. V. Gangnes, "Cosmic Ray Intensity above the Atmosphere at the Geomagnetic Equator," in *Proceedings of the Echo Lake Cosmic Ray Symposium*, 23 to 28 June 1949 (Washington: Office of Naval Research, Nov. 1949), pp. 199-204; J. A. Van Allen, "An Improved Upper Limit to the Primary Cosmic-Ray Intensity at Geomagnetic Latitude 41° N," ibid., pp. 195-98; idem, "Transition Effects of Primary Cosmic Radiation in Lead, Aluminum, and the Atmosphere," ibid., pp. 95-102; S. F. Singer, "The Specific Ionization of the Cosmic Radiation above the Atmosphere," *Physical Review* 76 (Sept. 1949): 701-02; S. E. Golian et al., "V-2 Cloud-Chamber Observation of a Multiply Charged Primary Cosmic Ray," *Physical Review* 75 (Feb. 1949): 524-25; H. E. Tatel and J. A. Van Allen, "Cosmic-Ray Bursts in the Upper Atmosphere," *Physical Review* 73 (Jan. 1948): 87-88; T. A. Bergstralh, "Photography from the V-2 Rocket at Altitudes Ranging up to 160 Kilometers," Naval Research Laboratory rpt. R-3083 (Washington, Apr. 1947); Delmar L. Crowson, "Cloud Observations from Rockets," *Bulletin of the American Meteorological Society* 30 (Jan. 1949): 17-22. See also the series of Upper Atmosphere Research Reports put out by the Naval Research Laboratory, Washington, starting in 1946.

8. Newell, *High Altitude Rocket Research*; Rocket Panel, "Pressures, Densities, and Temperatures in the Upper Atmosphere," *Physical Review* 88 (Dec. 1952): 1027-32.

9. Homer E. Newell, Jr., *Sounding Rockets* (New York: McGraw-Hill Book Co., 1959).

10. James A. Van Allen, L. W. Fraser, and J. F. R. Lloyd, "Aerobee Sounding Rocket—A New Vehicle for Research in the Upper Atmosphere," *Science* 108 (31 Dec. 1948): 746; James A. Van Allen, John W. Townsend, Jr., and Eleanor C. Pressly, "The Aerobee Rocket," chap. 4 of Newell, *Sounding Rockets*; William R. Corliss, *NASA Sounding Rockets, 1958-1968*, NASA SP-4401 (Washington, 1971), pp. 18-21.

11. Milton W. Rosen, *The Viking Rocket Story* (New York: Harper & Bros., 1955); Newell, "Viking," chap. 13 in Newell, *Sounding Rockets*.

12. Megerian, minutes of panel, rpt. 14, 28 Jan. 1948.

13. Corliss, *NASA Sounding Rockets*, pp. 79-84.

14. Constance McLaughlin Green and Milton Lomask, *Vanguard: A History* (Washington: Smithsonian Institution Press, 1971), pp. 35-56.

15. Newell, *Sounding Rockets*; Corliss, *NASA Sounding Rockets*.

16. J. A. Van Allen and M. B. Gottlieb, "The Inexpensive Attainment of High Altitude with Balloon-launched Rockets," in *Rocket Exploration of the Upper Atmosphere*, ed. R. L. F. Boyd and M. V. Seaton (Oxford: Pergamon Press; New York: Interscience Publishers, 1954), pp. 53-64; James A. Van Allen, "Balloon-Launched Rockets for High-Altitude Research," chap. 9 in Newell, *Sounding Rockets*.

17. Megerian, minutes of panel, rpt. 1, 27 Feb. 1946; rpt. 3, 24 Apr. 1946; rpt. 6, 5 Sept. 1946; rpt. 15, 18 Mar. 1948; and ibid., Executive Session, 29 Apr. 1957.

18. Ibid., Attendance Lists; ibid., rpt. 1, 27 Feb. 1946, and rpt. 44, 31 May 1956.

19. C. N. Warfield, "Tentative Tables for the Properties of the Upper Atmosphere," NACA TN 1200 (Washington, Jan. 1947). Also, National Advisory Committee for Aeronautics, Panel on the Upper Atmosphere of the Committee on Aerodynamics, minutes, 4 Mar. 1946, 24 June 1946, 2 May 1947, 17 Sept. 1948, 5 Oct. 1950, 23 Oct. 1951; all mimeographed.

20. Eugene M. Emme, *Aeronautics and Astronautics: An American Chronology of Science and Technology in the Exploration of Space, 1915-1960* (Washington: NASA, 1961), p. 54.

21. C. S. Piggot, Exec. Dir., RDB Committee on Geophysical Sciences, to E. H. Krause, Chairman, V-2 Upper Atmosphere Panel, 20 Nov. 1947, in Megerian, minutes of panel, rpt. 13, 29 Dec. 1947, encl. D.

22. Megerian, minutes of panel, rpt. 24, 20 Apr. 1950.

23. Homer E. Newell, Jr., "The Challenge to United States Leadership in Rocket Sounding of the

Upper Atmosphere" (Washington: Naval Research Laboratory, 28 Aug. 1957), bound mimeographed typescript, pp. 3–4.
24. Col. H. N. Toftoy to Commanding General, White Sands Proving Ground, in Megerian, minutes of panel, rpt. 13, 29 Dec. 1947, encl. E.
25. Boyd and Seaton, *Rocket Exploration.*
26. Megerian, minutes of panel, rpt. 34, 29, 30 Jan. 1953, and rpt. 35, 29 Apr. 1953.
27. Ibid., rpt. 36, 7 Oct. 1953, and rpt. 37, 4 Feb. 1954; L. V. Berkner, ed., *Manual on Rockets and Satellites,* in *Annals of the International Geophysical Year,* 6 (London: Pergamon Press, 1958): 54–55.
28. Megerian, minutes of panel, rpt. 37, 4 Feb. 1954; rpt. 39, 9 Sept. 1954; and rpt. 40, 3 Feb. 1955.
29. Berkner, *Rockets and Satellites,* p. 55.
30. Megerian, minutes of panel, rpt. 39, 9 Sept. 1954.
31. James A. Van Allen, ed., *Scientific Uses of Earth Satellites* (Ann Arbor: University of Michigan Press, 1956); Megerian, minutes of panel, rpt. 43, 26, 27 Jan. 1956, Attendance List.
32. Megerian, minutes of panel, early October 1957 through February 1958. See specifically: Committee on the Occupation of Space files for October and November 1957; rpt. 48, 13–14 Nov. 1957; rpt. 49, 6 Dec. 1957; Executive Committee Report, Jan. 1958.
33. Ibid., rpt. 49, 6 Dec. 1957.
34. Space Flight Technical Committee of the American Rocket Society, "A National Space Flight Program," *Astronautics* 3 (Jan. 1958): 21–28.
35. Congress, Subcommittee of the Joint Committee on Atomic Energy of the United States, *Outer Space Propulsion by Nuclear Energy,* hearings, 85th Cong., 2d sess., 22, 23 Jan. and 6 Feb. 1958, pp. 149–73.
36. Megerian, minutes of panel, rpt. 49, 6 Dec. 1957.
37. Ibid.
38. W. H. Pickering to Homer E. Newell, Jr., 27 May 1960, in NASA History Office files.
39. Megerian, minutes of panel, rpt. 1960-1, 17 Feb. 1960.

Chapter 5

1. Niels H. de V. Heathcote and Angus Armitage, "The First International Polar Year (1882–1883)," *Annals of the International Geophysical Year,* 1 (London: Pergamon Press, 1959): 6 (hereafter cited as *IGY Annals*); V. Laursen, "The Second International Polar Year," *IGY Annals,* 1:211.
2. G. Breit and M. A. Tuve, "A Test of the Existence of the Conducting Layer," *Physical Review* 28 (Sept. 1926): 554–75.
3. Sir Harold Spencer Jones, "The Inception and Development of the International Geophysical Year," *IGY Annals,* 1:383. The reader desiring to delve further into the IGY will find in the *Annals* details of the planning and many of the results of the IGY. Very readable accounts of IGY are given by Sydney Chapman, *IGY: Year of Discovery* (Ann Arbor: Univ. of Michigan Press, 1959); Walter Sullivan, *Assault on the Unknown: The International Geophysical Year* (New York: McGraw-Hill Book Co., 1961, 1971); Ronald Fraser, *Once around the Sun* (New York: Macmillan, 1957); and Alexander Marshak, *The World in Space* (New York: Thomas Nelson & Sons, 1958).
4. Sydney Chapman, President of CSAGI, Brussels, 28 Jan. 1957, *IGY Annals,* 1:3.
5. L. V. Berkner, ed., *Manual on Rockets and Satellites,* in *IGY Annals,* 6:54–55. See also app. 7.1.
6. R. Cargill Hall, "Early U.S. Satellite Proposals," *Technology and Culture* 4 (Fall, 1963): 410–34; William H. Pickering, "History of the Juno Cluster System," in *Astronautical Engineering and Science,* ed. Ernst Stuhlinger et al. (New York: McGraw-Hill Book Co., 1963), pp. 204–05.
7. S. F. Singer, "Research in the Upper Atmosphere with Sounding Rockets and Earth Satellite Vehicles," *Journal of the British Interplanetary Society* 11 (1952): 61–73; idem, "A Minimum Orbital Instrumented Satellite—Now," ibid. 13 (1954): 74–79; idem, "Astrophysical Measurements from an Earth Satellite," chapter in *Rocket Exploration of the Upper Atmosphere,* ed. R. L. F. Boyd and M. J. Seaton (Oxford: Pergamon Press; New York: Interscience Publishers, 1954), p. 369; idem, "Studies of a Minimum Orbital Unmanned Satellite of the Earth (MOUSE)," pt. I, "Geophysical and Astrophysical Applications," *Astronautica Acta* 1 (1955): 171–84.
8. L. V. Berkner, CSAGI Reporter for Rockets and Satellites, in the introduction to *IGY Annals,* 6:1–2.
9. Ibid.
10. Ibid.

11. Constance McLaughlin Green and Milton Lomask, *Vanguard: A History* (Washington: Smithsonian Institution Press, 1971), pp. 34–56.
12. Berkner, *Rockets and Satellites*, pp. 283–84. See also app. 7.1
13. Berkner, *Rockets and Satellites*, p. 2. Also author's contemporaneous notes (hereafter referred to as author's notebooks [NF28] in NASA History Office).
14. Author's notebooks (NF28).
15. Ibid. The quotation is approximate.
16. Mary Stone Ambrose, "The National Space Program: Phase I: Passage of the 'National Aeronautics and Space Act of 1958,'" M.A. thesis (Washington: American Univ., July 1960), pp. 39–44, 154.
17. Alison Griffith, *The Nasa Act: A Study of the Development of Public Policy* (Washington: Public Affairs Press, 1962); Enid Curtis Bok Schoettle, "The Establishment of NASA," in *Knowledge and Power: Essays on Science and Government*, ed. Sanford A. Lakoff (New York: Free Press, 1966); Arthur L. Levine, *The Future of the U.S. Space Program* (New York: Praeger, 1975); Ambrose, "The National Space Program."

Chapter 6

1. Homer E. Newell, Jr., *High Altitude Rocket Research* (New York: Academic Press, 1953), pp. 111–42; the Rocket Panel, "Pressures, Densities, and Temperatures in the Upper Atmosphere," *Physical Review* 88 (Dec. 1952): 1027–32.
2. B. Haurwitz, "The Physical State of the Upper Atmosphere," reprinted from *Journal of the Royal Astronomical Society of Canada*, Oct. 1936–Feb. 1937 (Toronto: Univ. of Toronto Press, 1937; with addition, 1941).
3. T. H. Johnson, "Cosmic Ray Intensity and Geomagnetic Effects," *Reviews of Modern Physics* 10 (Oct. 1938): 193–244.
4. Fred L. Whipple, "Meteors and the Earth's Upper Atmosphere," *Reviews of Modern Physics* 15 (Oct. 1943): 246–64.
5. S. K. Mitra, *The Upper Atmosphere* (Calcutta: Royal Asiatic Society of Bengal, 1947), 519.
6. Ibid., pp. 5–8.
7. Ibid., p. 511.
8. B. Stewart, in *Encyclopaedia Britannica*, 9th ed., 16 (1882): 181.
9. A. E. Kennelly, "On the Elevation of the Electrically-Conducting Strata of the Earth's Atmosphere," *Electric World and Engineering* 39 (Mar. 1902): 473; O. Heaviside, "Telegraphy-Theory," *Encyclopaedia Britannica*, 10th ed., 33 (1902): 213–18.
10. E. V. Appleton and M. A. F. Barnett, "Local Reflection of Wireless Waves from the Upper Atmosphere," *Nature* 115 (March 1925): 333–34; idem, "On Some Direct Evidence for Downward Atmospheric Reflection of Electric Rays," *Proceedings of the Royal Society*, A 109(1925): 621–41.
11. G. Breit and M. A. Tuve, "A Test of the Existence of the Conducting Layer," *Physical Review* 28 (Sept. 1926): 554–75.
12. Mitra, *Upper Atmosphere*, chap. 6.
13. Ibid., pp. 257–61.
14. Ibid., p. 512.
15. Ibid., pp. 328–30.
16. See, for example, C. Størmer, "Twenty-five Years' Work on the Polar Aurora," *Terrestrial Magnetism and Atmospheric Electricity* 35 (1930): 193–208.
17. Newell, *High Altitude Rocket Research*, pp. 237–45.
18. See also I. S. Bowen, R. A. Millikan, and H. V. Neher, "New Light on the Nature and Origin of the Incoming Cosmic Rays," *Physical Review* 53 (June 1938): 855–61.
19. M. Schein, W. P. Jesse, and E. O. Wollan, "The Nature of the Primary Cosmic Radiation and the Origin of the Mesotron," *Physical Review* 59 (Apr. 1941): 615.
20. Mitra, *Upper Atmosphere*, pp. 77–87.
21. Haurwitz, *Upper Atmosphere*, p. 79; p. 8.
22. Mitra, *Upper Atmosphere*, p. 87.
23. Whipple, "Meteors and the Earth's Upper Atmosphere."
24. Mitra, *Upper Atmosphere*, p. 146.
25. Ibid., p. 151.
26. Ibid., p. 518.

27. Ibid., pp. 515–19.
28. Homer E. Newell, Jr., *Sounding Rockets* (New York: McGraw-Hill Book Co., 1959), pp. 37–43.
29. Homer E. Newell, Jr., and Leonard N. Cormier, eds., *First Results of IGY Rocket and Satellite Research*, vol. 12, pts. 1 and 2, in *Annals of the International Geophysical Year* (London: Pergamon Press, 1960).
30. W. A. Baum, F. S. Johnson, J. J. Oberly, C. C. Rockwood, C. V. Strain, and R. Tousey, "Solar Ultraviolet Spectrum to 88 Kilometers," *Physical Review* 70 (Nov. 1946): 781–82.
31. Herbert Friedman, "The Sun's Ionizing Radiations," chapter in *Physics of the Upper Atmosphere*, ed. J. A. Ratcliffe (New York: Academic Press, 1960), pp. 133–218.
32. J. J. Hopfield and H. E. Clearman, Jr., "The Ultraviolet Spectrum of the Sun from V-2 Rockets," *Physical Review* 73 (Apr. 1948): 877–84.
33. E. Durand, J. J. Oberly, and R. Tousey, "Analysis of the First Rocket Ultraviolet Spectra," *Astrophysical Journal* 109 (Jan. 1949): 1–16.
34. W. A. Rense, "Intensity of Lyman-Alpha Line in the Solar Spectrum," *Physical Review* 91 (15 July 1953): 299–302; idem, "Solar Ultraviolet Radiation and Its Effect on the Earth's Upper Atmosphere," in *Advances in Space Research*, ed. T. M. Tabanera et al. (London: Pergamon Press, 1964), pp. 275–76.
35. Rense, "Solar Ultraviolet and Its Effect on the Earth's Upper Atmosphere," pp. 278–79.
36. Newell, *High Altitude Rocket Research*, p. 161; Friedman, "The Sun's Ionizing Radiations," pp. 168–78.
37. F. S. Johnson, J. D. Purcell, and R. Tousey, "Measurements of the Vertical Distribution of Atmospheric Ozone from Rockets," *Journal of Geophysical Research* 56 (Dec. 1951): 583–94; F. S. Johnson et al., "Direct Measurements of the Vertical Distribution of Atmospheric Ozone to 70 Kilometers Altitude," *Journal of Geophysical Research* 57 (June 1952): 157–76; J. A. Van Allen and J.J. Hopfield, "Preliminary Report on Atmospheric Ozone Measurements from Rockets," in *L'étude optique de l'atmosphère terrestre: communications présentées au colloque international tenu à l'Institut d'Astrophysique de l'Université de Liège les 3 et 4 septembre 1951* (Louvain: Imprimere Ceuterick, 1952), pp. 179–83.
38. Newell, *High Altitude Rocket Research*, pp. 111–42; idem, "The Upper Atmosphere Studied by Rockets and Satellites," chapter in Ratcliffe, *Physics of the Upper Atmosphere*, pp. 74–101.
39. W. G. Stroud, W. Nordberg, W. R. Bandeen, F. L. Bartman, and P. Titus, "Rocket Grenade Measurements of Temperature and Winds in the Mesosphere over Churchill, Canada," in *Space Research: Proceedings of the First International Space Science Symposium, Nice, 1960*, ed. H. K. Kallmann-Bijl (Amsterdam: North-Holland Publishing Co., 1960), pp. 117–43.
40. Friedman, "The Sun's Ionizing Radiations," pp. 208–14.
41. The Rocket Panel, "Pressures, Densities, and Temperatures in the Upper Atmosphere."
42. Newell, "The Upper Atmosphere Studied by Rockets and Satellites," pp. 74–97.
43. Ibid., pp. 102–11.
44. Newell, *High Altitude Rocket Research*, pp. 203–08.
45. J. W. Townsend, Jr., "Radiofrequency Mass Spectrometer for Upper Air Research," *Reviews of Scientific Instruments* 23 (1952): 538–41.
46. Newell, "The Upper Atmosphere Studied by Rockets and Satellites," pp. 112–18. E. B. Meadows and J. W. Townsend, Jr., "IGY Rocket Measurements of Arctic Atmospheric Composition Above 100 Km," in *Space Research*, ed. Kallmann-Bijl, pp. 175–98. C. Y. Johnson, "Aeronomic Measurements," in *Advances in Space Research*, ed. Tabanera et al., pp. 295–317.
47. Newell, "The Upper Atmosphere Studied by Rockets and Satellites," pp. 119–20.
48. Ibid., pp. 108–11.
49. A. V. Gangnes, J. F. Jenkins, Jr., and J. A. Van Allen, "The Cosmic Ray Intensity above the Atmosphere," *Physical Review* 75 (Jan. 1949): 57–69.
50. G. J. Perlow et al., "Rocket Determination of the Ionization Spectrum of Charged Cosmic Rays at λ = 41° N," *Physical Review* 88 (Oct. 1952): 321–25.
51. L. H. Meredith, M. B. Gottlieb, and J. A. Van Allen, "Direct Detection of Soft Radiation Above 50 Kilometers in the Auroral Zone," *Physical Review* 97 (1 Jan. 1955): 201; J. A. Van Allen, "Rocket Measurement of Soft Radiation," in *First Results of IGY Rocket and Satellite Research*, ed. Newell and Cormier, pp. 646–50.
52. Paper presented by Van Allen at joint meeting of National Academy of Sciences and Physical Society on 1 May 1958. See also, J. A. Van Allen, G. H. Ludwig, E. C. Ray, and C. E. McIlwain, "The Observation of High Intensity Radiation by Satellites 1958 α and γ," in *First Results of IGY Rocket and Satellite Research*, ed. Newell and Cormier, pp. 671–81.

53. J. E. Kupperian et al., "Far Ultraviolet Radiation in the Night Sky," in *First Results of IGY Rocket and Satellite Research*, ed. Newell and Cormier, pp. 619-22; J. E. Kupperian et al., "Rocket Astronomy in the Far Ultraviolet," ibid., pp. 622-26; Herbert Friedman et al., *Space Astronomy: A New Era in the Making*, a series of articles reprinted from *Astronautics & Aeronautics* 7 (Mar. and May 1969), pp. 34-75.

Chapter 7

1. Alison Griffith, *The National Aeronautics and Space Act: A Study of the Development of Public Policy* (Washington: Public Affairs Press, 1962), pp. 100-01.
2. Ibid., pp. 17-18.
3. R. Cargill Hall, "Early U.S. Satellite Proposals," *Technology and Culture* 4 (Fall 1963): 410-34.
4. Homer E. Newell, Jr., *Sounding Rockets* (New York: McGraw-Hill Book Co., 1959); Milton W. Rosen, *The Viking Rocket Story* (New York: Harper and Brothers, 1955); Constance McLaughlin Green and Milton Lomask, *Vanguard: A History* (Washington: Smithsonian Institution Press, 1971); William H. Pickering, "History of the Juno Cluster System," chap. 12 in *Astronautical Engineering and Science*, ed. Ernst Stuhlinger et al. (New York: McGraw-Hill Book Co., 1963); Homer E. Newell, Jr., *Guide to Rockets, Missiles, and Satellites* (New York: McGraw-Hill Book Co., 1958).
5. "Introduction to Outer Space," rpt. prepared by panel of President's Science Advisory Committee, endorsed by PSAC, and issued by the White House with presidential endorsement 26 Mar. 1958.
6. Griffith, *The Space Act*, p. 9.
7. Ibid., p. 11.
8. William G. Stroud, handwritten notes on 14 Feb. 1958 meeting of Rocket and Satellite Research Panel (copy in NF41, NASA History Office).
9. Alex Roland, *Research by Committee: A History of the National Advisory Committee for Aeronautics*, NASA SP-4103, comment ed. (Washington, 1979).
10. Ibid.
11. NACA, "A National Research Program for Space Technology," multilith (Washington, 14 Jan. 1958).
12. Senate Special Committee on Space and Aeronautics, *Compilations of Materials on Space and Aeronautics*, no. 2, 85th Cong., 2d sess., committee print, 1958, pp. 293-94; NACA, "A Program for Expansion of NACA Research in Space Flight Technology with Estimates of the Staff and Facilities Required" (Washington, 10 Feb. 1958).
13. Robert L. Rosholt, *An Administrative History of NASA, 1958-1963*, NASA SP-4101 (Washington, 1966), p. 9.
14. Arthur L. Levine, "United States Aeronautical Research Policy, 1915-1958," doctoral dissertation, Columbia Univ. (1963), p. 155.
15. Griffith, *The Space Act*, p. 44.
16. Griffith, *The Space Act*; Enid Curtis Bok Schoettle, "The Establishment of NASA," in *Knowledge and Power: Essays on Science and Government*, ed. Sanford A. Lakoff (New York: Free Press, 1966); Arthur L. Levine, *The Future of the U.S. Space Program* (New York: Praeger Publishers, 1975); Mary Stone Ambrose, "The National Space Program: Phase I: Passage of the 'National Aeronautics and Space Act of 1958,'" M. A. thesis, American University (July 1960).
17. Griffith, *The Space Act*, p. 12.
18. Senate Committee on Space and Aeronautics, *Compilations*, no. 2, pp. 308-09.
19. House Committee on Science and Astronautics, *A Chronology of Missiles and Astronautic Events*, H. rpt. 67, 87th Cong., 1st sess., 8 Mar. 1961, pp. 41-42.
20. Ibid., p. 44.
21. Griffith, *The Space Act*, p. 14.
22. Ibid., chap. 6.
23. Ibid., pp. 65-74.
24. Ibid., pp. 75-96; Rosholt, *Administrative History of NASA*, pp. 12-15.
25. Griffith, *The Space Act*, p. 45.
26. Ibid., p. 1.
27. NASA, Office of the General Counsel, "National Aeronautics and Space Act of 1958, as Amended, and Related Legislation" (Washington, 1 July 1969), p. 1.

28. Ibid., passim.
29. Ibid., p. 4.
30. Reorganization Plan No. 4 of 1965, effective 27 July 1965 (30 Federal Register 9353, 28 July 1965, 79 Stat. 384).
31. NASA General Counsel, "NASA Act," p. 29.
32. Rosholt, *Administrative History of NASA*, p. 40.

Chapter 8

1. Author's notebooks, 16 Dec. 58, NF28.
2. Robert L. Rosholt, *An Administrative History of NASA, 1958–1963*, NASA SP-4101 (Washington, 1966), pp. 37–116.
3. Ibid., pp. 38–40.
4. Ibid., app. B, p. 332.
5. NACA, "A National Program for Space Technology," multilith (Washington: NACA, 14 Jan. 1958).
6. Rosholt, *Administrative History of NASA*, fig. 3-2, fig. 3-3.
7. T. Keith Glennan to Monte Wright, 15 July 1978, comments on draft Newell manuscript, NF40.
8. Homer E. Newell, conference report on meeting of Space Science Panel of the President's Science Advisory Committee, 18 Dec. 1959, NF12(173).
9. Rosholt, *Administrative History of NASA*, p. 341; facing p. 344, chart no. 1.
10. Bruce K. Byers, *Destination Moon: A History of the Lunar Orbiter Program*, NASA TM X-3487 (Washington, Apr. 1977), pp. 25–36.
11. Edward C. Ezell and Linda Neuman Ezell, *On Mars: Exploration of the Red Planet, 1958–1978*, NASA SP-4212, in press.
12. House Committee on Science and Astronautics, *Astronautical and Aeronautical Events of 1962*, 88th Cong., 1st sess., committee print, 12 June 1963, p. 203.
13. Author's notebooks, 18 Feb. 1960, NF28.
14. William R. Corliss, *The Interplanetary Pioneers*, vol. 1, Summary, NASA SP-278 (Washington, 1972), pp. 3–4.
15. Rosholt, *Administrative History of NASA*, fig. 3-3.
16. Joseph A. Shortal, *A New Dimension: Wallops Island Flight Test Range, the First Fifteen Years*, NASA RP-1028 (Washington, 1978).
17. Rosholt, *Administrative History of NASA*, p. 47.
18. Alfred Rosenthal, *Venture into Space: Early Years of Goddard Space Flight Center*, NASA SP-4301 (Washington, 1968), pp. 28–29.
19. Rosholt, *Administrative History of NASA*, p. 47.
20. American Geophysical Union, "Report of Committee on Geodetic Applications of Artificial Satellites" (Washington, 1 Sept. 1958), encl. in John A. O'Keefe to Homer E. Newell, 20 June 1978; NF40 (Newell Files, Box 40, in National Archives Federal Records Center; see Bibliographical Essay for description of files and explanation of NF designations); John O'Keefe, "Geodetic Significance of an Artificial Satellite," app. to ARS Space Flight Committee's "Utility of an Earth Satellite," pp. 75–76; O'Keefe and C. D. Batchelor, "Perturbations of a Close Satellite by the Equatorial Ellipticity of the Earth, *Astronomical Journal* 62(1957): 183; O'Keefe, "An Application of Jacobi's Integral to the Motion of an Earth Satellite," *Astronomical Journal* 62 (1957): 203; idem. "Geodesy Comes of Age with Vanguard," *Astronautics* 2 (Aug. 1957); 71–73, 92; F. L. Whipple and J. A. Hynek, "A Research Program Based on the Optical Tracking of Artificial Earth Satellites," *Proceedings of the Institute of Radio Engineers* 44 (June 1956), 760–64.
21. Minutes, NACA Executive Committee, Washington, 20 Feb. 1958, in NASA History Office.
22. "Recommendations to the NASA Regarding a National Civil Space Program," typescript rpt. of NACA Special Committee on Space Technology, 28 Oct. 1958, in NASA History Office.
23. Loyd S. Swenson, Jr., James M. Grimwood, and Charles C. Alexander, *This New Ocean: A History of Project Mercury*, NASA SP-4201 (Washington, 1966), pp. 76, 91–106.
24. Eugene M. Emme, *Aeronautics and Astronautics: An American Chronology of Science and Technology in the Exploration of Space, 1915–1960* (Washington: NASA, 1961), pp. 102–03.
25. Senate Committee on Aeronautical and Space Sciences, *NASA Authorization for Fiscal Year 1960*, hearings, pt. 1, *Scientific and Technical Presentations*, 86th Cong., 1st sess., 7–10 Apr. 1959, pp. 127–226.

26. Working papers summarizing space science program, 17 Jan. 1959, NF11(170).
27. "Areas of Research in the NASA Space Sciences Program," mimeographed NASA document, 48 pp., 10 Feb. 1959, NF2(33).
28. "The United States National Space Sciences Program," mimeographed NASA document, 10 Feb. 1959, NF2(33).
29. "National Space Sciences Program," mimeographed NASA document, 16 Apr. 1959, NF2(33).
30. Homer E. Newell, Jr., "Report on the Second Meeting of the Committee on Space Research Held at The Hague, 12–14 March 1959," NASA lithoprint typescript [Mar. 1959], encl. 7, NF8(127).
31. "NASA Program Planning in Space Sciences," mimeographed NASA document, Apr. 1960, NF2(33).
32. Homer E. Newell, "The NASA Space Science Program," mimeographed NASA document, 20 Mar. 1961, NF1.
33. Senate Committee on Aeronautical and Space Sciences, *NASA Authorization for Fiscal Year 1968*, hearings, 90th Cong., 1st sess., pt. 1, 18–20 Apr. 1967, p. 366.
34. Glennan to Wright, 15 July 1978, NF40.
35. Meeting of Harold Urey, Robert Jastrow, John O'Keefe, and Newell, 16 Jan. 1959, author's notebook, NF28.
36. Henry Stauss, mimeographed rpt. of meeting of NASA Space Sciences Steering Committee with selected consultants, 22 May 1961, NF11(160).
37. Author's notebook, 13 Dec. 1961, describes a discussion with Hugh Odishaw, executive director of the Space Science Board, and Ross Peavey of the Board staff on plans for the coming summer study, NF28.
38. Homer E. Newell, "Review of Progress toward NASA Announced Goals," typescript, 13 pp., 11 Jan. 1962, NF11(168).

Chapter 9

1. Robert L. Rosholt, *An Administrative History of NASA, 1958–1963*, NASA SP-4101 (Washington, 1966), pp. 71–179.
2. H. E. Newell, conference report on NASA-DoD meeting, 4 May 1959, J. B. Macauley to Newell, 26 Jan. 1959; Hugh Dryden to W. M. Holaday, 28 July 1959. A description of the NASA-DoD Space Science Committee is given in the NASA memorandum Newell to Alfred Hodgson, 20 Oct. 1959. NF12(173).
3. Rosholt, *Administrative History of NASA*, pp. 172–73.
4. Unmanned Spacecraft Panel, "Joint NASA-DoD Space Sciences Program Document," Oct. 1960, NF6(106).
5. Herbert F. York and G. Allen Greb, "Strategic Reconnaissance," *Bulletin of the Atomic Scientists* (Apr. 1977), pp. 33–42.
6. Conference between Weather Bureau and NASA representatives, 12 Feb. 1960, author's notebooks, NF28.
7. "Ecological Surveys from Space," NASA SP-230 (Washington, 1970).
8. Eugene M. Emme, *Aeronautics and Astronautics: An American Chronology of Science and Technology in the Exploration of Space, 1915–1960* (Washington: NASA, 1961), p. 130.
9. Ibid., *Satellite Geodesy, 1958–1964*, NASA SP-94 (Washington, 1966), pp. 125–28.
10. Hugh L. Dryden to Hugh Odishaw, 23 Nov. 1960; Odishaw to Dryden, 22 Dec. 1960; NF6(101).
11. House Committee on Science and Astronautics, *1964 NASA Authorization*, hearings before Subcommittee on Space Sciences and Advanced Research and Technology, 88th Cong., 1st sess., 6 Mar.–8 May 1963, p. 1493. See also Jerome D. Rosenberg, "National Geodetic Satellite Program." 29 Oct. 1964, and NASA Congressional Study Questionnaire Concerning the Geodetic Satellite Policy Board, Feb. 1963, NF40.
12. Author's notebook, 25 Jan. 1960, NF28.
13. Newell to Ira H. Abbott, 1 Mar. 1960, NF12(174).
14. Dryden to Alan Waterman, 20 Oct. 1959, NF6(109).
15. Dryden to Douglas Cornell, 20 Oct. 1959, NF6(109).
16. Newell, conference report on NAS meeting at Boulder, Colo., 9 Dec. 1959, NF4(67).
17. Newell to Harry Goett, 10 Dec. 1963, and succeeding correspondence relative to creation of the Space Science Data Center in Office of Space Science and Applications files, NASA. In particular,

see undated typescript paper, "NASA Space Science Data Center," which states the center was officially established in Apr. 1964, OSSA Space Science Data Center files for 1963–1965.

18. NASA Management Issuance 37-1-1, "Establishment and Conduct of Space Science Program—Selection of Scientific Experiments," typescript NASA document, 15 Apr. 1960, NASA History Office files.

19. NMI 37-1-2, "Membership of Space Sciences Steering Committee and Subcommittees," 27 May 1960, NASA History Office files, shows only NASA membership; the outside consultants were added later.

20. By way of illustration see Newell, report on meeting of Space Science Panel of the President's Science Advisory Committee, at Inst. of Radio Engineers, New York, 18 Dec. 1959; also Newell, "Briefing for Space Science Panel of the PSAC," Washington, 2 Apr. 1962, NF12(173), (180); T. Keith Glennan to Monte Wright, 15 July 1978, comments on draft Newell MS., NF40; Newell to Dryden, 18 FEb. 1960, NF12(174).

21. Newell, Briefing for Space Science Panel, 2 Apr. 1962, NF12(180).

22. NASA Management Issuance 7100.1, "Conduct of Space Science Program—Selection and Support of Scientific Investigations and Investigators," 29 Apr. 1964, p. 8.

23. Lloyd Berkner to George Kistiakowsky, 13 Nov. 1959, NF7(112).

24. John F. Clark, memorandum to files, 2 Mar. 1960, NF6(105).

25. Berkner to Kistiakowsky, 13 Nov. 1959, NF7(112).

26. Glennan to Kistiakowsky, 3 Dec. 1959, NF7(112).

27. T. L. K. Small, "The Nature and Scope of the NASA University Program," NASA SP-731 (Washington, 1965), pp. 11-12. Also, W. A. Greene, "Step Funded REsearch Grants," *Bioscience* 18 (Dec. 1968): 1133-36.

28. NASA Management Issuance 7100.1, "Conduct of Space Science Program—Selection and Support of Scientific Investigations and Investigators, 29 Apr. 1964, p. 8.

29. Newell to Abe Silverstein, 16 Aug. 1960, NF12(176).

30. Homer E. Newell, Jr., *High Altitude Rocket Research* (New York: Academic Press, 1953), pp. 221-29.

31. Alex Roland, *Research by Committee: A History of the National Advisory Committee for Aeronautics*, NASA SP-4103, comment ed. (Washington, 1979).

32. Naval Research Laboratory Upper Atmosphere Research Reports, and Rocket Research Reports. For examples see: H. E. Newell, Jr. and J. W. Siry, eds., "Upper Atmosphere Research Report No. II," NRL rpt. R-3030 (Washington: Naval Research Laboratory, 30 Dec. 1946); K. Watanabe, J. D. Purcell, and R. Tousey, "Upper Atmosphere Research Report XII: Direct Measurements of Solar Extreme Ultraviolet and X-rays from rockets by Means of a CaSo₄: Mn Phosphor," NRL rpt. 3733 (18 Sept. 1950); Milton W. Rosen and James M. Bridger, "Rocket Research Report No. I: The Viking No. 1 Firings," NRL rpt. 3583 (19 Dec. 1949); Milton W. Rosen and James M. Bridger, "Rocket Research Report No. XIX: The Viking 10 Firings," NRL rpt. 4513 (5 May 1955).

33. As an illustration, see R. L. Heacock et al., "Ranger VII," pt. 2, "Experimenters' Analyses and Interpretations," Tech. Rpt. 32-700 (Pasadena: Jet Propulsion Laboratory, California Institute of Technology, 10 Feb. 1965).

34. As an illustration, see *Communications of the Lunar and Planetary Laboratory*, Vol. 10, pt. 1 (Tucson: Univ. of Arizona, 1973), which shows the series as it was well along in its history.

35. Author's notebook, 11 May 1960, NF38.

36. H. E. Newell, reprot of telephone convsersation with Lloyd Berkner, 19 May 1960, NF12(175).

37. "International Programs," a summary prepared by the Office of International Programs, NASA, 1962: see first page dated 31 Jan. 1962. Alternatively, see any of the international program summaries issued semiannually by NASA.

38. See *Publications of the Goddard Space Flight Center 1959-1962*, vol. 1, *Space Sciences*, and succeeding issues.

39. Author's notebook, 1 Sept. 1959, NF38.

40. Roland, *Research by Committee:*

41. *Rosholt, Administrative History of NASA*, pp. 128-29, 222.

42. Robert Jastrow, ed., *The Exploration of Space* (New York: The Macmillan Co., 1960).

43. Robert Jastrow and Gordon MacDonald to Lloyd Berkner, 10 Dec. 1959, NF4(66).

44. H. E. Newell, "A Home for Planetary Science," *Transactions, American Geophysical Union* 41 (Sept. 1960): 407-09; Bruno Rossi, "Scientific Results of Experiments in Space," ibid., pp. 410-29; Robert Jastrow and Gordon J. F. MacDonald, "Highlights of the Planetary Sciences Program," ibid., pp. 430-34.

45. Arnold W. Frutkin, *International Cooperation in Space* (Englewood Cliffs, N. J.: Prentice Hall, 1965).
46. Newell, "Report on the Second Meeting of the Committee on Space Research Held at the Hague, 12–14 March 1959," NASA lithoprint typescript [Mar. 1959], encl. 7, NF8(127).
47. Ibid., encl. 9.

Chapter 10

1. Homer E. Newell, Jr., *Sounding Rockets* (New York: McGraw-Hill Book Co., 1959); William R. Corliss, *NASA Sounding Rockets, 1958–1968*, NASA SP-4401 (Washington, 1971).
2. William H. Pickering, "History of the Juno Cluster System," chap. 12 in *Astronautical Engineering and Science*, ed. Ernst Stuhlinger et al. (New York: McGraw-Hill Book Co., 1963).
3. Loyd S. Swenson, Jr., James M. Grimwood, and Charles C. Alexander, *This New Ocean: A History of Project Mercury*, NASA SP-4201 (Washington, 1966), pp. 21, 22, 341.
4. Constance McLaughlin Green and Milton Lomask, *Vanguard: A History*, NASA SP-4202 (Washington, 1970; Smithsonian Institution Press, 1971).
5. Ibid., p. 255.
6. Swenson et al., *This New Ocean*, chap. 13.
7. Green and Lomask, *Vanguard*, chap. 9.
8. As an illustration, see rpt. to Launch Vehicle Panel of Aeronautics and Astronautics Coordinating Board, "Intermediate-Class Launch Vehicles for Future DoD/NASA Manned Missions," prepared by Ad Hoc Joint DoD/NASA Vehicle Working Group, typescript, Nov. 1968.
9. Robert A. Goddard, "A Method of Reaching Extreme Altitudes," in *Smithsonian Miscellaneous Collections*, vol. 71, no. 2 (Washington: Smithsonian Institution, Dec. 1919).
10. NASA, "A National Space Vehicle Program: A Report to the President" (Washington, 27 Jan. 1959); also Senate Committee on Aeronautical and Space Sciences, *Investigation of Governmental Organization for Space Activities*, hearings before Subcommittee on Governmental Organization for Space Activities, 86th Cong., 1st sess., 24 Mar.–7 May 1959, pp. 12–24.
11. Wernher von Braun and Frederick I. Ordway III, *The History of Rocketry and Space Travel* (New York: Thomas Y. Crowell Co., 1966, rev. 1969). See also William R. Corliss, *Scientific Satellites*, NASA SP-133 (Washington, 1967), pp. 261–67.
12. Courtney G. Brooks, James M. Grimwood, Loyd S. Swenson, Jr., *Chariots for Apollo: A History of Manned Lunar Spacecraft*, NASA SP-4205 (Washington, 1979).
13. Roger E. Bilstein, *Stages to Saturn: A Technological History of the Apollo/Saturn Launch Vehicles*, NASA SP-4206, in press.
14. Robert L. Rosholt, *An Administrative History of NASA, 1958–1963*, NASA SP-4101 (Washington, 1966), pp. 72–76, 110, 213. In February 1962 NASA and the Department of Defense signed an agreement that neither agency would undertake development of a space launch vehicle without first obtaining the written acknowledgment of the other. House Committee on Science and Astronautics, *Astronautical and Aeronautical Events of 1962*, 88th Cong., 1st sess., committee print, 12 June 1963, p. 21.
15. House Committee on Science and Astronautics, *Review of the Soviet Space Program*, 90th Cong., 1st sess., committee print, 1967, pp. 5–10.
16. Arnold W. Frutkin, *International Cooperation in Space* (Englewood Cliffs, N.J.: Prentice-Hall, 1965), pp. 134–37.
17. Ibid., p. 45; *Europe in Space* (Paris: European Space Research Organization, 1974), pp. 180–81.
18. Air Force Systems Command, *Air Force Eastern Test Range*, a descriptive booklet prepared by Office of Information, Hqs., Air Force Eastern Test Range, Patrick Air Force Base, Florida; Senate Committee on Aeronautical and Space Sciences, *NASA Scientific and Technical Programs*, hearings, 87th Cong., 1st sess., 28 Feb.–1 Mar. 1961, pp. 455–66; Henry R. Van Goey, "Western Test Range Operations Handbook" (Lompoc, Cal.: Kennedy Space Center, WTR, Jan. 1968); Helen T. Wells, Susan H. Whitely, and Carrie E. Karegeannes, *Origins of NASA Names*, NASA SP-4402 (Washington, 1976), pp. 149–50.
19. Joseph A. Shortal, *A New Dimension: Wallops Island Flight Test Range, the First Fifteen Years*, NASA RP-1028 (Washington, 1978).
20. Senate Committee on Aeronautical and Space Sciences, *Soviet Space Programs, 1966–1970*, staff report, 92nd Cong., 1st sess. (9 Dec. 1971), pp. 126–28.

21. R. R. Teeter, "Foreign Space Launch Vehicle Performance Document," rpt. to NASA BMI-NLVP-DD-4(74) (Columbus, Ohio: Battelle Memorial Institute Columbus Laboratories, 8 Apr. 1974), pp. 30–37.

22. Frutkin, *International Cooperation*, pp. 54–57, 145–46.

23. William R. Corliss, *Scientific Satellites*; idem, "NASA Spacecraft," in the series America in Space/The First Decade (Washington: NASA, 1 Oct. 1968); idem, "Putting Satellites to Work," ibid.; idem, "Man in Space," ibid.; John E. Naugle, *Unmanned Space Flight* (New York: Holt, Rinehart & Winston, 1965); Max Faget, *Manned Space Flight* (New York: Holt, Rinehart & Winston, 1966); William K. Widger, Jr., *Meteorological Satellites* (New York: Holt, Rinehart & Winston, 1966); Brooks et al., *Chariots for Apollo*; Wells et al., *Origins of NASA Names*; James J. Haggerty, Jr., *Spacecraft* (Washington: National Science Teachers Association, 1961).

NASA Facts: B-62, "Orbiting Solar Observatory"; C-62, "Ariel: First International Satellite"; E-10-62, "The Explorer Satellites"; F-12-62, "Alouette—Canada's First Satellite"; G-12-62, "Project Relay"; A-1-63, "Explorer XVI: The Micrometeoroid Satellite"; vol. 2, no. 1, "Interplanetary Explorer Satellites"; vol. 2, no. 7, "Nimbus"; vol. 2, no. 8, "Manned Space Flight: Projects Mercury and Gemini"; vol. 2, no. 9, "Mariner IV"; vol. 2, no. 10, "Biosatellites"; vol. 2, no. 12, "TIROS"; vol. 2, no. 13, "Orbiting Geophysical Observatory"; vol. 2, no. 14, "Project Syncom"; vol. 2, no. 15, "Pegasus"; vol. 3, no. 2, "Project Ranger"; vol. 3, no. 4, "Explorer XXIX (The Geodetic Explorer)"; vol. 3, no. 7, "Orbiting Solar Observatory"; vol. 4, no. 1, "Gemini Pictorial"; NF-31, vol. 4, no. 3, "The Pioneer Spacecraft"; NF-32, vol. IV, no. 4, "Lunar Orbiter"; NF-35, vol. 4, no. 6, "Surveyor"; NF-23, vol. 4, no. 7, "Manned Space Flight: Apollo"; NF-39/2-68, "Mariner Spacecraft: Planetary Trailblazers"; NF-43/1-72, "Skylab" (all Washington) ; Newell Files, box 40.

Geophysics and Astronomy in Space Exploration, NASA SP-13; *Lunar and Planetary Sciences in Space Exploration*, NASA SP-14; *Celestial Mechanics and Space Flight Analysis*, NASA SP-15; *Data Acquisition from Spacecraft*, NASA SP-16; *Control, Guidance, and Navigation of Spacecraft*, NASA SP-17; *Bioastronautics*, NASA SP-18; *Chemical Rocket Propulsion*, NASA SP-19; *Nuclear Rocket Propulsion*, NASA SP-20; *Power for Spacecraft*, NASA SP-21; *Electric Propulsion for Spacecraft*, NASA SP-22; *Aerodynamics of Space Vehicles*, NASA SP-23; *Gas Dynamics for Space Exploration*, NASA SP-24; *Plasma Physics and Magnetohydrodynamics in Space Exploration*, NASA SP-25; *Laboratory Techniques in Space Environment Research*, NASA SP-26; *Materials for Space Operations*, NASA SP-27; *Structures for Space Operations*, NASA SP-28 (all Washington, Dec. 1962).

24. Samuel Glasstone, *Sourcebook on the Space Sciences* (New York: D. Van Nostrand Co., 1965), p. 31; Senate Committee on Aeronautical and Space Sciences, *NASA Authorization for Fiscal Year 1961*, hearings, 86th Cong., 2d sess., pt. 1, 28–30 Mar. 1960, p. 312.

25. Lloyd Berkner to George Kistiakowsky, 13 Nov. 1959, NF7(112); H. E. Newell, memo to file, conference rpt. of meeting of Space Science Panel of President's Science Advisory Committee at Institute of Radio Engineers, New York, 18 Dec. 1959, NF12(173).

26. Charles D. Benson and W. D. Compton, *A History of Skylab*, in process; Edward C. Ezell and Linda N. Ezell, *The Partnership: A History of the Apollo-Soyuz Test Project*, NASA SP-4209 (Washington, 1978).

27. Joseph F. Saunders, ed., *The Experiments of Biosatellite II*, NASA SP-204 (Washington, 1971).

28. Newell, conference rpt. on meeting of Space Science Panel, 18 Dec. 1959, NF12(173).

29. Senate Committee on Aeronautical and Space Sciences, *NASA Authorization for Fiscal Year 1976*, hearings, 94th Cong., 1st sess., pt. 2, 18, 19 Mar. 1975, pp. 394–401.

30. Senate Committee, *Soviet Space Programs, 1966–1970*, pp. 198–209.

31. William R. Corliss, *Histories of the Space Tracking and Data Acquisition Network (STADAN), the Manned Space Flight Network (MSFN), and the NASA Communications Network (NASCOM)*, NASA CR-140390 (Washington, June 1974).

32. Corliss, *A History of the Deep Space Network*, NASA CR-151915 (Washington, 1 May 1976).

33. Corliss, *Histories of STADAN, MSFN, and NASCOM*.

34. "Areas of Research in the NASA Space Sciences Program," 10 Feb. 1959, typescript; "The United States National Space Sciences Program," 10 Feb. 1959, typescript; typescript paper on plans for a biosciences program, 30 Mar. 1959; "NASA Program Planning on Space Sciences." Apr. 1960, typescript; all NF2(33); Senate Committee on Aeronautical and Space Sciences, *NASA Authorization for Fiscal Year 1961*, hearings, 86th Cong., 2d sess., pt. 2, *Scientific and Technical Aspects of NASA Program*, 30 June 1960, pp. 575–632.

35. Senate Committee, *NASA Authorization for Fiscal Year 1961*, pt. 2, *Scientific and Technical Aspects*

of *NASA Program*, 1960, p. 786; idem, *NASA Scientific and Technical Programs*, 1961, pp. 34-35; Green and Lomask, *Vanguard*.

36. Eugene M. Emme, *Aeronautics and Astronautics: An American Chronology of Science and Technology in the Exploration of Space, 1915-1960* (Washington: NASA, 1961), p. 102.
37. NASA, "Historical Pocket Statistics" (Washington, Jan. 1972), p. A-13.
38. Newell, memo for files, 16 Dec. 1960, NF12(176).
39. J. A. Simpson to Newell, 30 Aug. 1961, NF13(193).
40. Berkner to Kistiakowsky, 13 Nov. 1959, encl. p. 6, NF7(112), among other matters raised the question of backups for space science flights. NASA's response to Berkner's questions was given in T. Keith Glennan, 3 Dec. 1959, NF7(112).
41. Glennan to Alan Waterman, 29 May 1959, NF11(171). Waterman, director of the National Science Foundation, no doubt prompted by the Space Science Board, had queried NASA as to agency policy on providing backups for its missions. See also Newell, conference rpt. on Space Science Board meeting, 11 Mar. 1960, typescript, NF6(106).
42. Green and Lomask, *Vanguard*, pp. 208-12.
43. Ibid., pp. 98-106. Author's notebook, 30 Sept.-5 Oct. 1957, NF(28), records Richard Porter, U.S. delegate to CSAGI conference on rockets and satellites at National Academy of Sciences, Washington, as saying, in presenting the U.S. National Report, that America expects from Vanguard "at least one success."
44. Roger E. Bilstein, *Stages to Saturn: A Technological History of the Apollo/Saturn Launch Vehicle*, NASA SP-4206, in press.
45. Charles S. Sheldon II, *United States and Soviet Progress in Space: Summary Data through 1972 and a Forward Look*, 73-69 SP (Washington: Library of Congress, 29 Jan. 1973), pp. 2-3.
46. Ibid., p. 3.
47. Senate Committee on Appropriations, *Independent Offices Appropriations, 1961*, hearings before Subcommittee, on H.R. 11776, 86th Cong., 2d sess., 1960, p. 276.
48. House Committee on Science and Astronautics, *1966 NASA Authorization*, hearings before the Subcommittee on Science and Applications, 89th Cong., 1st sess., pt. 3, 4-26 Mar. 1965, pp. 803-06.
49. Brooks et al, *Chariots for Apollo*; Senate Committee, *Soviet Space Programs, 1966-1970*, pp. 230-32.
50. House Committee on Science and Astronautics, *Investigation into Apollo 204 Accident*, hearings, 90th Cong., 1st sess., vols. 1-3, 10 Apr.-10 May 1967.
51. NASA Historical Div., *Astronautics and Aeronautics, 1968: A Chronology on Science, Technology, and Policy*, NASA SP-4010 (Washington, 1969), p. 157.
52. Idem, *Astronautics and Aeronautics, 1964*, NASA SP-4005 (1965), p. 135.
53. Merton J. Peck and Frederic M. Scherer, *The Weapons Acquisition Process: An Economic Analysis* (Boston: Harvard, Grad. Sch. of Bus. Admin., Div. of Res., 1962), p. 22.
54. D. D. Wyatt to NASA administrator, "NASA Cost Projections," 10 Apr. 1969, NF40.
55. Hans Mark to Wernher von Braun, 22 Apr. 1971, NF40.
56. House Committee, *1966 NASA Authorization*, pp. 529, 1090-91.
57. Newell, conference rpt., Space Science Panel, 18 Dec. 1959, NF12(173).
58. R. Cargill Hall, *Lunar Impact: A History of Project Ranger*, NASA SP-4210 (Washington, 1977), pp. 252-54.
59. NASA chart on Centaur mission assignments, 27 Apr. 1962, NF2(32).
60. NASA, Office of Space Science and Applications, "Atlas Centaur Development Flight," Mission Operation Rpt. S-891-63-01 (Washington, 22 Nov. 1963), pp. 8-9.
61. House Committee on Science and Astronautics, *Centaur Program*, hearings before Subcommittee on Space Science and Applications, 87th Cong., 2d sess., 15, 18 May 1962, and *Centaur Launch Vehicle Development Program*, H. rpt. 1959, 2 July 1962.
62. Arnold Levine, *Managing NASA in the Apollo Era*, in process.
63. Ibid.
64. Ibid.
65. Ibid.
66. Wyatt to NASA admin., "NASA Cost Projections."
67. Arnold Levine, *Managing NASA*.
68. NASA, OSSA Monthly Status Reports (Washington: Office of Space Science and Applications, NASA, Nov. 1966 . . .); NASA Issuance NHB 2340.2, "OSSA/OART Project Management Information and Control System (MICS)" (Washington, 1 Nov. 1966), formalized publication of reviews of the status of projects which had been conducted monthly for the past three years.

69. Arnold Levine, *Managing NASA.*
70. NASA, "NASA Record of Performance" in "NASA Pocket Statistics," (Washington, Apr. 1976), p. 5.
71. NASA, "Historical Pocket Statistics" (Washington, Jan. 1972), pp. A-9 to A-11; ibid. (Jan. 1976), pp. C-3, C-5.

Chapter 11

1. *Significant Achievements in Satellite Meteorology, 1958–1964,* NASA SP-96 (Washington, 1966); William K. Widger, Jr., *Meteorological Satellites* (New York: Holt, Rinehart and Winston, 1966).
2. *Ecological Surveys from Space,* NASA SP-230 (Washington, 1970); Richard S. Williams, Jr., and William D. Carter, eds., *ERTS-1: A New Window on Our Planet,* Geological Survey Professional Paper 929 (Washington: U.S. Geological Survey, 1976).
3. John O'Keefe, "Geodetic Significance of an Artificial Satellite," appendix to the A[merican] R[ocket] S[ociety] Space Flight Committee, "Utility of an Artificial Unmanned Earth Satellite," *Jet Propulsion* 25 (Feb. 1955): 71–78; John O'Keefe and C. D. Batchelor, "Perturbations of a Close Satellite by the Equatorial Ellipticity of the Earth," *Astronomical Journal* 62 (1957): 183.
4. A. V. Gangnes, L. F. Jenkins, Jr., and J. A. Van Allen, "The Cosmic Ray Intensity above the Atmosphere," *Physical Review* 75 (Jan. 1949): 57–69.
5. J. A. Van Allen and S. F. Singer, "On the Primary Cosmic Ray Spectrum," *Physical Review* 78 (June 1950): 819; erratum, ibid., 80 (Oct. 1950): 116.
6. J. A. Van Allen, "Rocket Measurements of Soft Radiation," in *Annals of the International Geophysical Year,* 12, pt. 2, ed. Homer E. Newell, Jr., and Leonard N. Cormier (London: Pergamon Press, 1960): 646–50.
7. Paper presented by Van Allen at joint meeting of National Academy of Sciences and American Physical Society on 1 May 1958. See also, Van Allen, G. H. Ludwig, E. C. Ray, and C. E. McIlwain, "The Observation of High Intensity Radiation by Satellites 1958 α and γ," *Annals of the IGY,* 12, pt. 2 (1960): 671–81.
8. S. K. Mitra, *The Upper Atmosphere* (Calcutta: Royal Asiatic Society of Bengal, 1947), pp. 27–29.
9. K. Birkeland, *Arch. Sci. Phys.* 1 (1896): 497; E. Brüche, "Some new theoretical and experimental results on the aurora polaris," *Terrestrial Magnetism and Atmospheric Electricity* 36 (1931): 41–52; C. Stormer, "Twenty-five Years' York on the Polar Aurora," *Terrestrial Magnetism and Atmospheric Electricity* 35 (1930): 193–208.
10. P. S. Epstein, "Note on the Nature of Cosmic Rays," *Proceedings of the National Academy of Sciences, U.S.* 16 (1930): 658–63; G. Lemaitre and M. S. Vellarta, "On Compton's Latitude Effect of Cosmic Radiation," *Physical Review* 43 (Jan. 1933): 87–91; G. Lemaitre, M. S. Vellarta, and L. Bouckert, "On the North-South Asymmetry of Cosmic Radiation," *Physical Review* 47 (Mar. 1935): 434–36; G. Lemaitre and M. S. Vellarta, "On the Geomagnetic Analysis of Cosmic Radiation," *Physical Review* 49 (May 1936): 719–26; R. A. Alpher, "Theoretical Geomagnetic Effects in Cosmic Radiation," *Journal of Geophysical Research* 55 (Dec. 1950): 437–71.
11. S. F. Singer, "A New Model of Magnetic Storms," *Transactions of the American Geophysical Union* 38 (1957): 175.
12. *Significant Achievements in Particles and Fields, 1958–1964,* NASA SP-97 (Washington, 1966).
13. Ibid.; John Naugle, *Unmanned Space Flight* (New York: Holt, Rinehart and Winston, 1965).
14. E. N. Parker, "Dynamics of the Interplanetary Gas and Magnetic Fields," *Astrophysical Journal* 132 (1960): 821–66; idem, "The Hydrodynamic Theory of Solar Corpuscular Radiation and Stellar Winds," *Astrophysical Journal* 132 (1960): 821–66.
15. K. I. Gringauz et al., "A Study of the Interplanetary Ionized Gas, High-Energy Electrons, and Corpuscular Radiation from the Sun by Means of the Three-Electrode Trap for Charged Particles on the Second Cosmic Rocket," *Doklady Akad. Nauk SSSR* 131 (1960): 1301; *Soviet Phys.-Doklady* 5 (1960): 361–64.
16. A. Bonetti, H. S. Bridge, et al., "Explorer 10 Plasma Measurements," *Journal of Geophysical Research* 68 (1963): 4017–63.
17. M. Neugebauer and C. W. Snyder, "Solar Plasma Experiment: Preliminary Mariner II Observations," *Science* 138 (1962): 1095–97; Snyder and Neugebauer, "Interplanetary Solar Wind Measurements by Mariner II," in *Space Research IV,* ed. P. Muller (Amsterdam: North-Holland Publishing Co., 1964), pp. 89–113; Snyder, Neugebauer, and U. R. Rao, "The Solar Wind Velocity and Its

Correlation with Cosmic Ray Variations and with Solar and Geomagnetic Activity," *Journal of Geophysical Research* 68 (1963): 6361-70; H. Bridge et al., "Preliminary Results of Plasma Measurements on IMP-A," in *Space Research V*, ed. D. G. King-Hele et al. (Amsterdam: North-Holland Publishing Co., 1965), pp. 969-78; "Initial Results from the First Interplanetary Monitoring Platform (IMP I)," I. G. Bulletin 84, in *Transactions of the American Geophysical Union* 45 (1964): 501.

18. S. C. Freden and R. S. White, "Protons in the Earth's Magnetic Field," *Physical Review Letters* 3 (1959): 9.

19. NASA, *Particles and Fields, 1958-1964*, pp. 17-19.

20. Ibid., pp. 19-29.

21. N. F. Ness, C. S. Scearce, and J. B. Seek, "Initial Results of the IMP I Magnetic Field Experiment." *Journal of Geophysical Research* 69 (1964): 3531-69.

22. Bridge et al., "Preliminary Results of Plasma Measurements on IMP-A"; "Initial Results from the First Interplanetary Monitoring Platform (IMP I)," I.G. Bulletin 84.

23. N. F. Ness, "The Magnetohydrodynamic Wake of the Moon," *Journal of Geophysical Research* 70 (1965): 517-34.

24. NASA, *Particles and Fields, 1958-1964*, p. 55.

25. Ibid., pp. 52-53.

26. B. J. O'Brien, "Lifetimes of Outer Zone Electrons and Their Precipitation into the Atmosphere," *Journal of Geophysical Research* 67 (1962): 3687.

27. NASA, *Particles and Fields, 1958-1964*, pp. 33-34.

28. Singer, "A New Model of Magnetic Storms."

29. Norman F. Ness to Homer Newell, 12 July 1978, comments on draft Newell MS., NF40.

30. Norman F. Ness et al., "Early Results from the Magnetic Field Experiment on Lunar Explorer 35," *Journal of Geophysical Research* 72 (Dec. 1967): 5769.

31. W. A. Heiskanen, "Geodesy," *Encyclopaedia Britannica*, 1967 ed, vol. 10, p. 134. Much of the historical summary of the next few paragraphs is based on Heiskanen's article.

32. Ibid., p. 34.

33. Myron Lecar, John Sorenson, and Ann Eckels, "A Determination of the Coefficient J of the Second Harmonic in the Earth's Gravitational Potential from the Orbit of Satellite 1958 β_2," ibid., pp. 181-97.

34. E. Buchar, "Determination of the Flattening of the Earth by Means of the Displacement of the Node of the Second Soviet Satellite (1957 β)," *Annals of the IGY*, 12, pt. 1 (1960): 174-76; Harvard College Observatory Announcement Card 1408, "Oblateness of the Earth by Artificial Satellites," ibid., 176.

35. S. W. Henriksen, "The Hydrostatic Flattening of the Earth," ibid., pp. 197-98.

36. W. M. Kaula, "The Shape of the Earth," in *Introduction to Space Science*, ed. Wilmot N. Hess and Gilbert D. Mead (New York: Gordon and Breach, 2d ed., 1968), p. 333.

37. J. A. O'Keefe, Ann Eckels, and R. K. Squires, "Pear-Shaped Component of the Geoid from the Motion of Vanguard I," *Annals of IGY*, 12, pt. 1 (1960): 199-201; Desmond King-Hele, "The Shape of the Earth," *Science* 192 (25 June 1976): 1293-1300.

38. W. M. Kaula, "Current Knowledge of the Earth's Gravitational Field from Optical Observations," in *Significant Achievements in Satellite Geodesy, 1958-1964*, NASA SP-94 (Washington, 1966), p. 33.

39. C. A. Wagner, "A Determination of Earth Equatorial Ellipticity from Seven Months of Syncom 2 Longitude Drift," in *Satellite Geodesy, 1958-1964*, pp. 93-97.

40. H. M. Dufour, "Geodetic Junction of France and North Africa by Synchronized Photographs Taken from Echo I Satellite," ibid., pp. 109-20.

41. Robert R. Newton, "Geometrical Geodesy by Use of Doppler Data," ibid., pp. 121-23.

42. Owen W. Williams, Paul H. Dishong, and George Hadgigeorge, "Results from Satellite (ANNA) Geodesy Experiments," ibid., pp. 129-38.

43. *Satellite Geodesy, 1958-1964*, p. 155; Wagner, "Earth Equatorial Ellipticity," ibid., p. 96.

44. "Report of the Working Group on the System of Astronomical Constants: Agenda and Draft Reports," International Astronomical Union, 12th General Assembly, 25 Aug.-3 Sept. 1964, reprinted in *Satellite Geodesy, 1958-1964*, pp. 101-06.

45. John O'Keefe to Homer Newell, 22 June 1978, comments on draft Newell MS., NF40; W. A. Heiskanen and F. A. Vening Meinesz, *The Earth and Its Gravity Field* (New York: McGraw-Hill Book Co., 1958), p. 72.

Chapter 12

1. Public Law 85-568, 85th Cong., 2d sess., 72 Stat. 426, Sec. 102, 29 July 1958.
2. Lloyd Berkner to George Kistiakowsky, 13 Nov. 1959, NF7(112).
3. Hugh Odishaw to T. Keith Glennan, 1 Aug. 1958, NF7(112).
4. Norriss S. Hetherington, "Winning the Initiative: NASA and the U.S. Space Science Program," *Prologue* 7 (Summer 1975): 99-107.
5. NASA, Space Science Div., "Areas of Research in the NASA Space Sciences Program," mimeographed, 48 pp., 10 Feb. 1959; idem, "The United States National Space Sciences Program," mimeographed, 10 Feb. 1959; NF2(33); Lloyd V. Berkner and Hugh Odishaw, *Science in Space* (New York: McGraw-Hill Book Co., 1961).
6. Berkner to Kistiakowsky, 13 Nov. 1959, NF7(112).
7. Glennan to Kistiakowsky, 3 Dec. 1959, NF7(112).
8. Homer E. Newell, "The Extent and Adequacy of Relationships between NASA and the Scientific Community," typescript, 17 June 1960, NF11 (161).
9. Donald Hornig to Hugh Dryden, 20 Oct. 1961, NF6(100).
10. Newell to Hornig, 15 Nov. 1961, NF6(100).
11. *A Review of Space Research*, National Academy of Sciences–National Research Council publication 1079 (Washington, 1962). Author's notebook, 17 June-10 Aug. 1962, NF28, gives a number of insights not provided by the formal academy rpt.
12. Alex Roland, rpt. of telephone conversation with Abe Silverstein, 1978, NF40.
13. L. V. Berkner to James E. Webb, 31 Mar. 1961, and Berkner to Alan Waterman, 31 Mar. 1961, both transmitting a Space Science Board paper entitled "Man's Role in the National Space Program," NF6(100); "Man's Role in the National Space Program," National Academy of Sciences release, 7 Aug. 1961, NF5(93).
14. P. H. Abelson, "Manned Lunar Landing," *Science* 140 (19 Apr. 1963); "Moon Plans Called Wasteful," article based on testimony of Dr. Philip A. Abelson, editor of *Science*, before Senate Committee on Aeronautical and Space Sciences, *Washington Star*, 10 June 1963.
15. *Christian Science Monitor*, Friday, 9 Apr. 1965, NF13(200).
16. NASA Historical Staff, *Astronautics and Aeronautics, 1964: A Chronology on Science, Technology, and Policy*, NASA SP-4005 (Washington, 1965), p. 356; *1965*, NASA SP-4006 (1966), pp. 5, 133, 299-300; *1966*, NASA SP-4007 (1967), p. 302; *1967*, NASA SP-4008 (1968), pp. 1, 230, 233-44.
17. Author's notebook, Jan. 1971, NF28.
18. Newell, "Report of the Subcommittee on Scientist Astronauts of SPAC," 8 Sept. 1975.
19. *A Review of Space Research*, NAS-NRC pub. 1079; author's notebook, 17 June-10 Aug. 1962, NF28.
20. Author's notebook, 20 June-16 July 1965, NF28.
21. *Priorities for Space Research 1971-1980*, rpt. of study by Space Science Board (Washington: National Academy of Sciences, 1971). See also author's notebooks for July 1970, NF28.
22. Harold C. Urey to Newell, 13 May 1963, NF13(1963).
23. Urey to Newell, 15 June 1965, NF14(201).
24. NASA Management Issuance 37-1-1, "Establishment and Conduct of Space Science Program—Selection of Scientific Experiments," typescript, 15 Apr. 1960, NASA Historical Archives.
25. NASA Management Issuance, NMI 7100.1, "Conduct of Space Science Program—Selection and Support of Scientific Investigations and Investigators," 29 Apr. 1964, app. C.
26. See, for example, NASA, "Opportunities for Participation in Space Flight Investigations" (Washington, Apr. 1967); also, same title, NASA NHB 8030.1A (Washington, Apr. 1967).
27. James E. Webb to Norman Ramsey, 14 Jan. 1966, NF13(188).
28. NASA Ad Hoc Science Advisory Committee, "Report to the Administrator," mimeographed, 15 Aug. 1966.
29. Ibid., p. 3; T. Keith Glennan to Monte Wright, 15 July 1978, comments on draft Newell MS., NF40.
30. NASA Management Instructions 1156.12, 1 May 1967, and NMI 1156.16, 15 Mar. 1968.
31. See, for example, Astronomy Missions Board, *A Long Range Program in Space Astronomy*, NASA SP-213 (Washington, July 1969).
32. Richard Nixon to the Vice President, the Secretary of Defense, the Acting Administrator, National Aeronautics and Space Administration, and the Science Adviser, 13 Feb. 1969.
33. "The Post-Apollo Space Program: Directions for the Future," Space Task Group rpt. to the president, Sept. 1969.

34. NASA Management Instruction 1156.20, 8 Mar. 1971.
35. President's Science Advisory Committee, *The Space Program in the Post-Apollo Period* (Washington: The White House, 1967).
36. June Merker, looseleaf notes on recommendations received over the years by NASA from advisory groups, NASA Administrator's Office files.
37. Hugh Odishaw to Monte Wright, 11 Aug. 1978, comments on draft Newell MS., NF40.
38. Norman Ness to Homer Newell, 12 July 1978, comments on draft Newell MS., p. 3., NF40.

Chapter 13

1. Alex Roland, *Research by Committee: A History of the National Advisory Committee for Aeronautics,* NASA SP-4103, comment ed. (Washington, 1979).
2. Jane Van Nimmen and Leonard C. Bruno with Robert L. Rosholt, *NASA Historical Data Book, 1958–1968,* vol. 1, *NASA Resources,* NASA SP-4012 (Washington, 1976), p. 540.
3. NASA, Office of General Counsel, *National Aeronautics and Space Act of 1958, As Amended, and Related Legislation* (Washington, 1 July 1969), p. 9.
4. Homer E. Newell, rpt. of conference between Glenn Seaborg and colleagues and Hugh Dryden and Newell, 1 May 1959, NF11(171).
5. Public Law 87-98, 87th Cong., 1st sess., 75 Stat. 216, 21 July 1961.
6. Newell, memo on NASA policy of support of research in universities and nonprofit organizations, 10 Nov. 1960, NF12(176).
7. Author's notebook, 22 and 23 June 1961, NF28.
8. Newell, conference rpt. on internal NASA meeting to discuss developing NASA-university relationships, 30 June 1961, NF12(177).
9. Author's notebook, 14 Aug. 1961, NF28.
10. W. A. Greene, "Step Funded Research Grants," *Bioscience* 18 (1968): 1133–36; T. L. K. Smull, *The Nature and Scope of the NASA University Program,* NASA SP-73 (Washington, 1965), pp. 11–12; NASA, Office of Grants and Research Contracts Brochure, *A Guide to NASA Policies and Procedures for Grants and Research Contracts,* Sept. 1965, p. 20: NF19(402).
11. Edwin Gilliland, "Meeting Manpower Needs in Science and Technology," President's Science Advisory Committee rpt., 12 Dec. 1962.
12. Smull, *NASA University Program,* pp. 23–28.
13. Van Nimmen et al., *NASA Resources,* p. 541.
14. John C. Honey to Newell, 7 Feb. 1962, NF2(41); also author's notebook, 7 Feb. 1962, NF28.
15. Sidney G. Roth, "A Study of NASA-University Relations," 30 June 1964, NF2(41).
16. D. J. Montgomery, "A Study of NASA-University Relations," 30 Nov. 1965, NF2(41).
17. Smull, *NASA University Program,* NF2(41).
18. *Summary Report on the NASA-University Program Review Conference, Kansas City, 1–3 March 1965,* NASA SP-81 (Washington, 1965), NF18(378).
19. Author's working file, NF18(378).
20. Martin Summerfield to Newell, 3 May 1965, NF13(200).
21. Author's notebook, 19 Oct. 1961, NF28.
22. Ibid., 20 Oct. 1961.
23. Ibid., 23 Oct. 1961.
24. Ibid., 26 Oct. 1961.
25. Ibid., 30 Oct. 1961.
26. Donald Holmes, memo to file, 3 July 1962, National Archives Rec. Grp. 255, Acc. 67A601, FRC Box 27 (1/17-51-1).
27. James E. Webb to Newell, 25 July 1962, Rec. Grp. 255, Acc. 67A601, FRC Box 27 (1/17-51-1); Smull, *NASA University Program,* pp. 36–37. See also Newell to Assoc. Admin., NASA, 23 July 1962.
28. Webb to Newell, 19 Feb. 1965, NF2(41).
29. Newell to Webb, 20 Apr. 1966, NF2(41).
30. See, for example, Robert Jastrow and A. G. W. Cameron, eds., *Origin of the Solar System* (New York: Academic Press, 1968); Jastrow and Malcolm H. Thompson, *Astronomy: Fundamentals and*

Frontiers (New York: John Wiley & Sons, 1972); Jastrow, *Red Giants and White Dwarfs* (New York: Harper & Row, 1967).

31. Jastrow to T. Keith Glennan, 11 Dec. 1958, NF2(35).
32. Hugh Dryden to Abe Silverstein, 24 Dec. 1958; NASA release, "NASA Announces Research Appointments Program," 11 Mar. 1959: NF2(35).
33. Newell to Hugh Dryden, 15 Dec. 1959, NF12(173).
34. Jastrow to Silverstein, 14 Oct. 1960, NF2(46); Jastrow to Silverstein, 13 Dec. 1960, NF2(46); Glennan to Silverstein, 14 Dec. 1960. See also Glennan's approval signature, 4 Jan. 1961, at bottom of supporting memo, Silverstein to Glennan, 3 Jan. 1961, NF2(46).
35. Van Nimmen et al., *NASA Resources*, p. 284.
36. Ibid., pp. 284–85.
37. Jastrow and Cameron, *Origin of the Solar System.*
38. Author's notebook, 20, 21 June 1963, NF28. This committee should not be confused with the Physical Sciences Committee later established under NASA's Space Program Advisory Council.
39. *Weekly compilation of Presidential Documents,* week ending Friday, 8 Mar. 1968, pp. 411–12; E. W. Quintrell to B. L. Kropp, 23 Dec. 1968, forwards from NASA to National Academy of Sciences the initial grant beginning 1 Oct. 1968 for operation of Lunar Science Institute by the academy, NASA University Affairs Office records, Lunar Science Institute 1969; Office of Recorder of Deeds, District of Columbia, Certificate of Incorporation of Universities Space Research Association, 12 Mar. 1969, plus attachments; NASA University Affairs Office records, Lunar Science Institute 1969. See also unsigned paper, "History of Lunar Science Institute (LSI) and University Space Research Association (USRA)," July 1971, in NASA University Affairs Office records, Lunar Science Institute, 1969.

Chapter 14

1. Alex Roland, *Research by Committee: A History of the National Advisory Committee for Aeronautics.* NASA SP-4103, comment ed. (Washington, 1979).
2. Jane Van Nimmen and Leonard C. Bruno with Robert L. Rosholt, *NASA Historical Data Book, 1958–1968,* vol. 1, *NASA Resources,* NASA SP-4012 (Washington, 1976), pp. 222–27, 279–95, 297–312, 377–410, 347–76, 259–67; Alfred Rosenthal, *Venture into Space: Early Years of the Goddard Space Flight Center,* NASA SP-4301 (Washington, 1968); Joseph A. Shortal, *A New Dimension: Wallops Island Flight Test Range, the First Fifteen Years,* NASA RP-1028 (Washington, 1978); Clayton Koppes, *History of the Jet Propulsion Laboratory,* in preparation at California Institute of Technology.
3. Van Nimmen et al., *NASA Resources,* pp. 386–87.
4. Ibid., p. 351.
5. Ibid., pp. 302–09.
6. Richard F. Hirsh, "A Search for Scientific Paradigms in Early Non-solar X-ray Astronomy, "NASA HHN-139 (NASA History Office, Aug. 1976).
7. Van Nimmen et al., *NASA Resources,* pp. 421–35.
8. See published proceedings of Lunar Science Conferences held annually for many years in Houston; for example, A. A. Levinson, ed., *Proceedings of the Apollo 11 Lunar Science Conference, Houston, Texas, January 5–8, 1970,* 3 vols. (New York: Pergamon Press, 1970).
9. NASA, "Historical Pocket Statistics" (Washington, Jan. 1976), pp. A88, A6–A9, A13–A15, A10–A11.
10. Ibid., pp. A20–A26; Helen T. Wells, Susan H. Whiteley, and Carrie E. Karegeannes, *Origins of NASA Names,* NASA SP-4402 (Washington, 1976), p. 15.
11. Van Nimmen et al., *NASA Resources,* p. 336.
12. House Committee on Science and Astronautics, *1963 NASA Authorization,* hearings before Subcommittee on Space Science, 87th Cong., 2d sess., 6–20 Mar. 1962, pp. 1959–66; *Centaur Program,* hearings, 15–18 May 1962, pp. 111–15.
13. Author's notebook, 5 Oct. 1962, NF28.
14. Harry J. Goett to Robert Seamans, 5 July 1963.
15. Proposed revision of NMI 37-1-1, "Conduct of Space Sciences Program—Selection of Scientific Investigations and Investigators," typescript, 26 July 1963, App. A, NF17(318).

16. Homer E. Newell to Goett, 5 Nov. 1963, NF12(183).
17. Author's notes for talk at Airlie House, Warrenton, Va., to NASA project managers, 9 Mar. 1964, NF12(184).
18. Author's notebook, 30 Dec. 1963, NF28.
19. Years later an analysis by D. D. Wyatt showed what had caused the concern of Associate Administrator Seamans and his staff. See, D. D. Wyatt to NASA Admin., "NASA Cost Projections," 10 Apr. 1969, NF40.
20. Robert C. Seamans, Jr., to Monte D. Wright, 3 Aug. 1978, NF40.
21. Ibid.

Chapter 15

1. William H. Pickering, "History of the Juno Cluster System," chap. 12 in Ernst Stuhlinger et al., eds., *Astronautical Engineering and Science from Peenemünde to Planetary Space* (New York: McGraw-Hill Book Co.. 1963), pp. 204–14; Jane Van Nimmen and Leonard C. Bruno with Robert L. Rosholt, *NASA Historical Data Book, 1958–1968*, vol. 1, *NASA Resources*, NASA SP-4012 (Washington, 1976), p. 456.
2. Van Nimmen et al., *NASA Resources*, p. 456.
3. Ibid., pp. 454–72; Clayton R. Koppes, *History of the Jet Propulsion Laboratory*, in preparation. See also R. Cargill Hall, *Lunar Impact: The History of Project Ranger*, NASA SP-4210 (Washington, 1977); Erasmus Kloman, *Unmanned Space Project Management: Surveyor and Lunar Orbiter*, NASA SP-4901 (Washington, 1972); Jet Propulsion Laboratory staff, *Mariner: Mission to Venus* (New York: McGraw-Hill Book Co., 1963); Stewart A. Collins, *The Mariner 6 and 7 Pictures of Mars*, NASA SP-263 (Washington, 1971); William K. Hartmann and Odell Raper with cooperation of Mariner 9 Science Experiment Team, *The New Mars: The Discoveries of Mariner 9*, NASA SP-337 (Washington, 1974).
4. W. H. Pickering to Homer E. Newell, 30 Aug. 1978, NF40.
5. T. Keith Glennan to Newell, 25 July 1978, comments on draft Newell MS., NF40.
6. NASA contract NASW-6 with California Institute of Technology, 1 May 1959, pt. 1, art. 1, sec. a; NASA contract NAS7-100 with Cal Tech, 1 Jan. 1962, pt. 1, art. 1, sec. (b.)(1.), files of Institutional Operations Div., Office of Management Operations, NASA Hq.
7. Glennan to Newell, 25 July 1978 and 2 Jan. 1979, NF40.
8. Stuhlinger et al., *Astronautical Engineering and Science*, p. 204.
9. Author's notebook, 12 Jan. 1959, NF28.
10. Jet Propulsion Laboratory contract 950056 with Hughes Aircraft Co., 1 Mar. 1961.
11. Glennan to Newell, 25 July 1978, NF40.
12. Richard E. Horner to William Pickering, 16 Dec. 1959, NF11(163), NF15(251); Abe Silverstein to Pickering, 21 Dec. 1959; Newell, trip rpt. on visit to Jet Propulsion Laboratory, 28 Dec. 1959, rpt. dated 30 Dec. 1959: NF11(163), NF15(251).
13. Pickering to Silverstein, 17 Dec. 1959, NF15(251).
14. Pickering to Newell, 22 Mar. 1960, NF11(163).
15. Glennan to Newell, 25 July 1978, NF40.
16. Author's notebook, 5 Oct. 1962, NF28.
17. JPL staff, *Mariner: Mission to Venus*; Irl Newlan, *First to Venus: The Story of Mariner II* (New York: McGraw-Hill Book Co., 1963); Hall, *Lunar Impact*, pp. 81–255.
18. Author's notebook, 5 Oct. 1962, NF28.
19. Author's notebook, 23 Oct. 1963, NF28.
20. Newell to Edgar M. Cortright, 24 Dec. 1963; Newell, rpt. of phone conversation with Pickering, 24 Dec. 1963: NF12(183).
21. Hall, *Lunar Impact*, p. 237.
22. George P. Miller to Newell, 5 Feb. 1964; Newell to Pickering, 14 Feb. 1964: NF12(184).
23. NASA Announcement 64-27, "Establishment of the Ranger VI Review Board, 3 Feb. 1964," JPLHF 2-1811; R. Cargill Hall, *Project Ranger: A Chronology*, JPL/HR-2 (Pasadena, Apr. 1971), p. 431.
24. House Committee on Science and Astronautics, *Investigation of Project Ranger*, hearings before Subcommittee on NASA Oversight, 88th Cong., 2d sess., 27 Apr.–4 May 1964.

25. Pickering to Newell, 30 Aug. 1978, NF40.
26. Author's notebook, 16 June 1964, NF28.
27. Newell to Pickering, 13 July 1964, NF12(185).
28. Newell to Pickering, 14 July 1964, NF12(185); Newell rpt. of conference with Lee DuBridge, 14 Sept. 1964, RG 255, Acc. 72A070, Box 9.
29. NASA contract NAS7-100 with California Institute of Technology, Mod. 10, approved 16 Dec. 1964; Inst. Ops. Div., NASA. See NASA Contract Briefing Memo, contract NAS7-100 with Cal Tech, 12 Jan. 1965, NF40.
30. Transcript of *Ranger 7* press conference, 31 July 1964, NF18(353); summary of presidential briefing on *Ranger 7*, 1 Aug. 1964, NF2(37); House of Representatives, *Congressional Record*, 88th Cong., 2d sess., 11 Aug. 1964, pp. 18370–73, NF18(353).
31. NASA, Office of Space Science and Applications, Lunar and Planetary Div., Surveyor Program, *Surveyor Program Results*, NASA SP-184 (Washington, 1969).
32. Gregg Mamikunian to Newell, 26 May 1966, NF14(202).
33. Draft Evaluation Report on Cal Tech/JPL, 16 June 1965, NF12(186).
34. John E. Naugle to Richard McCurdy et al., 27 Sept. 1971, with encl.: "JPL Performance Evaluation for the Period 1 January through 30 June 1971," NF40; a complete copy is in files of Office of Tracking and Data Acquisition.

Chapter 16

1. Jane Van Nimmen and Leonard C. Bruno with Robert L. Rosholt, *NASA Historical Data Book, 1959–1968*, vol. 1, *NASA Resources*, NASA SP-4012 (Washington, 1976), p. 540; Robert L. Rosholt, *An Administrative History of NASA, 1958–1963*, NASA SP-4101 (Washington, 1966), pp. 124–27, 342–44.
2. Richard S. Johnston, Lawrence F. Dietlein, and Charles A. Berry, *Biomedical Results of Apollo*, NASA SP-368 (Washington, 1975); J. W. Dyer, tech. ed., *Biosatellite Project: Historical Summary Report* (Moffett Field, Calif.: Ames Research Center, Dec. 1969), chap. 9; Richard S. Young, "Biological Experiments in Space," *Space Science Reviews*, 8(1968): 665–89; Joseph F. Saunders, *The Experiments of Biosatellite II*, NASA SP-204 (Washington, 1971).
3. Clayton S. White and Otis O. Benson, Jr., eds., *Physics and Medicine of the Upper Atmosphere: A Study of the Aeropause* (Albuquerque: Univ. of New Mexico Press, 1952).
4. NASA, "Historical Pocket Statistics" (Washington, Jan. 1976), pp. A-52, A-88, 12, 52, 56, 66; Dyer, *Biosatellite Project*; Young, "Biological Experiments in Space"; T. O. Paine to Homer Newell, 16 Oct. 1978, comments on Newell draft MS., p. 18–22, NF40; Dyer, *Biosatellite Project*, chap. 9.
5. Colin S. Pittendrigh, Wolf Vishniac, and J. P. T. Pearman, *Biology and the Exploration of Mars*, National Academy of Sciences Publication 1296 (Washington, 1966); Cyril Ponnamperuma, ed., *Chemical Evolution of the Giant Planets* (New York: Academic Press, 1976); idem, *Exobiology* (Amsterdam & London: North-Holland Publishing Co., 1972).
6. Gerald A. Soffen et al., in *Scientific Results of the Viking Project*, reprinted from *Journal of Geophysical Research* 82, no. 28 (30 Sept. 1977).
7. NASA, "Historical Pocket Statistics," Jan. 1976, pp. A-37, A-64, 73; William H. Pickering, "Mariner 4's Flight to Mars," *Astronautics and Aeronautics*, Oct. 1965, p. 21; Walter Sullivan. "Mariner 4's Final Photos Depict a Moonlike Mars," *New York Times*, 30 July 1965, pp. 1, 9; Earl C. Gottschalk, "Chemist Denies Impossibility of Life on Mars," *St. Louis Post Dispatch*, 17 Nov. 1965; William K. Hartmann and Odell Raper, *The New Mars: The Discoveries of Mariner 9*, NASA SP-337 (Washington, 1974), p. 166.
8. Author's notebook, 31 July 1962, NF28.
9. President's Science Advisory Committee, Space Science and Technology Panel, "The Biomedical Foundations of Manned Space Flight" (Washington, Nov. 1969.).
10. Space Science Board, *Space Biology* (Washington: National Academy of Sciences, 1970); Space Science Board minutes, 13 Jan. 1970; also author's notebook, 13 Jan. 1970, NF28.
11. Author's notebook, 13 Nov. 1969, NF28.
12. Homer E. Newell to Phillip Handler, 12 Aug. 1969, NF24.
13. H. Bentley Glass, ed., *Life Sciences in Space* (Washington: National Academy of Sciences, 1970), especially pp. 4–6.
14. Newell to George M. Low, 9 Nov. 1970, NF25.
15. Telephone conversation, Newell with Bentley Glass, 13 Nov. 1970, NF29.

Chapter 17

1. Robert L. Rosholt, *An Administrative History of NASA, 1958–1963*, NASA SP-4101 (Washington, 1966), p. 185; Norriss S. Heatherington, "Winning the Initiative: NASA and the U.S. Space Science Program," *Prologue* 7, no. 2 (1975): 99–107.
2. House Committee on Science and Astronautics, *1966 NASA Authorization*, hearings before Subcommittee on Space Science and Applications, 89th Cong., 1st sess., rpt. 3, 4–26 Mar. 1965, pp. 856–59.
3. NASA Historical Div., *Astronautics and Aeronautics, 1967: A Chronology on Science, Technology, and Policy*, NASA SP-4008 (Washington, 1969), pp. 21–27.
4. Idem, *Astronautics and Aeronautics, 1965*, NASA SP-4006 (1966), p. 174.
5. Idem, *Astronautics and Aeronautics, 1966*, NASA SP-4007 (1967), pp. 52, 59.
6. Idem, *Astronautics and Aeronautics, 1968*, NASA SP-4010 (1969), p. 26.
7. Ibid., pp. 212–13.
8. Manned Spacecraft Center, *Analysis of Apollo 8 Photography and Visual Observations*, NASA SP-201 (Washington, 1969).
9. NASA, *Astronautics and Aeronautics, 1968*, p. 299; *1969*, NASA SP-4014 (1970), p. 167.
10. Space Task Group rpt. to the President, "The Post-Apollo Space Program: Directions for the Future" (Washington, Sept. 1969), apps. A, B.
11. Author's notebook, 4 Aug. 1969, NF28.
12. Summary of President's Science Advisory Committee rpt. to President's Space Task Group, "The Next Decade in Space," Sept. 1969, p. 1; also published as Executive Office of the President, "Summary," *The Next Decade in Space: A Report of the Space Science and Technology Panel of the President's Science Advisory Committee* (Washington, March 1970), p. i.
13. NASA, *Astronautics and Aeronautics, 1970*, NASA SP-4015 (1972), pp. 193–94.
14. NASA, "Goals and Objectives for America's Next Decades in Space," Sept. 1969, and "America's Next Decades in Space: A Report for the Space Task Group," Sept. 1969; President's Science Advisory Committee, "The Next Decade in Space," Sept. 1969. Much of the Department of Defense's contribution was classified.
15. Telcon, Homer E. Newell to Thomas O. Paine, 9 July 1969, NF29.
16. Space Task Group, "Post-Apollo Program," pp. 21, 22.
17. NASA, *Astronautics and Aeronautics, 1970*, pp. 77–79.
18. Paine to Newell, 16 Oct. 1978, comments on draft Newell MS., NF40.
19. Author's notebook, June 1970.
20. NASA, *Astronautics and Aeronautics, 1970*, p. 330.
21. PSAC rpt. to Space Task Group, "The Next Decade in Space."
22. NASA, *Astronautics and Aeronautics, 1970*, p. 303.
23. James E. Webb to Robert Gilruth, 26 Feb. 1963; Gilruth to Webb, 12 Mar. 1963: NF13(196).
24. Meeting of Robert Seamans with Raymond Bisplinghoff, George Mueller, and Homer Newell, 7 Sept. 1965, author's notebook.
25. Paine to Newell, 16 Oct. 1978, NF40.
26. Author's notebook, 27 Dec. 1969, NF28.

Chapter 18

1. Eugene M. Emme, *Aeronautics and Astronautics: An American Chronology of Science and Technology in the Exploration of Space, 1915–1960* (Washington: NASA, 1961), p. 104.
2. Arnold W. Frutkin, *International Cooperation in Space* (Englewood Cliffs, N.J.: Prentice-Hall, 1965), pp. 142–45.
3. Ibid.
4. United Nations, General Assembly Res. 1721 (XVI: 20 Dec. 1961).
5. Frutkin, *International Cooperation*, pp. 144–45.
6. Author's notes, 28 May–9 June 1962, NF28; also miscellaneous papers on meeting of Scientific and Technical Subcommittee of UN Committee on the Peaceful Uses of Outer Space, 28 May–9 June 1962, NF4(80).
7. Frutkin, *International Cooperation*, pp. 145–46.

8. Author's notes, 14 Aug. 1968 ff., NF28.
9. Frutkin, *International Cooperation*, pp. 85–86, 37.
10. Homer E. Newell, Jr., "Report on the Second Meeting of the Committee on Space Research Held at The Hague, 12–14 March 1959," [Mar. 1959], NF8(127).
11. Ibid., encl. 9.
12. See, for example, Hilde Kallmann Bijl, ed., *Space Research: Proceedings of the First International Space Science Symposium, Nice, January 11–16, 1960* (Amsterdam: North-Holland Publishing Co., 1960).
13. International Council of Scientific Unions, Res. 10 (1961), NF6(98).
14. COSPAR Res. 1 (1962), NF6(98).
15. Author's notes, 20 May 1963, NF28; also papers on meeting of Scientific and Technical Subcommittee of CPUOS, Geneva, 14–31 May 1963, NF4(78 & 79).
16. Author's notes, 21 May 1963, NF28. See also ibid.
17. Morton Werber, *Objectives and Models of the Planetary Quarantine Program*, NASA SP-344 (Washington, 1975), pp. 1–5, 101; Charles R. Phillips, *The Planetary Quarantine Program, Origins and Achievements, 1956–1973*, NASA SP-4902 (Washington, 1974); both NF40.
18. NASA, *International Programs*, booklet issued periodically by Office of International Programs (later International Affairs), 1962 ed.
19. NASA, *International Programs*, Jan. 1973 ed.
20. NASA, *International Programs*, 1962 ed., first page, and Jan. 1973, p. ii.
21. Ibid., pp. 23–46.
22. Ibid., p. 8.
23. Ibid., pp. 14–16.
24. Ibid., p. 2.
25. G. Breit and M. A. Tuve, "A Test of the Existence of the Conducting Layer," *Physical Review* 28 (Sept. 1926): 554–75.
26. Hideo Itokawa, "Japanese Sounding Rockets—Kappa and Sigma," in Homer E. Newell, Jr., *Sounding Rockets* (New York: McGraw-Hill Book Co., 1959), chap. 16; idem, "Developments of Project Kappa in 1959–60," *Proceedings of the Second International Symposium on Rockets and Astronautics, Tokyo, 1960* (Tokyo, 1961), pp. 146–157.
27. Homer E. Newell, "Report of Conferences with Japanese Scientists and Government Officials during Week of 22 May 1960," pp. 3–8 and 20–26, NF4(76).
28. Ibid., pp. 11–13.
29. NASA, *International Programs*, Jan. 1973 ed., p. 37.
30. NASA Historical Office, *Astronautics and Aeronautics, 1970: Chronology on Science, Technology, and Policy*, NASA SP-4015 (Washington, 1972), pp. 48–49.
31. NASA, *International Programs*, Jan. 1973 ed., p. 67.
32. Ibid., pp. 6–7.
33. Ibid., p. 66.
34. Ibid., pp. 47–55.
35. Ibid., p. 58.
36. See, for example, A. A. Levinson, ed., *Proceedings of the Apollo 11 Lunar Science Conference, Houston, Texas, January 5–8, 1970*, 3 vols. (New York: Pergamon Press, 1970). Similar proceedings were published for succeeding annual lunar science conferences.
37. Frutkin, *International Cooperation*, pp. 89–91.
38. President John F. Kennedy to Chairman Nikita S. Khrushchev, 22 Feb. 1962; Frutkin, *International Cooperation*, p. 92.
39. Kennedy to Khrushchev, 7 Mar. 1962; Frutkin, *International Cooperation*, p. 121, n. 11; Khrushchev to Kennedy, 20 Mar. 1962; Frutkin, *International Cooperation*, p. 123, n. 12.
40. Hugh Dryden and Anatoly Blagonravov, "Summary of Understandings," typescript, 8 June 1962, NF6(97) and NF4(80); *Department of State Bulletin*, 24 Dec. 1962, pp. 962 ff; Frutkin, *International Cooperation*, p. 96.
41. Frutkin, *International Cooperation*, pp. 96–97.
42. Ibid., p. 104.
43. Ibid., pp. 100–01.
44. Arnold Frutkin, "Notes on Soviet Deficiencies in COSPAR, Warsaw, Poland, June 3–11, 1963," typescript, [21 June 1963], NF9(139).
45. Frutkin, *International Cooperation*, pp. 119–20, 174–77.

46.. Senate Committee on Aeronautical and Space Sciences, *NASA Authorization for Fiscal Year 1968*, hearings, 90th Cong., 1st sess., pt. 1, 18–20 Apr. 1967, p. 57.
47. Author's notes, 14 Feb. 1966 ff., NF28.
48. Frutkin, *International Cooperation*, pp. 133–41.
49. NASA, "Historical Pocket Statistics," Jan. 1975, p. A-82.
50. NASA Historical Office, *Astronautics and Aeronautics, 1973*, NASA SP-4018 (Washington, 1975), pp. 76, 268.
51. Idem, *Astronautics and Aeronautics, 1967*, NASA SP-4008 (1968), pp. 257, 333; *1968*, NASA SP-4010 (1969), p. 10.
52. Thomas O. Paine to M. V. Keldysh, 31 July 1970; NASA, Office of International Affairs, "US/USSR Cooperation in Space Research," 1 July 1970, rev. 1 1970, NF40.
53. Edward C. Ezell and Linda N. Ezell, *The Partnership: A History of the Apollo-Soyuz Test Project*, NASA SP-4209 (Washington, 1978); NASA, *Astronautics and Aeronautics, 1970*, pp. 334–35, 347, 351, 413; idem, *1971*, NASA SP-4016 (1972), pp. 171, 177, 243.
54. NASA, *Astronautics and Aeronautics, 1971*, p. 10.

Chapter 19

1. R. M. Goody, "Weather on the Inner Planets," *New Scientist* (June 1973), pp. 602–05.
2. *Significant Achievements in Space Communications and Navigation, 1958–1964*, NASA SP-93 (Washington, 1966), pp. 53–54.
3. *Significant Achievements in Satellite Geodesy, 1958–1964*, NASA SP-94 (Washington, 1966).
4. See, for example, Nicholas M. Short et al., eds., *Mission to Earth: Landsat Views the World*, NASA SP-360 (Washington, 1976); Richard S. Williams, Jr., and William D. Carter, eds., *ERTS-1: A New Window on Our Planet*, Geological Survey Professional Paper 929 (Washington: USGS, 1976).
5. Charles Townes to James C. Fletcher, 13 Aug. 1971 (National Academy of Sciences, Space Science Board files).
6. Short et al., *Mission to Earth*; Williams and Carter, *ERTS-1: A New Window on Our Planet*.
7. C. B. Spann (NASA contracting officer) to National Academy of Sciences, 20 Dec. 1971; National Academy of Sciences, *News Report* 22, no. 6 (June–July 1972): 2.
8. R. A. Minzner and W. S. Ripley, "The ARDC Model Atmosphere, 1956," Air Force Surveys in Geophysics, TN-56-204, no. 86, 15 pp. (Bedford, Mass.: Air Force Cambridge Research Center, Dec. 1956). See also, The Rocket Panel, "Pressures, Densities, and Temperatures in the Upper Atmosphere," *Physical Review* 88 (Dec. 1952): 1027–32; and, H. Kallman-Bijl et al., *COSPAR International Reference Atmosphere, 1961* (Amsterdam: North-Holland Publishing Co., 1961).
9. R. Cargill Hall, "Early U.S. Satellite Proposals," *Technology and Culture* 4 (Fall 1963).
10. J. Spriggs, ed., "Practical Results from the NASA Space Program," app. 11 to a statement on NASA's space science and applications program presented to Senate Committee on Aeronautical and Space Sciences, 20 Apr. 1967, reprinted by NASA, Washington, 1967.
11. Vannevar Bush, *Science, the Endless Frontier: A Report to the President* (Washington, 1945).
12. Daniel S. Greenberg, *The Politics of Pure Science* (New York: New American Library, 1967, rev. 1971).
13. Spriggs, "Practical Results from the NASA Space Program."
14. William K. Widger, Jr., *Meteorological Satellites* (New York: Holt, Rinehart & Winston, 1966).
15. Leonard Jaffe, *Communications in Space* (New York: Holt, Rinehart & Winston, 1966).
16. John Young to Homer Newell, 22 June 1978, comments on draft Newell MS., NF40.

Chapter 20

1. By way of illustration, see Ivan I. Mueller, *Introduction to Satellite Geodesy* (New York: Frederick Ungar Publishing Co., 1964); William M. Kaula, *Theory of Satellite Geodesy* (Waltham, Mass.: Blaisdell Publishing Co., 1966); G. Mamikunian and M. H. Briggs, eds., *Current Aspects of Exobiology* (Pasadena: Jet Propulsion Laboratory, 1965); Colin S. Pittendrigh et al., eds., *Biology and the Exploration of Mars* (Washington: National Academy of Sciences, 1966); Elie A. Shneour and

Eric A. Otteson, compilers, *Extraterrestrial Life: An Anthology and Bibliography* (Washington: National Academy of Sciences, 1966); Robert J. Mackin, Jr., and Marcia Neugebauer, eds., *The Solar Wind* (Pasadena: Jet Propulsion Laboratory, 1966); Wilmot N. Hess and Gilbert D. Mead, eds., *Introduction to Space Science* (New York: Gordon and Breach, 2d ed., 1968); Donald J. Williams and Gilbert D. Mead, eds., *Magnetospheric Physics* (Washington: American Geophysical Union, 1969); Siegfried J. Bauer, *Physics of Planetary Ionospheres* (New York, Heidelberg, Berlin: Springer-Verlag, 1973); Edward G. Gibson, *The Quiet Sun*, NASA SP-303 (Washington, 1973); Riccardo Giacconi and Herbert Gursky, eds., *X-ray Astronomy* (Dordrecht-Holland: D. Reidel Publishing Co., 1974); Stuart Ross Taylor, *Lunar Science: A Post-Apollo View* (New York: Pergamon Press, 1975); Nicholas M. Short, *Planetary Geology* (Englewood Cliffs, N.J.: Prentice-Hall, 1975); R. Grant Athay, *The Solar Chromosphere and Corona: Quiet Sun* (Dordrecht-Holland: D. Reidel Publishing Co., 1976).

2. Hess and Mead, eds, *Introduction to Space Science*; Lloyd V. Berkner and Hugh Odishaw, eds., *Science in Space* (New York: McGraw-Hill Book Co., 1961).

3. Homer E. Newell, "A New Laboratory—How to Work in It," address before American Physical Society, Washington, 29 Apr. 1965.

4. Newell, "NASA's Space Science and Applications Program," statement to Senate Committee on Aeronautical and Space Sciences, 20 Apr. 1967.

5. Homer E. Newell and Leonard Jaffe, "Impact of Space Research on Science and Technology," *Science* 157 (7 July 1967): 29–39.

6. George Gamov, *One Two Three—Infinity* (New York: The New American Library, 1947), pp. 253–314; idem, *The Birth and Death of the Sun* (ibid., 1952); Fred Hoyle, *Frontiers of Astronomy* (ibid., 1955); D. W. Sciama, *The Unity of the Universe* (Garden City, N.Y.: Doubleday & Co., 1959).

7. Giacconi and Gursky, *X-ray Astronomy*.

8. Pittendrigh et al., *Biology and the Exploration of Mars*.

9. Bernard H. Chovitz, "Geodesy," in *Collier's Encyclopedia*, 10 (New York: Crowell-Collier Educational Corp., 1972): 629–38.

10. John O'Keefe to Newell, 22 June 1978, comments on draft Newell MS., NF40; Chovitz, "Geodesy," p. 638; Short, *Planetary Geodesy*, pp. 72–75.

11. Charles Y. Johnson, "Basic E and F Region Aeronomy," presented at Defense Nuclear Agency Symposium, Stanford Research Institute, Aug. 1971; A. P. Willmore, "Exploration of the Ionosphere from Satellites," *Journal of Atmospheric and Terrestrial Physics* 36 (Dec. 1974): 2255–86.

12. S. A. Bowhill, "Investigation of the Ionosphere by Space Techniques," *Journal of Atmospheric and Terrestrial Physics* 36 (Dec. 1974): 2240.

13. H. Friedman, "Solar Ionizing Radiation," ibid., p. 2252.

14. Willmore, "Exploration of the Ionosphere from Satellites," p. 2279.

15. Bauer, *Physics of Planetary Ionospheres*, p. 6.

16. W. R. Bandeen and S. P. Maran, eds., *Possible Relationships between Solar Activity and Meteorological Phenomena*, GSFC symposium, 7–8 Nov. 1973, NASA SP-366 (Washington, 1975).

17. *Significant Achievements in Space Science 1965*, NASA SP-136, (Washington, 1967), pp. 40–42.

18. Gerard P. Kuiper, ed., *The Atmospheres of the Earth and Planets*, 2d ed. (Chicago: Univ. of Chicago Press, 1952); Gerard P. Kuiper and Barbara Middlehurst, eds., *Planets and Satellites* (ibid., 1961).

19. Short, *Planetary Geology*, pp. 249, 282–87.

20. Ibid., p. 285.

21. Ibid., p. 292.

22. Bauer, *Physics of Planetary Ionospheres*, pp. 194–203.

23. Short, *Planetary Geology*, p. 291.

24. Robert L. Carovillano et al., eds., *Physics of the Magnetosphere* (Dordrecht-Holland: D. Reidel Publishing Co., 1968); B. M. McCormac, ed., *Earth's Magnetospheric Processes* (ibid., 1972); idem, *Magnetospheric Physics* (ibid., 1974); V. Formisano, ed., *The Magnetospheres of the Earth and Jupiter* (ibid., 1975); Syun-Ichi Akasofu, *Physics of Magnetospheric Substorms* (ibid., 1977).

25. Formisano, *Magnetospheres of Earth and Jupiter*; K. Knott and B. Battrick, eds., *The Scientific Satellite Programme during the International Magnetospheric Study* (Dordrecht-Holland: D. Reidel Publishing Co., 1976).

26. E. N. Parker, "Solar Wind Interaction with the Geomagnetic Field," in *Magnetospheric Physics*, ed. Williams and Mead, pp. 3–9; idem, "Dynamical Properties of the Magnetosphere," in *Physics of the Magnetosphere*, ed. Carovillano et al., pp. 3–64.

27. Formisano, *Magnetospheres of Earth and Jupiter*.
28. Zdenek Kopal, *The Solar System* (London: Oxford Univ. Press, 1972), p. 15.
29. *Significant Achievements in Planetology 1958–1964*, NASA SP-99 (Washington, 1966), p. 43.
30. Ibid., pp. 44–46; *Significant Achievements in Space Science 1965*, pp. 139–40.
31. N. F. Ness et al., "The Magnetic Field of Mercury," pt. 1, *Journal of Geophysical Research* 80 (July 1975): 2708–16.
32. Y. C. Whang, "Magnetospheric Magnetic Field of Mercury," *Journal of Geophysical Research* 82 (Mar. 1977): 1024–30.
33. Short, *Planetary Geology*, p. 291.
34. Formisano, *Magnetospheres of Earth and Jupiter*.
35. Jet Propulsion Laboratory Staff, *Mariner Mission to Venus* (New York: McGraw-Hill Book Co., 1963).
36. Short, *Planetary Geology*, pp. 242–81; Stewart A. Collins, *The Mariner 6 and 7 Pictures of Mars*, NASA SP-263 (Washington, 1971); William K. Hartmann and Odell Raper, *The New Mars: The Discoveries of Mariner 9*, NASA SP-337 (Washington, 1974).
37. Short, *Planetary Geology*, p. 284; Hartmann and Raper, *The New Mars*, pp. 38–42.
38. Short, *Planetary Geology*, pp. 284–90.
39. Ibid., p. 285; NASA release 76-153, 10 Sept. 1976.
40. Paul D. Lowman, Jr., "The Geologic Evolution of the Moon," *The Journal of Geology* 80 (Mar. 1972): 125–26; Taylor, *Lunar Science*; Short, *Planetary Geology*, pp. 196–240.
41. Yosio Nakamura et al., "Deep Lunar Interior Inferred from Recent Seismic Data," *Geophysical Research Letters* (July 1974): 137–40.
42. Short, *Planetary Geology*, pp. 287–89.
43. Ibid., pp. 255–59.
44. Paul D. Lowman, Jr., "Crustal Evolution in Silicate Planets: Implications for the Origin of Continents," *Journal of Geology* 84 (Jan. 1976): 1–26.
45. Short, *Planetary Geology*, p. 292.
46. Ibid., pp. 231–32.
47. *Viking 1: Early Results*, NASA SP-408 (Washington, 1976), pp. 59–63; American Geophysical Union, *Scientific Results of the Viking Project*, collection of reprints from the *Journal of Geophysical Research* (Washington, 1977).
48. Hartmann and Raper, *The New Mars*, p. 65.
49. Robert J. Davis, William A. Deutschman, and Katherine L. Haramundanis, *The Celescope Catalog of Ultraviolet Stellar Observations* (Washington: Smithsonian Institution, 1973).
50. Bruno Rossi, "X-ray Astronomy," in "Discoveries and Interpretations: Studies in Contemporary Scholarship," 2, *Daedalus*, Fall 1977 (issued as vol. 106, no. 4, of the *Proceedings of the Academy of Arts and Sciences*), pp. 37–58.
51. Riccardo Giacconi, introduction to *X-ray Astronomy*, ed. Giacconi and Gursky, p. 6.
52. Ibid., pp. 8–11.
53. Ibid., pp. 11–12.
54. T. A. Chubb and H. Friedman, "Glimpsing the Hidden X-ray Universe," *Astronautics and Aeronautics* 7 (Mar. 1969): 50–55.
55. Giacconi, introduction to *X-ray Astronomy*, pp. 15–19.
56. Giacconi, "X-ray Sky," chap. in *X-ray Astronomy*, pp. 155–68.
57. Riccardo Giacconi, "Progress in X-ray Astronomy," talk presented at Thirty-fourth Richtmeyer Memorial Lecture of American Association of Physics Teachers, Anaheim, Calif., 30 Jan. 1975, Preprint Series No. 304, Center for Astrophysics, Cambridge, Mass., pp. 8–19.
58. Ibid., p. 21.
59. Gursky and Ruffini, *Neutron Stars, Black Holes and Binary X-ray Sources*, app. 1, pp. 259–317.
60. Henry Norris Russell, Raymond Smith Dugan, and John Quincy Stewart, *Astronomy*, 2 vols. (New York: Ginn and Co., 1927); Edward G. Gibson, *The Quiet Sun*, NASA SP-303 (Washington, 1973).
61. Giuseppe Vaiana and Wallace Tucker, "Solar X-ray Emission," chapter in *X-ray Astronomy*, ed. Giacconi and Gursky, pp. 171–78; Gibson, *The Quiet Sun*, pp. 7–30.
62. Leo Goldberg, in a foreword to Gibson's *The Quiet Sun*.
63. Richard Tousey, "Some Results of Twenty Years of Extreme Ultraviolet Solar Research," *The Astrophysical Journal* 149 (Aug. 1967): 239–52 plus plates; Kenneth G. Widing, "Solar Research from Rockets and Satellites," *Astronautics and Aeronautics* 7 (Mar. 1969): 36–43.
64. Tousey, "Twenty Years of Ultraviolet Solar Research," plates 19–28.

65. Vaiana and Tucker, "Solar X-ray Emission."
66. Ibid., p. 197.
67. Ibid., pp. 183-84.
68. Ibid., pp. 187-90.

Chapter 21

1. Meeting in Abe Silverstein's office between NASA and Weather Bureau representatives, author's notebooks, 12 Feb. 60, NF28; Hugh Dryden to F. Reicheldoerfer, 6 May 60, NF12(176); Dryden and Luther Hodges, "Basic Agreement between the United States Department of Commerce and the National Aeronautics and Space Administration concerning Meteorological Satellite Systems," 30 June 1964, NF40.
2. Leonard Jaffe, *Communications in Space* (New York: Holt, Rinehart and Winston, 1966), pp. 107-13, 122-40, 152-60; House Committee on Science and Astronautics, *Communications Satellites*, hearings, 87th Cong., 1st sess., 1, 8 May-13 July 1961; pt. 2, 14 July-10 Aug. 1961; Senate Committee on Aeronautical and Space Sciences, *Communications Satellites: Technical, Economic, and International Developments*, staff rpt., 87th Cong., 2d sess., 25 Feb. 1962. Senate Committee on Aeronautical and Space Sciences, *Communications Satellite Legislation*, hearings, 87th Cong., 2d sess., 27 Feb.-7 Mar. 1962; Senate Committee on Commerce, *Communications Satellite Incorporators*, hearings, 88th Cong., 1st sess., 11 Mar. 1963; Senate Committee on Aeronautical and Space Sciences, *NASA Authorization for Fiscal Year 1964*, pt. 1, *Scientific and Technical Programs*, 88th Cong., 1st sess., 24-30 Apr. 1963, pp. 422-24.
3. Robert L. Rosholt, *An Administrative History of NASA, 1958-1963*, NASA SP-4101 (Washington, 1966), p. 130.
4. Working papers of NASA's space science div. summarizing space science program, 17 Jan. 59, NF11(170); space science div., "Areas of Research in the NASA Space Sciences Program," mimeographed, 48 pp., 10 Feb. 1959; idem, "The United States National Space Sciences Program," mimeographed, 10 Feb. 1959; idem, "National Space Sciences Program" mimeographed, 16 Apr. 1959; idem, "NASA Program Planning in Space Sciences," mimeographed, Apr. 1960, all NF2(33); Homer E. Newell, "The NASA Space Sciences Program," mimeographed, 20 Mar. 1961, NF1.
5. Note space science div. paper, "NASA Space Sciences Long Range Program," Oct. 1959, whereas three-quarters of a year later a similar document was entitled "Long Range Thinking in Space Sciences," July 1960; NASA History Office files.
6. "Long Range Thinking in Space Sciences," Oct. 1960; "NASA Program Planning in Space Sciences," Nov. 60; and "NASA Program Planning in Space Sciences," Nov. 61; all in NASA History Office files.
7. H. E. Newell to Space Science Steering Committee et al., 11 Sept. 1962, NASA History Office files.
8. See, for example, "Prospectus," NASA Office of Space Sciences, Nov. 62, NASA History Office files.
9. See, for example, *The Budget of the United States Government for the Fiscal Year Ending June 30, 1964* (Washington, 1963), Append. pp. 781-87. See also the 5-vol. *"Budget Estimates: Fiscal Year 1964,"* which NASA prepared and distributed to the Congressional committees, giving extensive detail on the budget request; these were used in the budget defense before Congress.
10. Jane Van Nimmen and Leonard Bruno with Robert L. Rosholt, *NASA Historical Data Book, 1958-1968*, vol. 1, *NASA Resources*, NASA SP-4012 (Washington, 1976), p. 118; also NASA, "Chronological History: Fiscal Year 1969 Budget Submission" (final rpt., 14 Oct. 1968); 1970 (final rpt. 5 Dec. 1969); 1971 (final rpt. 11 June 1971); 1972 (final rpt. 30 Aug. 1971); 1973 (final rpt. 5 Sept. 1972); all in NASA History Office files.
11. Van Nimmen, et al., *NASA Historical Data Book*, pp. 118, 134-49; *Aeronautics and Space Report of the President, 1976 Activities* (Washington: NASA, 1977), p. 107; *Report to the Congress from the President of the United States: United States Aeronautics and Space Activities, 1966* (Washington: National Aeronautics and Space Council, 31 Jan. 1967), p. 166; "NASA Historical Pocket Statistics," Jan. 1974, pp. D-4, D-5.
12. Senate Committee on Aeronautical and Space Sciences, *Scientists' Testimony on Space Goals*, hearings, 88th Cong., 1st sess., 10-11 June 1963, p. 3.
13. House Committee on Science and Astronautics, *Review of the Soviet Space Program*, 90th Cong., 1st sess., rpt. prepared by Science Policy Research Div., Legislative Reference Service, Library of

Congress, 1967, pp. 11–45. The referenced material compares Soviet and U.S. space accomplishments through the summer of 1967, to which time, despite the highly successful Gemini program, the preponderance of U.S. achievements was provided by the unmanned space science and applications programs.

14. For reductions in Ranger, see: H. E. Newell to W. H. Pickering, 12 July 1963, JPL Historical Files 2-190; E. M. Cortright, memo for record, "Some Comments on the NASA Reorientation of the Ranger Programs," 15 July 1963, JPL Historical Files 2-719; Newell to Pickering, 13 Dec. 1963, JPL Historical Files 2-196. For retrenchments on Surveyor, see author's notebooks, 12 Dec. 1965, NF28. A year later 3 more Surveyors were eliminated from the program: NASA release 66-318, 13 Dec. 1966.

15. NASA release 65-380. Also, see author's notebook, 12 Dec. 1965, NF28.

16. James E. Webb to the President, 30 Nov. 1962, NF40.

17. Senate Committee on Aeronautical and Space Sciences, *NASA Authorization for Fiscal Year 1970*, hearings, 91st Cong., 1st sess., 1–9 May 1969, pp. 505, 528, 555–58.

18. By way of illustration, see House Committee on Science and Astronautics, *1965 NASA Authorization*, hearings before Subcommittee on Space Science and Applications, 88th Cong., 2d sess., pt. 3, 17 Feb.–3 Mar. 1964.

19. J. A. Van Allen to Hon. Walter Mondale, 29 June 1970, in *Congressional Record—Senate*, 6 July 1970, p. S10606; T. Gold to Sen. Walter Mondale, 3 July 1970, ibid., pp. S10608–09. See also Thomas Gold, "Machines, Not Men, in Space," *New York Times Magazine*, 22 Aug. 1971, pp. 14–30.

20. NASA, "Chronological History: Fiscal Year 1970 Budget Submission," through "Fiscal Year 1973 Budget Submission," (Washington, 1969-1973), in NASA History Office files.

21. House Committee on Science and Technology, *Operational Cost Estimates: Space Shuttle*, rpt. prepared by Subcommittee on Space Science and Applications, 94th Cong., 2d sess., committee print, Dec. 1976; *Space Shuttle*, NASA SP-407 (Washington, 1976); NASA, "Environmental Impact Statement: Space Shuttle Program," draft, July 1977.

22. John V. Becker, "The X-15 Program in Retrospect," *Raumfahrtforschung*, Mar.-Apr. 1969, pp. 45–53.

23. Jay Miller, "The X-Series," *Aerophile* 1, no. 2, Mar.-Apr. 1977, pp. 80-2, NF40.

24. For example, physicists Herbert Friedman, Lewis Branscomb, and Charles Townes; astronomer Gerard Kuiper; and engineers Raymond Bisplinghoff and H. Guyford Stever. See telephone conversation between Newell and Dale Grubb, 4 Jan. 1971, and telephone conversation between Newell and Gerard Kuiper, 8 June 1971, NF29.

25. Space Science Board, *Priorities for Space Research 1971-1980* (Washington: National Academy of Sciences, 1971).

26. Author's notebook, July 1970, NF28.

27. Ibid.

28. Ibid., 24 Oct. 1972.

29. Space Science Board, *Scientific Uses of the Space Shuttle* (Washington: National Academy of Sciences, 1974).

Index

Appendixes, Bibliographic Essay, and Source Notes are not included in the index.

The Author

Homer Edward Newell was born in Holyoke, Massachusetts, 11 March 1915. During the years of his primary and secondary education in Holyoke's public school system, the city was undergoing a substantial transition. Formerly known as "Paper City," Holyoke was losing its foremost position in paper manufacture to the Midwest, while its considerable textile industry was similarly losing out to mills in the South. Newell's early interest in science was fostered by his paternal grandfather's extensive technical library, by competent and interested teachers in Holyoke High School, and by a home chemistry laboratory, the nucleus of which had been contributed by a local sulfite wood pulp mill.

At Harvard Newell studied science courses including mineralogy, physics, and astronomy and pursued a major in mathematics to an A.B. and an A.M.T. from Harvard and a Ph.D. from the University of Wisconsin in 1940. He then taught mathematics at the University of Maryland until 1944, when he joined the U.S. Naval Research Laboratory. From 1946, as a mathematician-turned-physicist, he was one of a group of scientists and engineers who used their World War II experience with missiles and radio communications to instrument rockets for high-altitude research and then to launch them at White Sands, New Mexico. In the fall of 1947 he became head of the Rocket Sonde Research Section (later Branch). In September 1955, when the Naval Research Laboratory was assigned the task of developing the Vanguard launch vehicle for the International Geophysical Year satellite program, he became Vanguard Science Program Coordinator. The National Aeronautics and Space Administration opened for business in October 1958; Newell joined its headquarters the same month and remained for 15 years. He guided the space science program through much of the 1960s and was NASA's Associate Administrator from 1967 until his retirement at the end of 1973.

In 1953 his book *High Altitude Rocket Research* was published. Six more books followed, as well as hundreds of articles on such subjects as space science, vector analysis, sounding rockets, missiles, and astronomy. Several of the books were directed specifically to young people. Since leaving NASA, Newell has retained this interest and is preparing a book for the young reader on mineralogy and rock-hounding.